AF805919

On the Class Number of Abelian Number Fields

Helmut Hasse

On the Class Number of Abelian Number Fields

Extended with Tables by Ken-ichi Yoshino and Mikihito Hirabayashi

Springer

Helmut Hasse
Hamburg, Germany

Translated by
Mikihito Hirabayashi
Kanazawa Institute of Technology
Ishikawa, Japan

The book is a translation of the German edition - two chapters are added as reprints.

ISBN 978-3-030-01510-7 ISBN 978-3-030-01512-1 (eBook)
https://doi.org/10.1007/978-3-030-01512-1

Library of Congress Control Number: 2018961405

Mathematics Subject Classification (2010): 11-XX, 11Rxx, 11R37, 11R29, 12-XX

Part II Chapter 4 reprinted with kind permission of Kanazawa Medical University from: Ken-ichi Yoshino and Mikihito Hirabayashi, On the Relative Class Number of the Imaginary Abelian Number Field I, Memoirs of the College of Liberal Arts, Vol. 9, December, p. 5–53, © Kanazawa Medical University 1981.
Part II Chapter 5 reprinted with kind permission of Kanazawa Medical University from: Ken-ichi Yoshino and Mikihito Hirabayashi, On the Relative Class Number of the Imaginary Abelian Number Field II, Memoirs of the College of Liberal Arts, Vol. 10, December, p. 33–81, © Kanazawa Medical University 1982. Corrected printing 2015

English translation of the German reprint published by Springer-Verlag Berlin, Heidelberg, 1985

© Springer Nature Switzerland AG 1952, 1985, 2019
This work is subject to copyright. All rights are reserved by the Publisher, whether the whole or part of the material is concerned, specifically the rights of translation, reprinting, reuse of illustrations, recitation, broadcasting, reproduction on microfilms or in any other physical way, and transmission or information storage and retrieval, electronic adaptation, computer software, or by similar or dissimilar methodology now known or hereafter developed.
The use of general descriptive names, registered names, trademarks, service marks, etc. in this publication does not imply, even in the absence of a specific statement, that such names are exempt from the relevant protective laws and regulations and therefore free for general use.
The publisher, the authors and the editors are safe to assume that the advice and information in this book are believed to be true and accurate at the date of publication. Neither the publisher nor the authors or the editors give a warranty, express or implied, with respect to the material contained herein or for any errors or omissions that may have been made. The publisher remains neutral with regard to jurisdictional claims in published maps and institutional affiliations.

This Springer imprint is published by the registered company Springer Nature Switzerland AG
The registered company address is: Gewerbestrasse 11, 6330 Cham, Switzerland

Foreword

Jacques Martinet has sketched the development of the main ideas in the theory of cyclotomic fields from the publication of Hasse's monograph up to the heydays of Iwasawa theory (for a polished proof of the Main Conjecture, see, e.g., the beautiful book [10]). The notorious conjectures (Vandiver, Greenberg) are still open, and progress has shifted somewhat to the non-abelian side. This note is devoted to the algorithmic part of the theory of cyclotomic fields.

As Hasse explained in his preface, he was not content with the state of the art in the theory of cyclotomic fields: Hilbert had erected a magnificent building of algebraic number theory, which had been crowned by Takagi's class field theory. On the other hand, the computation of the main invariants of number fields, such as the unit group and the class number, was not feasible except for fields of small degree and very small discriminant. In Hasse's opinion, it was unsatisfactory to have such a great tool as the analytic class number formula while having to admit that it was pretty useless for actually computing class numbers.

Hasse's goal was to investigate the class number formula from an arithmetic point of view; even showing that the number h^- provided by the class number formula is an integer is quite non-trivial. He then showed that the class number formula may be used for computing the relative class numbers of all cyclotomic fields of conductor ≤ 100.

Hasse's monograph served as a blueprint for the corresponding book on abelian extensions of complex quadratic number fields by Curt Meyer [45]. The whole situation in this case is much more complicated, and Meyer's book is a lot harder to read than Hasse's. What is missing are the many tables that Hasse provided in his book and in particular the beautiful diagrams of the subfields involved.

Class Number Formulas

The statement that the analytic class number formula for quadratic number fields was first proved by Dirichlet must be taken with a grain of salt, since Dirichlet worked with binary quadratic forms. This fact makes translating his result on the class number formula for Dirichlet number fields $\mathbb{Q}(i, \sqrt{m})$ (or, as he would have said, for binary quadratic forms with complex coefficients and determinant m) a highly non-trivial task, since his class groups of forms correspond to certain ring class groups in the modern sense. The generalizations of his class number formula to more general biquadratic number fields provided by Eisenstein [15], Bachmann [5] and Amberg [1] are so difficult to translate into the language of number fields that it is hard to say whether their results are correct or not (Hasse alludes to this state of affairs in his Introduction).

Hilbert [27] and Herglotz [26] proved these results in the language of number fields, and the most profound study of the class number formula in multiquadratic number fields is the little-known thesis by Värmon [66], which contains most of the results that were later rediscovered independently by other authors.

Computation of Class Numbers

Already Gauss had computed extended tables of class numbers of binary quadratic forms, but he used his theory of composition, a method that is superior to the class number formula for computations by hand. Even after Dirichlet's proof of the class number formula in the quadratic case, the theory of quadratic forms remained the favourite computational tool.

Similar techniques were not available when Kummer started investigating the class groups of cyclotomic fields. He proved that the class number h of the field $K = \mathbb{Q}(\zeta_p)$ of p-th roots of unity admits a factorization $h_p = h_p^+ h_p^-$, where h_p^+ is the class number of the maximal real subfield $K^+ = \mathbb{Q}(\zeta_p + \zeta_p^{-1})$, and where h_p^- is an integer that can be computed explicitly. The relative class number $h_p^- = h(K)/h(K^+)$ is numerically accessible, and the plus class number is essentially the index of the group of cyclotomic units inside the full unit group.

Using the techniques provided by Kummer, C.G. Reuschle [55] (1812–1875), a teacher at the Gymnasium in Stuttgart, computed generators of the principal ideals of small norms in cyclotomic number fields $\mathbb{Q}(\zeta_m)$ for all prime values $m < 100$ as well as for several composite values of $m \leq 120$. Reuschle's correspondence with Kummer was published by Folkerts and Neumann [17].

After the publication of Hasse's book, Schrutka von Rechtenstamm [61] published extensive tables of relative class numbers of cyclotomic fields in 1964. D.H. Lehmer [36] computed many minus class numbers of cyclotomic fields; his results were extended considerably in [18]. Yoshino and Hirabayashi [69] expanded Hasse's tables, including the diagrams for the subfields (see Part II of the present translation).

The first calculations of h_m^+ in some non-trivial cases were done by van der Linden [39], who used Odlyzko's bounds for computing h_p^+ for all primes $p \leq 163$ (in some cases, he had to assume GRH).

Great advances were made by R. Schoof [59] (see also [33] and [22]), who was able to determine factors of class numbers of real cyclotomic fields, which very likely coincide with the actual class numbers.

Stéphane Louboutin has written a wealth of articles on the computation of relative class numbers of CM fields using analytic means and has used these techniques for classifying abelian (and non-abelian) CM fields with small class numbers; see, e.g., [40].

Class numbers of cyclotomic fields showed up in investigations of Catalan's equation $x^m - y^n = 1$. It is not difficult to reduce the statement that this equation does not have any non-trivial solutions to the case $x^p - y^q = 1$, where p and q are distinct odd prime numbers. It can be shown (see [7, 46]) that if this equation has a non-trivial solution, then $p \mid h_q^-$ and $q \mid h_p^-$. The conjecture that the only non-trivial solution of the Catalan equation in natural numbers is $3^2 - 2^3 = 1$ was obtained by P. Mihailescu using Stickelberger's theorem (see [6, 60] for expositions of the proof).

For an overview on the determination of class numbers using the p-adic class number formula, see the recent thesis by Zhang [70].

Parity of Class Numbers

Hasse proved in Theorem 3.45 that the class number of $\mathbb{Q}(\zeta_n)$ is odd if and only if h_n^- is odd. The conjecture that h_p^- is odd if p and $q = \frac{p-1}{2}$ are both prime emerged in the work of Davis [13] and Estes [14]. This conjecture was proved if 2 is a primitive root modulo q by Estes; see Stevenhagen [63] for a generalization and other useful references. Gras [21] proved an important duality result between groups of totally positive and primary cyclotomic units; a different proof based on the Main Conjecture was given recently by Ichimura [30]. Yoshino [68] proved that the class number of $\mathbb{Q}(\zeta_n)$ is even if $n \not\equiv 2 \bmod 4$ is divisible by 4 distinct prime factors.

For the state of the art concerning the parity of plus class numbers, and the determination of the 2-class group of real abelian fields, see the recent thesis by Verhoek [67].

Structure of Class Groups

Already Kummer [34] investigated the structure of some minus class groups of cyclotomic fields by studying the action of the Galois group on class groups; in this way, he was able to show that the minus class group of $\mathbb{Q}(\zeta_{29})$, which has order 8, is elementary abelian. Subsequently, Kummer's methods were refined by Tateyama [64], Horie and Ogura [29] and many other authors.

Kummer's result that $p \mid h_p^+$ implies that $p \mid h_p^-$ was proved algebraically by Hecke [24], who proved that the p-rank e^- of the minus part of $\mathrm{Cl}_p(\mathbb{Q}(\zeta_p))$ and the corresponding rank e^+ of the plus part satisfy $e^- \geq e^+$. Generalizing results by Arnold Scholz [57], who compared the 3-ranks of the ideal class groups of $\mathbb{Q}(\sqrt{m}\,)$ and $\mathbb{Q}(\sqrt{-3m}\,)$, Leopoldt [38] obtained strong bounds between individual pieces of the plus and the minus class groups of cyclotomic fields.

Class Number 1 Problems

Kummer conjectured in 1851 that the minus class number h_p^- of $\mathbb{Q}(\zeta_p)$, where p is prime, is asymptotically equal to

$$G(p) = \frac{p^{\frac{p+3}{4}}}{2^{\frac{p-3}{2}}\pi^{\frac{p-1}{2}}}.$$

Ankeny and Chowla could prove that

$$\log \frac{h_p^-}{G(p)} = o(\log p),$$

which implies that $h_p^- = 1$ for only finitely many primes numbers p. Granville [20] showed that Kummer's conjectures are not compatible with conjectures in analytic number theory that are believed to be true.

Masley [41] (see also [42]) determined all cyclotomic fields $\mathbb{Q}(\zeta_m)$ with class number 1 and found that all of them satisfy $m \leq 84$. A useful result in this connection was proved by Horie [28]: if $K \subseteq L$ are complex abelian number fields, then $h_K^- \mid 4h_L^-$; this was generalized to arbitrary CM fields by Okazaki [52].

In this book, Hasse discusses Weber's result that the class numbers of the fields L_n of 2^{n+2}-th roots of unity are all odd in Sect. 3.16. Harvey Cohn [12] pointed out that the maximal real subfields K_n of L_n seem to have class number 1. This is easy to prove for $K_1 = \mathbb{Q}(\sqrt{2}\,)$ and $K_2 = \mathbb{Q}(\sqrt{2+\sqrt{2}}\,)$, and it follows from Reuschle's tables that $h(K_3) = 1$. Cohn proved that the fields K_n either have class number $= 1$ or ≥ 257. He adds the remark, "We still have obtained no evidence to doubt" that the class numbers of the fields K_n are all trivial.

Fukuda and Komatsu [19] improved Cohn's result and showed that the class numbers $h(K_n)$ are not divisible by any prime number $< 10^9$. Bauer and van der Linden showed that $h(K_4) = h(K_5) = 1$; Miller [47, 48] proved that $h(K_6) = 1$ and, assuming GRH, that $h(K_7) = 1$. In addition, he conjectures that all subfields $K_{p,n}$ of the cyclotomic p-extensions for any prime p and any n have class number 1. Buhler, Pomerance and Robertson [8] have shown that the Cohen–Lenstra heuristics predict that $h^+(p^n) = h^+(p)$ for almost all primes p and all integers n, where $h^+(p^n)$ is the class number of the maximal real subfield of the field of p^n-th roots of unity, and they remark, "It is possible that there are no exceptional primes" p at all.

Hilbert Class Fields

The fields $\mathbb{Q}(\zeta_p)$ with $p \leq 19$ have class number 1, and $L = \mathbb{Q}(\zeta_{23})$ has class number 3 coming from the quadratic subfield $K = \mathbb{Q}(\sqrt{-23})$. In particular, we get the Hilbert class fields of L and K by adjoining a root of the polynomial $f(x) = x^3 - x + 1$ with discriminant -23.

In a similar way, we can construct unramified cubic extensions for many other cyclotomic fields. The smallest example of a quadratic number field with class number 5 was treated by Hasse [23]. Nowadays, such calculations can be performed routinely using the methods described in Henri Cohen's book [11].

A few other known examples of unramified abelian extensions of cyclotomic fields are given by the following table.

Δ	h	f	$\lvert \operatorname{disc} F \rvert$
47	5	$x^5 - 2x^4 + 2x^3 - x^2 + 1$	47^2
79	5	$x^5 - 2x^4 + 3x^2 - 2x + 1$	79^2
71	7	$x^7 - 2x^6 + 2x^5 + x^3 - 3x^2 + x - 1$	71^3
29	8	$x^8 - 4x^7 + 8x^6 - 6x^5 + 2x^4 + 6x^3 - 3x^2 + x + 3$	29^6
31	9	$x^9 - x^7 - 2x^6 + 3x^5 + x^4 + 2x^3 - x^2 + x - 3$	31^6

The field F of degree 17 whose compositum with $K = \mathbb{Q}(\zeta_{64})$ is the Hilbert class field of K was computed by Noam Elkies [16]: F is generated by a root of the polynomial

$$\begin{aligned} f(x) = x^{17} - 2x^{16} + 8x^{13} + 16x^{12} - 16x^{11} + 64x^9 - 32x^8 \\ - 80x^7 + 32x^6 + 40x^5 + 80x^4 + 16x^3 - 128x^2 - 2x + 68 \end{aligned}$$

and has discriminant $\lvert \operatorname{disc} F \rvert = 2^{79}$.

Families of unramified extensions of cyclotomic number fields were constructed by Arnold Scholz ([58]; see also [37]). An investigation of Metsänkylä's results [44] on prime factors of the minus class number of cyclotomic fields from a class field theoretical point of view might be a rewarding project.

The main investigation of unramified abelian extensions of cyclotomic fields was done in connection with p-class groups of $\mathbb{Q}(\zeta_p)$. Kummer had already shown that these p-class groups are governed by the divisibility of Bernoulli numbers by p, and some of his results may be interpreted as an explicit construction of p-class fields of $\mathbb{Q}(\zeta_p)$.

The precise connection between the divisibility of Bernoulli numbers and certain pieces of the p-class group of $\mathbb{Q}(\zeta_p)$ was investigated by Pollaczek [53], Morishima

[50, 51] and Vandiver [65], but the clearest exposition and the most complete results concerning this correspondence were given by Herbrand [25]. For $K = \mathbb{Q}(\zeta_p)$, set $G = \mathrm{Gal}(K/\mathbb{Q})$, $A = \mathrm{Cl}(K)/\mathrm{Cl}(K)^p$ and $\sigma_a(\zeta) = \zeta^a$. Let

$$A_i = \{c \in A : \sigma_a(c) = c^{a^i} \text{ for all } \sigma_a \in G\}.$$

Then A_0 is the part of A fixed by the Galois group, and we have

$$A = A_0 \oplus A_1 \oplus \cdots \oplus A_{p-2}.$$

It is easy to see that $A_0 = A_1 = 0$. Herbrand proved that if $A_i \neq 0$ for some odd index i, then $p \mid B_{p-i}$, which refines Kummer's result that if $p \mid h_p^-$, then p divides some Bernoulli number. Since the plus part of the class group is the sum of the A_k with even index, Vandiver's conjecture states that $A_k = 0$ for all even $0 \leq k \leq p-3$. Herbrand proved that if $p \mid B_{p-i}$, then $A_i \neq 0$ if Vandiver's conjecture holds, and Ribet [56] (see also Mazur's beautiful survey [43]) succeeded in eliminating Vandiver's conjecture using modular forms.

In recent years, it was discovered how to use algebraic K-theory to prove results about the A_i. Kurihara [35] was able to show that $A_{p-3} = 0$, and Soulé [62] showed that $A_{p-n} = 0$ for odd values of n satisfying $\log p > n^{224n^4}$.

The triviality of certain pieces of A is also related to a conjecture of Ankeny, Artin and Chowla [2–4], according to which the fundamental unit $\varepsilon = t + u\sqrt{p}$ for primes $p \equiv 1 \bmod 4$ satisfies $p \nmid u$. Kiselev [31], Carlitz [9] and Mordell [49] proved independently that this is equivalent to $p \nmid B_{(p-1)/2}$. This conjecture was verified for all primes $p < 2 \cdot 10^{11}$ in [54].

For a survey on the fascinating connections between values of zeta functions and algebraic K-theory, see Kolster [32].

Jagstzell, Germany

Franz Lemmermeyer

References

1. E.J. Amberg, Über den Körper, dessen Zahlen sich rational aus zwei Quadratwurzeln zusammensetzen. Diss. Zürich 1897
2. N.C. Ankeny, E. Artin, S. Chowla, The class-number of real quadratic number fields. Ann. Math. **56**, 479–493 (1952)
3. N.C. Ankeny, S. Chowla, A note on the class number of real quadratic fields. Acta Arith. **6**, 145–147 (1960)
4. N.C. Ankeny, S. Chowla, A further note on the class number of real quadratic fields. Acta Arith. **7**, 271–272 (1962)
5. P. Bachmann, Zur Theorie der complexen Zahlen. J. Reine Angew. Math. **67**, 200–204 (1867)

6. Y.F. Bilu, Y. Bugeaud, M. Mignotte, *The Problem of Catalan* (Springer, Berlin, 2014)
7. Y. Bugeaud, G. Hanrot, Un nouveau critère pour l'équation de Catalan. Mathematika **47**, 63–73 (2000)
8. J. Buhler, C. Pomerance, L. Robertson, Heuristics for class numbers of prime-power real cyclotomic fields, in *High Primes and Misdemeanours: Lectures in Honour of the 60th Birthday of Hugh Cowie Williams*, ed. by A. van der Poorten (AMS, Providence, 2003)
9. L. Carlitz, Note on the class number of real quadratic fields. Proc. Am. Math. Soc. **4**, 535–537 (1953)
10. J. Coates, R. Sujatha, *Cyclotomic Fields and Zeta Values* (Springer, Berlin, 2006)
11. H. Cohen, *Advanced Topics in Computational Number Theory* (Springer, Berlin, 2000)
12. H. Cohn, A numerical study of Weber's real class number calculation. Numer. Math. **2**, 347–362 (1960)
13. D. Davis, Computing the number of totally positive circular units which are squares. J. Number Theory **10**, 1–9 (1978)
14. D.R. Estes, On the parity of the class number of the field of q-th roots of unity. Rocky Mountain J. Math. **19**, 675–682 (1989)
15. G. Eisenstein, Über die Anzahl der quadratischen Formen in den verschiedenen complexen Theorien. J. Reine Angew. Math. **27**, 311–316 (1844); *Mathematische Werke I*, Chelsea, New York, 1975, 89–94
16. N. Elkies, Question 172148: degree 17 number fields ramified only at 2 (2015), https://mathoverflow.net/questions/172148/degree-17-number-fields-ramified-only-at-2/
17. M. Folkerts, O. Neumann, *Der Briefwechsel zwischen Kummer und Reuschle: Ein Beitrag zur Geschichte der algebraischen Zahlentheorie* (Erwin Rauner Verlag, 2006)
18. G. Fung, A. Granville, H.C. Williams, Computation of the first factor of the class number of cyclotomic fields. J. Number Theory **42**, 297–312 (1992)
19. T. Fukuda, K. Komatsu, Weber's class number problem in the cyclotomic $\mathbb{Z}_2$-extension of $\mathbb{Q}$, III. Int. J. Number Theory **7**, 1627–1635 (2011)
20. A. Granville, On the size of the first factor of the class number of a cyclotomic field. Invent. Math. **100**, 321–338 (1990)
21. G. Gras, Critère de parité du nombre de classes des extensions abéliennes réeles de $\mathbb{Q}$ de degré impair. Bull. Soc. Math. France **103**, 177–190 (1975)
22. T. Hakkarainen, On the computation of class numbers of real abelian fields. TUCS Technical Report no. 770, May 2006
23. H. Hasse, Über den Klassenkörper zum quadratischen Zahlkörper mit der Diskriminante −47. Acta Arith. **9**, 419–434 (1964)
24. E. Hecke, Über nicht-reguläre Primzahlen und den Fermatschen Satz. Nachr. Akad. Wiss. Göttingen 420–424 (1910)
25. J. Herbrand, Sur les classes des corps circulaires. J. Math. Pures Appl. **11**, 417–441 (1932)

26. G. Herglotz, Über einen Dirichletschen Satz. Math. Z. **12**, 255–261 (1922)
27. D. Hilbert, Über den Dirichletschen biquadratischen Zahlkörper. Math. Ann. **45**, 309–340 (1894); Ges. Abh. I, 24–52
28. K. Horie, On a ratio between relative class numbers. Math. Z. **211**, 505–521 (1992)
29. K. Horie, H. Ogura, On the ideal class groups of imaginary abelian fields with small conductor. Trans. Am. Math. Soc. **347**, 2517–2532 (1995)
30. H. Ichimura, On a duality of Gras between totally positive and primary cyclotomic units. Math. J. Okayama Univ. **58**, 125–132 (2016)
31. A.A. Kiselev, An expression for the number of classes of ideals of real quadratic fields by means of Bernoulli numbers (Russian). Dokl. Akad. Nauk (N.S.) **61**, 777–779 (1948)
32. M. Kolster, K-theory and arithmetic, in *Contemporary Developments in Algebraic K-Theory*, ed. by M. Karoubi et al. Proceedings of the School and Conference on Algebraic K-theory and Its Applications (2003), ICTP, Trieste (2002)
33. Y. Koyama, K. Yoshino, Prime divisors of real class number of the real p^rth cyclotomic field and characteristic polynomials attached to them. RIMS Kokyuroku Bessatsu B12, 149–172 (2009)
34. E.E. Kummer, Über die Irregularität von Determinanten. Monatsber. Akad. Wiss. Berlin 1853, 194–200; *Collected Papers,* Vol. I, 539–545
35. M. Kurihara, Some remarks on conjectures about cyclotomic fields and K-groups of $\mathbb{Z}$. Compositio Math. **81**, 223–236 (1992)
36. D.H. Lehmer, Prime factors of cyclotomic class numbers. Math. Comput. **31**, 599–607 (1977)
37. F. Lemmermeyer, Ideal class groups of cyclotomic number fields. III. Acta Arith. **131**, 255–266 (2008)
38. H.W. Leopoldt, Zur Struktur der ℓ-Klassengruppe galoisscher Zahlkörper. J. Reine Angew. Math. **199**, 165–174 (1958)
39. F. van der Linden, Class number computations of real abelian number fields. Math. Comput. **39**, 693–707 (1982)
40. S. Louboutin, Computation of relative class numbers of imaginary abelian number fields. Exp. Math. **7**, 293–303 (1998)
41. J. Masley, On the class number of cyclotomic fields. Diss. Princeton Univ., 1972
42. J. Masley, H. Montgomery, Cyclotomic Fields with unique factorization. J. Reine Angew. Math. **286–287**, 248–256 (1976)
43. B. Mazur, How can we construct abelian Galois extensions of basic number fields? Bull. Am. Math. Soc. **48**, 155–209 (2011)
44. T. Metsänkylä, Über den ersten Faktor der Klassenzahl des Kreiskörpers. Ann. Acad. Sci. Fenn. Ser. A I **416**, 7–48 (1967)
45. C. Meyer, *Die Berechnung der Klassenzahl Abelscher Körper über quadratischen Zahlkörpern.* (Akademie-Verlag, Berlin, 1957)
46. P. Mihailescu, On the class groups of cyclotomic extensions in presence of a solution to Catalan's equation. J. Number Theory **118**, 123–144 (2006)

47. J.C. Miller, Class numbers of totally real number fields. Ph.D. thesis, Rutgers Univ., 2015
48. J.C. Miller, Class numbers in cyclotomic $\mathbb{Z}_p$-extensions. J. Number Theory **150**, 47–73 (2015)
49. L.J. Mordell, On a Pellian equation conjecture. Acta Arith. **6**, 137–144 (1960)
50. T. Morishima, Über die Einheiten und Idealklassen des Galoisschen Zahlkörpers und die Theorie der Kreiskörper der l^{ν}-ten Einheitswurzeln. Jpn. J. Math. **10**, 83–126 (1933)
51. T. Morishima, *Über die Theorie der Kreiskörper der l^{ν}-ten Einheitswurzeln. II.* Jpn. J. Math. **11**, 225–240 (1934)
52. R. Okazaki, Inclusion of CM-fields and divisibility of relative class numbers. Acta Arith. **92**, 319–338 (2000)
53. F. Pollaczek, Über die irregulären Kreiskörper der l-ten und l^2-ten Einheitswurzeln. Math. Z. **21**, 1–38 (1924)
54. A.J. van der Poorten, H.J.J. te Riele, H.C. Williams, Computer verification of the Ankeny-Artin-Chowla Conjecture for all primes less than 100.000.000.000. Math. Comput. **70**, 1311–1328 (2001); Corr. ibid. **72** (2003), 521–523
55. C.G. Reuschle, *Tafeln complexer Primzahlen, welche aus Wurzeln der Einheit gebildet sind* (Königl. Akademie der Wissenschaften, Berlin, 1875)
56. K. Ribet, A modular construction of unramified p-extensions of $\mathbb{Q}(\mu_p)$. Invent. Math. **34**, 151–162 (1976)
57. A. Scholz, Über die Beziehung der Klassenzahlen quadratischer Körper zueinander. J. Reine Angew. Math. **166**, 201–203 (1932)
58. A. Scholz, Minimaldiskriminanten algebraischer Zahlkörper. J. Reine Angew. Math. **179**, 16–21 (1938)
59. R. Schoof, Class numbers of real cyclotomic fields of prime conductor. Math. Comput. **72**, 913–937 (2003)
60. R. Schoof, *Catalan's Conjecture* (Springer, Berlin, 2008)
61. G. Schrutka von Rechtenstamm, Tabelle der (Relativ)-Klassenzahlen der Kreiskörper, deren ϕ-Funktion des Wurzelexponenten (Grad) nicht größer als 256 ist. Abh. Deutsch. Akad. Wiss. Berlin (1964)
62. C. Soulé, Perfect forms and the Vandiver conjecture. J. Reine Angew. Math. **517**, 209–221 (1999)
63. P. Stevenhagen, Class number parity for the pth cyclotomic field. Math. Comput. **63**, 773–784 (1994)
64. K. Tateyama, On the ideal class groups of some cyclotomic fields. Proc. Jpn. Acad. **58**, 333–335 (1982)
65. H.S. Vandiver, On the composition of the group of ideal classes in a properly irregular cyclotomic field. Monatsh. Math. Phys. **48**, 369–380 (1939)
66. J. Värmon, *Über abelsche Körper, deren alle Gruppeninvarianten aus einer Primzahl l bestehen, und über abelsche Körper als Kreiskörper.* Akad. Abh. Lund, 1925
67. H.W. Verhoek, Class bumber parity of real fields of prime conductor. Master thesis Univ. Leiden, 2006

68. K. Yoshino, Class number parity for cyclotomic fields. Proc. Am. Math. Soc. **126**, 2589–2591 (1998)
69. K. Yoshino, M. Hirabayashi, On the relative class number of the imaginary abelian number field I, II. Mem. Coll. Liberal Arts, Kanazawa Medical Univ. **9**, 5–53 (1981); **10**, 33–81 (1982)
70. Y. Zhang, *p*-adic verification of class number computations. Ph.D. thesis Univ. Sydney, 2013

Translator's Preface

In 1952, Helmut Hasse published a book entitled *Über die Klassenzahl abelcher Zahlkörper (On the Class Number of Abelian Number Fields),* and in 1985, this book was reprinted with a second preface, supplemental bibliography and some corrections made by Jacques Martinet. This book is organized in to two parts: Part I is a translation of that reprint from 1985 with additional corrections, while Part II is an extension of Hasse's relative class numbers table.

In the original book, Hasse set a goal for himself: to deal with the abelian number fields in a systematic and structure-invariant way as freely as one deals with the quadratic fields. Since then, even though substantial progress in number theory has been made, there continue to remain many problems related to the contents of this book, especially problems of units and the class number of abelian number fields. To investigate such problems, as Hasse wrote, numerical examples are absolutely vital and so this book justifies its value.

To extend this book, we added Part II in which we included the tables of relative class numbers and unit indices of imaginary abelian number fields with conductor f, $100 < f \leq 200$. These tables are continuations of Hasse's table. They are reprints from our papers:

On the Relative Class Number of the Imaginary Number Field I, Mem. Coll. Liberal Arts, Kanazawa Medical Univ. **9** (1981), 5–53: part II, **10** (1982), 33–81.

(These two papers are referred to as Paper I and Paper II, respectively.) In fact, the generating characters, the types of the groups of characters, the numbers w of roots of unity, the unit indices Q and the relative class numbers h^* are tabulated and the subfields are displayed by diagram in the same way as Hasse's, except for the columns of the unit indices where we add the number of Hasse's Theorem by which we determined the value $Q = 1$ or 2.

Hasse calculated the relative class number of abelian number fields K by using the formula

$$h^* = Qw \prod_{\psi} N_{\psi}(\Theta(\psi)).$$

In fact, he calculated the contribution $N_\psi(\Theta(\psi))$ of representatives ψ and the unit indices Q and made the table of these values.

Our calculations in Part II were made, different from Hasse's, by starting with cyclic fields and applying the values to the formulas for non-cyclic fields obtained in Paper I. As for the unit index Q, we first determined the value of Q by Hasse's Theorems. After the completion of Paper I, we found some properties of unit index, which allowed us to easily determine the values in many cases in Paper II.

Looking back on the two tables, we realize we could have also used the contributions appearing in the table of Schrutka von Rechtenstamm instead of our calculation of determinant. At any rate, there would remain problems: how to determine the unit index and to draw the Hasse diagram of subfields of a cyclotomic field in general.

After the publication of Paper I and Paper II, we obtained some method of determining the value of the unit index of the composites of two, three and four imaginary quadratic fields. As for the product of contributions, $\prod_\psi N_\psi(\Theta(\psi))$, we obtained many kinds of determinantal formulas for an imaginary abelian number field, some of which were useful for actual calculation of the relative class number.

Nowadays, there are many tables and databases of invariants of number fields as well as softwares for number theory not only in publication but also available on the Internet. For example, LMFDB, Paris/GP, Magma and KANT are very convenient and indispensable for the study of number theory. However, it seems that almost all such tables contain no Hasse diagram and that the considered fields are restricted to special types of fields. In our tables in Part II, we treat all the imaginary abelian number fields with conductors f, $100 < f \leq 200$, and display the fields by Hasse diagram. Visual diagrams help us to understand the relations of subfields.

Recently, John C. Miller[1] has determined the class number of the maximal real subfield of some cyclotomic fields with conductor f, $100 < f \leq 200$. Considering the works of others so far, we may expect that in the near future we will determine the class number and the class group of all the abelian fields listed in the tables in Part II : not only the imaginary fields denoted by bullet points ○ but also the real fields by bullet points ●.

In the final section of Part II, in an updated version of the references, I include bibliographies on Hasse unit indices of CM fields, calculations of relative class numbers and determinantal expressions of relative class numbers. These bibliographies substantially depend on the ones that Ken Yamamura has made, in which he describes the lists with more detail.

Throughout the book, I have made some remarks in footnotes to assist the reader. I have also corrected errors and misprints in both the original version and the reprint. Some trivial misprints, especially in the tables, are corrected without remark.

There are many people I would like to thank. First of all, I thank the staff of Springer-Verlag for searching for the copyright holder of the book, for agreeing to

[1] J.C. Miller, Real Cyclotomic Fields of Prime Conductor and Their Class Numbers, Mathematics of Computations **84** (2015), 2459–2469.

publish this translation, for checking and reforming the English of the manuscript, for completing the tables in TeX and for engaging Franz Lemmermeyer to write the scholarly foreword. I would like to thank Jacques Martinet for accepting the reprint of his preface and for writing his own translation of that preface. I would also like to express my warmest thanks to Franz Lemmermeyer for advising me to add the papers with tables published at Kanazawa Medical University, for informing me of an error in the book and for writing the scholarly foreword. I have to thank Kanazawa Medical University for permitting us to reprint the papers with tables in Part II. I would like to express my thanks to Kôji Uchida and to Lawrence C. Washington for giving us (K. Yoshino and me) information about Theorem 29 and also to Hendrik W. Lenstra Jr. and to Guntram Schrutka von Rechtenstamm for sending us counterexamples to Theorem 29 during or after the publication of Part II. Additionally, I would like to thank Ken Yamamura for providing me the Tex files of the bibliographies of the three topics and for giving me the comments on the lists and Tetsuya Taniguchi for giving me information about the Internet bibliographies and software for calculation of relative class numbers. I would like to thank Hiroshi Fujisaki and Toru Nakahara, who gave me useful comments on the translation, and Katsuya Miyake for reading through the manuscript and giving me many invaluable comments. I should thank Atsuki Umegaki for the effort of locating the copyright holder. I also thank Ken-ichi Yoshino, who has studied number theory together with me for a long time, for giving me advice on the tables and for accepting the reprints of the tables in Part II. Finally, I wish to express my sincere thanks to my teacher, Yoshiomi Furuta, who introduced me to Hasse's book as a textbook in the graduate course at Kanazawa University, for his constant encouragement during the translation.

Nonoichi, Japan Mikihito Hirabayashi

Preface to the Second Edition

The aim of this book is clearly explained by Hasse: based on calculations at $s = 1$ of residues of zeta functions, to give for abelian fields procedures for computing the class number and a system of fundamental units. It is well known that knowledge of the residue at $s = 1$ of a number field K allows one to calculate the product $h_k R_K$ of the class number h_K and the regulator R_K of K. It amounts to the same as considering the value at $s = 1$ of the "L-function" L_K of K (the quotient of the zeta functions of K and $\mathbb{Q}$), or the value at $s = 0$ of a derivative (of a convenient order) of this function, a viewpoint that yields more handsome formulae. When K is abelian, the decomposition of L_K as a product of Artin L-functions $L(s, \chi)$ of irreducible characters takes a simple form: all characters have degree 1 and can be viewed as Dirichlet characters, so that the calculation of $L(1, \chi)$ or derivatives $L^{(k)}(0, \chi)$ reduces to the calculation of a finite sum. We obtain in this way a formula for $h_K R_k$ (Sects. 1.1–1.6). This is the starting point of Hasse's program, about which I would like to point out the following two facts: use as often as possible analytic proofs and take great care of numerical applications.

I shall comment below on the main results of the book and on progress obtained during the last 30 years, thanks in particular to the work of K. Iwasawa and H.W. Leopoldt. Iwasawa's theory and the related theory of p-adic L-functions (cf. [1]) lie at the core of recent developments. This is a wide subject, in the details of which it was not possible to enter.

Due to its major historical importance, it seemed desirable to provide the mathematics community with the original form of Hasse's book. This second edition reproduces up to some minor corrections the 1952 edition (the letter P then refers to this preface). Besides the comments below I have added a short reference list. More details can be found in the long reference list of L.C. Washington's book *Introduction to Cyclotomic Fields* [7][2]; see also the reference list of S. Lang [5]. I have restricted myself to this short list, either because the sources looked especially

[2]Second edition: 1997.

important, or because they threw light on some interesting aspects of the theory, or simply because they did not occur in [7].

I thank J. Coates, V. Ennola, R. Gillard, G. Gras, B. Gross, M. Hirabayashi, H.W. Lenstra, B. Mazur, B. Oriat, J.-P. Serre and L.C. Washington for their comments on a preliminary version of this preface.

I also thank the editors at Akademie-Verlag and Springer-Verlag who gave me the opportunity of writing this preface.

Real Abelian Fields (Sects. 2.1–2.12)

These are the fields all the characters of which are *even* (i.e. one has $\chi(-1) = +1$). Hasse tries to get rid of regulators that show up in Sect. 1.5, formula (1.5.3a), by proving equalities of the form $c_K h_K = [E_K : H_K]$, in which H_K stands for a subgroup of finite index in the group E_K of units of K constructed using explicit units of the cyclotomic field having the same conductor as K, and c_K is a convenient rational number. He succeeds in some cases, obtaining two types of formulae. In the first series (Sect. 2.5), the coefficient c_K, denoted by g_K, depends on the ramification, and his formulae apply to fields in which a unique prime ideal lies above each ramified prime. In the second series, there appears a coefficient denoted by c_G that solely depends on the Galois group G of $K/\mathbb{Q}$. They apply every time G is cyclic, with indeed $c_G = 1$ in this case (Theorem 2.9, Sect. 2.11).

Leopoldt constructed in [26] convenient cyclotomic unit groups thanks to which he was able to extend in full generality the formulae above. He also proved formulae that put in evidence factors h_χ attached to non-trivial, $\mathbb{Q}$-irreducible characters of the Galois groups, which give the class number formulae the form $h_K = (Q_K^+/Q_G)\prod_\chi h_\chi$, where Q_G is an integer depending only on G and Q_K^+ is an integer that can be calculated once E_K is known, the prime divisors of which are a priori known; these are analogues of formulae proved by Hasse for relative class numbers (see Relative Classes: The Hasse Index (Sects. 3.2–3.8)). Actually, there are several possible choices for the group of cyclotomic units. Gillard [16] proved formulae involving smaller factors Q_G (even sometimes with non-trivial denominators). Other formulae have been proved by Sinnot (see Relative Classes and Stickelberger Elements).

Using the methods above to construct extended table requires important computational tools, which were not available to Hasse when he wrote his book. One must first have good a priori upper bounds for the index $[E_K : H_K]$ of the cyclotomic unit group. Such upper bounds have been given by G. and M.-N. Gras in [15]. One can then give a computer the task of making the necessary *dévissages* to construct a basis for E_K from a basis for H_K. This is how M.-N. Gras constructed her extended tables for degrees 3 and 4. Tables for degree 6 were then constructed by S. Mäki. The tables for degrees 3 to 6 have then been enlarged by Ennola, Mäki and Turunen.[3] It would be interesting to improve the bounds in [15] for degrees other than 2, 3, 4 and 6.

The structure of class number formulae suggests comparing the structure *as G-modules* of the class group and of the quotient group E_K/H_K. A precise conjecture comparing the p-components of the Jordan–Hölder sequences of these groups for primes p that do not divide the order of G is stated in [15, p. 127]. This conjecture has been reduced by R. Greenberg to classical problems in the Iwasawa theory and consequently has been proved in various cases; I shall return to this question in Iwasawa's Theory below.[4]

Relative Classes: The Hasse Index (Sects. 3.2–3.8)

As an error occurs in Sect. 3.7, Theorem 3.29, it is necessary to study closely this index. Let K be an imaginary abelian field, or more generally, a *CM* field, which is a totally complex quadratic extension of a totally real subfield K_0. Calculating the value at $s = 1$ (or at $s = 0$) of L_K/L_{K_0} involves the quotient R_K/R_{K_0}, hence the (finite) index $[E_K : E_{K_0}]$. Denote by μ_K the group of roots of unity of K, and let $Q_K = [E_K : \mu_K E_{K_0}]$ be the *Hasse index of* K. The Herbrand quotient of E_K for the action of $g := \mathrm{Gal}(K/K_0)$ is equal to $2^{[K_0:\mathbb{Q}]-1}$ (the calculation is easy, because E_K/E_{K_0} is a finite group). This implies the following alternative, in which N denotes the norm from K to K_0: either $N(E_K) = E_{K_0}^2$, and then $H^1(g, E_K) = \mu_K/\mu_K^2$ is of order 2 and $Q_K = 1$, or $N(E_K)$ contains $E_{K_0}^2$ to index 2, and then $H^1(g, E_K) = \{1\}$ and $Q_K = 2$. The following theorem easily follows from the results above.

Theorem 29′ *If $\tilde{K}$ is an imaginary subfield of K and if the index of $\mu_{\tilde{K}}$ in μ_K is* odd, *then $Q_{\tilde{K}}$ divides Q_K.*

Making use of the results proved in Sect. 3.8, one can easily check that for $\tilde{K} = \mathbb{Q}(\sqrt{-1}, \sqrt{34})$ and $K = \tilde{K}(\sqrt{2})$, we have $Q_{\tilde{K}} = 2$ and $Q_K = 1$ (the orders of $\mu_{\tilde{K}}$ and μ_K are 4 and 8, respectively). In this example, due to Lenstra, the fields K and $\tilde{K}$ both have conductor 136. Index calculations of Auxiliary table: values of the unit indices, pp. 247–250, are actually correct and can be proved using when necessary Theorem 29′ above or Theorem 3.25 of Sect. 3.7 instead of Hasse's Theorem 3.29.

[3]On real cyclic sextic fields, *Math. Comp.* 45 (1985), 591–611.

[4]Gras' conjecture follows from the "Main Conjecture" as stated in [1], which is now a theorem. See B. Mazur and A. Wiles, Class fields of abelian extensions of $\mathbb{Q}$, *Invent. Math.* **76** (1984), 179–330; see also A. Wiles, On p-adic representations for totally real fields, *Ann. Math.* **123** (1986), 407–456.

Relative Classes: Questions of Integrality (Sects. 3.1 and 3.9–3.15)

Let K be a CM field of degree 2 over its maximal real subfield K_0. Write its class number as a product $h_K = h^* h_{K_0}$. Then h^* is an integer: this easily follows from Class Field Theory, interpreting h^* as a relative class number, and Hasse gives an analytic proof of this result when K is an abelian field. Using Leopoldt's [4] generalized Bernoulli numbers $B_n(\chi)$ (previously considered in 1890 by A. Berger—see [7]), we may rewrite formula (3.9.1) of Sect. 3.9 as

$$h^* = Q_K\, w \prod_{\chi(-1)=-1} \left(\left(-\frac{1}{2}\right) B_1(\chi)\right).$$

Putting together factors associated with $\mathbb{Q}$-conjugate characters, we write h^* as a product of rational numbers, which is indeed integral except for some characters, the conductors of which are prime powers (Theorem 3.34, Sect. 3.15). These properties of generalized Bernoulli numbers are better understood in terms of Kubota–Leopoldt's p-adic L-functions [24]: we are led to calculate the valuation of $\frac{1}{2} L_p(0, \psi)$, where ψ denotes the image of χ under Leopoldt's reflection (*Spiegelung*). Details can be found in [13], the introduction of which contains a review of the various ways of defining p-adic L-functions, and in [36].

Parity of the Class Number and Signature of the Unit Group (Sect. 2.6)

Various authors have studied the relationship between the parity of the class number and the signature of the group of cyclotomic units (or of the whole unit group—there is no need to limit oneself to abelian fields), and the results of Sect. 2.6 have been generalized in various settings. Details on this problem can be found in [33] and in the six references quoted there.

Results Related to Invariant Classes (Sects. 3.16–3.20)

A well-known formula allows us to compute the number of classes that are invariant by the Galois group of a cyclic extension. The general formula appears in Chevalley's thesis, but the case of a prime degree, which suffices for many applications, was known to Takagi. The results of Sects. 3.16–3.20 can be easily proved using this formula (which also often proves useful for the calculation of the Hasse index). It should be noted that Theorems 3.36 and 3.37 of Sect. 3.16 had been previously generalized by Furtwängler to regular primes [10], by arguments that follow from Hilbert's Theorem 94. One should also note that Hasse prefers to make use of the theory of genera, which yields quotients of the class group, rather than that of invariant classes, which yields subgroups.

Relative Classes: Numerical Results

Schrutka von Rechtenstamm has published a table that gives for fields K of conductor f such that $\varphi(f) \leq 256$ the contributions (*Beiträge*) associated with classes of $\mathbb{Q}$-conjugacy of odd characters of K. One deduces from this table the relative class number h^* of cyclotomic fields with conductor as above, but not that of the other abelian fields, by lack of knowledge of the Hasse index.

Table 3.2 has been extended by Hirabayashi and Yoshino ([19] for $101 \leq f \leq 150$ and [20] for $151 \leq f \leq 200$). A few errors in [19], due to the use of Hasse's statement of Theorem 3.29, which concern conductors 104, 112, 120 and 136, have been corrected in [20].

Among other numerical results let us cite computations relative to cyclotomic fields carried out by Wagstaff in connection with Fermat's Last Theorem; see [7]. See also [42], which contains a list of subfields that may imply the irregularity of primes < 125,000 considered by Wagstaff.

In calculations concerning conductor 25, which Hasse chose as an illustration of Sect. 3.16, the three formulae on pages 159–160 must be replaced by those displayed below:

$$\text{(i) } \Theta(\psi_2) = \frac{1+\psi_2(2)(1+\zeta^3)}{1-\zeta}\,; \qquad \text{(ii) } \Theta(\psi_2) = \frac{1+i\zeta^2(1+\zeta^3)}{1-\zeta}\,;$$

$$\text{(iii) } N(\Theta(\psi_2)) = \frac{1+\zeta^4(1+\zeta^3)^2}{(1-\zeta)^2} = \frac{2+\zeta^4+2\zeta^2}{(1-\zeta)^2} = \frac{1+\zeta^2}{1-\zeta}\,.$$

The end of Sect. 3.16 is correct.

Relative Classes and Stickelberger Elements

Generalizing Iwasawa's paper [23], Sinnot has obtained for abelian fields formulae interpreting the class number of a real field or the relative class number of an imaginary field as a group index, in [40] for cyclotomic fields, and then in [41] for the general case. In the real case, he makes use of unit groups. In the relative case, he considers the group ring $\mathbb{Z}[G]$ ($G = \mathrm{Gal}(K/\mathbb{Q})$) and makes use of the index of an ideal S of $\mathbb{Z}[G]$ (the *Stickelberger ideal*) in a submodule of $\mathbb{Z}[G]$.

Sinnot's theorem quoted above is one of the numerous applications of "Stickelberger elements" to the theory of abelian fields. They play an important rôle, in particular in the construction of p-adic L-functions; see [1, §§**2–4**].

For another interpretation of the relative class number as a group index, see C.-G. Schmidt [37, 38].

Small Class Numbers

Important results concerning imaginary abelian fields have been proved. First, we must cite the theorem of Heegner–Stark–Baker, which asserts that there are exactly 9 imaginary quadratic fields with class number 1, those which were known to Gauss (see [43]); the problem of class number 2 has also been solved (see [31] and [44]), and it seems that recent results of Gross and Zagier, together with previous work of Goldfeld [17], yield an effective solution for any class number.[5] Of course, that there exist only finitely many imaginary quadratic fields having a given class number is an old result of Heilbronn (and is also a consequence of the Brauer–Siegel Theorem).

Thanks to the work of Masley, Montgomery and Uchida, we also know the cyclotomic fields having a given (not too large) class number; in particular, there are exactly 30 cyclotomic fields with class number 1.

Lower Bounds of Discriminants

Thanks to the work of Stark and Odlyzko, completed by Serre and Poitou, relying on analytic tools, we now have sharp lower bounds of discriminants of number fields (in absolute value, and for any given signature) much better than those previously obtained using geometric methods. We now have tables due to Odlyzko (unpublished,[6] but excerpts were published in [29]) and to Diaz y Diaz.[7] Using these lower bounds, we can obtain upper bounds for the degrees of unramified extensions of number fields of a not-too-large discriminant. This method can be refined using Galois action, following the procedure described by Masley in [30]. This method has been applied to many abelian fields. In particular, van der Linden [28] has obtained results that look out of scope of methods based on the determination of cyclotomic

[5]See J. Oesterlé, Nombres de classes des corps quadratiques imaginaires, *Astérisque* **121–122** (1985), 309–323; see also D. Goldfeld, Gauss's class number problem for imaginary quadratic fields, *Bull. Amer. Math. Soc.* **13** (1985), 23–37.

[6]However, see A.M. Odlyzko, Bounds for discriminants and related estimates for class numbers, regulators and zeros of zeta functions: A survey of recent results, *Séminaire (now Journal) de Théorie des Nombres de Bordeaux* **2** (1990), 119–141.

[7]F. Diaz y Diaz, Tables minorant la racine n-ième du discriminant d'un corps de degré n, *Publ. Math. Orsay* **80, 6** (1980), 59 pp.

units in the case of maximal real subfields of cyclotomic fields. Unfortunately, this method only applies to finitely many fields.

Iwasawa's Theory

This is the domain in which the most important developments in the theory of abelian fields arise. The starting point is the fundamental paper of Iwasawa ([21]; see also [39] and [22]), in which he proves that in a $\mathbb{Z}_p$-extension K_∞ of a number field K, the order of the p-component of the class group of K_n (the subextension of degree p^n in K_∞/K) is of the form p^{e_n}, where $e_n = \mu p^n + \lambda n + \nu$ for n large enough and μ, λ, ν are integers with $\mu, \lambda \geq 0$. The case when K_∞/K is the cyclotomic $\mathbb{Z}_p$-extension is particularly important: on the one hand, some invariants attached to K_∞/K account for various properties of K itself (cf. [1]); on the other hand, and this is directly connected with the subject of Hasse's book, the fields K_n are abelian whenever K is. Here we cite a few contributions to the theory that brought answers to famous conjectures or had important applications. To the work by Iwasawa referred to above, by Kubota and Leopold cited in Relative Classes: Questions of Integrality (Sects. 3.1 and 3.9–3.15) and by Greenberg cited in Real Abelian Fields (Sects. 2.1–2.12), we add the following five references:

- The proof by Ferrero and Greenberg [11] that the zeros at $s = 0$ of the Kubota-Leopoldt L-functions (if any) are simple.
- The proof by Ferrero and Washington ([12]; see also [7, Ch. 7] and [4, Ch. 11]) that Iwasawa's μ-invariant vanishes when K_∞/K is the cyclotomic $\mathbb{Z}_p$-extension and K is abelian.
- The proof by Washingtom [45] that the exponent of a prime ℓ in the class groups of the K_n inside the cyclotomic $\mathbb{Z}_p$-extension of an abelian field ($\ell \neq p$) is bounded.
- The study by Friedman of cyclotomic extensions of abelian fields having a Galois group isomorphic to a product $\mathbb{Z}_{p_1} \times \cdots \times \mathbb{Z}_{p_s}$.
- Finally, the announced proof by Mazur and Wiles of the proof of the Main Conjecture (see Coates [8, §**5**]) for cyclotomic fields. This fundamental result, which extends to the previous work by Ribet [35] and Wiles [46], and which is expected to extend to all abelian fields,[8] gives important information on the action of Galois groups on class groups; we meet again this way the problems considered in Real Abelian Fields (Sects. 2.1–2.12).

Talence, France
April 28, 1983

Jacques Martinet

[8]Done, see footnote 4, p. xxi.

References

(a) Books and Expository Articles

1. J. Coates, p-adic L-functions and Iwasawa's theory, in *Algebraic Number Fields*, ed. by A. Fröhlich (Academic Press, London, 1977), pp. 269–353
2. K. Iwasawa, *Lectures on p-adic L-Functions*. Annals of Mathematical Studies, vol. 74 (Princeton Univ. Press, Princeton, 1972)
3. S. Lang, *Cyclotomic Fields*. Graduate Texts in Mathematics, vol. 59 (Springer, New York, 1978)
4. S. Lang, *Cyclotomic Fields II*. Graduate Texts in Mathematics, vol. 69 (Springer, New York, 1980)
5. S. Lang, Units and class groups in number theory and algebraic geometry. Bull. Am. Math. Soc. **6**, 253–316 (1982)
6. K. Ribet, *Fonctions L p-adiques et théorie d'Iwasawa*. (Notes by Ph. Satgé.) Publications Math. Orsay 79.01 (1979)
7. L.C. Washington, *Introduction to Cyclotomic Fields*. Graduate Texts in Mathematics, vol. 83 (Springer, New York, 1982)

(b) Articles

8. J. Coates, The work of Mazur and Wiles on cyclotomic fields, in *Sém. Bourbaki*. Lecture Notes in Mathematics, vol. 901 (Springer, Berlin, 1981), 23 pp.
9. E. Friedman, Ideal class groups in basic $\mathbb{Z}_{p_1} \times \cdots \times \mathbb{Z}_{p_s}$ extensions of number fields. Invent. Math. **65**, 425–440 (1982)
10. Ph. Furtwängler, Über die Klassenzahlen der Kreisteilungskörper. J. Reine Angew. Math. **140**, 29–32 (1911)
11. B. Ferrero, R. Greenberg, On the behavior of p-adic L-functions at $s = 0$. Invent. Math. **50**, 91–102 (1978)
12. B. Ferrero, L.C. Washington, The Iwasawa invariant μ_p vanishes for abelian number fields. Ann. Math. **109**, 377–395 (1979)
13. G. Gras, Sur la construction des fonctions L p-adiques abéliennes. Sém. D.P.P., Paris exp. nu **22**, 20 pp. (1978–1979)
14. G. Gras, Canonical divisibilities of values of p-adic L-functions, in *Journées Aritmétiques 1980*, ed. by J.V. Armitage. Lecture Notes of the London Mathematical Society, vol. 56 (Cambridge Univ. Press, Cambridge, 1982)
15. M.-N. Gras, Calcul du nombre de classes et des unités des extensions abéliennes réelles de $\mathbb{Q}$. Bull. Sci. Math. **101**, 97–129 (1977)
16. R. Gillard, Remarques sur les unités cyclotomiques et les unités elliptiques. J. Number Theory **11**, 21–48 (1979)

17. D.M. Goldfeld, The class number of quadratic fields and the conjectures of Birch and Swinnerton-Dyer. Ann. Scuola Nor. Sup. Pisa, Cl. Sc. **3**, 624–663 (1976)
18. R. Greenberg, On p-adic L-functions and cyclotomic fields II. Nagoya Math. J. **67**, 139–158 (1977)
19. M. Hirabayashi, K. Yoshino, On the relative class number of the imaginary abelian number field I. Mem. Coll. Libr. Arts Kawasawa Med. Univ. **9**, 5–53 (1981)
20. M. Hirabayashi, K. Yoshino, On the relative class number of the imaginary abelian number field II. Mem. Coll. Libr. Arts Kawasawa Med. Univ. **10**, 33–81 (1982)
21. K. Iwasawa, On Γ-extensions of algebraic number fields. Bull. Am. Math. Soc. **65**, 183–226 (1959)
22. K. Iwasawa, On $\mathbb{Z}_\ell$-extensions of algebraic number fields. Ann. Math. **98**, 246–326 (1973)
23. K. Iwasawa, A class number formula for cyclotomic fields. Ann. Math. **76**, 171–179 (1962)
24. T. Kubota, H.W. Leopoldt, Eine p-adische Theorie des Zetawerte, Teil I: Einführung der p-adischen Dirichletschen L-Functionen. J. Reine Angew. Math. **214–215**, 328–339 (1964)
25. D.S. Kubert, The 2-divisibility of the class number and the Stickelberger ideal. J. Reine Angew. Math. **369**, 192–218 (1986) (Instead of "*Bull. Soc. Math. France*, to appear")
26. H.W. Leopoldt, Über Einheitengruppe und Klassenzahl reller abelscher Zahlkörper. Abh. Deutsch. Akad. Wiss, Kl. Math. **1953**(2) (1954)
27. H.W. Leopoldt, Eine Verallgemeinerung der Bernoullischen Zahlen. Abh. Math. Sem. Univ. Hamburg **22**, 131–140 (1958)
28. F.J. van der Linden, Class number computations of real abelian number fields. Math. Comput. **39**, 693–707 (1982)
29. J. Martinet, Petits discriminants des corps de nombres, in *Journées Aritmétiques 1980*, ed. by J.V. Armitage. Lecture Notes of the London Mathematical Society, vol. 56 (Cambridge Univ. Press, Cambridge, 1982)
30. J. Masley, Odlyzko bounds and class number problems, in *Algebraic Number Fields*, ed. by A. Fröhlich (Academic Press, London, 1977), pp. 465–474
31. H.L. Montgomery, P.J. Weinberger, Notes on small class numbers. Acta Arith. **24**, 529–542 (1973–1974)
32. B. Mazur, A. Wiles, Class fields of abelian extensions of $\mathbb{Q}$. Invent. Math. **76**, 179–330 (1984)
33. B. Oriat, Relation entre les 2-groupes des classes d'idéaux aux sens ordinaire et restreint de certains corps de nombres. Bull. Soc. Math. France **104**, 301–307 (1976)
34. G. Poitou, Sur les petits discriminants. Sém. D.P.P., Paris, exp. nu. **6**, 17 pp. (1976–1977)
35. K. Ribet, A modular construction of unramified p-extensions of $\mathbb{Q}(\mu_p)$. Invent. Math. **34**, 151–162 (1976)

36. K. Ribet, p-Adic L-functions attached to characters of p-power order. Sém. D.P.P., Paris, exp. nu. **9**, 8 pp. (1977–1978)
37. C.-G. Schmidt, Größencharactere und Relativklassenzahl abelscher Zahlkörper. J. Number Theory **11**, 128–159 (1979)
38. C.-G. Schmidt, On ray class annihilators of cyclotomic fields. Invent. Math. **66**, 215–230 (1982)
39. J.-P. Serre, Classes des corps cyclotomiques (d'après K. Iwasawa). Sém. Bourbaki, exp. nu. **174**, 11 pp. (1958)
40. W.M. Sinnot, On the Stickelberger ideal and the circular units of a cyclotomic field. Ann. Math. **108**, 107–134 (1978)
41. W.M. Sinnot, On the Stickelberger ideal and the circular units of an abelian field. Invent. Math. **62**, 181–234 (1980)
42. K. Selucky, L. Skula, Irregular imaginary fields. Arch. Math. **2**, Scripta Fac. Sci. Nat. Ujep Brunensis (= Brno) XVII (1981), 95–112
43. H.M. Stark, On the "gap" in a theorem of Heegner. J. Number Theory **1**, 16–27 (1969)
44. H.M. Stark, On complex quadratic fields with class-number two. Math. Comput. **29**, 289–302 (1975)
45. L.C. Washington, The non p-part of the class number in a cyclotomic $\mathbb{Z}_p$-extension. Invent. Math. **49**, 87–97 (1979)
46. A. Wiles, Modular curves and the class group of $\mathbb{Q}(\zeta_p)$. Invent. Math. **58**, 1–35 (1980)

Preface to the First Edition

Algebraic number theory, since Gauss put it on the stage, has developed into a mighty edifice of theory—by the hands of great masters of the last and present centuries (the nineteenth and twentieth centuries)—which is nowadays essentially self-contained and abundant with general theorems, dominant methodical viewpoints and profound structural insights. The first stage of this development was summarized by Hilbert in his famous report [2][9] on the theory of algebraic number fields. This report, Zahlbericht, presents in the first two parts the general foundation of the theory and then deals in the subsequent three parts with the three types of algebraic number fields in detail: quadratic fields, cyclotomic fields and Kummer fields. From today's point of view, the last three parts of Hilbert's Zahlbericht are special cases of the general theory of relative abelian number fields. They introduce the second stage of the development to which Hilbert himself gave an impetus with his audacious concept of class fields and the main theorems of class field theory. Following Hilbert's Zahlbericht, I summarized this second stage, the theory of relative abelian number fields, in a three-part report [1],[10] where the theory of class fields has developed into a fully general form and has been applied to the derivation of general reciprocity laws.

In this full development that has been led by general theoretic, structural, methodical and systematic aspects, there appears, however, a peculiar desire of every serious number theorist for explicit control of treated objects to execute numerical examples in the background. Today, when one asks a number theorist for what type of algebraic number fields he could explain the legitimacy of the general theory by means of explicit materials of general structure invariants of treated type of fields, or also as preparation of explanation of the legitimacy, he could obtain, for example, only an integral basis, the discriminant, a system of fundamental units and the class number by means of systematic structure invariants; then, if he is honest, the answer generally would be: only for quadratic fields. Today, only in quadratic

[9]In the following, this report will be cited as "Zahlbericht" for short.

[10]In the following, this report will be cited as "Klassenkörperbericht" for short.

fields would every number theorist, and probably also many mathematicians of other branches, feel quite at home such that in those fields they could deal with the concept of general theory as they like; whereas, for fields of higher degree, the freedom of movement (treatment) is, at least, severely restricted by the complete and sovereign control of general theory. In fact, we are not short of approaches corresponding to control of fields different from quadratic fields and not short of numerical examples that are added to use for the explanation of general legality here and there. For most of such approaches, however, there is a lack of uniformity, a lack of systematic representation and, above all, the sense that one has to describe the treated type of fields not by contingent determining materials—for example, the coefficients of a generating polynomial, even if they are arbitrarily selected or normalized through any condition of reduction—but by structure invariants, such as discriminants, associated class groups, conductors and characters. And, in the numerical examples, ad hoc made-up tricks and tentative guesses, more or less, prevail against systematic calculation processes. Indeed, one could make a constructive process of general theory to obtain an integral basis of treated types of fields, say, absolute cyclic fields; nevertheless, in this way one cannot obtain in general any excellent integral basis constructed by structure-invariant determinative pieces of these types of fields. The requirement of profoundly strenuous number theorists is not satisfied with *a list of all cases*, which seems to be the final aim of numerous North American works in such and similar cases. For the *complete solutions* that often seem to be sufficiently plain to us, we are mostly short of incorporation of treated objects in the general theory and (short) of meaningful control of special cases by substantial structure invariants in this theory. One would be reminded of certain types of works from the descriptive natural sciences[11] or from the prehistory, which have actually gathered and registered all the materials most faithfully but failed to bring them in a gathering order or system, to interpret them from the general point of view and to draw essential from unessential, legality from random.

As for the execution of numerical examples mentioned many times so far, I would like here to prevent the misunderstanding that I have given such examples of general theorems unjustified significance. Numerical examples are frankly inserted in many textbooks, works and speeches on number theory, and general research is carried out really with examples. Experience teaches us that the reader or the audience, if he is not excessively conscientious, easily passes over these examples and strives against the general threads of the research. He would rather seek examples by himself and carry them out; this is in the right because the significance of numerical examples, after all, does not lie in a perfectly shaped and informed calculation and its result but in the activity that is requisite for the execution of the examples. The study of general theory develops the potential of ability, intellectual power and strength to the treated subjects. The execution of examples is a touchstone that one could have had the theory as one's own and that one has sovereignly mastered the theory; the

[11](Translator's remark) The descriptive natural sciences consist of the branches of natural science except mathematics, physics and chemistry.

execution is also a test of the obtained power and strength. The practice of this ability, the execution of this power and the application of this strength induce a full feeling of joy and satisfaction when the example is computed by oneself, but not when the example has already been prepared and obtained. Here I only raise a question as to what extent the correspondence between theory and examples holds completely, in general, for every way of mathematical activity and then for the development of mathematical theory: there might be many things to say on this correspondence. It is true that in physics lectures on experimental physics are given in conjunction with practical courses, where the receptive learning is consolidated in its own activity. In number theory, the execution of numerical examples has a completely corresponding role. Moreover, the examples under the control of research mathematicians are considered just as the experiments for physicists, that is, one of the main means of discovery of new laws. Hereby it is clear why and why not I put high value on the explicit command of general theory up to the execution of numerical examples; in my own works, such examples are introduced only for actual reasons that seem to be imperative.

The sense that the explicit control of objects in all details should keep pace with the general growth of theory existed in a very distinctive way by Gauss and later, above all, by Kummer. Especially, Kummer has produced plenty of supplementary research in this direction to his general theory of ideal numbers. But this sense has gradually disappeared under the overwhelming influence that Hilbert has had on the wide developments of number theory. It is completely typical for Hilbert's position of focusing on the generalization and conceptualization, as well as on existence and structure, that in his Zahlbericht [2] he dismissed all the research and results of Kummer and others, who dealt with explicit control of objects, by giving short references and indications without explaining them one by one; Hilbert has, in fact, replaced the systematic and logical proof methods of Kummer—which are more numerical, constructive and hence easily accessible by explicit execution—by more conceptual, numerically difficult and hardly controllable conclusive methods. Dedekind also participated decisively in this development with his very conceptional and deeply axiomatic method; whereas, in the another direction, the more constructive methods of Kronecker and Hensel have extended the tradition of Kummer. These methods will hardly prevail against the overwhelming influence of Hilbert until, then and even today, the decisive results of these methods have offered us the groundwork of the return of the organized balance of both directions. It would not be self-evident that we misunderstand or diminish the great merit that Hilbert has acquired: excessively impressive generality, conceptional clearness and perfect-in-form simplicity by his unwavering and consistent adhesion to his above-mentioned attitude towards the upward development of algebraic number theory. This is clearly due to his great success and to everything that could contribute to the completion of the architecture by the generation after him who follow his spirit and methods. Standing myself at the final stage of this development, I am always aware clearly and impressively that along with all the splendid beauty and striking greatness we are still short of something essential that one could really feel at home in the constructed house. One must come to know most accurately one's individual floors and one's individual rooms in their peculiar nature and their relation, and one must learn how to freely move about in the floors. To this purpose, it seems to me requisite, by returning from concept to object, to make the setting contrary to the approach of Hilbert come back more into its own in explicitness and in detail to the numeration. Being led by this viewpoint, in the second part of my report on the class field (Klassenkörperbericht [1]), which is concerned with general reciprocity laws, I have given comparatively broad space to the explicit formulas for these laws that connect with the researchers before Hilbert. With this larger work I adopt this viewpoint to another object again.

I have set myself a goal, if possible, to make extensively a class of absolute abelian number fields, at least the class of absolute cyclic fields, accessible in a systematic and structure-invariant way so that one can deal with such fields as freely as quadratic fields. As a touchstone for achieving this goal, it may be important to make tables for these fields on the grounds of methods, formulas and results to be developed in a systematic procedure such as Sommer [4] who has included tables for quadratic fields in his famous textbook. For every number theorist who wishes to acquire numerical examples by himself—for the general laws acquired so far from the theory of algebraic number fields and for any eventual broader laws—such tables would provide supremely valuable hand instruments, as so are Sommer's tables today, whether they would be used to make the theory familiar or to find clues to new laws and relations in experimental ways. So far, in this direction there exists only the table of complex prime numbers of Reuschle [3] computed by Kummer's works, which is, however, unsatisfactory in spite of the ample quantity of gathered materials in the table, because it does not contain the essential components, namely, fundamental units, class numbers and class groups.

The present work, as a first contribution to the above-mentioned broad aim, shall develop the foundation of systematic calculations of the class numbers of absolute abelian number fields. Moreover, this book is considered as a supplement to Hilbert's Zahlbericht [2] and to my Klassenkörperbericht [1]. In addition to the viewpoint of using general class number formula for actual calculation of class numbers, there appear accordingly the viewpoints of reporting the results obtained by Kummer and others on the class numbers of cyclotomic fields and other types of fields, of generalizing these results to arbitrary abelian number fields and of shedding light on them from the standpoint of the general class field theory.

Besides proper results on the class numbers of absolute abelian number fields, the present work also contains plenty of algebraic and number-theoretic details, which are attractive by themselves: partly incorporated into the proofs and partly emphasized individually, such as a special generalization of the famous decomposition of the group determinant of an abelian group into linear factors, theorems on the behaviour of ramification of an absolute abelian number field over its maximal real subfield and a new side piece of Gauss' lemma on the quadratic residue. Contrary to many observations on the other side, I have always thought that the classical number theory, which has goal-setting and methodology directing to the reality of natural numbers, is today thoroughly capable of further active and productive development on the inherent ground and also that we need not seek new subsistence for the impulse of arithmetic activity in the domain of abstract algebra or in topology, set theory and axiomatics. As in music, after the periods of Romanticism and post-Romanticism indulging in heroic and demonic works as well as in boldest fantasy, one would again vigorously remember today, with all the joys of these creative works, the fountainhead of pure and simple musicality of the ancient masters; also in number theory, which is controlled by the laws of harmony perhaps more than any other branch of mathematics, it seems to me that there appears the back deliberation, that is, what great masters who have established the theory had in mind as real figures. The present work may be lively evidence of my view and lead the way to broader research on this trodden field.

Göttingen, Germany
August 1945

Helmut Hasse

I am greatly indebted to Miss G. Beyer, Mr. H.W. Leopoldt and Mr. C. Meyer for many invaluable comments by perusal of the manuscript and the corrections, to Mr. R. Schwarzenberger for his help in executing some very complicated calculations in the table and to the publishing company for their willing acceptance of my various requests in the printing.

Hamburg, Germany
November 1951

Helmut Hasse

References[12]

1. H. Hasse, Bericht über neuere Untersuchungen und Probleme aus der Theroie der algbraichen Zahlkörper. Part I: Klassenörpertheorie. Jahresbericht D.M.-V. **35** (1926); H. Hasse, Bericht über neuere Untersuchungen und Probleme aus der Theroie der algbraichen Zahlkörper. Part Ia: Beweise zu Teil I. Jahresbericht D.M.-V. **36** (1927); H. Hasse, Bericht über neuere Untersuchungen und Probleme aus der Theroie der algbraichen Zahlkörper. Part II: Reziprozitätsgesetz. Jahresbericht D.M.-V. Supplemental Ed. **6** (1930). Cited as "Klassenkörperbericht." [Introduction, **1.2**, **1.3**, **2.3**, **2.12**, **3.1**, **3.2**, **3.4**, **3.5**, **3.9**, **3.19**]
2. D. Hilbert, Die Theorie der algebraischen Zahlkörper. Jahresbericht D.M.-V. **4** (1897). Cited as "Zahlbericht." [Preface, Introduction, **1.2**, **1.6**, **2.9**, **3.2**, **3.5**, **3.8**, **3.19**]
3. C.G. Reuschle, *Tafeln komplexer Primzahlen, whelche aus Wurzeln der Einheit gebildet sind.* Berlin (1875) [Preface, Appendix]
4. J. Sommer, *Vorlesungen über Zahlentheorie. Einführung in die Theorie der algebraischen Zahlkörpern* (Druck Und Verlag, Leipzig-Berlin, 1907) [Preface, **3.8**]

[12]The bold-typed numbers attached in the square brackets denote the sections of this book in which the individual works are cited.

Contents

List of Theorems

Part I

Introduction

It is basically known that for a general abelian number field[1] a corresponding class number formula can be derived, as Dirichlet [7, 8][2] obtained for the quadratic forms (quadratic fields) and then Kummer [22, 24, 26, 27][3] for the cyclotomic fields.[4] In the literature there appear the general formulas of Fuchs [10], Beeger [3–5], and Gut [11], except the short comment on the existence of the one of Kummer [22] at page 111 f. I derive briefly these formulas in Chap. 1 (of this book) by application of modern number-theoretic methods and calculation techniques. Here I would like to assume (that the reader is acquainted with) the residue formula for the Dedekind zeta function, which lies at the basis of this derivation and is precisely found in Hilbert's Zahlbericht [19], and the product formula for the Dirichlet L-functions, which is treated in my Klassenkörperbricht [12].

So far, one has always taken a viewpoint that the class number formulas, which flow from the analytic spring but are essentially arithmetic, would be the final goal and that the derivation of such formulas would be everything. However, simple examples of quadratic fields have already shown that the formula obtained in the first form is barely or completely not suitable for the actual arithmetic calculation of the class number, because, first, in given cases these calculations would require excessively large numerical computation and, second, the formulas contain analytic elements originating from the derivation, which are strange to the methods of numerical calculations of number theory. The process that one has to quote tables of logarithms and trigonometric functions cannot be regarded in the real sense as

[1]Throughout this book, by an *abelian number field* I mean an abelian extension over the rational number field; it is also called an absolute abelian number field.

[2]See for this also Zahlbericht [19], §§**79, 86**, and Hasse [15], §**18**.

[3]See for this also Zahlbericht [19], §§**117, 118**, and Hasse [15], §**18**.

[4]Throughout this book, by a *cyclotomic field* I mean only the full number field generated by the roots of unity of fixed order, not its subfield, an abelian number field.

the arithmetic calculation of the class number. Consequently one has given up the application of the formula to actual calculation of the class number and instead has recommended the tentative test process applicable to an arbitrary algebraic number field that relies on the famous theorem: in every divisor class of an algebraic number field of degree n with r_2 pairs of the complex-conjugates and with absolute value d of the discriminant, there exists an integral divisor $\mathfrak{a}$ with $N(\mathfrak{a}) \leq \left(\frac{4}{\pi}\right)^{r_2} \frac{n!}{n^n} \sqrt{d}$.[5] Although by much effort one has obtained a beautiful explicit formula for the class number, one has renounced and then given up its actual application; what a pathetic viewpoint for number theorists who are conscious of its strength! For number theorists the real task—namely, setting the formula to be accessible to arithmetic calculation—starts by removing tools not familiar to them and by driving ones indispensable so far. How we tackle and manage this task I [13] have been able to show briefly as an example of real quadratic fields with prime discriminant. Subsequently it was my student Bergström [6] who managed to extend this method to all real quadratic fields. Moreover I was once able to give first a solution to the task for real cyclic cubic and biquadratic fields, which I will explain in another place [16].

Beside the above-mentioned practical viewpoint of actual calculations of the class number, there also exists the theoretical viewpoint that imperatively reminds every sincere number theorist of broad arithmetic research on the obtained class number formula. The formula gives the value of a number and then that of a positive rational integer with definite conceptual meaning. However, in the appearance of the form obtained first, the formula shows neither the integrality nor the positiveness of the found expression, let alone reveals in a pure arithmetic way that this expression is the number of the divisor classes of the abelian number field in question or in some way has to do with this number. Then, for number theorists, there has emerged the task of entering into the arithmetic structure as deeply as possible to research these problems. From the solution of the last extensive question to a direct pure arithmetic proof of the general class number formula, we are, even today, just as far removed as from a pure arithmetic proof of Dirichlet's theorem on the prime numbers in a prime residue class.[6] To the contrary, on the first restricted problem-setting we will solve really, though with some difficulty, the problem of direct proof of the integrality of the obtained expression of the class number (treated in Chap. 2 at least in the cyclic case and treated in Chap. 3 perfectly), and we can feed back the problem on positiveness to the remaining problem of positiveness of the special case of quadratic fields; certainly, this remaining problem, which has already been emphasized and also marked as deep-rooted in the previous literature, has to be left unsolved.

[5] See, for example, Hasse [14], §**30**, (c), III′.

[6] Supplement by the proofreader (in 1951): In the meantime there appear elementary proofs of this theorem by A. Selberg and by H. Zassenhaus, independently to each other: An elementary proof of Dirichlet's theorem about primes in an arithmetic progression, *Ann. of Math.* (2) **50** (1949); Über die Existenz von Primzahlen in arithmetischen Progressionen, *Comm. Math. Helvetici* **22** (1949).

In Chaps. 2 and 3 we fully engage in arithmetic investigation on the above-mentioned methods and goal setting for the two factors of the analytically obtained expressions for the class number, which were already made prominent by Kummer. Kummer [22–25, 27, 28] has already conducted such arithmetic investigations with the analytically obtained class number formulas for the special case of cyclotomic fields, the case treated by himself. At that time Kronecker [20, 21] also participated in these investigations. Afterward, in a corresponding way, Weber [30, 31] has proved interesting theorems on the class number and the units of the cyclotomic field with 2-power order. Finally, what further belong to these investigations are the famous theorem of Dirichlet [9][7] on the class number of special bicyclic biquadratic fields in relation to the class numbers of its quadratic subfields, which later Hilbert [18] could prove indeed pure-arithmetically, and also the generalizations to such arbitrary fields given by Bachmann [2], Amberg [1], and Herglotz [17].[8] All these results, as already remarked in my Preface, have been referenced in Hilbert's Zahlbericht [19] only briefly and without hint of proof. Collecting the proofs from original works is tremendously arduous, because first they are greatly scattered, second they are not always based on the same concept of class as today, third some deduction methods are incomplete, incorrect, or completely defect and finally, in the situation of development of algebraic number theory at that time, sufficiently general, systematic and simplified concepts and methods were not at one's disposal as they are today; we are now ready for, for example, the forms of field theory, the theory of abelian groups and its characters, Galois theory, the Hilbert theory of Galois field and class field theory. Therefore, I take the generalization to arbitrary abelian number fields and the penetration by the modern number-theoretic methods, which is here in the foreground, as a welcome opportunity to return to this research that is attractive and older from a pure number-theoretic point of view in this reported form at every given place in a little more detail than it would be demanded as the proper course of investigation. The matter is certainly not a complete report on all the research in this direction even only of Kummer. Rather I would take only the research for which I could see the possibility of the generalization to arbitrary abelian number fields or at least to definite types of abelian number fields.

References[9]

1. E.J. Amberg, Über den Körper, dessen Zahlen sich rational aus zwei Quadratwurzeln zusammensetzen. Diss. Zürich 1897 [Introduction, **3.8**]

[7] See for this also Zahlbericht [19], §**87**.

[8] See for this also the more progressed research in the direction of Vårmon [29], into which I could not enter in this book.

[9] The bold-typed numbers attached in the square brackets denote the sections of this book in which the individual works are cited.

2. P. Bachmann, Zur Theorie der komplexen Zahlen. J. f. d. reine und angew. Math. **67** (1867) [Introduction, **3.8**]
3. N.G.W.H. Beeger, Über die Teilkörper des Kreiskörpers $K(\zeta_{l^n})$. Proc. Akad. Wet. Amsterdam **21** (1919) [Introduction]
4. N.G.W.H. Beeger, Bestimmung der Klassenzahl der Ideale aller Unterkörper des Kreiskörpers der ζ_m, wo m durch mehr als eine Primzahl teilbar ist. Proc. Akad. Wet. Amsterdam **22** (1920) [Introduction]
5. N.G.W.H. Beeger, Berichting zu vorstehender Arbeit. Proc. Akad. Wet. Amsterdam **23** (1922) [Introduction]
6. H. Bergström, Die Klassenzahlformel für reelle quadratische Zahlkörper mit zuammengesetzter Diskriminante als Produkt verallgemeinerter Gaußscher Summen. J. f. d. reine und angew. Math. **186** (1944) [Introduction, **1.3**]
7. G.L. Dirichlet, Sur l'usage des séries infinies dans la théorie des nombres. J. f. d. reine und angew. Math. **18** (1838) = Collected Works **1**, 357–374 [Introduction]
8. G.L. Dirichlet, Recherches sur diverses applications de l'analyse infinitésimale à la théorie des nombres. J. f. d. reine und angew. Math. **19** (1839) = Collected Works **1**, 411–496. [Introduction]
9. G.L. Dirichlet, Recherches sur les formes quadratiques à coefficients et à indéterminées complexes. J. f. d. reine und angew. Math. **24** (1842) = Collected Works **1**, 533–618 [Introduction, **3.8**]
10. L. Fuchs, Über die aus Einheiteswurzeln gebildeten complexen Zahlen von periodischem Verhalten, insbespmdere die Bestimmung der Klassenzahl derselben. J. f. d. reine und angew. Math. **65** (1866) = Collected Works **1**, 69–109 [Introduction]
11. M. Gut, Die ζ-Funktion, die Klassenzahl und die Kroneckersche Grenzformel eines beliebigen Kreiskörpers. Comm. Math. Helvetici **1** (1929) [Introduction]
12. H. Hasse, Bericht über neuere Untersuchungen und Probleme aus der Theroie der algbraichen Zahlkörper. Part I: Klassenörpertheorie. Jahresbericht D.M.-V. **35** (1926); Bericht über neuere Untersuchungen und Probleme aus der Theroie der algbraichen Zahlkörper. Part Ia: Beweise zu Teil I. Jahresbericht D.M.-V. **36** (1927); Bericht über neuere Untersuchungen und Probleme aus der Theroie der algbraichen Zahlkörper. Part II: Reziprozitätsgesetz. Jahresbericht D.M.-V. Supplemental Ed. **6** (1930). Cited as "Klassenkörperbericht." [Introduction, **1.2, 1.3, 2.3, 2.12, 3.1, 3.2, 3.4, 3.5, 3.9, 3.19**]
13. H. Hasse, Produktformeln für verallgemeinerte Gaußsche Summen und ihre Anwendung auf die Klassenzahlformel für reelle quadratische Zahlkörper. Math. Zeitschr. **46** (1940) [Introduction]
14. H. Hasse, *Zahlentheorie.* Berlin (1949) [Introduction, **2.9, 3.1, 3.2, 3.5, 3.9**]
15. H. Hasse, *Vorlesungen über Zahlentheorie* (American Mathematical Society, Berlin 1950) [Introduction, **1.4**]
16. H. Hasse, Arithmetische Bestimmung von Grundeinheit und Klassenzahl in zyklischen kubischen und biquadratischen Zahlkörper. Abh. Deutsche Akad. Wiss. 1948 No.2 (1950) [Introduction]
17. G. Herglotz, Über einen Dirichletschen Satz. Math. Zeitschr. **12** (1922) [Introduction, **3.8**]
18. D. Hilbert, Über den Dirichletschen biquadratischen Zahlkörper. Math. Ann. **45** (1894) [Introduction, **3.8**]
19. D. Hilbert, Die Theorie der algebraischen Zahlkörper. Jahresbericht D.M.-V. **4** (1897). Cited as "Zahlbericht." [Preface, Introduction, **1.2, 1.6, 2.9, 3.2, 3.5, 3.8, 3.19**]
20. L. Kronecker, Bemerkung über dek Klassenzahl der aus Einheitswurzeln gebildeten komplexen Zahlen. Monatsber. Akad. d. Wissensch. Berlin (1863) = Collected Works **1**, 123–131 [Introduction, **1.6, 3.7**]
21. L. Kronecker, Auseinandersetzung einiger Eigenschaften der Klassenzahl idealer komplexer Zahlen. Monatsber. Akad. d. Wissensch. Berlin (1870) = Collected Works **1**, 271–282 [Introduction, **1.6, 3.1**]

22. E. Kummer, Bestimmung der Anzahl nicht-äquivalenter Klassen für die aus λ-ten Wurzeln der Einheit gebildeten komplexen Zahlen und die idealen Faktoren derselben. J. f. d. reine und angew. Math. **40** (1850) [Introduction, **1.6**, **3.20**]
23. E. Kummer, Zwei besondere Untersuchungen über die Klassenzahl und über die Einheiten der aus λ-ten Wurzeln der Einheit gebildeten komplexen Zahlen. J. f. d. reine und angew. Math. **40** (1850) [Introduction, **3.19**]
24. E. Kummer, Sur la théorie des nombres complexes composés de racines de l'unité et de nombres entiers. J. de Math. **16** (1851) (French summary laying the foundation of the theory of ideal numbers from J. f. d. reine und angew. Math. **35** (1847), as well as three works cited above) [Introduction, **1.6**, **3.19**, **3.20**, Appendix]
25. E. Kummer, Über die Irregularität der Determinanten. Monatsber. Akad. d. Wissensch. Berlin (1853) (Extract from the letter to Dirichlet) [Introduction, **3.1**, **3.15**]
26. E. Kummer, Über die Klassenanzahl der aus n-ten Einheitenwurzeln gebildeten komplexen Zahlen. Monatsber. Akad. d. Wissensch. Berlin (1861) [Introduction, **3.16**]
27. E. Kummer, Über die Klassenanzahl der aus zusammengesetzten Einheitenwurzeln gebildeten idealen komplexen Zahlen. Monatsber. Akad. d. Wissensch. Berlin (1863) [Introduction, **1.6**, **3.10**, **3.12**, **3.15**, **3.20**, Appendix]
28. E. Kummer, Über eine Eigenschaft der Einheiten der aus den Wurzeln der Gleichung $\alpha^\lambda = 1$ gebildeten komplexen Zahlen und über den zweiten Faktor der Klassenzahl. Monatsber. Akad. d. Wissensch. Berlin (1870) [Introduction, **3.19**, **3.20**]
29. J. Vårmon, *Über Abelscher Körper, deren alle Gruppeninvarianten aus einer Primzahl l bestehen, und über Abelscher Körper als Kreiskörper.* Akad. Abhandl. Lund 1925 [Introduction]
30. H. Weber, Theorie der Abelschen Zahlkörper. Acta Math. **8** (1886) [Introduction, **2.1**, **2.6**, **2.10**, **3.16**]
31. H. Weber, *Lehrbuch der Algebra*, vol. 2, 2nd edn. (American Mathematical Society, Braunschweig, 1899), pp. 219–223 [Introduction, **2.1**, **2.6**, **2.10**, **3.16**]

Chapter 1
The Generalized Class Number Formulas

1.1 Abelian Number Fields as Class Fields

I begin with a brief summary of the essential class-field-theoretic facts on abelian number fields for all the rest of this book, most of which, as we will see, can be found in detail in my Klassenzahlbericht [2], or arise as special cases of abelian number fields by general theorems on the class field. Here I restrict myself to main facts. At given places later in this work, I will quote some facts in more detail. I will use, without newly defining at each place, the notations introduced in this summary and moreover some terms introduced later in the process of deriving generalized class number formulas.

Let K be an abelian number field of degree n over the rational number field P. It is the class field corresponding to a rational congruence group H of index n. Let f be the conductor of H, which is also called the conductor of K. Then H consists of all the numbers of the group of prime residue classes mod f such that the n classes mod H consist of the numbers of n cosets of the group of all prime residue classes mod f and such that this partition cannot be described in the region of the prime residue group mod any proper divisor of f. A property of class field K says that the type of decomposition of a rational prime p into primes of K depends only on the class to which the prime p belongs in the definite way of the decomposition law of the class field. Therefore, for this property of the class field K, the problem is not the congruence group H itself, the principal class, but rather the class group mod H.

The field K is a subfield of the cyclotomic field P_f of f-th roots of unity ζ_f^x; in fact, it is the field invariant to the group of automorphisms $\zeta_f \to \zeta_f^a$ with a in H.[1] Consequently, K consists of all rational symmetric functions of ζ_f^a with coefficients

[1](Translator's remark) In the following "a in H" means that the element a mod f of $(\mathbf{Z}/f\mathbf{Z})^\times$ is contained in H.

© Springer Nature Switzerland AG 2019
H. Hasse, *On the Class Number of Abelian Number Fields*,
https://doi.org/10.1007/978-3-030-01512-1_1

of rational numbers where ζ_f is a fixed primitive f-th root of unity and a runs over a system of representatives of the residue classes mod f in H. As one could show, the Gaussian cyclotomic period $\sum_{\substack{a \bmod f \\ a \text{ in } H}} \zeta_f^a$ suffices for a generator of K. The Galois group $\mathfrak{G}$ of K is isomorphic to the class group mod H. This isomorphism is given by the representation $\zeta_f \to \zeta_f^s$ of the automorphism S from $\mathfrak{G}$ where s runs over the representatives of the classes mod H. Moreover, we have to investigate the group X of characters of $\mathfrak{G}$ isomorphic to $\mathfrak{G}$, which are also called characters of K. These characters are regarded as the ones of the class group mod H by means of the isomorphism and also as the ones of the group of the residue classes mod f.[2] Let $f(\chi)$ be the conductor of a character χ, so that χ is a primitive character mod $f(\chi)$. Therefore f is the least common multiple of the $f(\chi)$, while the product of the $f(\chi)$ is the absolute value of the discriminant of K.

From now on we will have to classify the field K into the two cases according to whether K is real or imaginary. When K is *imaginary*, $n = 2n_0$ is even and K is of degree 2 over its maximal real subfield K_0 of degree n_0, the field invariant by the automorphism $\zeta_f \to \overline{\zeta_f} = \zeta_f^{-1}$ (transform induced by complex-conjugate). The Galois group $\mathfrak{G}_0$ of K_0 is the factor group of $\mathfrak{G}$ by the subgroup generated by this automorphism. The group X of n characters of $\mathfrak{G}$ is divided into two classes: the subgroup X_0 of characters χ_0 of K_0 characterized by $\chi_0(-1) = 1$ and its coset consisting of characters χ_1 characterized by $\chi_1(-1) = -1$. The χ_0 are the characters of K_0. The χ_1 are the characters of K that are not of K_0; we call χ_1 *characters of* K/K_0 for short. The congruence group H does not contain -1; in addition to the class containing -1 and the ones containing $-a$, the group H goes on to the congruence group H_0 corresponding to K_0. The classes mod H_0 consist of the classes mod H by the union of opposite classes $\pm s$ mod H. When K is *real*, then $K = K_0$, so $\mathfrak{G} = \mathfrak{G}_0$, $X = X_0$ and always $\chi(-1) = 1$, and further $H = H_0$ and -1 is contained in H and hence $\pm s$ belong to the same class mod H.

Corresponding to the purpose of this book that we seek the arithmetic determination of the class number of K by actual numerical procedures, we attribute the present properties and statements of K, as much as possible, to the rational congruence group H characterizing the field K or rather to the class group mod H by virtue of the above-mentioned and outlined class-field-theoretic facts. At the same time just as well, we can describe this group H by means of characters of residue classes in the group X of K, by which the basically lying congruence group H itself can be represented as the set of a with $\chi(a) = 1$ for all χ in X. Relating to the characters is more convenient than to the classes for the following reasons. First, by the ordering theorem of class field theory, the subfields K', K'', ... of K correspond to the congruence groups H', H'', ... over H in the inverse order ($K' \subseteqq K''$ is synonymous with $H' \supseteqq H''$); on the one hand, for the corresponding subgroups X', X'', ... of X (that is, for the character groups of K', K'', ...), there is the direct order relation $X' \subseteqq X''$. The transform from K' to K'' is therefore described by means of the class group as a group-theoretic partition of the classes

[2](Translator's remark) Throughout this book we suppose that the characters χ of K are primitive.

mod H' into the ones mod H''; on the other hand, the partition of characters is rather simpler because it involves just the addition to X' of characters in X'' not contained in X'. Second, the characters, as the functions on the group of the full prime residue classes mod f, are generally treated more conveniently than the individual classes mod H that are, in each case, composed only of definite prime residue classes mod f. This complete and simplified action of introducing characters instead of classes appeared already in the classical proof of the famous theorem of Dirichlet on the prime numbers in a prime residue class.

The decomposition law of the class field for K is described by the character group X in the following way: for the inertia field K_T for p, the character group X_T consists of characters χ in X with $p \not| f(\chi)$, or with $\chi(p) \neq 0$, by letting, as usual, $\chi(x) = 0$ in general for any integer x not coprime to $f(\chi)$. For the decomposition field K_Z for p, the character group X_Z consists of characters χ in X with $\chi(p) = 1$. Consequently the numbers that characterize the type of decomposition for p in K are determined, namely, the ramification index $e_p = [K : K_T]$, the residue class degree $n_p = [K_T : K_Z]$, and the number $r_p = [K_Z : P]$ of prime divisors of p in K, which are also described in the character group X by the group indices: $e_p = [X : X_T]$, $n_p = [X_T : X_Z]$ and $r_p = [X_Z : 1]$.

1.2 The Analytic Class Number Formula

The starting point for the analytic determination of the class number h of K is, on the one hand, the formula[3]

$$\zeta_K(s) = \prod_{\chi} L(s, \chi), \tag{1.2.1}$$

by which the Dedekind zeta function of K is represented by the product of Dirichlet L-functions attached to the characters χ of K, and on the other hand, the formula[4]

$$\lim_{s\to 1}(s-1)\zeta_K(s) = \left\{ \begin{array}{ll} \dfrac{2^n hR}{w\sqrt{d}} = \dfrac{2^{n-1}hR}{\sqrt{d}}, & \text{if } K \text{ is real} \\[2ex] \dfrac{(2\pi)^{n_0} hR}{w\sqrt{d}}, & \text{if } K \text{ is imaginary} \end{array} \right\} \tag{1.2.2}$$

for the residue at $s = 1$ of the Dedekind zeta function of K. Here R means the regulator of K; that is, when K is real, R is the absolute value of the determinant

[3]See Klassenkörperbericht [2], Part I, §**8**, Theorem 14.

[4]See Zahlbericht [7], §§**25**, **26**.

formed by a system of fundamental units $\varepsilon_1, \ldots, \varepsilon_{n-1}$ of K,

$$R = \left\| \log |\varepsilon_\nu^S| \right\| \quad \begin{pmatrix} \text{row index} & S \neq 1 \text{ in } \mathfrak{G} \\ \text{column index} & \nu = 1, \ldots, n-1 \end{pmatrix}, \tag{1.2.3a}$$

and when K is imaginary, R is the absolute value of the determinant formed by a system of fundamental units $\varepsilon_1, \ldots, \varepsilon_{n_0-1}$ of K,

$$R = \left\| \log |\varepsilon_{\nu_0}^{S_0}|^2 \right\| = \left\| 2 \log |\varepsilon_{\nu_0}^{S_0}| \right\| \begin{pmatrix} \text{row index} & S_0 \neq 1 \text{ in } \mathfrak{G}_0 \\ \text{column index} & \nu_0 = 1, \ldots, n_0 - 1 \end{pmatrix}. \tag{1.2.3b}$$

Here and later on, w means the number of roots of unity in K. It is characterized by class field theory as follows: for even f, w is the greatest divisor v of f such that $a \equiv 1 \pmod{v}$ holds for all a in H; for odd f, the corresponding fact holds for $w/2$. Since K always has the square roots of unity ± 1, w is always even; if K is real, then $w = 2$.

Taking $s = 1$ in (1.2.1) and noting that $L(s, 1) = \zeta(s)$ (the Riemann zeta function) has residue 1 at $s = 1$, one obtains

$$\lim_{s \to 1} (s-1)\zeta_K(s) = \prod_{\chi \neq 1} L(1, \chi),$$

and hence one obtains by (1.2.2) the analytic class number formula

$$hR = \left\{ \begin{array}{ll} \dfrac{\sqrt{d}}{2^{n-1}} \displaystyle\prod_{\chi \neq 1} L(1, \chi), & \text{if } K \text{ is real} \\[2ex] \dfrac{w\sqrt{d}}{(2\pi)^{n_0}} \displaystyle\prod_{\chi \neq 1} L(1, \chi), & \text{if } K \text{ is imaginary} \end{array} \right\}. \tag{1.2.4}$$

By this formula the determination of the class number h, or rather of the product hR of the class number h and the regulator R, is reduced to the calculation of the product of Dirichlet L-functions with $\chi \neq 1$ at $s = 1$. The values of these functions are the conditionally convergent infinite series defined by natural number order,

$$L(1, \chi) = \sum_{m=1}^{\infty} \frac{\chi(m)}{m}, \tag{1.2.5}$$

which are the so-called Dirichlet L-series.

1.3 Product Formulas for the Conductors and for the Gaussian Sums

For the calculation of the product of the L-series $L(1, \chi)$ along with the product formula for the conductors $f(\chi)$ of characters χ of K, which has already been explained in Sect. 1.1, a corresponding product formula for the Gaussian sums

$$\tau(\chi) = \sum_{x \bmod f(\chi)} \chi(x)\zeta_{f(\chi)}^{x} \tag{1.3.1}$$

associated to them is required, where, from now on,

$$\zeta_{f(\chi)} = e^{\frac{2\pi i}{f(\chi)}}$$

always denotes the analytically normalized $f(\chi)$-th root of unity. When one attaches importance to these two formulas, one could prove them pure-arithmetically.[5] However, it seems to me that, in the present situation where the analytic methods are playing the crucial role anyway, the elegant proof analytically given by Hecke [5, 6] takes a best position. His proof is produced from the product formula (1.2.1) by comparing the functional equation for $\zeta_K(s)$ with that for the $L(s, \chi)$, and it is informed in my Klassenkörperbericht [2] for arbitrary relative abelian number fields.[6] For the simplest spacial case of absolute abelian number fields, the proof can briefly be described once again in the following way.

The functional equation for $\zeta_K(s)$ states that by the transformation $s \to 1 - s$, the function

$$\left\{ \begin{array}{ll} d^{\frac{s}{2}} \left(\pi^{-\frac{s}{2}}\Gamma\left(\frac{s}{2}\right)\right)^{n} \zeta_K(s), & \text{if } K \text{ is real} \\ d^{\frac{s}{2}} \left(\pi^{-\frac{s}{2}}\Gamma\left(\frac{s}{2}\right)\right)^{n_0} \left(\pi^{-\frac{s}{2}}\Gamma\left(\frac{1+s}{2}\right)\right)^{n_0} \zeta_K(s), & \text{if } K \text{ is imaginary} \end{array} \right\}$$

takes factor 1. (The function is invariant under the transformation $s \to 1 - s$.)[7] The functional equation for $L(s, \chi)$ states that by the transforms $s \to 1 - s$ and $\chi \to \overline{\chi}$, the function

$$\left\{ \begin{array}{ll} f(\chi)^{\frac{s}{2}} \left(\pi^{-\frac{s}{2}}\Gamma\left(\frac{s}{2}\right)\right) L(s, \chi), & \text{if } \chi(-1) = 1 \\ f(\chi)^{\frac{s}{2}} \left(\pi^{-\frac{s}{2}}\Gamma\left(\frac{1+s}{2}\right)\right) L(s, \chi), & \text{if } \chi(-1) = -1 \end{array} \right\}$$

[5] Regarding the conductors, see Hasse [3]; regarding the Gaussian sums, see, for example, Bergström [1].

[6] See Klassenkörperbericht [2], Part I, §9.

[7] (Translator's remark) This means the following: let $\xi(s)$ be the function defined in the parentheses { }. Then the functional equation $\xi(s) = 1 \cdot \xi(1 - s)$ holds.

takes factor $\tau(\chi)/\sqrt{\chi(-1)f(\chi)}$ (of absolute value 1) with positive real or positive imaginary[8] quadratic root.[9] By dividing the functional formula for $\zeta_K(s)$ by the product of those for $L(s,\chi)$ in accordance with the product formula (1.2.1), it immediately follows that the function $\left(d/(\prod_\chi f(\chi))\right)^{\frac{s}{2}}$ takes factor $(\prod_\chi \sqrt{\chi(-1)f(\chi)})/\prod_\chi \tau(\chi)$ by the transformation $s \to 1-s$ in our way of writing this functional equation.[10] This indeed comes true only when the base of the exponential function[11] and the factor take value 1. Thus one obtains the following two product formulas:

$$\prod_\chi f(\chi) = d, \tag{1.3.2}$$

$$\prod_\chi \tau(\chi) = \left\{ \begin{array}{ll} \left(\prod_\chi f(\chi)\right)^{\frac{1}{2}} = d^{\frac{1}{2}}, & \text{if } K \text{ is real} \\ i^{n_0}\left(\prod_\chi f(\chi)\right)^{\frac{1}{2}} = i^{n_0} d^{\frac{1}{2}}, & \text{if } K \text{ is imaginary} \end{array} \right\}. \tag{1.3.3}$$

Here one can restrict these products over $\chi \neq 1$, because $f(1) = 1$ and $\tau(1) = 1$.

1.4 Calculation of *L*-series

For the calculation of the product of L-series (1.2.5) at $s = 1$, first the individual factor $L(s,\chi)$ $(\chi \neq 1)$ for $s > 1$ is reformed elementarily.[12] As long as we deal with only an individual factor $L(s,\chi)$, we take for short f instead of $f(\chi)$ and ζ instead of $\zeta_{f(\chi)}$. As shown in the definition (1.3.1) of Gaussian sums and also all through in the following, we mean by $\sum_{x \bmod f}$ the summation over a system of prime residues x mod f, and by $\sum_{\pm x \bmod f}$ the summation over a half system of prime residues mod f, i.e., a system such that the $\pm x$ mod f constitute the whole

[8](Translator's remark) "Positive imaginary" means that the imaginary part is positive.

[9](Translator's remark) This means the following: let $\xi(s,\chi)$ be the function defined in the parentheses { } and let $W(\chi)$ be $\tau(\chi)/\sqrt{\chi(-1)f(\chi)}$. Then the functional equation $\xi(s,\chi) = W(\chi)\cdot\xi(1-s,\overline{\chi})$ holds. The printed factor $\sqrt{\chi(-1)f(\chi)}/\tau(\chi)$ in the original book should be read as $\tau(\chi)/\sqrt{\chi(-1)f(\chi)}$.

[10](Translator's remark) The printed factor $(\prod_\chi \tau(\chi))/\prod_\chi \sqrt{\chi(-1)f(\chi)}$ in the original book should be read as $(\prod_\chi \sqrt{\chi(-1)f(\chi)})/\prod_\chi \tau(\chi)$.

[11](Translator's remark) For the exponential function $\left(d/(\prod_\chi f(\chi))\right)^{\frac{s}{2}}$, the base is $d/(\prod_\chi f(\chi))$.

[12]See Hasse [4], §**18, 2**.

system of prime residues mod f.[13] The indication $(x, f) = 1$ in the summation is tacitly taken into consideration all throughout by the already-used general setting $\chi(x) = 0$ for $(x, f) \neq 1$.

Then it holds that

$$L(s, \chi) = \sum_{m=1}^{\infty} \frac{\chi(m)}{m^s} = \sum_{z \bmod f} \chi(z) \sum_{\substack{n \equiv z \pmod f \\ n \geqq 1}} \frac{1}{n^s}$$

$$= \sum_{z \bmod f} \chi(z) \sum_{n=1}^{\infty} \frac{1}{n^s} \frac{1}{f} \sum_{x \bmod f} \zeta^{(z-n)x}$$

$$= \frac{1}{f} \sum_{x \bmod f} \sum_{z \bmod f} \chi(z) \zeta^{zx} \sum_{n=1}^{\infty} \frac{\zeta^{-nx}}{n^s}.$$

From Abel's limit theorem for the Dirichlet series, it follows that

$$L(1, \chi) = -\frac{1}{f} \sum_{x \bmod f} \tau_x(\chi) \log(1 - \zeta^{-x}) = -\frac{\tau(\chi)}{f} \sum_{x \bmod f} \overline{\chi}(x) \log(1 - \zeta^{-x}),$$

where the notation[14]

$$\tau_x(\chi) = \sum_{z \bmod f} \chi(z) \zeta^{zx}$$

and the relation $\tau_x(\chi) = \overline{\chi}(x)\tau(\chi)$, which is obtained by an elementary theory of Gaussian sums,[15] are used; the logarithm has the principal value with imaginary part between $-\pi i$ and πi (here more precisely between $-\frac{\pi i}{2}$ and $\frac{\pi i}{2}$). For further calculation of the sum

$$S(\chi) = \sum_{x \bmod f} \chi(x) \log(1-\zeta^{-x}) = \frac{1}{2} \sum_{x \bmod f} \chi(x)(\log(1-\zeta^{-x}) + \chi(-1)\log(1-\zeta^{x})),$$

the two cases $\chi(-1) = 1$ and $\chi(-1) = -1$ have to be classified.

<u>(a) $\chi(-1) = 1$</u>

[13]Remark that the case $f = 2$, which is meaningless by this definition, never arises. More precisely, the conductor $f(\chi)$ of a character χ contains no factor of prime number 2 or factor of at least power 2^2, because for odd conductor f_0 the group of the prime residues mod $2f_0$ coincides with that of mod f_0. We must keep the last remark in mind at many places in the following process.

[14](Translator's remark) In the sum $\sum_{x \bmod f} \chi(z)\zeta^{zx}$ of the original book, the variables x mod f should be z mod f.

[15]For this see Hasse [4], §**20**, **1**.

Since $1-\zeta^x$ and $1-\zeta^{-x}$ are complex-conjugate to each other, so are the principal values of their logarithms. Then there arises

$$S(\chi) = \frac{1}{2}\sum_{x \bmod f} \chi(x)(\log(1-\zeta^{-x}) + \log(1-\zeta^x)) = \frac{1}{2}\sum_{x \bmod f} \chi(x)\log|1-\zeta^x|^2$$
$$= \sum_{x \bmod f} \chi(x)\log|1-\zeta^x| = 2\sum_{\pm x \bmod f} \chi(x)\log|1-\zeta^x|$$

and hence[16]

$$L(1,\chi) = 2\frac{\tau(\chi)}{f(\chi)}\sum_{\pm x \bmod f}\left(-\overline{\chi}(x)\log|1-\zeta^x_{f(\chi)}|\right). \tag{1.4.1a}$$

(b) $\chi(-1) = -1$

The principal values of logarithms of $1-\zeta^x$ and $1-\zeta^{-x}$ arise from the expressions

$$1-\zeta^x = 1-e^{\frac{2\pi ix}{f}} = \left(e^{\frac{-\pi ix}{f}} - e^{\frac{\pi ix}{f}}\right)e^{\frac{\pi ix}{f}} = -2i\sin\frac{\pi x}{f}\cdot e^{\frac{\pi ix}{f}}$$
$$= 2\sin\frac{\pi x}{f}\cdot e^{\pi i\left(\frac{x}{f}-\frac{1}{2}\right)},$$
$$1-\zeta^{-x} = 1-e^{-\frac{2\pi ix}{f}} = \left(e^{\frac{\pi ix}{f}} - e^{-\frac{\pi ix}{f}}\right)e^{-\frac{\pi ix}{f}} = 2i\sin\frac{\pi x}{f}\cdot e^{-\frac{\pi ix}{f}}$$
$$= 2\sin\frac{\pi x}{f}\cdot e^{-\pi i\left(\frac{x}{f}-\frac{1}{2}\right)}.$$

For the representatives $x \bmod f$, take the system of the least positive prime residues mod f (the summation over this residue system will be denoted by $\sum^+_{x \bmod f}$), and then in this expression it holds that $2\sin\frac{\pi x}{f} > 0$ and $\left|\frac{x}{f}-\frac{1}{2}\right| < \frac{1}{2}$. Consequently $2\sin\frac{\pi x}{f} > 0$ are the common absolute values of logarithms of $1-\zeta^{\pm x}$, and $\pm\pi i\left(\frac{x}{f}-\frac{1}{2}\right)$ are its imaginary parts lying between $-\pi i$ and πi (precisely between $-\frac{\pi i}{2}$ and $\frac{\pi i}{2}$), as these values correspond to the principal values of these logarithms.

---

[16](Translator's remark) In the summand in (1.4.1a) in the original book published in 1985, "$-\chi(x)\log|1-\zeta^x_{f(\chi)}|$" should be "$-\overline{\chi}(x)\log|1-\zeta^x_{f(\chi)}|$." In the first version published in 1952, it is printed correctly.

Then there arises

$$S(\chi) = \frac{1}{2} \sum_{x \bmod f} \chi(x)(\log(1-\zeta^{-x}) - \log(1-\zeta^{x}))$$

$$= -\pi i \sum_{x \bmod f}{}^{+} \chi(x)\left(\frac{x}{f} - \frac{1}{2}\right) = \frac{\pi}{i}\frac{1}{f} \sum_{x \bmod f}{}^{+} \chi(x)x,$$

and hence

$$L(1,\chi) = \frac{\pi}{i}\frac{\tau(\chi)}{f(\chi)}\frac{1}{f(\chi)} \sum_{x \bmod f}{}^{+} (-\overline{\chi}(x)x). \tag{1.4.1b}$$

1.5 The Arithmetic Class Number Formula

By the results (1.4.1a) and (1.4.1b), one obtains immediately the following formulas for the product of the regulator and the class number through the product construction (1.2.4) and with attention to the product formulas (1.3.2) and (1.3.3)[17]:

$$hR = \prod_{\chi \neq 1} \sum_{\pm x \bmod f(\chi)} \left(-\chi(x) \log|1-\zeta_{f(\chi)}^{x}|\right), \quad \text{if } K \text{ is real,} \tag{1.5.1a}$$

$$hR = \prod_{\chi_0 \neq 1} \sum_{\pm x \bmod f(\chi_0)} \left(-\chi_0(x) \log|1-\zeta_{f(\chi_0)}^{x}|\right) \cdot \frac{w}{2} \prod_{\chi_1} \frac{1}{f(\chi_1)}$$

$$\sum_{x \bmod f(\chi_1)}{}^{+} (-\chi_1(x)x)\,, \text{ if } K \text{ is imaginary.} \tag{1.5.1b}$$

The formula (1.5.1a) would not need to be especially quoted at all, because the formula (1.5.1b) holds also for real K, where no χ_1 arises and $w = 2$, so that the second factor could be abolished. The first factor in (1.5.1b) is, according to (1.5.1a),

[17]Formally the next formulas (1.5.1a) and (1.5.1b) arise at first with complex-conjugate character $\overline{\chi}$ in the place of χ. It would be more adequate theoretically to maintain the original form with $\overline{\chi} = \chi^{-1}$ throughout the work, like how in group theory one replaces a so-called group number $\sum_S \chi(S)S$ by the more adequate form $\sum_S \chi^{-1}(S)S$, so that, namely, by the group representation $S \to \chi(S)$ of corresponding χ for the individual summand it holds that $\chi^{-1}(S)S \to 1$. However, for our arithmetic investigation this formal-algebraic viewpoint is not so important, and hence in this book we will take formally, as a basis, a little smoother form of formulas. Nevertheless for our investigation formal simplification is important when we regard the minus sign of the individual summand as being attached to the factor of $\chi(x)$, and we do not put the sign in front of the sum or even entirely extract it from the product.

the product $h_0 R_0$ of the class number and the regulator of the maximal real subfield K_0 of K.

Now K_0 has the same rank $n_0 - 1$ of units as that of K; a system of fundamental units of K_0 is also a system of independent units of K of maximal rank. However, its regulator constructed in K is different from (the regulator) R_0 constructed in K_0 and rather has value $2^{n_0-1} R_0$ from (1.2.3a) and (1.2.3b). The quotient

$$Q = \frac{2^{n_0-1} R_0}{R} \tag{1.5.2}$$

is a natural number, that is, the index of the subgroup consisting of units of K_0 and its product with roots of unity in K in the group of all the units of K. We call this number Q the *unit index of* K/K_0,[18] which we have to deal with in detail later. Remark here that the index of the group of sheer units of K_0 (without factors of roots of unity in K) in the group of units of K is not Q itself, but the product $Q\,\frac{w}{2}$.

By writing the formula (1.5.1b) from this arithmetic point of view, we have the following result:

General Class Number Formula for an Abelian Number Field

$$\boxed{h = \frac{\prod_{\chi_0 \neq 1} \sum_{\pm x \bmod f(\chi_0)} \left(-\chi_0(x) \log|1 - \zeta_{f(\chi_0)}^x|\right)}{R_0} \cdot Qw \prod_{\chi_1} \frac{1}{2f(\chi_1)} \sum_{x \bmod f(\chi_1)}^{+} (-\chi_1(x)x)} \tag{1.5.3}$$

The meaning of the notation appearing here is once again summarized:

χ	*the characters of* K (regarded as characters of the rational congruence group mod H with conductor f corresponding to K), which are divided into
χ_0	*the characters of the maximal real subfield* K_0 *of* K $(\chi_0(-1) = 1)$,
χ_1	*the characters of* K/K_0 $(\chi_1(-1) = -1)$,
$f(\chi)$	*their conductors,*
$\sum_{\pm x \bmod f(\chi)}$	*the summation over a half system of prime residues* $x \bmod f(\chi)$,
$\sum_{x \bmod f(\chi)}^{+}$	*the summation over the system of the least positive prime residues* $x \bmod f(\chi)$,

[18](Translator's remark) The unit index Q of K/K_0 is also called the *Hasse index of* K or the *Hasse unit index of* K.

$\zeta_{f(\chi)} = e^{\frac{2\pi i}{f(\chi)}}$ *the analytically normalized primitive $f(\chi)$-th root of unity,*
R_0 *the regulator of K_0,*
Q *the unit index of K/K_0,*
w *the number of roots of unity in K.*

The two factors in (1.5.3) separated by the multiplication dot are called the second and the first factor of the class number, respectively. According to the construction of K from the sub-construction K_0 and the over-construction K/K_0, it would be more natural that these factors are numbered conversely. After all it seems to me that such a schematic numbering is irrelevant in general. For the first factor

$$h_0 = \frac{\prod_{\chi_0 \neq 1} \sum_{\pm x \bmod f(\chi_0)} \left(-\chi_0(x) \log |1 - \zeta^x_{f(\chi_0)}| \right)}{R_0} \tag{1.5.3a}$$

in my ordering, further naming is not necessary with regard to the meaning than the *class number of* K_0. For the second factor

$$h^* = Qw \prod_{\chi_1} \frac{1}{2f(\chi_1)} \sum_{x \bmod f(\chi_1)}^{+} (-\chi_1(x)x) \tag{1.5.3b}$$

in my ordering, I am willing to name it the *relative class number of* K/K_0, because by formula (1.5.3) it is the quotient of the class number h of K by h_0 of K_0. In the special case of $K = K_0$, where no character χ_1 exists, one has correspondingly that $h^* = 1$ from the derivation.

1.6 Preliminary Remarks on the Arithmetic Structure of the Two Class Number Factors

For the expression (1.5.3a) it is clear from the interpretation of the class number h_0 of K_0 that the expression is a positive rational integer. By the explanation in the Introduction, there arises a task to make clear at least the rational integrality of the expression (1.5.3a) also from its form. For this purpose one must give a system of independent units of K_0 with maximal rank $n_0 - 1$ whose regulator is just equal to the numerator of (1.5.3a). This is simultaneously the first step toward the actual arithmetic calculation of the class number h_0 of K_0 from the formula (1.5.3a) and a test to evaluate how much larger this system of units is than that of fundamental units of K_0, i.e., how much is the index of the subgroup generated by such units and their inverses in the group of units of K.[19] For the special case of $K = P_p$, the cyclotomic field with prime conductor p, Kummer [10, 11][20] has already given such a system of units by describing the product in the numerator of (1.5.3a) as a group determinant of the cyclic group $\mathfrak{G}_0$ of K_0 and by transforming this group determinant into a regulator. However, this process cannot be transferred to the general case at all, because it essentially makes use of two special conditions: that the Galois group $\mathfrak{G}_0$ is cyclic and that the conductor $f_0 = p$ is a prime number and hence all individual conductors $f(\chi_0) = p$ have the same value. In Chap. 2 I will return to the problem raised here in detail. For a real cyclic field I give a system of units of the desirable type and thereby make sure of the integrality of the expression (1.5.3a) for h_0 in appearance. For a more general class of real abelian number fields, I succeed only in the problem on a system of units whose regulator is a definite integral multiple of the product in the numerator of (1.5.3a). Hence for this class of fields the arithmetic calculation of the class number is actually executable, however its integrality cannot be made clear.

For the relative class number h^* of K/K_0 defined by (1.5.3b), as I will show in Chap. 3, one can deduce from the class field theory of the relative quadratic extension K/K_0 that h is divisible by h_0 and then h^* is a positive integer. Moreover one obtains an arithmetic interpretation of h^* as the order of a definite subgroup of the class group of K, namely, the group of ideal classes of K whose relative norms fall in the principal class of K_0. In addition, by virtue of the genus theory of K/K_0, the relative class number h^* is divisible by a definite power 2^γ ($\gamma \geqq 0$), the genus factor. By the explanation in the Introduction, there arises here again a task to make clear at least the integrality of h^* directly from the form of expression (1.5.3b). For the special case of $K = P_{p^\rho}$, the cyclotomic field with prime-power conductor p^ρ, Kummer [10–12] has established the integrality of h^* in this latter way. For the cyclotomic field $K = P_f$ with composite[21] conductor f, Kummer [12] could prove only that $2^{r-2}h^*$ is an integer where r is the number of distinct primes dividing f.[22] Successively Kronecker [8] has also ascertained the integrality of h^* itself for the cyclotomic field with composite conductor f, as he proved the divisibility of h by h_0 by the consideration that amounts to the class-field-theoretic deducing way mentioned so far.[23] For the general case I make clear

[19] Or also, how much is the index of the subgroup generated by absolute values of such units in the group of absolute values of units of K.

[20] See also Zahlbericht [7], §**117**.

[21] Here and later on, I always use the word *composite* to mean "composed of several distinct prime-powers."

[22] By Kummer the factor 2^{r-1} reads for 2^{r-2}. Notice that Kummer's first class number factor in the latter case (of composite conductor) is just $\frac{1}{2}h^*$, because Kummer has taken as a basis the classification in the narrow sense (norm-positive principal divisors). This distinction could be made neither in imaginary K nor in real K_0 in the former case; in the latter case, as Kronecker [8] has carried out, K_0 has the class number $2h_0$ in the narrow sense.

[23] In the proof of this divisibility, Kummer [10, p. 115] has assumed tacitly as self-evident that no ideal class in K_0 distinct from the principal class falls into the principal ideal of K. Kronecker [8]

the integrality of h^* in Chap. 3 also directly from the form of expression (1.5.3b) by generalizing and refining Kummer's deducing way. The transformation undertaken here for the individual factor in this expression gives at the same time considerable simplification for numerical calculation of the relative class number h^* of K/K_0 by means of formula (1.5.3b). To my regret I have not yet achieved the proof of the existence of the genus factor 2^γ in h^* directly from formula (1.5.3b) in the corresponding way, not even in the simplest special case of an imaginary quadratic field, where the genus factor is just 2^{r-1}.

As for the positiveness of h_0 and h^*, considering the pairs of factors with complex-conjugate characters in expressions (1.5.3a) and (1.5.3b), in which only the factors of quadratic characters are isolated, one knows that the problem of positiveness returns to the quadratic case. This is a completely corresponding fact to the non-vanishing of L-functions in the proof of Dirichlet's Prime Number Theorem. And as the non-vanishing for the remaining quadratic characters has not yet been able to be proved in an elementary way,[24] the positiveness for the quadratic field is also not yet known directly in an elementary way. For an imaginary quadratic field it holds that

$$h = h^* = \frac{w}{2}\frac{1}{f}\sum_{x \bmod f}{}^{+}\left(-\left(\frac{-f}{x}\right)x\right) = \frac{w}{2}\frac{\sum_b{}^+ b - \sum_a{}^+ a}{f},$$

where the numbers $x = a,\ b$ run over the least positive prime residue classes mod f with $\left(\frac{-f}{a}\right) = 1$ and $\left(\frac{-f}{b}\right) = -1$. The positiveness of $\sum_b{}^+ b - \sum_a{}^+ a$ is a deep-rooted statement on the distribution of the a's and the b's in the system of residues mod f, for which no one knows so far—except by the indirect and essentially analytic proof of the class number formula—any direct proof. For a real quadratic field it holds that

$$h = h_0 = \frac{\log \prod_{\pm x \bmod f}{}^{+}\left(\zeta_{2f}^{x} - \zeta_{2f}^{-x}\right)^{-\left(\frac{f}{x}\right)}}{\log \varepsilon} = \frac{\log \dfrac{\prod_{\pm b}{}^{+}\left(\zeta_{2f}^{b} - \zeta_{2f}^{-b}\right)}{\prod_{\pm a}{}^{+}\left(\zeta_{2f}^{a} - \zeta_{2f}^{-a}\right)}}{\log \varepsilon},$$

where ε is the fundamental unit (uniquely normalized $\varepsilon > 1$) and the numbers $x = a,\ b$ run over a half system of the least prime positive residue classes mod f with $\left(\frac{-f}{a}\right) = 1$ and $\left(\frac{-f}{b}\right) = -1$. The positiveness of the latter expression, i.e., the

pointed out this defect and also proved that Kummer's assumption actually holds true for a cyclotomic field. By a later attempt to prove in a completely quite general way the divisibility of the class number of an algebraic number field by that of every subfield, Kronecker [9] strangely made the same mistake in a hidden form. Today we know by class field theory that Kronecker's assumption does not in general hold true (see the discussion with Theorem 3.40), and we also know a sufficient condition for the correctness; see footnote 11, p. 69, in Sect. 3.1.

[24] See, however, the supplement of the proofreader (1951) in the Introduction.

positiveness of another expression

$$\log \prod_{\pm b}^{+} \sin \frac{\pi b}{f} - \log \prod_{\pm a}^{+} \sin \frac{\pi a}{f}$$

is also a deep-rooted statement on the distribution of the a's and the b's in the system of prime residues mod f, for which no one knows any direct proof.

References[25]

1. H. Bergström, Die Klassenzahlformel für reelle quadratische Zahlkörper mit zuammengesetzter Diskriminante als Produkt verallgemeinerter Gaußscher Summen. J. f. d. reine und angew. Math. **186** (1944) [Introduction, **1.3**]
2. H. Hasse, Bericht über neuere Untersuchungen und Probleme aus der Theroie der algbraichen Zahlkörper. Part I: Klassenörpertheorie. Jahresbericht D.M.-V. **35** (1926); H. Hasse, Bericht über neuere Untersuchungen und Probleme aus der Theroie der algbraichen Zahlkörper. Part Ia: Beweise zu Teil I. Jahresbericht D.M.-V. **36** (1927); H. Hasse, Bericht über neuere Untersuchungen und Probleme aus der Theroie der algbraichen Zahlkörper. Part II: Reziprozitätsgesetz. Jahresbericht D.M.-V. Supplemental Ed. **6** (1930). Cited as "Klassenkörperbericht." [Introduction, **1.2, 1.3, 2.3, 2.12, 3.1, 3.2, 3.4, 3.5, 3.9, 3.19**]
3. H. Hasse, Führer, Diskriminate und Verzweigungskörper relativ-abelscher Zahlkörper. J. f. d. reine und angew. Math. **162** (1930) [**1.3**]
4. H. Hasse, *Vorlesungen über Zahlentheorie* (American Mathematical Society, Berlin, 1950) [Introduction, **1.4**]
5. E. Hecke, Über die L-Funktionen und den Dirichletschen Primzahlsatz für einen beliebigen Zahlkörper. Nachr. Kgl. Gel. d. Wissensch. Göttingen (1917) [**1.3**]
6. E. Hecke, Über eine neue Art von Zetafunktionen und ihre Beziehungen zur Verteilung der Primahlen. Math. Z. **1** (1918); E. Hecke, Über eine neue Art von Zetafunktionen und ihre Beziehungen zur Verteilung der Primahlen. Math. Z. **4** (1920) [**1.3**]
7. D. Hilbert, Die Theorie der algebraischen Zahlkörper. Jahresbericht D.M.-V. **4** (1897). Cited as "Zahlbericht" [Preface, Introduction, **1.2, 1.6, 2.9, 3.2, 3.5, 3.8, 3.19**]
8. L. Kronecker, Bemerkung über dek Klassenzahl der aus Einheitswurzeln gebildeten komplexen Zahlen. Monatsber. Akad. d. Wissensch. Berlin (1863) = Collected Works **1**, 123–131 [Introduction, **1.6, 3.7**]
9. L. Kronecker, Auseinandersetzung einer Einschaften der Klassenzahl idealer komplexer Zahlen. Monatsber. Akad. d. Wissensch. Berlin (1870) = Collected Works **1**, 271–282 [Introduction, **1.6, 3.1**]
10. E. Kummer, Bestimmung der Anzahl nicht-äquivalenter Klassen für die aus λ-ten Wurzeln der Einheit gebildeten komplexen Zahlen und die idealen Faktoren derselben. J. f. d. reine und angew. Math. **40** (1850) [Introduction, **1.6, 3.20**]

[25] The bold-typed numbers attached in the square brackets denote the sections of this book in which the individual works are cited.

11. E. Kummer, Sur la théorie des nombres complexes composés de racines de l'unité et de nombres entiers. J. de Math. **16** (1851) (French summary laying the foundation of the theory of ideal numbers from Journ. f. d. reine und angew. Math. **35** (1847), as well as three works cited above) [Introduction, **1.6**, **3.19**, **3.20**, Appendix]
12. E. Kummer, Über die Klassenanzahl der aus zusammengesetzten Einheitenwurzeln gebildeten idealen komplexen Zahlen. Monatsber. Akad. d. Wissensch. Berlin (1863) [Introduction, **1.6**, **3.10**, **3.12**, **3.15**, **3.20**, Appendix]

Chapter 2
The Arithmetic Structure of the Class Number Formula for Real Fields

In this second chapter we always assume that an abelian number field K we consider is real and hence coincides with its maximal real subfield K_0. Then the class number h of K is given by the formula (1.5.1a):

$$\boxed{hR = \prod_{\chi \neq 1} \sum_{\pm x \bmod f(\chi)} \left(-\chi(x) \log |1 - \zeta_{f(\chi)}^{x}| \right)} \tag{0}$$

2.1 Plan of Investigation

In the individual factors of the product (0), the summation quantity x does not always run over the same residue system; rather the system of summation as well as the summands $-\chi(x) \log |1 - \zeta_{f(\chi)}^{x}|$ depends on the characters χ over which the product is extended. From this situation, which makes the treatment of the product difficult, one can get free *formally* in a quite simple way when one introduces the decomposition

$$1 - \zeta_{f(\chi)}^{x} = \prod_{y \bmod \frac{f}{f(\chi)}} \left(1 - \zeta_{f}^{x + y f(\chi)} \right),$$

where y runs over the full residues mod $\frac{f}{f(\chi)}$.[1] Since here the equation

$$\chi(x) = \chi(x + y f(\chi))$$

[1](Translator's remark) In the following we should always pay attention to whether the summation as well as the product is taken over a system of *prime* residues or over a system of *full* residues.

© Springer Nature Switzerland AG 2019

H. Hasse, *On the Class Number of Abelian Number Fields*,
https://doi.org/10.1007/978-3-030-01512-1_2

holds, one obtains the transformation

$$hR = \prod_{\chi \neq 1} \sum_{\pm x \bmod f}{}^{*} \left(-\chi(x) \log|1-\zeta_f^x|\right), \tag{2.1.1}$$

where the star here means that—different from the ones in (0) and in all our other formulas—the summation is not restricted on a half system of *prime* residues x mod f, but is extended over a half system of *full* residues x mod f (then containing the non-prime residue classes mod f). Of course, it may very likely occur that, though x is prime to $f(\chi)$ and so $\chi(x) \neq 0$, x is not prime to f. For a cyclotomic field, Kummer has given his final formula in the form of (2.1.1) (and, in fact, not only for the second factor treated here, but also in the corresponding form for its first factor of the class number). However, for our purpose, from the merely formal transformation (2.1.1) of the class number formula (0), nothing could be obtained; on the contrary, the way would rather go backward, because the terms x mod f not coprime to f and with $\chi(x) \neq 0$ in the further actual algebraic transformation would hinder the process.

In the following, starting from the beginning formula (0), we will deal with such real algebraic transformations in two different ways. Each of these two transformations represents the definite multiples of hR by natural number factors g_K and $c_{\mathfrak{G}}$ determined in invariant ways by the field K and its Galois group $\mathfrak{G}$, respectively,[2] as determinants with elements of logarithms of absolute values of units of K. In the first case, $g_K \neq 0$ does not necessarily hold indeed. For the special class of the fields K for which $g_K \neq 0$ holds, however, a representation of $g_K hR$ arises directly as the regulator of a system of units of K, or—as we will call it for short—an *arithmetic representation of* $g_K h$. In the second case, $c_{\mathfrak{G}} \neq 0$ always holds indeed; however, only for a special class of fields K the obtained determinant is the regulator of a system of units of K, so that again for a special class of fields K an *arithmetic representation of* $c_{\mathfrak{G}} h$ is obtained. In both cases explicit expressions of the factors g_K and $c_{\mathfrak{G}}$ are given, which are also interesting of themselves and which particularly tell us what the state of the field K must be when the factors take value 1. For only the fields K for which the latter case occurs, the purpose of the present research is completely accomplished, while otherwise the obtained systems of units of K are g_K and $c_{\mathfrak{G}}$ times higher than those of the fundamental units of K, respectively. Nevertheless in such cases one can also call them *arithmetic representations of* h, because with these aids one can determine the class number h purely arithmetically in every given case.

We will obtain an *arithmetic representation of* h *in the narrow sense* ($g_K = 1$ as well as $c_{\mathfrak{G}} = 1$) only for cyclic real fields K—in fact, only for a special class of fields by the first transformation and for all such fields by the second transformation. Moreover, by the first transformation we will obtain an *arithmetic representation of* h *in the wide sense* ($g_K \neq 0$) for a definite class of real abelian number fields

[2](Translator's remark) This means that the factors g_K and $c_{\mathfrak{G}}$ are invariants of K and its Galois group $\mathfrak{G}$, respectively.

K that contains also non-cyclic fields—they are characterized by the restriction on the prime-decomposition of a divisor of discriminant[3]—while we will not be able to utilize the second transformation to derive an arithmetic representation of h for non-cyclic fields.

Hereafter for the required completion of results of this chapter, as I believe, it would be advantageous to quote the method that Weber [4, 5] has developed for the proof of his theorem stating that the class number h_0 of $K_0 = P_{2^\rho,0}$ for the cyclotomic field P_{2^ρ} of 2-power conductor 2^ρ is odd.[4] Actually in the present work I will deduce the theorem of Weber in a considerably simple way from the result of the first transformation; however, I hope in a further work that we could substantially exploit the method of Weber's proof to complete the results of this chapter of the present work.

2.2 The First Way of Transformation

Our first way of transformation of the class number formula (0) is distinguished from the transformation leading to (2.1.1) in the way that we ensure that there remain only the summands corresponding to a half system of *prime* residues modulo f in the individual sums of which the product (0) is composed. For the transformation some number factors[5] are added to the sums, which all together yield the already-mentioned number factor g_K. In regard to later application of the same method, we proceed here in a little more general way than what is necessary for the present purpose by deriving beforehand a general summation formula that underlies this transformation.

We consider the sums[6]

$$S_f(\chi) = \sum_{\pm x \bmod f} \chi(x) A_f(x)$$

corresponding to characters $\chi \neq 1$ of K, where f is a multiple of $f(\chi)$ and $A_f(x)$ is an expression depending only on the residue class x mod f on which we will later require further two formal conditions. We seek a reduction of the sum $S_f(\chi)$ with any multiple f of $f(\chi)$ to the corresponding sum $S_{f(\chi)}(\chi)$ with the smallest

---

[3](Translator's remark) See Theorem 2.1.

[4]Weber has proved this theorem and also the corresponding theorem on the relative class number of $P_{2^\rho}/P_{2^\rho,0}$ (see below Theorem 3.36!) in order to apply the theorem to his new proof of Kronecker's fundamental theorem that any abelian number field is contained in a cyclotomic field.

[5](Translator's remark) These number factors are $\prod_{p|f}(1-\chi(p))$.

[6](Translator's remark) We should always pay attention to the region over which a summation we consider is extended; this summation $\sum_{\pm x \bmod f} \chi(x)A_f(x)$ is extended over a half system of *prime* residues mod f, while the following summation $\sum_{y \bmod p}$ is over the system of *full* residues mod p.

multiple $f(\chi)$ itself. For this purpose we continue to separate prime factors p from f step by step to attain $f(\chi)$. Accordingly, let

$$f = f_0\, p,$$

where f_0 is also a multiple of $f(\chi)$. Then we can put the relation

$$x \equiv x_0 + y f_0 \pmod{f}$$

between the summation quantities x mod f of $S_f(\chi)$ and x_0 mod f_0 of $S_{f_0}(\chi)$ with y mod p. Here we have to divide the multiple $f = f_0\, p$ into two cases: $p \mid f_0$ and $p \nmid f_0$. In the case of $p \mid f_0$, x runs over a half system of prime residues mod f when x_0 runs over a half system of prime residues mod f_0 and y runs over a system of full residues mod p. Then the simple relation

$$S_f(\chi) = \sum_{\pm x_0 \bmod f_0} \chi(x_0) \sum_{y \bmod p} A_{f_0 p}(x_0 + y f_0) \tag{2.2.1a}$$

arises. In the case of $p \nmid f_0$, however, x runs over a half system of prime residues mod f and also over non-prime residues mod f when x_0 and y run over as above; that is, for every x_0 exactly one such class x^* mod f is given by

$$x^* \equiv x_0 + y^* f_0 \equiv x_0' p \pmod{f}$$

with uniquely determined y^* mod p. Hereby x_0' with x_0 runs over a half system of prime residues mod f_0, and it holds that

$$\chi(x^*) = \chi(x_0) = \chi(x_0')\chi(p).$$

Then the slightly complicated relation

$$S_f(\chi) + \chi(p) \sum_{\pm x_0' \bmod f_0} \chi(x_0') A_{f_0 p}(x_0' p) = \sum_{\pm x_0 \bmod f_0} \chi(x_0) \sum_{y \bmod p} A_{f_0 p}(x_0 + y f_0) \tag{2.2.1b}$$

arises. From now on we assume that the expression $A_f(x)$ has the following two properties for a multiple f_0 of $f(\chi)$:

$$\left\{ \begin{array}{ll} \sum_{y \bmod p} A_{f_0 p}(x_0 + y f_0) & = A_{f_0}(x_0) + \alpha \\ A_{f_0 p}(x_0' p) & = A_{f_0}(x_0') \end{array} \right\}, \tag{2.2.2}$$

where α is independent of x_0. Since, due to $\chi(-1) = 1$ and $\chi \neq 1$, the equations

$$\sum_{\pm x_0 \bmod f_0} \chi(x_0) = \frac{1}{2} \sum_{x_0 \bmod f_0} \chi(x_0) = 0$$

hold, the reductions[7]

$$\begin{aligned} S_f(\chi) &= \qquad S_{f_0}(\chi) \qquad && \text{for } p \mid f_0, \\ S_f(\chi) &= (1 - \chi(p)) S_{f_0}(\chi) && \text{for } p \nmid f_0 \end{aligned}$$

follow from relations (2.2.1a) and (2.2.1b). By separating gradually prime factors p from f to attain $f(\chi)$, one therefore obtains the entire reduction:

$$S_f(\chi) = \prod_{p \mid \frac{f}{f(\chi)}} (1 - \chi(p)) \cdot S_{f(\chi)}(\chi) = \prod_{p \mid f} (1 - \chi(p)) \cdot S_{f(\chi)}(\chi), \tag{2.2.3}$$

where the product is extended over the distinct prime divisors p of f; the primes p with $p \mid f(\chi)$ that are missing in the first product of (2.2.3) can be rounded up, because for these p it holds that $\chi(p) = 0$. This is the above-noticed general summation formula; it holds under condition (2.2.2) on the expression $A_f(x)$ in the considered sum $S_f(\chi)$. As is clear from the proof, one needs to use condition (2.2.2) only for the prime numbers $p \mid \frac{f}{f(\chi)}$.

For our present purpose we especially have to set

$$A_f(x) = -\log |1 - \zeta_f^x|.$$

Then condition (2.2.2) (with $\alpha = 0$) is satisfied and the summation formula (2.2.3) yields

$$\prod_{p \mid f} (1 - \chi(p)) \cdot \sum_{\pm x \bmod f(\chi)} (-\chi(x) \log |1 - \zeta_{f(\chi)}^x|) = \sum_{\pm x \bmod f} (-\chi(x) \log |1 - \zeta_f^x|),$$

where now f is the least common multiple of all $f(\chi)$ and it should be the conductor of K. Bringing this into the class number formula (0), as a preliminary result of our first transformation, one obtains the formula

$$g_K hR = \prod_{\chi \neq 1} \sum_{\pm x \bmod f} (-\chi(x) \log |1 - \zeta_f^x|) \tag{2.2.4}$$

[7] (Translator's remark) Here we recall our conditions: $\chi \neq 1$, $f(\chi) \mid f_0$ and $f = f_0 p$.

with the number factor

$$g_K = \prod_{p|f} \prod_{\chi \neq 1} (1 - \chi(p)). \tag{2.2.5}$$

Different from the sum in (2.1.1), the summation in the right-hand side of (2.2.4) is indeed restricted to a half system of *prime* residues mod f; for this reason, the number factor g_K given by (2.2.5) is actually added in the left-hand side of (2.2.4).

2.3 The Number Factor g_K

Before continuing to transform the obtained formula (2.2.4), we will enter into details of the factor g_K appearing in the formula and being given by (2.2.5). This number factor is determined by K in the invariant way. It is a rational and non-negative integer; the former is obvious from the symmetry of the algebraic-conjugate characters, and the latter is recognized by the combination of pairs of factors with complex-conjugate characters.[8] Also from the form we will deduce a rational and integral representation of g_K, by which we can see the arithmetic meaning of g_K.

This representation arises by the remark that $1 - \chi(p)$ are the values of the contributions $1 - \chi(p)/p^s$ of the product expansion of $1/L(s, \chi)$ for prime numbers p at $s = 0$. By the product formula (1.2.1) taken as the starting point in Chap. 1, which actually holds p-contribution-wise and is proved,[9] (it holds that) $\prod_{\chi \neq 1}(1 - \chi(p))$ are the values of the contribution of p to the product expansion of $\zeta(s)/\zeta_K(s)$ at $s = 0$ and hence

$$\prod_{\chi \neq 1} (1 - \chi(p)) = \left. \frac{\left(1 - \frac{1}{p^{n_p s}}\right)^{r_p}}{1 - \frac{1}{p^s}} \right|_{s=0} = \left\{ \begin{matrix} 0, & \text{if } r_p > 1 \\ n_p, & \text{if } r_p = 1 \end{matrix} \right\}, \tag{2.3.1}$$

where n_p denotes the residue class degree of p in K and r_p the number of the prime divisors $\mathfrak{p}$ of p in K. Consequently there arises by (2.2.5) the rational and integral representation of g_K:

$$g_K = \left\{ \begin{matrix} 0, & \text{if } r_p > 1 & \text{for at least one } p|f \\ \prod_{p|f} n_p, & \text{if } r_p = 1 & \text{for all } p|f \end{matrix} \right\}. \tag{2.3.2}$$

[8](Translator's remark) When $\chi = \overline{\chi}$, it obviously holds that $1 - \chi(p) \geqq 0$.

[9]See, for example, Klassenkörperbericht [1], Part I, §8, Proof of Theorem 14.

From this representation there follow two criteria for the question of when $g_K \neq 0$ or $g_K = 1$ occurs, which are particularly interesting for our application[10]:

Theorem 2.1 *$g_K \neq 0$ holds if and only if every prime divisor p of the discriminant of K is not decomposed in K, i.e., (it is decomposed) of the form $p \cong \mathfrak{p}^{\frac{n}{n_p}}$ with a prime divisor $\mathfrak{p}$ of K of degree n_p; then $g_K = \prod_{p|d} n_p$ holds actually.*

Theorem 2.2 *$g_K = 1$ holds if and only if every prime divisor p of the discriminant of K is totally ramified in K, i.e., (it is decomposed) of the form $p \cong \mathfrak{p}^n$ with a prime divisor $\mathfrak{p}$ of K of degree 1.*

The property of prime decomposition in K appearing in these criteria can be led back to the property of characters of K given in its formulation at the end of Sect. 1.1 by the decomposition theorem of the class field. Therefore the following two criteria arise:

Theorem 2.1′ *$g_K \neq 0$ holds if and only if $\chi(p) \neq 1$ holds for every prime divisor p of the conductor of K and for all characters $\chi \neq 1$ of K; then $g_K = \prod_{p|f} n_p$ holds actually, where n_p is determined such that the character values $\chi(p) \neq 1$ of p are n_p-th roots of unity.*

Theorem 2.2′ *$g_K = 1$ holds if and only if $\chi(p) = 0$, i.e., $p|f(\chi)$ holds for every prime divisor p of the conductor of K and for all characters $\chi \neq 1$ in K.*

One also reads directly from (2.2.5) the criterion in Theorem 2.1′ (except the data of the value of g_K). The criterion in Theorem 2.2′ emphasizes strictly the special case in which for all $\chi \neq 1$ the condition $(x, f(\chi)) = 1$ is tantamount to $(x, f) = 1$, where, during the derivation of formula (2.1.1) from (0), one restricts the summation over a half system of prime residues mod f and hence one can deduce directly (2.2.4) with $g_K = 1$.

This latter special case, which is particularly important in the following, can also be characterized in another way. We consider, for this purpose, the unique decomposition

$$\chi = \prod_{p|f} \chi_p$$

of a character χ mod H into components χ_p whose conductors are powers of prime divisors p of f; then certainly the components χ_p themselves are not always characters mod H, but are characters of residue classes mod f in any case—more precisely, mod p^ρ, where p^ρ are the p-components of f. If $f = p^\rho$ is a prime-power, then the condition in question—that $(x, f(\chi)) = 1$ is tantamount to $(x, f) = 1$ for all $\chi \neq 1$—is clearly satisfied. If f is composite, then it is necessary

[10] In Theorems 2.1 and 2.2 we replace the conductor f by the absolute value d of the discriminant of K (on the basis of the conductor-discriminant theorem of class field theory), because this seems more natural in this relation.

for the condition that for every $\chi \neq 1$ all the components χ_p have the same order as χ—i.e., the same order is also the order of χ—because every power of χ, as well as χ itself, is also a character mod H. Under this assumption, select χ as a character mod H of maximal order m, and let

$$\chi' = \prod_{p|f} \chi'_p$$

be any another character mod H of order m', which divides m. We select a prime $p \neq 2$ of f, which is possible by the compositeness of f. Since the prime residue classes mod p^ρ as well as its character group is cyclic for $p \neq 2$, then for χ and χ' the p-component of the form $\chi'\chi^{-\frac{m}{m'}k}$ (with some definite k coprime to m') vanishes. Hence by the above-mentioned fact it must necessarily be the principal character and then $\chi' = \chi^{\frac{m}{m'}k}$ holds. Consequently, for the composite f, the field K is necessarily cyclic. For the anticipated case of $f = p^\rho$, the field K is cyclic, because for $p \neq 2$ the group of the prime residue classes mod f and for $p = 2$ the factor group by the subgroup generated by ± 1 mod 2^ρ (on which only the real field K depends) are cyclic. Thus it has been found as a necessary condition for the considered special case that K is cyclic and that a generating character χ of K has p-components of the same order as χ itself for all $p|f$. This condition is obviously also sufficient.

We will name as a *pure character* a character χ of order n and with conductor f such that for every prime divisor p of f the component χ_p (of χ) is also of order n. As a result of the consideration above, we have, in addition to the supplement of Theorem 2.2′, the following criterion:

Theorem 2.2″ *$g_K = 1$ holds if and only if K is cyclic and a generating character of K is pure.*

This criterion can be applied particularly to cyclic fields of prime order and moreover, as already emphasized in the proof, to (real) cyclic fields of prime-power conductor.

2.4 Introduction to Cyclotomic Units

From now on we proceed to transform the class number formula (2.2.4) obtained in Sect. 2.2.

In the individual sums of which the product is composed, we first decompose the summations that are now throughout over some half system of prime residues x mod f, corresponding to the class partition mod H, into the summation over a half system of residues a mod f of H and over a system of representatives of the classes

$s \bmod H$. Hence we obtain by (2.2.4)

$$g_K hR = \prod_{\chi\neq 1} \sum_{\pm x \bmod f} \left(-\chi(x) \log |1 - \zeta_f^x|\right)$$

$$= \prod_{\chi\neq 1} \sum_{s \bmod H} \left(-\chi(s) \sum_{\substack{\pm a \bmod f \\ a \text{ in } H}} \log |1 - \zeta_f^{sa}|\right)$$

$$= \prod_{\chi\neq 1} \sum_{s \bmod H} \left(-\chi(s) \log \prod_{\substack{\pm a \bmod f \\ a \text{ in } H}} |1 - \zeta_f^{sa}|\right).$$

For further arithmetic treatment of this formula, it is practical to bring the absolute value of the right-hand side into a symmetric form of x and $-x$:

$$|1-\zeta_f^x| = |\zeta_{2f}^x - \zeta_{2f}^{-x}| = \sqrt{(1-\zeta_f^x)(1-\zeta_f^{-x})} \quad \text{(positive square root)}. \qquad (2.4.1)$$

Then it follows that

$$g_K hR = \prod_{\chi\neq 1} \sum_{s \bmod H} \left(-\chi(s) \log \prod_{\substack{\pm a \bmod f \\ a \text{ in } H}} \sqrt{(1-\zeta_f^{sa})(1-\zeta_f^{-sa})}\right). \qquad (2.4.2)$$

In this formula there appear the absolute values of conjugates of the integer of the cyclotomic field P_{2f},

$$\lambda = \prod_{\substack{\pm a \bmod f \\ a \text{ in } H}} \sqrt{-(1-\zeta_f^a)(1-\zeta_f^{-a})} = \prod_{\substack{\pm a \bmod f \\ a \text{ in } H}} (\zeta_{2f}^a - \zeta_{2f}^{-a}),$$

which we now have to investigate in detail. The number λ, whose indefiniteness of the sign (positive or negative) of the (pure-imaginary) quadratic root corresponds to the first product shown, is determined only up to undetermined sign: the sign can be normalized on the basis of the second product by taking a fixed half system of residues $a \bmod f$ from H and moreover a fixed normalization mod $2f$ of this half system.[11] The square

$$\lambda^2 = \prod_{\substack{\pm a \bmod f \\ a \text{ in } H}} \left(-(1-\zeta_f^a)(1-\zeta_f^{-a})\right)$$

[11](Translator's remark) We take a under the condition that a is coprime to $2f$.

lies in P_f and belongs to the considered field K on account of the invariance by the automorphisms $\zeta_f \to \zeta_f^a$ of P_f/K. Moreover λ^2 does not depend on the choice of half system a mod f from H, so that λ^2 is a fixed number of K determined by H in the invariant way.

For the construction of formula (2.4.2), the automorphisms $\zeta_{2f} \to \zeta_{2f}^s$ take a basic position when one expresses the quadratic roots of (2.4.1) as absolute values of the numbers of P_{2f}. For even conductor f these are the automorphisms $\zeta_{2f} \to \zeta_{2f}^s$ that provide the automorphisms S of the subfield K (of P_{2f}) for the numbers of K. For odd conductor f, where $P_{2f} = P_f$, this holds only when the representatives s of classes mod H are odd; and so the s are normalized to be coprime to $2f$, which can occur by the choice between the pairs s and $s + f$.[12] Hence, in every case the transformations $\zeta_{2f} \to \zeta_{2f}^s$ generate the automorphisms S of K. For the conjugates of λ generated by the transformations, even when the number λ itself does not lie in K in general but only λ^2 lies in K, we will safely use the notation λ^S, and we let

$$\lambda^S = \prod_{\substack{\pm a \bmod f \\ a \text{ in } H}} \sqrt{-(1-\zeta_f^{sa})(1-\zeta_f^{-sa})} = \prod_{\substack{\pm a \bmod f \\ a \text{ in } H}} (\zeta_{2f}^{sa} - \zeta_{2f}^{-sa}). \tag{2.4.3}$$

Here, as that of λ itself, the signs of λ^S remain undecided, or rather, the signs depend on the choice of the half system of a mod f from H and their normalization mod $2f$. For our present purpose these signs do not matter, because in formula (2.4.2) only the absolute values $|\lambda^S|$ appear.

For further calculation the quotients

$$\eta_S = \frac{\lambda}{\lambda^S} = \prod_{\substack{\pm a \bmod f \\ a \text{ in } H}} \frac{\zeta_{2f}^{a} - \zeta_{2f}^{-a}}{\zeta_{2f}^{sa} - \zeta_{2f}^{-sa}} \tag{2.4.4}$$

will be introduced. These numbers η_S belong to the considered field K. In fact, if λ is already contained in K, then this is clear. If λ^2 is contained in K (and if λ is not in K), then λ generates a relative quadratic field over K. Since the field (the quadratic extension) is contained in P_{2f} and is absolute abelian, it is also generated by each one of the conjugates λ^S. By the uniqueness of the generating radical except for square of elements of the ground field, η_S also belongs to K in this case. One also recognizes that η_S belongs to K by the fact that the representation (2.4.4) for the element η_S is invariant under the automorphisms of P_{2f}/K—$\zeta_{2f} \to \zeta_{2f}^a$ and $\zeta_{2f} \to \zeta_{2f}^{-1}$—and (moreover) when f is even, $\zeta_{2f} \to \zeta_{2f}^{1+f} = -\zeta_{2f}$.[13] On account

[12] We take this odd-numbered normalization of the representatives s mod f of classes mod H for odd f in the following execution and also throughout in the further process of this book; this assumption will not always be noted particularly.

[13] For odd f, in this conclusion, the half representatives system a mod f of H is thought to be normalized to be odd, which can occur by the following remark.

of the invariance of these representations of the replacements $a \to -a$ and $a \to a + f$ for individual representatives a, the elements η_S do not depend on the choice of the half system a mod f from H. By transition to another representatives system of classes s mod H, they (the elements η_S) change at most their signs. Therefore the absolute values

$$|\eta_S| = \frac{|\lambda|}{|\lambda^S|} = \prod_{\substack{\pm a \bmod f \\ a \text{ in } H}} \sqrt{\frac{(1-\zeta_f^a)(1-\zeta_f^{-a})}{(1-\zeta_f^{sa})(1-\zeta_f^{-sa})}} \quad \text{(with positive quadratic root),}$$

which alone are important first of all, are numbers of K determined in the invariant way by H and by automorphisms S of K, whereas the numbers η_S themselves are determined by H and S up to their signs. Since all the $1-\zeta_f^x$ with integers x coprime to f are associate with each other, the η_S are units; they are called *cyclotomic units of K*. For composite f, as is well known, the $1 - \zeta_f^x$ are already units, so the λ^2 are already units of K. However we will not have to make use of this fact in this chapter, but we will use it in the proof of Theorem 3.26 in Sect. 3.7.

2.5 The First Arithmetic Representation of the Class Number

On the basis of (2.4.3), we can express the class number formula (2.4.2) in the form

$$g_K hR = \prod_{\chi \neq 1} \sum_S \left(-\chi(S) \log |\lambda^S|\right), \tag{2.5.1}$$

where S—as in the rest of this book—runs over the elements of the Galois group $\mathfrak{G}$ of K. Therefore we have come back to the original meaning of the χ as the characters of $\mathfrak{G}$.

Now, as is well known, for the indeterminate $u(S)$ that corresponds to an element S, the linear factor decomposition of the group determinant of $\mathfrak{G}$ holds:

$$\prod_\chi \sum_S \chi(S)u(S) = |u(ST^{-1})| \quad \begin{pmatrix} S \text{ row index} \\ T \text{ column index} \end{pmatrix}. \tag{2.5.2}$$

If one wants to remove the linear factor $\sum_S u(S)$ corresponding to $\chi = 1$ from the product, one needs to replace only some row (or column) by 1, because all the rows (or columns) have the same sum $\sum_S u(S)$. One replaces, for example, the column corresponding to $T = 1$ by 1 and subtracts the row corresponding to $S = 1$ from all other rows, then one obtains the product formula

$$\prod_{\chi \neq 1} \sum_S \chi(S)u(S) = |u(ST^{-1}) - u(T^{-1})|_{S,T \neq 1}. \tag{2.5.3}$$

This formula permits us to describe the product of the right-hand side of (2.5.1) as determinant: for this purpose one only has to carry out the specialization $u(S) = -\log|\lambda^S|$, by which the equations

$$u(ST^{-1}) - u(T^{-1}) = -\log|\lambda^{ST^{-1}}| + \log|\lambda^{T^{-1}}| = \log\frac{|\lambda^{T^{-1}}|}{|\lambda^{ST^{-1}}|} = \log|\eta_S^{T^{-1}}|$$

hold from (2.4.4). Hence by (2.5.1) there arises the determinant representation

$$g_K hR = \left|\log|\eta_S^{T^{-1}}|\right|_{S,T\neq 1}. \tag{2.5.4}$$

The determinant appearing in this way is the regulator of the system of cyclotomic units η_S of K. Therefore it holds that

$$g_K h = \frac{R(\eta_S)}{R}. \tag{2.5.5}$$

As long as $g_K \neq 0$—see Theorems 2.1 and 2.1′ in Sect. 2.3[14]—we have therefore obtained an arithmetic representation of the class number of K in the wide sense. Precisely, by summarizing the above in detail once again, the obtained result is stated as follows:

Theorem 2.3 *Let K be a real abelian number field of degree n with the property that every prime p dividing the discriminant of K is not decomposed in K, that is, (it is decomposed) of the form $p \cong \mathfrak{p}^{\frac{n}{n_p}}$ in K with a prime divisor $\mathfrak{p}$ of K of degree n_p, or—what is the same—that $\chi(p) \neq 1$ for every prime $p \mid f$ and for every $\chi \in X$, $\chi \neq 1$, where the number n_p is determined such that the $\chi(p) \neq 0$ are n_p-th roots of unity.*

Then $\prod_{p\mid f} n_p$ times the class number h of K is equal to the index of the subgroup generated by the absolute values of the cyclotomic units,

$$|\eta_S| = \prod_{\substack{\pm a \bmod f \\ a \text{ in } H}} \sqrt{\frac{(1-\zeta_f^a)(1-\zeta_f^{-a})}{(1-\zeta_f^{sa})(1-\zeta_f^{-sa})}}$$

($s \bmod f$ runs over a system of representatives of the classes $\bmod H$), in the group of absolute values of units of K.

By considering formula (2.5.5) from the viewpoint of the theory of structure of the group of units of K, the formula above yields the general fact that the index of the subgroup given in Theorem 2.3 is a multiple of the number g_K. In the case of

[14](Translator's remark) "die obigen Sätze 1, 1" in the original book published in 1985 should be "die obigen Sätze 1, 1′."

$g_K = 0$, this means that the cyclotomic units of K are not independent; in the case of $g_K \neq 1$, surely the units do not construct a system of fundamental units of K.

Now for formula (2.5.4), notice that, in the case of $g_K \neq 0$, according to the positiveness of $g_K hR$, the determinant appearing in the right-hand side is also positive by fixed coupling of rows and columns (of the same order), and then the determinant could be written without the absolute sign of the general regulator. This positiveness is a generalization of the deep-rooted description on the class distribution of the system of prime residues mod f for a real quadratic field, which was emphasized in Sect. 1.6. By the remarks there, however, this generalization is not deeper than the description of positiveness of the special quadratic case.

Especially when K is cyclic, one can replace in (2.5.4), in (2.5.5) and in Theorem 2.3 the system of cyclotomic units η_S, which are assigned by automorphisms S of K and, for instance, are not conjugate to each other, by a system of units with the same height that consists of conjugates η^S of *one* cyclotomic unit η, that is, the system

$$\eta = \pm\eta_Z,\ \eta^Z = \pm\eta_{Z^2}\eta_Z^{-1},\ \ldots,\ \eta^{Z^{n-2}} = \pm\eta_{Z^{n-1}}\eta_{Z^{n-2}}^{-1}\ (\eta^{Z^{n-1}} = \pm\eta_{Z^{n-1}}^{-1}), \tag{2.5.6}$$

where Z is a generating automorphism of K. Conversely, in this case

$$\eta_Z = \pm\eta,\ \eta_{Z^2} = \pm\eta^{1+Z},\ \ldots,\ \eta_{Z^{n-1}} = \pm\eta^{1+Z+\cdots+Z^{n-2}}\ (1 = \pm\eta^{1+Z+\cdots+Z^{n-1}}). \tag{2.5.6$'$}$$

We will call η *a generating cyclotomic unit of* K. The undetermined signs of these formulas come from the fact that in general the η_S are invariantly determined only up to sign. In the system η_S the $n-1$ signs are independent of each other, while in the system η^S they (the signs of η^S) are connected to the undetermined sign of η. One can normalize this (the sign of η) as positive, just as we could normalize the signs of η_S as positive by the determination (of sign). However, the normalization is not structure-invariant, insofar as the positiveness is not transformed to those of conjugates in general. Positive numbers would appear there only in the beginning, but not necessarily in the further Eqs. (2.5.6) and (2.5.6′) of transformation.

By the statement above, in the cyclic case, the formulas

$$g_K hR = \left|\log|\eta^{ST^{-1}}|\right|_{S,T\neq 1} = \left|\log|\eta^{Z^{\mu-\nu}}|\right|_{\mu,\,\nu\not\equiv 0\ (\mathrm{mod}\ n)}, \tag{2.5.4z}$$

$$g_K h = \frac{R(\eta^S)}{R} \tag{2.5.5z}$$

follow from (2.5.4), (2.5.5), and hence by Theorem 2.3 there arises the following:

Theorem 2.3z *Let K be a real cyclic field satisfying the condition quoted in Theorem 2.1.*

Then $\prod_{p|f} n_p$ times the class number h of K is equal to the index of the subgroup generated by absolute values of conjugates of a generating cyclotomic unit

$$\eta = \prod_{\substack{\pm a \bmod f \\ a \text{ in } H}} \sqrt{\frac{(1-\zeta_f^a)(1-\zeta_f^{-a})}{(1-\zeta_f^{za})(1-\zeta_f^{-za})}}$$

(z mod f is a representative of a generating class mod$H$*) in the group of absolute values of all units of K.*

Finally we further summarize (the results above) and emphasize a result obtained as a special case of $g_K = 1$—see Theorems 2.2, 2.2′, and 2.2″ in Sect. 2.3.

Theorem 2.4 *Let K be a real abelian number field of degree n with the property that every prime p dividing the discriminant of K is totally ramified in K, that is, (it is decomposed) of the form $p \cong \mathfrak{p}^n$ in K with a prime divisor $\mathfrak{p}$ of K of degree* 1, *or—what is the same—that K is a real cyclic field with the property that a generating character χ of K is pure.*

Then the class number h of K is equal to the index of the subgroup generated by the absolute values of conjugates of a generating cyclotomic unit

$$\eta = \prod_{\substack{\pm a \bmod f \\ a \text{ in } H}} \sqrt{\frac{(1-\zeta_f^a)(1-\zeta_f^{-a})}{(1-\zeta_f^{za})(1-\zeta_f^{-za})}}$$

(z mod f is a representative of a generating class mod$H$*) in the group of absolute values of all units of K.*

By this theorem, for a special kind of fields K, an arithmetic representation of the class number in the narrow sense is given. This kind of fields contains especially real cyclic fields of prime degree and real abelian number fields with prime-power conductor.

One recognizes that not only in the formulation but also in the actual situation Theorem 2.3 is much more general than Theorems 2.3z and 2.4, both of which treat only cyclic fields, since there exist non-cyclic fields with the property quoted in Theorem 2.3: for example, the real bicyclic biquadratic fields

$$K = P(\sqrt{p}, \sqrt{p'}) \text{ with primes } p, p' \equiv 1 \pmod 4 \text{ and } \left(\frac{p}{p'}\right) = \left(\frac{p'}{p}\right) = -1,$$

where p' can be replaced by 2 and then $(\frac{p}{2}) = (\frac{2}{p}) = -1$ is to be demanded. For these fields $g_K = 4$ holds clearly, and hence surely the cyclotomic units do not form a system of the fundamental units of K.

Examples of $g_K = 0$ are supplied, for example, by the real bicyclic biquadratic fields

$$K = P(\sqrt{p}, \sqrt{p'}) \text{ with primes } p, p' \equiv 1 \pmod 4 \text{ and } \left(\frac{p}{p'}\right) = \left(\frac{p'}{p}\right) = 1,$$

where p' can be also replaced by 2 and then $(\frac{p}{2}) = (\frac{2}{p}) = 1$ is to be demanded. Then for these fields the cyclotomic units are not independent.

2.6 The Theorem of Weber and Its Generalization

As already remarked at the end of Sect. 2.1, Weber [4, 5] has proved that the maximal real subfield $P_{2^\rho,0}$ of the cyclotomic field P_{2^ρ} has odd class number h_0. In the following by virtue of the results in Sect. 2.5, I will develop a new and considerably simple proof of the theorem of Weber and of an additional statement on the units of $P_{2^\rho,0}$, which Weber had obtained in the same context, and (I will) subsequently give a generalization of this theorem for a certain kind of real cyclic fields. It is quite surprising that Weber, who also started from the arithmetic representation of the class number of $P_{2^\rho,0}$ given in Theorem 2.4, had not seen this theoretically, as well as numerically, tremendous simple proof but had first taken aim at the basis of research entering very broadly and deeply into the structure of the group of units of $P_{2^\rho,0}$, which itself is very interesting.

The basic idea of our proof is the following concluding method well-known for a real quadratic field K. Let ε be the fundamental unit of K (with the unique normalization $\varepsilon > 1$) and η the cyclotomic unit of K (with the unique normalization $\eta > 0$); then by (2.5.4) the relation

$$\eta = \varepsilon^h$$

holds. Hence, if the establishment $N(\eta) = -1$ succeeds, then $N(\varepsilon)^h = -1$ follows, so that h is odd and also $N(\varepsilon) = -1$.

Next we will generalize this concluding method to an arbitrary real abelian field K. In this process we cannot also treat simply the sign of the norm, rather in its place we have to apply a more general invariant construction that is defined as follows. Let us denote, in general, by

$$\operatorname{sgn} \alpha^S = (-1)^{\sigma(\alpha^S)}$$

the signature of a number $\alpha \neq 0$ of K; it is a vector with n terms corresponding to elements S of $\mathfrak{G}$, whose exponent vectors $\sigma(\alpha^S)$ are the point here. Now let $\eta_0, \eta_1, \ldots, \eta_{n-1}$ be any system of n terms of units of K, and consider the determinant of the signature exponents of η_ν:

$$\Sigma(\eta_\nu) \equiv |\sigma(\eta_\nu^S)| \pmod 2 \quad \begin{pmatrix} S \text{ row index} \\ \nu \text{ column index} \end{pmatrix}. \tag{2.6.1}$$

This is a formation similar to the regulator $R(\eta_\nu)$. Except that the numbers of rows and columns of $\Sigma(\eta_\nu)$ are greater by one than those of the regulator, this determinant is different from the regulator in the situation that the signatures of exponents $\sigma(\eta_\nu^S)$, which formally behave like logarithms, stand in the place of logarithms $\log|\eta_\nu^S|$. We will call this determinant the *regulatrix* of a system of units η_ν for short. The relation $\Sigma(\eta_\nu) \not\equiv 0 \pmod 2$ means that the signatures of η_ν are independent. For a system of fundamental units of K with supplement -1,[15] $\varepsilon_0 = -1, \varepsilon_1, \ldots, \varepsilon_{n-1}$, the regulatrix

$$\Sigma(\varepsilon_\nu) \equiv \Sigma \pmod 2$$

is an invariant of K, which takes a position of the regulator $R(\varepsilon_\nu) = R$, called the *regulatrix of* K. For a real quadratic field it holds that

$$\Sigma \equiv \begin{vmatrix} 1 & \sigma(\varepsilon) \\ 1 & \sigma(\varepsilon') \end{vmatrix} \equiv \sigma(\varepsilon') - \sigma(\varepsilon) \equiv \sigma(N(\varepsilon)) \pmod 2,$$

then $\Sigma \equiv 0$ or $1 \pmod 2$ according as $N(\varepsilon) = 1$ or -1. In general, the relation $\Sigma \not\equiv 0 \pmod 2$, which corresponds to $N(\varepsilon) = -1$ (in the case of a real quadratic field), means that the signatures of the fundamental units ε_ν are independent, that is, there exist units with arbitrary given signature in K.

From now on we consider the cyclotomic units η_S of K; here, differently than in Sect. 2.5, we actually include the trivial cyclotomic unit $\eta_1 = \pm 1$ corresponding to $S = 1$, and in fact, we also put $\eta_1 = -1$ corresponding to $\varepsilon_0 = -1$. By giving attention to this last correspondence, one sees that the relation (2.5.5) between the regulators produces the corresponding relation between the regulatrices:

$$\Sigma(\eta_S) \equiv g_K h\, \Sigma \pmod 2. \tag{2.6.2}$$

Hereby there arises the following generalization of the above-mentioned concluding method for a real quadratic field:

Theorem 2.5 *Let K be a real abelian number field.*

If the signatures of the cyclotomic units are independent, then the class number h of K is odd and also the signatures of the fundamental units of K are independent.

By (2.6.2) one sees moreover that the assumption of this theorem can hold only for the fields K with odd invariant g_K, and especially with $g_K \neq 0$.

[15]For brevity, henceforth, so far as it is the matter of signatures, we will always understand the concept of a system of fundamental units of K with supplement -1. In fact, with this supplement the system is really sufficient for the composition of all the units of K.

Particularly, if K is cyclic and if a generating cyclotomic unit η of K has the property $N(\eta) = -1$, then by (2.5.6) and (2.5.6′) the system of the conjugates η^S is equivalent to the system of the cyclotomic units η_S with respect to signatures, so that the congruence

$$\Sigma(\eta^S) \equiv g_K h\, \Sigma \pmod 2 \tag{2.6.2z}$$

also holds and the system η_S in Theorem 2.5 can be replaced by the system η^S. The regulatrix of the system η^S is transformed by (2.6.1) and by the group-determinant formula (2.5.2) to

$$\Sigma(\eta^S) \equiv \prod_{\chi} \sum_{S} \chi(S)\sigma(\eta^S) \pmod 2. \tag{2.6.3}$$

To establish the independence of the signatures of η^S, one has to show accordingly that all the linear factors $\sum_S \chi(S)\sigma(\eta^S)$ are prime to 2. Particularly the linear factor corresponding to $\chi = 1$ provides the relation $\sum_S \sigma(\eta^S) \equiv \sigma(N(\eta)) \equiv 1 \pmod 2$[16] and so $N(\eta) = -1$, which we already had to assume for the application of Theorem 2.5 at any rate. In accordance with this, we have the following:

Theorem 2.5z *Let K be a real cyclic field.*

If for a generating cyclotomic unit η and for all characters χ of K all the sums $\sum_S \chi(S)\sigma(\eta^S)$ over the signatures of exponents $\sigma(\eta^S)$ are prime to 2, then the class number h of K is odd and the signatures of the fundamental units of K are independent.

By (2.6.2z) one sees once again that the assumption of this theorem can hold only for the fields K with odd invariant g_K, and especially with $g_K \neq 0$.

For the special cyclic field $K = P_{2^\rho,0}$ of degree $n = 2^{\rho-2}$, the situation is particularly simple, because for the field $\chi(S) \equiv 1 \pmod{\mathfrak{z}}$ holds for all $S \in \mathfrak{G}$, where $\mathfrak{z}$ is the unique prime divisor of order 1 of (the prime) 2 in the field $P_n = P_{2^{\rho-2}}$ (which is generated by the values) of characters χ of K. Therefore, here for all the sums the congruences

$$\sum_S \chi(S)\sigma(\eta^S) = \sum_S \sigma(\eta^S) = \sigma(N(\eta)) \pmod{\mathfrak{z}} \tag{2.6.4}$$

hold, so that in this special case the establishment $N(\eta) = -1$ is sufficient for drawing the deduction assigned in Theorem 2.5z. This establishment indeed can be easily provided. A generating cyclotomic unit of $P_{2^\rho,0}$ is

$$\eta = \frac{\zeta_{2^{\rho+1}} - \zeta_{2^{\rho+1}}^{-1}}{\zeta_{2^{\rho+1}}^{z} - \zeta_{2^{\rho+1}}^{-z}},$$

[16](Translator's remark) $\sum_S \sigma(\eta_S)$ in the original book should be $\sum_S \sigma(\eta^S)$.

where z is a basis element of the prime residue class group mod 2^ρ modulo the subgroup $\pm 1 \bmod 2^\rho$,[17] and so, for example, $z \equiv 1 + 2^2 \pmod{2^\rho}$. The conjugates of η are

$$\eta^{Z^\nu} = \frac{\zeta_{2^{\rho+1}}^{z^\nu} - \zeta_{2^{\rho+1}}^{-z^\nu}}{\zeta_{2^{\rho+1}}^{z^{\nu+1}} - \zeta_{2^{\rho+1}}^{-z^{\nu+1}}} \quad (\nu \bmod 2^{\rho-2}).$$

With attention to $z^{2^{\rho-2}} \equiv 1 + 2^\rho \pmod{2^{\rho+1}}$, (we see that) there actually appears

$$N(\eta) = \eta^{1+Z+\dots+Z^{2^{\rho-2}-1}} = \frac{\zeta_{2^{\rho+1}} - \zeta_{2^{\rho+1}}^{-1}}{\zeta_{2^{\rho+1}}^{1+2^\rho} - \zeta_{2^{\rho+1}}^{-1-2^\rho}} = -1.$$

By the statement above we have therefore obtained Weber's result:

Theorem 2.6 *For the maximal real subfield $P_{2^\rho,0}$ of the cyclotomic field P_{2^ρ}, the class number h_0 is odd and the signatures of the fundamental units are independent.*[18]

The same concluding method (as above) can be applied to more general real cyclic fields, namely, whenever the following two conditions are satisfied:

(a) The prime number 2 is decomposed into prime divisors $\mathfrak{z}$ of degree 1 in the field P_n of a generating character χ of K (allowed to be ramified).
(b) For a generating cyclotomic unit η it holds that $N(\eta) = -1$.

Hence, namely, by (a) the congruence (2.6.4) holds for all the divisors $\mathfrak{z}$ of 2, so that by (b) the assumption of Theorem 2.5z is satisfied. Consequently the assertion of Theorem 2.5z follows, and moreover, as already mentioned, the validity of the fact follows:

(c) The invariant g_K is odd.

Condition (a) is satisfied by the famous decomposition theorem of cyclotomic fields exactly when the degree n is a power of 2, $n = 2^\nu$ (and then, in fact, the prime 2 also has a unique prime divisor of degree 1 in P_n). Regarding condition (b), one has

$$N(\eta) = \eta^{1+Z+\dots+Z^{n-1}} = \prod_{\substack{\pm a \bmod f \\ a \text{ in } H}} \frac{\zeta_{2f}^{a} - \zeta_{2f}^{-a}}{\zeta_{2f}^{z^n a} - \zeta_{2f}^{-z^n a}},$$

[17](Translator's remark) z is a representative of a generator of the factor group of the prime residue class group mod 2^ρ by the subgroup $\pm 1 \bmod 2^\rho$.

[18]Weber [4, 5] formulated the last fact in the following equivalent way: *In $P_{2^\rho,0}$ every totally positive unit is square.*

where z is a representative of a basic class (a generating class) mod H. For odd f the number z is normalized to be odd by the remarks in Sect. 2.4, and we also suppose that the half system a mod f from H is normalized to be odd (then as a half system a mod $2f$ from H). We consider the reduction

$$z^n a \equiv (-1)^{\delta_a} a' \pmod{f},$$

where the a' are a permutation of the a and there appear the signature exponents δ_a mod 2 corresponding to a. For odd f one has moreover, by the encountered normalization,

$$z^n a \equiv (-1)^{\delta_a} a' \pmod{2f};$$

on the other hand, for even f there exists in a slightly different way the extended reduction

$$z^n a \equiv (-1)^{\delta_a} a'(1+\kappa_a f) \equiv (-1)^{\delta_a}(1+f)^{\kappa_a} a' \pmod{2f},$$

with extended exponents κ_a mod 2 corresponding to a. Thereby it holds apparently that[19]

$$N(\eta) = (-1)^{\sum_a \delta_a} \qquad \text{for odd } f,$$

$$N(\eta) = (-1)^{\sum_a \delta_a + \sum_a \kappa_a} \text{ for even } f.$$

By the construction of the product over the half system a mod f from H, one now confirms easily the validity of the following relations[20]:

$$(-1)^{\sum_a \delta_a} = -1 \text{ or } 1 \text{ according as } f \text{ is a power of a prime } p \neq 2 \text{ or not,}$$

$$(-1)^{\sum_a \kappa_a} = -1 \text{ or } 1 \text{ according as } f \text{ is a power of the prime } p = 2 \text{ or not.}$$[21]

Then it follows that

$$N(\eta) = -1 \text{ or } 1 \text{ according as } f \text{ is a power of a prime or a composite.}$$

Condition (b) is therefore satisfied only when $f = p^\rho$ is a power of a prime.

[19](Translator's remark) In the following summations $\sum_a$, $\pm a$ mod f runs over H, as in the product $\prod_{\pm a \bmod f,\, a \text{ in } H}$ above.

[20]They (the relations) are valid in any case where f is the conductor and z is a representative of a basic class mod H for a real cyclic filed K. To this, see also a little more detailed and quite similar concluding method in the proof of the lemma in Sect. 3.11.

[21](Translator's remark) When we consider κ_a, we suppose that f is even.

By summing up the above, for a real cyclic field K the two conditions (a) and (b) are satisfied exactly when the degree $n = 2^\nu$ is a power of 2 and the conductor $f = p^\rho$ is a power of a prime p. As was emphasized above, one recognizes also directly by Theorem 2.2″ that the condition (c) holds in this case; by Theorem 2.2″, moreover, $g_K = 1$ holds for $f = p^\rho$.

For $p = 2$, since the prime residue class group mod 2^ρ has the cyclic factor group by the subgroup ± 1 mod 2^ρ, the field $K = P_{2^\rho,0}$ of degree $n = 2^{\rho-2}$ already anticipated in Theorem 2.6 is the unique real cyclic field with conductor $f = 2^\rho$.

For $p \neq 2$ by $n = 2^\nu$, it holds necessarily that $\rho = 1$ and hence $f = p$ and also $p \equiv 1 \pmod{2^\nu}$, and by

$$\chi(-1) = \zeta_{2^\nu}^{\frac{p-1}{2}} = (-1)^{\frac{p-1}{2^\nu}} = 1$$

moreover $p \equiv 1 \pmod{2^{\nu+1}}$. For every prime number p with this property, there exists exactly one real cyclic field K with conductor $f = p$ and of degree $n = 2^\nu$, namely, the subfield of degree 2^ν of the cyclic cyclotomic field P_p of degree $p - 1$.

Thus, by the already-mentioned fact, as a generalization of Weber's Theorem 2.6, the following theorem has been proved.

Theorem 2.7 *For a real cyclic field K of 2-power degree $n = 2^\nu$ and with prime-power conductor $f = p^\rho$, that is, except for $P_{2^\rho,0}$, moreover for the subfield of degree 2^ν of the cyclotomic field P_p with prime number $p \equiv 1 \pmod{2^{\nu+1}}$, the class number h is odd and the signatures of fundamental units are independent.*

For the special case of $n = 2$ this is the famous fact from the genus theory of a real quadratic field $K = P(\sqrt{f})$ that for prime-power conductor f (then $f = 2^3$ or $f = p \equiv 1 \pmod 4$), the class number is odd and $N(\varepsilon) = -1$.

2.7 Generalized Group Matrix

Our second way of transformation of the class number formula (0), to which we will devote ourselves from now on, is based on a generalization of the linear factor decomposition (2.5.2) of the group determinant of an abelian group $\mathfrak{G}$, which is itself interesting to us. While the ordinal group matrix $(u(ST^{-1}))$ is constructed by a single system of indeterminates $u(S)$ that correspond to elements S of $\mathfrak{G}$, we consider a similarly constructed matrix with elements of a system of several indeterminates $u_\chi(S)$ that correspond (also) to characters χ of $\mathfrak{G}$ and among the indeterminates certain conditions should hold further.

First, the system $u_\chi(S)$ should depend not only on a character χ itself but also on a Frobenius class[22] of the group X of the characters of $\mathfrak{G}$. Then it should hold

[22](Translator's remark) In an abelian group G a Frobenius class consists of an element a of G and of its powers a^μ with exponent μ, where μ is coprime to the order of a.

that

$$u_\chi(S) = u_{\chi^\mu}(S) \quad \text{for} \ \ (\mu, n_\chi) = 1, \tag{2.7.1a}$$

where n_χ denotes the order of χ. Second, the indeterminates $u_\chi(S)$ of such a class should depend only on the class modulo the subgroup $\mathfrak{H}_\chi$ of $\mathfrak{G}$ corresponding to the χ to which S belongs.[23] Then it should hold that

$$u_\chi(S) = u_\chi(S') \quad \text{for} \ \ \chi(S) = \chi(S'). \tag{2.7.1b}$$

The subgroup $\mathfrak{H}_\chi$ and also its index n_χ in $\mathfrak{G}$ depend only on the class of χ.

In the following, ψ always runs over a system of representatives of the Frobenius classes of X so that one obtains all characters χ of $\mathfrak{G}$ in the form $\chi = \psi^\nu$ uniquely where ν always runs over a system of prime residues mod n_ψ. Since the powers $\chi = \psi^\nu$ are just the entire system of algebraic-conjugates of ψ, one can also define the ψ as a system of representatives of non-algebraic-conjugate characters of K. Further for every ψ let us select a fixed representative T_ψ of a generating element of the cyclic factor group $\mathfrak{G}/\mathfrak{H}_\psi$ of order n_ψ so that the value of the character $\psi(T_\psi)$ is a primitive n_ψ-th root of unity.

The subject to be treated here is the calculation of the determinant of the square matrix

$$U_\mathfrak{G} = (u_\psi(ST_\psi^{-k})) \quad \begin{pmatrix} \text{row index} & S \\ \text{column indices} & \psi, k \end{pmatrix}, \tag{2.7.2}$$

where the row index S runs over the elements of the group $\mathfrak{G}$, the first column index ψ runs over a system of representatives of the Frobenius classes of X and the second column index k runs over some system of $\varphi(n_\psi)$ residue classes mod n_ψ such that the roots of unity $\varphi(T_\psi)^k$ construct an integral basis of the cyclotomic field P_{n_ψ}, for example, for $k = 0, 1, \ldots, \varphi(n_\psi) - 1$. Since by the unique expression $\chi = \psi^\nu$, the number $\sum_\psi \varphi(n_\psi) = n$ of columns is equal to the order of $\mathfrak{G}$, $U_\mathfrak{G}$ is actually a square matrix and has a determinant.

If $\mathfrak{G}$ is cyclic of order p, p a prime number, then X has only two Frobenius classes: one corresponding to a generating character and the other to the principal character. Correspondingly one has two systems of indeterminates: the first system $u(S)$ consists of p different indeterminates and the second $v(S) = v$ is reduced to one indeterminate. Then the matrix $U_\mathfrak{G}$ is obtained from the ordinary matrix $(u(ST^{-1}))$ by replacing the column corresponding to $T = 1$ by the constant column $v(S) = v$. Exactly the determinant of such a modified group matrix played a roll in our first transformation (for arbitrary abelian groups)—see (2.5.2), (2.5.3), and (2.5.4)—and, actually by the specialization of indeterminates, $u(S) = -\log|\lambda^S|$ and $v = 1$.

[23](Translator's remark) Here and in what follows, we let $\mathfrak{H}_\chi = \{S \in \mathfrak{G};\ \chi(S) = 1\}$.

We now clarify the constructing law of the above-mentioned generalization of the group matrix by the explicit exercise of the matrix $U_{\mathfrak{G}}$ for the first three abelian groups of non-prime order. For the cyclic group of order 4, we have[24]

$$U_{\mathfrak{G}} = \begin{pmatrix} u_0 & u_3 & v_0 & w \\ u_1 & u_0 & v_1 & w \\ u_2 & u_1 & v_0 & w \\ u_3 & u_2 & v_1 & w \end{pmatrix}.$$

For the bicyclic group of order 4 (Klein's four group), we have

$$U_{\mathfrak{G}} = \begin{pmatrix} u_0 & v_0 & w_0 & z \\ u_0 & v_1 & w_1 & z \\ u_1 & v_0 & w_1 & z \\ u_1 & v_1 & w_0 & z \end{pmatrix}.$$

For the cyclic group of order 6, we have

$$U_{\mathfrak{G}} = \begin{pmatrix} u_0 & u_5 & v_0 & v_2 & w_0 & z \\ u_1 & u_0 & v_1 & v_0 & w_1 & z \\ u_2 & u_1 & v_2 & v_1 & w_0 & z \\ u_3 & u_2 & v_0 & v_2 & w_1 & z \\ u_4 & u_3 & v_1 & v_0 & w_0 & z \\ u_5 & u_4 & v_2 & v_1 & w_1 & z \end{pmatrix}.$$

Here the systems of indeterminates corresponding to the individual classes of X are, respectively, denoted by special letters, and the ordering of elements of $\mathfrak{G}$ is just described in an easy and clear way of corresponding numbering.

We once more particularly remark that one cannot obtain the ordinary group matrix by specialization (equality-setting of indeterminates) from $U_{\mathfrak{G}}$ surely,

---

24(Translator's remark) Let $\mathfrak{G} = \langle T \rangle$ be the cyclic group of order 4 generated by T and $X = \langle \chi \rangle$, its character group generated by χ. Putting

$$u_i = u_\chi(T^i) = u_{\chi^3}(T^i) \quad \text{for } i = 0, 1, 2, 3; \;\; v_0 = u_{\chi^2}(1) = u_{\chi^2}(T^2), \;\; v_1 = u_{\chi^2}(T) = u_{\chi^2}(T^3);$$
$$w = u_1(T^i) \quad \text{for } i = 0, 1, 2, 3,$$

and taking $T_\chi = T,\ T_{\chi^2} = T,\ T_1 = 1$, we have

$$U_{\mathfrak{G}} = (u_\psi(T^i T_\psi^{-k}))_{i=0,1,2,3;\ \psi=\chi,\chi^2,1, k=0,1,\ldots,\varphi(n_\psi)-1} = \begin{pmatrix} u_0 & u_3 & v_0 & w \\ u_1 & u_0 & v_1 & w \\ u_2 & u_1 & v_0 & w \\ u_3 & u_2 & v_1 & w \end{pmatrix}.$$

because $U_{\mathfrak{G}}$ always contains a constant column corresponding to the principal character, but the ordinary group matrix does not. Therefore the question is not to generalize the linear factor decomposition (2.5.2) of the ordinary group determinant in the strict sense, but to introduce a new determinant formula in a similar form.

2.8 Linear Factor Decomposition of Generalized Group Determinant

For the determinant of the generalized matrix $U_{\mathfrak{G}}$ defined in (2.7.2), as a generalization of the formula (2.5.2) for the ordinary group determinant,[25] the following linear factor decomposition holds:

$$|U_{\mathfrak{G}}| = \pm c_{\mathfrak{G}} \prod_{\chi} \sum_{S \bmod \mathfrak{H}_{\chi}} \chi(S)\, u_{\chi}(S), \tag{2.8.1}$$

as we will prove in the following. Here $c_{\mathfrak{G}}$ is a natural number determined by $\mathfrak{G}$, which will be given by the expression[26]

$$c_{\mathfrak{G}} = \prod_{p|n} p^{\frac{1}{2}\sum_{p^{\mu}|n}\left(q(\frac{n}{p^{\mu}})-\frac{n}{p^{\mu}}\right)}, \tag{2.8.2}$$

where $q(m)$ denotes generally the number of solutions of $X^m = 1$ in $\mathfrak{G}$.

One knows that this number $c_{\mathfrak{G}}$ is a rational integer by the fact that for a divisor m of the order of the group $\mathfrak{G}$, by the basic theorem of a finite abelian group, on the one hand the inequality $q(m) \geqq m$ holds and on the other hand $q(m)$ and m have the same prime factors, from which follows $q(m) \equiv m \pmod 2$.

It is the nature of things that the sign in (2.8.1) remains undetermined, because, different from the case of the ordinary group matrix, no couple between rows and columns of the matrix $U_{\mathfrak{G}}$ is fixed.

To prove formula (2.8.1) we multiply the matrix $U_{\mathfrak{G}}$ from the left-hand side by the transpose of the character matrix

$$X = (\chi(S)) \quad \begin{pmatrix} \text{row index} & S \\ \text{column index} & \chi \end{pmatrix}$$

---

[25](Translator's remark) As mentioned at the end of Sect. 2.7, the following formula (2.8.1) does not seem to be a generalization of (2.5.2) in the strict sense.

[26](Translator's remark) The number n of the right-hand side of (2.8.2) is the order of the group $\mathfrak{G}$. The factor $c_{\mathfrak{G}}$ will be defined by (2.8.5) later in this section.

and construct the matrix product[27]

$$X'U_{\mathfrak{G}} = \left(\sum_{S} \chi(S) u_{\psi}(S T_{\psi}^{-k})\right) \left(\begin{matrix}\text{row index} & \chi \\ \text{column indices} & \psi, k\end{matrix}\right). \tag{2.8.3}$$

The elements of this matrix product are calculated by the decomposition of the summation over the elements of $\mathfrak{G}$ into the ones corresponding to the subgroups $\mathfrak{H}_{\psi}$ and to the factor group $\mathfrak{G}/\mathfrak{H}_{\psi}$ in the following way:

$$\begin{aligned}\sum_{S \text{ in } \mathfrak{G}} \chi(S)\, u_{\psi}(S T_{\psi}^{-k}) &= \sum_{A \text{ in } \mathfrak{H}_{\psi}} \sum_{\nu=0}^{n_{\psi}-1} \chi(A T_{\psi}^{\nu})\, u_{\psi}(A T_{\psi}^{\nu-k}) \\ &= \sum_{A \text{ in } \mathfrak{H}_{\psi}} \chi(A) \sum_{\nu=0}^{n_{\psi}-1} \chi(T_{\psi}^{\nu})\, u_{\psi}(T_{\psi}^{\nu-k});\end{aligned}$$

the last equality is obtained by giving attention to the property (2.7.1b) presupposed for the indeterminates. Here, as is well known, it holds that

$$\sum_{A \text{ in } \mathfrak{H}_{\psi}} \chi(A) = \left\{\begin{matrix} \dfrac{n}{n_{\psi}} & \text{for } \chi = \psi^{i} \\ 0 & \text{for } \chi \neq \psi^{i} \end{matrix}\right\} \quad (i \text{ arbitrary, not necessarily prime to } n_{\psi}),$$

and in the first case[28]

$$\begin{aligned}\sum_{\nu=0}^{n_{\psi}-1} \chi(T_{\psi}^{\nu})\, u_{\psi}(T_{\psi}^{\nu-k}) &= \sum_{\nu \bmod n_{\psi}} \psi^{i}(T_{\psi}^{\nu})\, u_{\psi}(T_{\psi}^{\nu-k}) \\ &= \psi(T_{\psi})^{ik} \sum_{\nu \bmod n_{\psi}} \psi^{i}(T_{\psi}^{\nu})\, u_{\psi}(T_{\psi}^{\nu}) \\ &= \psi(T_{\psi})^{ik} \sum_{S \bmod \mathfrak{H}_{\psi}} \psi^{i}(S)\, u_{\psi}(S).\end{aligned}$$

Therefore

[27](Translator's remark) X' is the transpose of X.

[28](Translator's remark) The first case is the one where $\chi = \psi^{i}$.

$$\sum_{S \text{ in } \mathfrak{G}} \chi(S)\, u_\psi(ST_\psi^{-k}) = \begin{cases} \dfrac{n}{n_\psi} \psi(T_\psi)^{ik} \sum\limits_{S \bmod \mathfrak{H}_\psi} \psi^i(S)\, u_\psi(S) & \text{for } \chi = \psi^i \\ 0 & \text{for } \chi \neq \psi^i \end{cases}. \tag{2.8.4}$$

Now, (in the matrix in (2.8.3)) let the row indices χ be arranged such that the elements in one Frobenius class always stand next to each other and such that a Frobenius class consisting of another powers follows after that. Let the first column indices ψ be arranged in the order of the sequences of classes as (the row indices) χ, and the second column indices k, the number of which is $\varphi(n_\chi)$, formally correspond to the $\varphi(n_\chi)$ elements of an individual class. In this way the thick-schema[29] for the rows and columns of the matrix product $X'U_\mathfrak{G}$, which is really filled with submatrices of size $(\varphi(n_\chi),\ \varphi(n_\psi))$, is determined. Then the second formula of formula (2.8.4) says that in the thick-schema at most on and under the diagonal the submatrices are different from null. Therefore the determinant of $X'U_\mathfrak{G}$ is decomposed into the product of determinants of submatrices on the diagonal, which is given by the first formula of (2.8.4) (for integers i prime to n_ψ). Hence, from (2.8.3) and (2.8.4) (by any row-order and column-order), there arises the formula[30]

$$|X|\,|U_\mathfrak{G}| = \pm \prod_\psi \left(|\psi(T_\psi)^{ik}| \cdot \prod_i \frac{n}{n_\psi} \sum_{S \bmod \mathfrak{H}_\psi} \psi^i(S)\, u_\psi(S) \right),$$

where the indices i in the determinant and in the product always run over a system of prime residues mod n_ψ. Consequently, from another combination of the products and from attention to the presupposed property (2.7.1a) for the indeterminates u_χ, it follows that

$$|X|\,|U_\mathfrak{G}| = \pm \prod_\psi \left(\left(\frac{n}{n_\psi}\right)^{\psi(n_\psi)} |\psi(T_\psi)^{ik}| \right) \cdot \prod_\chi \sum_{S \bmod \mathfrak{H}_\chi} \chi(S)\, u_\chi(S),$$

29(Translator's remark) For a matrix A we partition A into submatrices A_{ij} as

$$A = \begin{pmatrix} A_{11} & A_{12} & \cdots & A_{1n} \\ A_{21} & A_{22} & \cdots & A_{2n} \\ & & \cdots & \\ A_{m1} & A_{m2} & \cdots & A_{mn} \end{pmatrix},$$

which we will call the *thick-schema* of A.

30(Translator's remark) In the product $\prod_\psi$ the quantity $|\psi(T_\psi)^{ik}|$ means here a determinant where the index i runs over $1 \leqq i \leqq n_\psi$, $(i,\ n_\psi) = 1$ and $k = 0,\ 1,\ \ldots,\ \varphi(n_\psi) - 1$.

and hence the asserted linear factor decomposition (2.8.1) has the number factor[31]

$$c_{\mathfrak{G}} = \pm |X|^{-1} \prod_{\psi} \left(\left(\frac{n}{n_\psi} \right)^{\psi(n_\psi)} |\psi(T_\psi)^{ik}| \right). \tag{2.8.5}$$

2.9 The Number Factor $c_{\mathfrak{G}}$

Now we have to convert the expression (2.8.5) of the number factor $c_{\mathfrak{G}}$ found previously into the asserted expression (2.8.2). For this it suffices to accomplish the conversion of the square $c_{\mathfrak{G}}^2$, because the sign of $c_{\mathfrak{G}}$ has already been fixed as positive (arbitrarily).

Since the $\psi(T_\psi)^k$ should construct a basis of the integers of the cyclotomic field P_{n_ψ} and whereas the index i runs over prime residues mod n_ψ and provides the automorphism of P_{n_ψ} as the i-th power of ψ, the square of the determinant is

$$|\psi(T_\psi)^{ik}|^2 = \sigma_{n_\psi} d_{n_\psi},$$

where d_{n_ψ} and σ_{n_ψ} denote the absolute value and the sign of the discriminant of P_{n_ψ}, respectively. By virtue of

$$X'X = \left(\sum_S \chi_1(S)\chi_2(S) \right) = \left(\sum_S \chi_1\chi_2(S) \right) \quad \begin{pmatrix} \chi_1 & \text{row index} \\ \chi_2 & \text{column index} \end{pmatrix},$$

it follows moreover that

$$|X|^2 = \sigma n^n,$$

where σ denotes the signature of the permutation $\chi \to \chi^{-1}$ of characters χ of $\mathfrak{G}$. Hence it holds by (2.8.5) that

$$c_{\mathfrak{G}}^2 = \sigma n^{-n} \cdot \prod_{\psi} \left(\frac{n}{n_\psi} \right)^{2\varphi(n_\psi)} \cdot \prod_{\psi} \sigma_{n_\psi} d_{n_\psi}. \tag{2.9.1}$$

On the one hand, it holds now that

$$\sigma_{n_\psi} = \left\{ \begin{matrix} 1 & \text{for } n_\psi = 1, 2 \\ (-1)^{\frac{1}{2}\varphi(n_\psi)} & \text{for } n_\psi \neq 1, 2 \end{matrix} \right\},$$

[31](Translator's remark) As remarked in the beginning of this section, the factor $c_{\mathfrak{G}}$ is not defined by the identity (2.8.2), but by this (2.8.5).

because in the first case the field P_{n_ψ} coincides with the rational field P, and in the second case the field P_{n_ψ} has $\frac{1}{2}\varphi(n_\psi)$ pairs of complex-conjugate characters. On the other hand, by the permutation $\chi \to \chi^{-1}$, every individual class of X is transformed to itself, and indeed the number $\varphi(n_\psi)$ of the classes is not equal to 1 and then the class of ψ has exactly $\frac{1}{2}\varphi(n_\psi)$ transformations provided that $n_\psi \neq 1, 2$. Hence summarizing the above, it follows that

$$\sigma \prod_\psi \sigma_{n_\psi} = 1,$$

so that the sign-factors in (2.9.1) collapse.[32] Since $\sum_\psi \varphi(n_\psi) = n$, it arises from (2.9.1) that

$$c_{\mathfrak{G}}^2 = \frac{n^n \prod_\psi d_{n_\psi}}{\prod_\psi n_\psi^{2\varphi(n_\psi)}}. \tag{2.9.2}$$

Now, as is well known,[33] the absolute value of the discriminant of the cyclotomic field P_{n_ψ} is

$$d_{n_\psi} = \left(\frac{n_\psi}{\prod_{p|n_\psi} p^{\frac{1}{p-1}}} \right)^{\varphi(n_\psi)}.$$

Here we can remove the exponent $\varphi(n_\psi)$ by taking the multiplication over all characters χ of $\mathfrak{G}$ instead of over the representatives ψ, and then we have[34] by (2.9.2)

$$c_{\mathfrak{G}}^2 = \prod_\chi \frac{n}{n_\chi \prod_{p|n_\chi} p^{\frac{1}{p-1}}}. \tag{2.9.3}$$

If $\mathfrak{G} = \mathfrak{G}_1 \cdot \mathfrak{G}_2$ is a direct product of $\mathfrak{G}$ by two groups $\mathfrak{G}_1$ and $\mathfrak{G}_2$ of coprime orders n_1 and n_2, respectively, then a character χ of $\mathfrak{G}$ has a corresponding direct

[32] Remark that the consideration of these signs in our proof is essential. Of course, the question here is not the unessential sign of $c_{\mathfrak{G}}$, but the sign of $c_{\mathfrak{G}}^2$, and so of the reality of $c_{\mathfrak{G}}$.

[33] See, for example, Zahlbericht [3], §§**96, 97**, and Hasse [2], §**27**, c), **1**.

[34] (Translator's remark) The formulas

$$\prod_\chi c_{\mathfrak{G}}^2 = \frac{n}{n_\chi \prod_{p|n_\chi} p^{\frac{1}{p-1}}}, \quad c_{\mathfrak{G}}^2 = \frac{n}{n_\chi \prod_{p|n_\chi} p^{\frac{1}{p-1}}}$$

in the original books published in 1952 and 1985, respectively, should be written as the formula in (2.9.3).

decomposition $\chi = \chi_1\chi_2$ by characters χ_1 of $\mathfrak{G}_1$ and χ_2 of $\mathfrak{G}_2$. Here it holds that $n = n_1 n_2$, $n_\chi = n_{\chi_1} n_{\chi_2}$, $\prod_{p|n_\chi} = \prod_{p|n_{\chi_1}} \cdot \prod_{p|n_{\chi_2}}$ and $\prod_\chi = (\prod_{\chi_1})^{n_2} \cdot (\prod_{\chi_2})^{n_1}$. Hence the reduction law for the number factor $c_{\mathfrak{G}}$ follows from (2.9.3):

$$c_{\mathfrak{G}} = c_{\mathfrak{G}_1}^{n_2} c_{\mathfrak{G}_2}^{n_1}.$$

A similar reduction law also holds for the asserted expression (2.8.2); since one has $\prod_{p|n} = \prod_{p_1|n_1} \cdot \prod_{p_2|n_2}$ and since $q_1(n_1) = n_1$ and $q_2(n_2) = n_2$ (which is obtained) by an easily understandable meaning of notation,[35] it holds that (for a prime p dividing n)[36]

$$\sum_{p^\mu|n} \left(q\left(\frac{n}{p^\mu}\right) - \frac{n}{p^\mu}\right) = \left\{\begin{array}{ll} n_2 \sum_{p_1^{\mu_1}|n_1} \left(q_1\left(\frac{n_1}{p_1^{\mu_1}}\right) - \frac{n_1}{p_1^{\mu_1}}\right) & \text{for } p = p_1,\ p_1|n_1 \\ n_1 \sum_{p_2^{\mu_2}|n_2} \left(q_2\left(\frac{n_2}{p_2^{\mu_2}}\right) - \frac{n_2}{p_2^{\mu_2}}\right) & \text{for } p = p_2,\ p_2|n_2 \end{array}\right\}.$$

Consequently it suffices to deduce formula (2.8.2) from (2.9.3) in the case of $\mathfrak{G}$ being an abelian group of prime-power order $n = p^\nu$.

In this case formula (2.9.3), which we have already proved, states[37]

$$c_{\mathfrak{G}}^2 = \prod_\chi c_\chi^2 \quad \text{with} \quad c_\chi^2 = \left\{\begin{array}{ll} \dfrac{p^\nu}{n_\chi} = p^\nu & \text{for } \chi = 1 \\ \dfrac{p^\nu}{n_\chi p^{\frac{1}{p-1}}} & \text{for } \chi \neq 1 \end{array}\right\}, \qquad (2.9.4)$$

[35](Translator's remark) Here, $q_1(k_1)$ and $q_2(k_2)$ denote the numbers of the solutions of $x^{k_1} = 1$ in $\mathfrak{G}_1$ and $x^{k_2} = 1$ in $\mathfrak{G}_2$, respectively.

[36](Translator's remark) The original formula of Hasse is

$$\sum_{p^\mu|n} \left(q\left(\frac{n}{p^\mu}\right) - \frac{n}{p^\mu}\right) = \left\{\begin{array}{ll} n_2 \sum_{p_1^{\mu_1}|n_1} \left(q_1\left(\frac{n_1}{p_1^{\mu_1}}\right) - \frac{n_1}{p_1^{\mu_1}}\right) & \text{for } p_1|n_1 \\ n_1 \sum_{p_2^{\mu_2}|n_2} \left(q_2\left(\frac{n_2}{p_2^{\mu_2}}\right) - \frac{n_2}{p_2^{\mu_2}}\right) & \text{for } p_2|n_2 \end{array}\right\}.$$

The formula in the corrections of Martinet of the second edition

$$\sum_{p^\mu|n} \left(q\left(\frac{n}{p^\mu}\right) - \frac{n}{p^\mu}\right) = n_2 \sum_{p_1^{\mu_1}|n_1} \left(q_1\left(\frac{n_1}{p_1^{\mu_1}}\right) - \frac{n_1}{p_1^{\mu_1}}\right) + n_1 \sum_{p_2^{\mu_2}|n_2} \left(q_2\left(\frac{n_2}{p_2^{\mu_2}}\right) - \frac{n_2}{p_2^{\mu_2}}\right)$$

holds as the sums over *all* prime-powers dividing n, n_1 and n_2, respectively, while our formula above is a summation over the powers of *one fixed* prime p.

[37](Translator's remark) The number c_χ is defined as a positive number by the second relation in (2.9.4).

and formula (2.8.2), which one still has to prove, states

$$c_{\mathfrak{G}}^2 = p^{\sum_{\kappa=0}^{\nu}(q(p^\kappa)-p^\kappa)}. \tag{2.9.5}$$

Now the orders n_χ of characters χ of $\mathfrak{G}$ in this considered case are powers of p, and from the isomorphism of the character group X to the group $\mathfrak{G}$ itself, it follows that for a given power p^κ $(\kappa \geqq 1)$ of a prime p there exist exactly $q(p^\kappa) - q(p^{\kappa-1})$ characters χ of order $n_\chi = p^\kappa$. Hence, taking

$$c_\chi = p^{\gamma_\chi}, \quad c_{\mathfrak{G}} = p^{\gamma_{\mathfrak{G}}},$$

one has by (2.9.4)

$$2\gamma_\chi = \left\{\begin{array}{ll} \nu - \kappa = \nu & \text{for the character } \chi = 1 \text{ with } \kappa = 0 \\ \nu - \kappa - \dfrac{1}{p-1} & \text{for each } q(p^\kappa) - q(p^{\kappa-1}) \text{ character } \chi \neq 1 \\ & \text{with fixed } \kappa \geqq 1 \end{array}\right\}$$

and hence

$$\begin{aligned} 2\gamma_{\mathfrak{G}} = \sum_\chi 2\gamma_\chi &= \nu + \sum_{\kappa=1}^{\nu}\left(\nu - \kappa - \frac{1}{p-1}\right)(q(p^\kappa) - q(p^{\kappa-1})) \\ &= \nu + \sum_{\kappa=0}^{\nu-1} q(p^\kappa) + \left(-\frac{1}{p-1}\right) q(p^\nu) - \left(\nu - \frac{1}{p-1}\right) q(1) \\ &\qquad\qquad \text{(partial summation!)} \\ &= \sum_{\kappa=0}^{\nu-1} q(p^\kappa) - \frac{p^\nu - 1}{p\ \ 1} \quad (\text{by } q(p^\nu) = p^\nu,\ q(1) = 1) \\ &= \sum_{\kappa-0}^{\nu-1}(q(p^\kappa) - p^\kappa) \ = \ \sum_{\kappa=0}^{\nu}(q(p^\kappa) - p^\kappa), \end{aligned}$$

as required by assertion (2.9.5).

Thus formulas (2.8.1) and (2.8.2) for the generalized group determinants have been completely proved.

We further emphasize the important fact for our application:

Theorem 2.8 *For a finite abelian group $\mathfrak{G}$, $c_{\mathfrak{G}} = 1$ if and only if $\mathfrak{G}$ is cyclic.*

This is because only in a cyclic group $q(p^\kappa) = p^\kappa$ always holds for the direct factors of p-power order (and it also holds generally always that $q(\frac{n}{p^\mu}) = \frac{n}{p^\mu}$).

2.10 The Second Way of Transformation

By the completely corresponding consideration, as it was conducted from (2.2.4) to (2.5.1) by our first way of transformation—see Sect. 2.4—starting from the initial class number formula (0), one obtains the formula

$$hR = \prod_{\chi \neq 1} \sum_{S \bmod \mathfrak{H}_\chi} \left(-\chi(S) \log |\lambda_\chi^S|\right), \tag{2.10.1}$$

where the numbers λ_χ of the subfields K_χ corresponding to the characters χ have the same meaning as the number λ for the full field K.[38] The subfield K_χ corresponding to χ is defined as the invariant field of the group $\mathfrak{H}_\chi$, and then its Galois group is the quotient group $\mathfrak{G}/\mathfrak{H}_\chi$, that is, it is cyclic of order n_χ with χ as the generating character.

Now the expression (2.10.1) has exactly the structure of expression (2.8.1) found in our generalized group determinant, and it is actually obtained by the specialization of indeterminates:

$$u_\chi(S) = -\log |\lambda_\chi^S| \quad (\chi \neq 1),^{39} \quad u_1(S) = 1.$$

For this specialization the prerequisite conditions (2.7.1a) and (2.7.1b) for (2.8.1) are satisfied with regard to the given meaning of the number λ_χ. Then the formula (2.8.1) is applicable. According to (2.7.2), it provides the expression of the absolute value of the determinant,

$$c_{\mathfrak{G}} hR = \left|\left|\log \left|\lambda_\psi^{ST_\psi^{-k}}\right|\right|\right| \quad \begin{pmatrix}\text{row index} & S \\ \text{column indices} & \psi, k\end{pmatrix}, \tag{2.10.2}$$

---

[38](Translator's remark) In Sect. 2.4, we defined λ as

$$\lambda = \prod_{\substack{\pm a \bmod f \\ a \text{ in } H}} \sqrt{-(1-\zeta_f^a)(1-\zeta_f^{-a})} = \prod_{\substack{\pm a \bmod f \\ a \text{ in } H}} (\zeta_{2f}^a - \zeta_{2f}^{-a}).$$

Here we define

$$\lambda_\chi = \prod_{\substack{\pm a \bmod f(\chi) \\ a \text{ in } H_\chi}} \sqrt{-\left(1-\zeta_{f(\chi)}^a\right)\left(1-\zeta_{f(\chi)}^{-a}\right)} = \prod_{\substack{\pm a \bmod f(\chi) \\ a \text{ in } H_\chi}} \left(\zeta_{2f(\chi)}^a - \zeta_{2f(\chi)}^{-a}\right),$$

where

$$H_\chi = \{a \bmod f(\chi)\ ;\ \chi(a) = 1\}.$$

[39](Translator's remark) In the original book, this "$\chi \neq 1$" was not given.

where ψ, T_ψ, and k have the meanings defined in Sect. 2.7, and the column corresponding to $\psi = 1$ is to be replaced by 1.[40] Hence, subtracting the row corresponding to $S = 1$ from the other rows, one obtains the formula analogous to (2.5.4),

$$c_{\mathfrak{G}} hR = \left| \left\| \log \left| \eta_{\psi,S}^{T_\psi^{-k}} \right| \right\|_{S,\psi \neq 1} \right| \quad \begin{pmatrix} \text{row index} & S \\ \text{column indices} & \psi, k \end{pmatrix}, \tag{2.10.3}$$

where $\eta_{\psi,S} = \lambda_\psi / \lambda_\psi^S$ is the cyclotomic units of the field K_ψ. Corresponding to the meaning of ψ, the fields K_ψ constitute just the system of all distinct cyclic subfields of K, and the elements T_ψ (of $\mathfrak{G}$) constitute the corresponding system of generating automorphisms.

However the determinant in (2.10.3) is not of the form of regulator $\| \log |\eta_S^T| \|_{S,T\neq 1}|$ of a system of units η_S of K of which we desire to conduct our research. Namely, on the one hand, the automorphisms T_ψ^{-k} corresponding to individual rows are not in general the system of all elements $T \neq 1$ of $\mathfrak{G}$, and on the other hand, instead of a single independent system of units η_S, one has, as a basis, several systems of units $\eta_{\psi,S}$ among which there exist relations.

In this respect there arise the following facts for the individual (elements) $\eta_{\psi,S}$. For fixed ψ the element $\eta_{\psi,S}$ (up to sign) depends only on the class of S mod $\mathfrak{H}_\psi$. For the principal class $\mathfrak{H}_\psi$ it holds that $\eta_{\psi,S} = \pm 1$; for the other classes mod $\mathfrak{H}_\psi$ the elements $\eta_{\psi,S}$ constitute a system of $n_\psi - 1$ cyclotomic units of K_ψ. Therefore the whole system of $\eta_{\psi,S}$ consists of $\sum_{\psi\neq 1}(n_\psi - 1)$ formally distinct genuine units[41] if one takes account of non-genuine units and of equality. This number is generally greater than $n - 1 = \sum_{\psi \neq 1} \varphi(n_\psi)$, the rank of units of K; only when all n_ψ are primes, that is, when K is composed of only cyclic fields of degree of one fixed prime, it (the sum $\sum_{\psi\neq 1}(n_\psi - 1)$) is equal to the rank of units of K. In general, there exist relations among the $\sum_{\psi\neq 1}(n_\psi - 1)$ formally distinct genuine units $\eta_{\psi,S}$, and there exist exactly relations when the above-mentioned special case is not under discussion. A series of such relations can be easily given. For a fixed $\psi \neq 1$ either $g_{K_\psi} = 0$ holds—in this case, as remarked after Theorem 2.3, there exists at least one relation among the $n_\psi - 1$ cyclotomic units $\eta_{\psi,S}$ of K_ψ—or $g_{K_\psi} \neq 0$ holds—in this case these $\eta_{\psi,S}$ are independent and for a cyclotomic unit $\eta_{\psi',S}$ of a proper subfield $K_{\psi'}$ of K_ψ (corresponding to the powers of the class of ψ of lower degree), a suitable power of $\eta_{\psi',S}$ is expressed by the units $\eta_{\psi,S}$. I have to leave undecided whether or not all the relations among the $\sum_{\psi\neq 1}(n_\psi - 1)$ formally distinct genuine units $\eta_{\psi,S}$ are exhausted by such relations.

To obtain an arithmetic expression of the class number of K from formula (2.10.3), we need eventually a deep profound research of the structure

[40](Translator's remark) The column corresponding to $\psi = 1$ consists of only 1 by the definition $u_1(S) = 1$.

[41](Translator's remark) *Genuine units* mean units different from roots of unity ± 1. In Sect. 1.5 we called genuine units *sheer units*.

of the group of units of K with regard to the groups of units of cyclic subfields K_ψ of K.[42] We have already seen at the end of Sect. 2.1 the initial steps to the research in the above-mentioned method of Weber [4, 5]. I hope that we could return later to this problem.

2.11 The Second Arithmetic Representation of the Class Number

If one restricts oneself to the special case of real cyclic fields, then the determinant expression (2.10.3) found in Sect. 2.10, or rather expression (2.10.2) in Sect. 2.10, on which expression (2.10.3) depends, can be transferred easily to an arithmetic representation of the class number.

When K is cyclic, on the one hand, as ascertained in Theorem 2.8, the number factor $c_{\mathfrak{G}}$ is equal to 1. On the other hand, let Z be a generating automorphism of K. In the determinant in (2.10.2), subtracting the row corresponding to $S = Z^{\nu+1}$ from the one to $S = Z^\nu$ for $\nu = 1, 2, \ldots, n-1$, one obtains, as an analogous expression to (2.5.4z), an expression of the absolute value of the determinant

$$hR = \left| \left\| \log \left| \eta_\psi^{ST_\psi^{-k}} \right| \right\|_{S,\psi\neq 1} \right| \quad \begin{pmatrix} \text{row index} & S \\ \text{column indices} & \psi, k \end{pmatrix}, \tag{2.11.1}$$

where

$$\eta_\psi = \eta_{\psi,Z}.^{43}$$

The absolute value of the determinant obtained in this way is the regulator of the system of $\sum_{\psi\neq 1}\varphi(n_\psi) = n-1$ units $\eta_\psi^{T_\psi^{-k}}$ of K. For a fixed ψ the unit η_ψ is a generating cyclotomic unit of K_ψ because Z also provides a generating

[42](Translator's remark) See Martinet's Preface to the Second Edition, I Real Abelian Fields of the present book.

[43](Translator's remark) As defined in Sect. 2.10, we let η_ψ be

$$\eta_\psi = \eta_{\psi,Z} = \frac{\lambda_\psi}{\lambda_\psi^Z},$$

where

$$\lambda_\psi = \prod_{\substack{\pm a \bmod f(\psi) \\ a \text{ in } H_\psi}} \sqrt{-(1-\zeta_{f(\psi)}^a)(1-\zeta_{f(\psi)}^{-a})} = \prod_{\substack{\pm a \bmod f(\psi) \\ a \text{ in } H_\psi}} (\zeta_{2f(\psi)}^a - \zeta_{2f(\psi)}^{-a})$$

with $H_\psi = \{a \bmod f(\psi)\,;\ \psi(a) = 1\}$.

automorphism of K_ψ. However the full set of all of the $n_\psi - 1$ conjugates $\eta_\psi^{T_\psi^{-k}}$ with $T_\psi^{-k} \neq 1$ does not enter in the system of units under discussion, but only does the subset of $\varphi(n_\psi)$ terms that are determined by the rule of selection for k given in Sect. 2.7; for example, by the one for $k = 0, 1, \ldots, \varphi(n_\psi) - 1$.

In the present cyclic case, the class representatives ψ are indeed characterized by the orders n_ψ, and these orders run over the divisors n' of n. By setting $K_\psi = K_{n'}$ and $\eta_\psi = \eta_{n'}$ for $n_\psi = n'$, formula (2.11.1) can also be written analogously to (2.5.4z) in the form

$$hR = \left|\left|\log|\eta_{n'}^{Z^{\mu-\nu'}}|\right|\right| \quad \begin{pmatrix} \text{row index } \mu \not\equiv 0 \pmod n \\ \text{column indices } n'|n, n' \neq 1, \text{ and} \\ \nu' = 0, 1, \ldots, \varphi(n') - 1 \end{pmatrix}, \tag{2.11.1'}$$

where $\eta_{n'}$ is a generating cyclotomic unit of the subfield $K_{n'}$ of degree n'. Thus we have the arithmetic expression of the class number analogous to (2.5.5z):

$$h = \frac{R(\eta_{n'}^{Z^{-\nu'}})}{R} \quad (n'|n,\ n' \neq 1 \text{ and } \nu' = 0, 1, \ldots, \varphi(n') - 1). \tag{2.11.2}$$

As a generalization of Theorems 2.3z and 2.4 that refers to special kinds of real cyclic fields, the following theorem holds by (2.11.1′) and (2.11.2):

Theorem 2.9 *Let K be an arbitrary real cyclic field of degree n and Z a generating automorphism of K. For a positive divisor n' of n, $n' \neq 1$, let $K_{n'}$ denote the subfield of K of degree n', and $\eta_{n'}$ a generating cyclotomic unit of $K_{n'}$. For $n' = 1$ let $\eta_{n'} = \eta_1 = -1$.*

Then the class number h of K is equal to the index of the subgroup generated by the conjugates $\eta_{n'}^{Z^{-\nu'}}$ of $\eta_{n'}$ for $\nu' = 0, 1, \ldots, \varphi(n') - 1$ (for n' being positive divisors of n) in the group of all units of K.[44]

To this result it is still unsatisfactory that the subset of conjugates selected from the set of whole conjugates of $\eta_{n'}$ is not determined in an invariant way. The setting assigned there, $\nu' = 0, 1, \ldots, \varphi(n') - 1$, is in fact merely the simplest way to realize the general selecting rule given in Sect. 2.7 by which the roots of unity $\zeta_{n'}^{\nu'}$ constitute an integral basis of the cyclotomic field $P_{n'}$; by changing Z to another generating automorphism, one obtains another realization of this rule.[45] But even if one keeps only the general rule in mind, the requirement of invariance is not yet satisfied, because it is not seen what this rule has to do with the behavior of $\eta_{n'}$ by the symbolic exponents with a polynomial of Z. Indeed, one has $\eta_{n'}^{Z^{n'}} = \eta_{n'}$; however,

[44](Translator's remark) $\eta_{n'}^{Z^{-\nu}}$ in the original book published in 1985 should be $\eta_{n'}^{Z^{-\nu'}}$.

[45]In the meantime, replacing Z by Z^{-1}, one can remove the ugly minus sign of ν' operating here and appearing also in (2.11.2) and Theorem 2.9, which seems to be suitable only for the proof.

the generator Z does not act as an exponent of $\eta_{n'}$ like as that of a primitive n'-th root of unity, that is, for the primitive (irreducible) cyclotomic polynomial $G_{n'}(t)$ of degree $\varphi(n')$, $\eta_{n'}^{G_{n'}(Z)}$ is different than ± 1 when n' is not a prime. Therefore, it does not directly become clear that different realizations of general selection rules lead to equivalent systems of units. At present I could not see any possibility to remove this defect of beauty.[46]

For a later application we will treat, in addition, the regulatrix of the system of cyclotomic units $\eta_{n'}^{Z^{-\nu'}}$ of the subfields $K_{n'}$ of K. First of all, the relation

$$\Sigma(\eta_{n'}^{Z^{-\nu'}}) \equiv h\Sigma \pmod 2, \tag{2.11.3}$$

which is analogous to (2.6.2z), is derived from the arithmetic expression (2.11.2) of the class number. Moreover, the linear factor decomposition of the regulatrix in question,

$$\Sigma(\eta_{n'}^{Z^{-\nu'}}) \equiv \prod_{\chi} \sum_{S \bmod \mathfrak{H}_\chi} \chi(S)\,\sigma(\eta_\chi^S) \pmod 2, \tag{2.11.4}$$

which is analogous to (2.6.3), is derived from the product formula (2.8.1) for the generalized group determinant and Theorem 2.8. From (2.11.3) and (2.11.4) we have a theorem analogous to Theorem 2.5z, whose special formulation we pass over because we shall not use it. Now here, nevertheless, for a later application we calculate the linear factors in (2.11.4).

The linear factor corresponding to $\chi = 1$ is simply $\sigma(-1) \equiv 1 \pmod 2$ because $\eta_1 = -1$, and hence the factor can be canceled in (2.11.4). For $\chi \neq 1$,

$$\eta_\chi^S = \prod_{\substack{\pm a \bmod f(\chi) \\ a \text{ in } \mathfrak{H}_\chi}} \frac{\zeta_{2f(\chi)}^{sa} - \zeta_{2f(\chi)}^{-sa}}{\zeta_{2f(\chi)}^{zsa} - \zeta_{2f(\chi)}^{-zsa}}.$$

Therefore it holds that

$$\sum_{S \bmod \mathfrak{H}_\chi} \chi(S)\,\sigma(\eta_\chi^S) \equiv \sum_{s \bmod \mathfrak{H}_\chi} \chi(S) \sum_{\substack{\pm a \bmod f(\chi) \\ a \text{ in } \mathfrak{H}_\chi}} \sigma\left(\frac{\zeta_{2f(\chi)}^{sa} - \zeta_{2f(\chi)}^{-sa}}{\zeta_{2f(\chi)}^{zsa} - \zeta_{2f(\chi)}^{-zsa}}\right) \pmod 2,$$

and so

$$\sum_{S \bmod \mathfrak{H}_\chi} \chi(S)\,\sigma(\eta_\chi^S) \equiv \sum_{\pm x \bmod f(\chi)} \chi(x)\,\sigma\left(\frac{\zeta_{2f(\chi)}^{x} - \zeta_{2f(\chi)}^{-x}}{\zeta_{2f(\chi)}^{zx} - \zeta_{2f(\chi)}^{-zx}}\right) \pmod 2.$$

[46](Translator's remark) See Martinet's references to R. Gillard's result in Preface to the Second Edition, I Real Abelian Fields of the present book.

The σ-value (sign exponent) appearing here is determined by virtue of the analytic expression

$$\frac{\zeta_{2f}^{x}-\zeta_{2f}^{-x}}{\zeta_{2f}^{zx}-\zeta_{2f}^{-zx}}=\frac{\sin 2\pi\frac{x}{2f}}{\sin 2\pi\frac{zx}{2f}}.$$

Since the function $\sin 2\pi\xi$ of ξ has period 1 and has the (same) sign of ξ for $0 < |\xi| < \frac{1}{2}$, the function $\sin 2\pi\frac{x}{2f}$ has the sign of the absolutely least residue of x mod $2f$. The exponent of this sign is generally denoted by $\delta_{2f}(x)$.[47] Then it holds that

$$\sigma\left(\frac{\zeta_{2f}^{x}-\zeta_{2f}^{-x}}{\zeta_{2f}^{zx}-\zeta_{2f}^{-zx}}\right)\equiv\delta_{2f}(x)-\delta_{2f}(zx) \pmod 2.$$

From this congruence, for the linear factor to be calculated for $\chi\neq 1$, it follows that

$$\sum_{S \bmod \mathfrak{H}_\chi}\chi(S)\,\sigma(\eta_\chi^S)\equiv\sum_{\pm x \bmod f(\chi)}\chi(x)\,(\delta_{2f(\chi)}(x)-\delta_{2f(\chi)}(zx)) \pmod 2.$$

Thus for the regulatrix in question there arises the expression

$$\Sigma(\eta_{n'}^{Z-\nu'})\equiv\prod_{\chi\neq 1}\sum_{\pm x \bmod f(\chi)}\chi(x)\,(\delta_{2f(\chi)}(x)-\delta_{2f(\chi)}(zx)) \pmod 2. \qquad (2.11.5)$$

2.12 Real Cyclic Biquadratic Fields

For an explanation of the results obtained by our two ways of transformation, we consider real cyclic biquadratic fields K.

Let χ be a generating character of (such a field) K. We consider χ to be decomposed into the components χ_p. Every such χ_p is either a biquadratic character or a quadratic one. We collect the biquadratic characters χ_p into a sub-product χ_0 and the quadratic characters χ_p into a sub-product ψ. Then we obtain a unique decomposition

$$\chi=\chi_0\psi$$

[47](Translator's remark) We define $\delta_{2f}(x)$ by

$$\delta_{2f}(x)=\begin{cases}1 \text{ if } -f<x<0,\\ 0 \text{ if } 0<x<f.\end{cases}$$

of χ into a pure biquadratic character χ_0 and a quadratic character ψ with the property that by the corresponding decomposition

$$f = f_0 g$$

of conductor f, the conductors f_0 of χ_0 and g of ψ are coprime to each other. Particularly, $\psi = 1$, $g = 1$, i.e., $\chi = \chi_0$, $f = f_0$ can also occur exactly when χ itself is a pure character.

The field K is uniquely and invariantly determined by giving the two characters χ_0 and ψ beforehand. Since K is supposed to be real, there exists the relation

$$\chi_0(-1) = \psi(-1).$$

The character χ_0 can be developed by the corresponding quadratic character χ_0^2 whose conductor $f_0^{(2)}$ is always a divisor of f_0. While ψ can be any arbitrary quadratic character, for the characters χ_0^2 only the quadratic characters in question whose components χ_{0p}^2 are squares of characters χ_{0p} are eligible. Let f_{0p} and $f_{0p}^{(2)}$ be the conductors of χ_{0p} and χ_{0p}^2, respectively; they are p-contributions (p-components) of f_0 and $f_0^{(2)}$, respectively. As is easily seen, it is necessary and sufficient for the eligibility that

$$f_{0p}^{(2)} = \left\{ \begin{array}{ll} p \equiv 1 \pmod 4 & \text{for } p \neq 2 \\ 2^3 & \text{for } p = 2 \end{array} \right\}.$$

For any such $f_{0p}^{(2)}$ there exists exactly one quadratic character in the form χ_{0p}^2 given by the quadratic residue symbol

$$\chi_{0p}^2(x) = \left\{ \begin{array}{ll} \left(\dfrac{x}{p}\right) \equiv x^{\frac{p-1}{2}} \pmod p & \text{for } p \neq 2 \\ \left(\dfrac{2}{x}\right) = (-1)^{\frac{x^*-1}{4}} & \text{for } p = 2 \end{array} \right\} \left(x^* = (-1)^{\frac{x-1}{2}} x \equiv 1 \pmod 4\right).$$

Accordingly, for $p \neq 2$ there exists exactly one pair and for $p = 2$ exactly two pairs—of the complex-conjugate biquadratic characters χ_{0p} given by

$$\chi_{0p}(x) = \left\{ \begin{array}{ll} \left(\dfrac{x}{\mathfrak{p}}\right)_4 \equiv x^{\frac{p-1}{4}} \pmod{\mathfrak{p}} & \text{for } p \neq 2; \\ & \text{then } \chi_{0p}(-1) = (-1)^{\frac{p-1}{4}} \\ \left(\dfrac{-1}{x}\right)^{\alpha} \left(\dfrac{1+i}{x}\right)_4^{\pm 1} = (-1)^{\alpha\frac{x-1}{2}} i^{\pm\frac{x^*-1}{4}} & \text{for } p = 2; \\ \text{(with } \alpha \bmod 2) & \text{then } \chi_{0p}(-1) = (-1)^{\alpha} \end{array} \right\},$$

where $\mathfrak{p}$ is one of the two complex-conjugate prime divisors of p in the 4-th cyclotomic field $P_4 = P(i)$ and the 4-th residue symbol is understood in this field.[48] Then it holds that

$$f_{0p} = \left\{ \begin{matrix} p \equiv 1 \pmod 4 & \text{for } p \neq 2 \\ 2^4 & \text{for } p = 2 \end{matrix} \right\}.$$

Therefore one obtains all possible $\chi = \chi_0\psi$ by letting ψ run over all quadratic characters and the principal character, and for every such ψ with conductor g (by letting χ_0 run over) all the products

$$\chi_0 = \prod_p \chi_{0p} \quad \text{with } p \nmid g$$

of biquadratic characters χ_{0p} of the assigned way for which the following holds:

$$\chi_0(-1) = \left\{ \begin{matrix} (-1)^{\sum_p \frac{p-1}{4}} & \text{when all } p \neq 2 \\ (-1)^{\alpha + \sum_{p\neq 2} \frac{p-1}{4}} & \text{when one } p = 2 \end{matrix} \right\} = \psi(-1).$$

Hence it holds that

$$f_0 = \prod_p f_{0p} = \left\{ \begin{matrix} \prod_p f_{0p}^{(2)} = f_0^{(2)} = \prod_p p & \text{for } 2 \nmid f_0^{(2)} \\ 2\prod_p f_{0p}^{(2)} = 2f_0^{(2)} = 2^4 \prod_{p\neq 2} p & \text{for } 2 \mid f_0^{(2)} \end{matrix} \right\}.$$

Each of the two complex-conjugate characters χ_0 leads to the complex-conjugate $\chi = \chi_0\psi$ and to the same field K. From a given entire overview of the fields K to be considered by their characters, we in fact need only a few for the subsequent process of the determination of class number; it seems interesting for us to send it (the overview) beforehand in detail like this and then one becomes aware of the object of consideration explicitly and looks over its extent.

The characters $\neq 1$ of K are the two complex-conjugate biquadratic characters χ and $\overline{\chi}$ with the corresponding subfield (being) K itself and the character $\chi^2 = \chi_0^2$ with the corresponding field (being) the quadratic subfield K_2 of K. The characters χ and $\overline{\chi}$ make no contribution to the expression (2.2.5) for the invariant g_K, because $\chi(p)$, $\overline{\chi}(p) = 0$ always holds for the primes $p|f$; to the contrary the character $\chi^2 = \chi_0^2$ makes contribution for all primes p dividing f and not dividing $f_0^{(2)}$, therefore for all $p|g$. Hereby there arises

$$g_K = \prod_{p|g} (1 - \chi^2(p)) = \prod_{p|g} (1 - \chi_0^2(p)). \tag{2.12.1}$$

[48]For this see Klassenkörperbericht [1], Part II, §**10**, (2), and §**19**, VI.

Hence it follows that

$$g_K = \left\{ \begin{array}{l} 0, \text{ when there exists at least a prime } p|g \text{ with } \chi_0^2(p) = 1 \\ 2^r, \text{ when } \chi_0^2(p) = -1 \text{ always holds for the } r\ (r \geqq 0) \text{ primes } p|g \end{array} \right\}. \tag{2.12.2}$$

As one sees, in general $g_K = 0$, while $g_K \neq 0$ is a special case where the coincidence of many conditions is required.

While the application of the result of our first transformation (Theorem 2.3z) is connected to the condition $g_K \neq 0$, the result of our second transformation (Theorem 2.9) is applicable to any case because it is really valid for an arbitrary real cyclic field. We start with an application of Theorem 2.9 and will compare afterward the result obtained there with the one obtained by application of Theorem 2.3z for the special case $g_K \neq 0$.

The system of units in Theorem 2.9 consists here of two generating conjugate cyclotomic units $\eta,\ \eta'$ of K and of a cyclotomic unit η_2 of K_2.[49] Since only from the 4-th roots of unity $1,\ i,\ i^2 = -1,\ i^3 = -i$ the integral basis $\pm 1,\ \pm i$ of P_4 can be selected, the unit η' is generated from η by one of the two generating automorphisms of K. Then it holds that

$$|\eta| = \prod_{\substack{\pm a \bmod f \\ \chi(a)=1}} \frac{|1-\zeta_f^a|}{|1-\zeta_f^{za}|}, \quad |\eta'| = \prod_{\substack{\pm a \bmod f \\ \chi(a)=1}} \frac{|1-\zeta_f^{za}|}{|1-\zeta_f^{z^2a}|}, \tag{2.12.3}$$

where $z \bmod f$ is represented by $\chi(z) = i$, for example. We here return from the symmetric expression of a and $-a$ of the cyclotomic units of P_{2f} themselves to the formal non-symmetric expression of their absolute values in P_f, because the latter is sufficient and convenient for the following process. For two of the further conjugates of η, it holds that

$$|\eta''| = \prod_{\substack{\pm a \bmod f \\ \chi(a)=1}} \frac{|1-\zeta_f^{z^2a}|}{|1-\zeta_f^{z^3a}|}, \quad |\eta'''| = \prod_{\substack{\pm a \bmod f \\ \chi(a)=1}} \frac{|1-\zeta_f^{z^3a}|}{|1-\zeta_f^{a}|},$$

[49](Translator's remark) Recall the definition of η: in Theorem 2.3z we define

$$\eta = \eta_Z = \prod_{\substack{\pm a \bmod f \\ a \text{ in } H}} \sqrt{\frac{(1-\zeta_f^a)(1-\zeta_f^{-a})}{(1-\zeta_f^{za})(1-\zeta_f^{-za})}}.$$

and so

$$|\eta''| = \prod_{\substack{\pm b \bmod f \\ \chi(b)=-1}} \frac{|1-\zeta_f^{b}|}{|1-\zeta_f^{zb}|}, \quad |\eta'''| = \prod_{\substack{\pm b \bmod f \\ \chi(b)=-1}} \frac{|1-\zeta_f^{zb}|}{|1-\zeta_f^{z^2 b}|}. \tag{2.12.4}$$

Moreover it holds that

$$|\eta_2| = \prod_{\substack{\pm a_0 \bmod f_0^{(2)} \\ \chi_0^2(a_0)=1}} \frac{|1-\zeta_{f_0^{(2)}}^{a_0}|}{|1-\zeta_{f_0^{(2)}}^{za_0}|}. \tag{2.12.5}$$

Among the five given different formal genuine units of K, except the trivial relation

$$|N(\eta)| = |\eta|\,|\eta'|\,|\eta''|\,|\eta'''| = 1,$$

there exists one more relation. The relation yields that a power of η_2 is expressed by conjugates of η. One obtains easily this expression by the general summation formula (2.2.3). By its application to the transformation from $f(\chi_0^2) = f_0^{(2)}$ to $f = f_0 g$ ($= f_0^{(2)} g$ or $2 f_0^{(2)} g$ according as $2 \nmid f_0^{(2)}$ or $2 \mid f_0^{(2)}$) in the logarithms of expression (2.12.5) for $|\eta_2|$, one has, returning to the antilogarithm, the relation

$$|\eta_2|^{\prod_{p|g}(1-\chi_0^2(p))} = \prod_{\substack{\pm x \bmod f \\ \chi_0^2(x)=1}} \frac{|1-\zeta_f^{x}|}{|1-\zeta_f^{zx}|}.$$[50]

By (2.12.1) the exponent of the left-hand side is the invariant g_K. In the product of the right-hand side, it holds that $\chi_0^2(x) = \chi^2(x)$ and x runs over a half system mod f of the solutions $x = a$ of $\chi(a) = 1$ and $x = b$ of $\chi(b) = -1$; hence by (2.12.3)

[50](Translator's remark) Indeed, $A_f(x) = \log|1-\zeta_f^x|$ satisfies the condition (2.2.2) with $\alpha = 0$. Hence

$$S_f(\chi) = \sum_{\pm x \bmod f} \chi(x) A_f(\chi) = \sum_{\pm x \bmod f} \chi(x) \log|1-\zeta_f^x|$$

has the reduction formula. Taking χ as χ_0^2 in this formula, we have

$$S_f(\chi_0^2) = \prod_{p|f}(1-\chi_0^2(p)) \cdot S_{f_0^{(2)}}(\chi_0^2),$$

which leads to this desired formula.

and (2.12.4) the product is equal to $|\eta||\eta''|$. Thus there arises the desired relation

$$|\eta_2|^{g_K} = |\eta|\,|\eta''| = |N_{K/K_2}(\eta)|,$$

and hence by (2.12.2)

$$|N_{K/K_2}(\eta)| = |\eta|\,|\eta''|$$
$$= \left\{\begin{array}{ll} 1, & \text{when there exists at least a prime } p|g \text{ with } \chi_0^2(p) = 1 \\ |\eta_2|^{2^r}, & \text{when } \chi_0^2(p) = -1 \text{ always holds for the } r\ (r \geqq 0) \text{ primes } p|g \end{array}\right\}. \tag{2.12.6}$$

In the first general case ($g_K = 0$), one has $|\eta||\eta''| = 1$, which is a non-trivial relation among the cyclotomic units $\eta,\ \eta',\ \eta'',\ \eta'''$ of K described after Theorem 2.3. Then the group generated by $\eta,\ \eta',\ \eta'',\ \eta'''$ alone has rank less than that of the full unit group of K, so that one relies on the cyclotomic unit η_2 of K_2 to determine the class number of K. One obtains here the arithmetic expression in the narrow sense obtained from Theorem 2.9:

$$h = \frac{R(\eta,\ \eta',\ \eta_2)}{R}. \tag{2.12.7}$$

In the latter special case ($g_K \neq 0$), the subgroup generated by the absolute values of $\eta,\ \eta',\ \eta'',\ \eta'''$ has index 2^r in the group generated by the absolute values of $\eta,\ \eta',\ \eta_2$, and one has, as well, the arithmetic expression obtained by Theorem 2.3z in the wide sense:

$$2^r h = \frac{R(\eta, \eta', \eta'')}{R}. \tag{2.12.8}$$

Especially, if $r = 0$, i.e., if χ is a pure biquadratic character, then this is also an arithmetic expression in the narrow sense (special case $g_K = 1$, Theorem 2.4).

References[51]

1. H. Hasse, Bericht über neuere Untersuchungen und Probleme aus der Theroie der algbraichen Zahlkörper. Part I: Klassenörpertheorie. Jahresbericht D.M.-V. **35** (1926).; H. Hasse, Bericht über neuere Untersuchungen und Probleme aus der Theroie der algbraichen Zahlkörper. Part Ia: Beweise zu Teil I. Jahresbericht D.M.-V. **36** (1927).; H. Hasse, Bericht über neuere

[51] The bold-typed numbers attached in the square brackets denote the sections of this book in which the individual works are cited.

Untersuchungen und Probleme aus der Theroie der algbraichen Zahlkörper. Part II: Reziprozitätsgesetz. Jahresbericht D.M.-V. Supplemental Ed. **6** (1930). Cited as "Klassenkörperbericht." [Introduction, **1.2**, **1.3**, **2.3**, **2.12**, **3.1**, **3.2**, **3.4**, **3.5**, **3.9**, **3.19**]
2. H. Hasse, *Zahlentheorie.* Berlin (1949) [Introduction, **2.9**, **3.1**, **3.2**, **3.5**, **3.9**]
3. D. Hilbert, Die Theorie der algebraischen Zahlkörper. Jahresbericht D.M.-V. **4** (1897). Cited as "Zahlbericht." [Preface, Introduction, **1.2**, **1.6**, **2.9**, **3.2**, **3.5**, **3.8**, **3.19**]
4. H. Weber, Theorie der Abelschen Zahlkörper. Acta Math. **8** (1886) [Introduction, **2.1**, **2.6**, **2.10**, **3.16**]
5. H. Weber, *Lehrbuch der Algebra*, vol. 2, 2nd edn. (American Mathematical Society, Braunschweig 1899), pp. 219–223 [Introduction, **2.1**, **2.6**, **2.10**, **3.16**]

Chapter 3
The Arithmetic Structure of the Relative Class Number Formula for Imaginary Fields

In this third chapter we always assume that an abelian number field K we consider is imaginary and hence quadratic over its maximal real subfield K_0. Therefore the class number h of K is the product

$$h = h_0 h^*,$$

where h_0 is the class number of K_0—we treated its arithmetic structure in Chap. 2—and h^* is the relative class of K/K_0; this is given by the formula (1.5.3b):

$$\boxed{h^* = Qw \prod_{\chi_1} \frac{1}{2f(\chi_1)} \sum_{x \bmod f(\chi_1)}^{+} (-\chi_1(x)x)\,.} \qquad (*)$$

3.1 Class-Field-Theoretic Proof of the Rational Integrality and the Arithmetic Meaning

We start with a class-field-theoretic proof of the rational integrality of the relative class number h^* of K/K_0 and with the arithmetic meaning of h^* given in the process. To do so we must make a scaffold of the proof of class-field-theoretic properties of the relative quadratic field K/K_0.[1] Later on I will use some of the notation introduced here without newly defining them explicitly.

The proof of the property of the class field K/K_0 is divided into two parts: the analytic part and the arithmetic part. In the analytic part it is established that for an arbitrary module $\mathfrak{m}_0$ the congruence group H_0 in K_0 corresponding to K

[1]For a detailed explanation of the following, see Klassenkörperbericht [5], Part I, §**6**, Theorems (B$''$), (B$'''$) and (B$''''$) as well as the corresponding proofs in Part Ia, especially in §**14**.

© Springer Nature Switzerland AG 2019

H. Hasse, *On the Class Number of Abelian Number Fields*,
https://doi.org/10.1007/978-3-030-01512-1_3

generated by the relative norms of ideals (of K) coprime to $\mathfrak{m}_0$ has index *at most* 2. In the arithmetic part it is deduced by the so-called fundamental inequality that this index is *at least* 2 if the module is taken as $\mathfrak{m}_0 = \mathfrak{d}_0 p_\infty$, where $\mathfrak{d}_0$ is the relative discriminant of K/K_0 and p_∞ is the product of infinite prime places (divisors) of K_0.[2] In summary of the above, it follows that H_0 mod $\mathfrak{d}_0 p_\infty$ has index *just* 2, the property of the class field K/K_0, and moreover that in the series of fundamental inequalities the equalities hold in every place, which yields additional arithmetic facts on the principal genus and ambiguous classes as well as on the expression of units of K_0 as the relative norms of numbers of K.

The arithmetic part of this proof, whose scaffold is the point here, depends essentially on the application of two homomorphisms $1 + J$ and $1 - J$ on the class group of K, where J denotes the generating automorphism of K/K_0 (transformation by complex-conjugate). In the following we describe the situations of applications of the two homomorphisms by a clearly arranged schematic figure; actually, we describe how the situation arises by the conclusion of the proof and by the expressions of the equalities in a series of fundamental inequalities. In the process, above all, the point for us is the indices of several congruence groups in K_0 and in K. In our scheme we put the group indices near the lines that connect the points denoting the congruence groups. We denote by arrows the homomorphisms $1 + J$ and $1 - J$.

With respect to the congruence groups, besides the groups Δ_0 and Δ of all divisor classes (of K_0 and K) and the principal classes[3] E_0 and E of K_0 and K, respectively, we have to consider the following:

1. Congruence groups in K_0:

H_0 relative norms of divisors of K,

$E_0^* = H_0 \cap E_0$ relative norms of divisors of K in the principal class of K_0,

N_0 relative norms of divisors in the principal class of K.

In the process only the divisors of K coprime to $\mathfrak{d}_0$ will be taken into consideration, and the indicated relative norms will be filled up with the entire ray class mod $\mathfrak{d}_0 p_\infty$. Groups H_0, E_0^*, and N_0 are then congruence groups mod $\mathfrak{d}_0 p_\infty$. We also note that Δ_0 and E_0 are defined by mod $\mathfrak{d}_0 p_\infty$, that is, with restriction to divisors coprime to $\mathfrak{d}_0$, as compounded with the ray class mod $\mathfrak{d}_0 p_\infty$.[4] For

[2]This module is the conductor of K/K_0. In the meaning of the concept of the conductor of general class field theory, throughout this work, we should actually have taken into account the infinite divisors p_∞ for the conductor f of K and $f(\chi)$ of χ, i.e., we must include p_∞ in f and in $f(\chi)$ when K is imaginary, that is, $\chi(-1) = -1$. In the special case of the base field P (the field of rational numbers), however, one manages without this slightly irksome burden of expressive way and of formulas; for these, see also Klassenkörperberichit [5], Part I, Exposition 11.

[3]When the meaning of the divisor group is not so the point, we write it as 1, or we omit it in the construction of factor groups (and we write them) simply like Δ/E and Δ_0/E_0.

[4]For this see Klassenkörperbericht [5], Part I, §3.

describing all of these congruence groups, instead of the classification by the ray class mod $\mathfrak{d}_0 p_\infty$, the rather larger classification by the classes mod N_0 is indeed sufficient, where the principal class N_0 is composed of all the principal divisors of K_0 that are produced by prime norm residues mod $\mathfrak{d}_0 p_\infty$ of K/K_0; we call these classes the *norm residue classes of* K/K_0. The relative norms of divisors coprime to $\mathfrak{d}_0$ from an ordinal class C of K always lie in the same norm residue class C_0 of K/K_0; we call C_0 the *relative norm of the class* C.

2. Congruence groups in K:

H classes with relative norms N_0,
H^* classes with relative norms in E_0 then in E_0^*,
A classes invariant by J.

Groups H, H^*, and A are congruence groups mod 1, that is, compounded by the ordinal classes.

On the basis of the results from the proofs of properties of the class field K/K_0, the relations among these congruence groups in K_0 and K are described schematically in Fig. 3.1.

Now we explain the schematic diagram in Fig. 3.1 and deduce from this diagram some conclusions that are interesting to us.

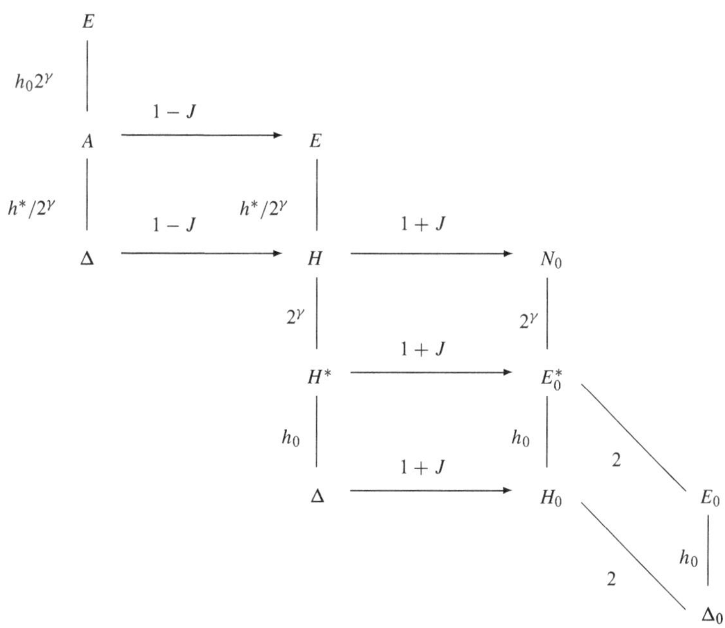

Fig. 3.1 ($\gamma = \delta + q^* - 1$) Relation among classes of K and K_0

H_0 is the congruence group mod $\mathfrak{d}_0 p_\infty$ in K_0 mentioned in the beginning of this book. The proof of the property of the class field K/K_0 establishes that the index is

$$[\Delta_0 : H_0] = 2. \tag{3.1.1}$$

For the further proof process of the class field theory, it arises moreover that H_0 has exactly the conductor $\mathfrak{d}_0 p_\infty$.[5] From this there follows the establishment crucial for our aim that the conductor of H_0 is not 1 in any case, because even if the relative discriminant is equal to 1, i.e., $\mathfrak{d}_0 = 1$—which is certainly possible, as we will see—the conductor contains still the product p_∞ of infinite divisors.[6] Along with H_0, the subgroups E_0^* and N_0 defined by mod $\mathfrak{d}_0 p_\infty$ also have exactly the conductor $\mathfrak{d}_0 p_\infty$.

As for the index of the norm residue principal class N_0, the value is, by definition, found to be[7]

$$[\Delta_0 : N_0] = h_0 2^{\delta+q^*}, \tag{3.1.2}$$

where δ is the number of distinct primes of K_0 dividing $\mathfrak{d}_0$ and q^* the number of independent units η_0^* of K_0 that are norm residues mod $\mathfrak{d}_0 p_\infty$ of K/K_0 with respect to squares ε_0^2 of units ε_0 of K_0[8]:

$$[\eta_0^* : \varepsilon_0^2] = 2^{q^*}. \tag{3.1.3}$$

These norm residue units η_0^* are proved to be actually relative norms of numbers of K.[9] Since $[\Delta_0 : E_0] = h_0$, it follows from (3.1.2) that

$$[E_0 : N_0] = 2^{\delta+q^*}. \tag{3.1.4}$$

As already emphasized, since $\mathfrak{d}_0 p_\infty \neq 1$, the ordinary principal class E_0 is not contained in H_0 of conductor $\mathfrak{d}_0 p_\infty$. Therefore, as for the intersection $E_0^* = H_0 \cap E_0$, there arise by (3.1.1)

$$[E_0 : E_0^*] = 2, \quad [H_0 : E_0^*] = h_0. \tag{3.1.5}$$

Hence the whole ordinary classes of K_0 are divided into two classes mod E_0^*, exactly one of which always contains the relative norms of classes of K. These last classes that contain relative norms of classes of K constitute a subgroup H_0/E_0^*, which is

---

[5]For this see Klassenkörperbericht [5], Part Ia, §**17**, Theorem 20.

[6]In this last fact, to the general theory of class field of the relative quadratic number field K/K_0, the special assumption states that K is imaginary while K_0 is real.

[7]See Klassenkörperbericht [5], Part Ia, §**14**, II.

[8]In regard to the abbreviated way of writing the following group indices, see Klassenkörperbericht [5], Part Ia, §**1**.

[9]See Klassenkörperbericht [5], Part Ia, §**14**, Theorem 15.

isomorphic to the ordinary class group Δ_0/E_0, of index 2 in the full class group Δ_0/E_0^*.

H is the principal genus and Δ/H is the group of genera of K/K_0. By the homomorphism $1 + J$ (relative norm transformation), the genus group Δ/H is isomorphically transformed to the subgroup H_0/N_0 of index 2 in the full norm residue class group Δ_0/N_0. In the process the factor group Δ/H^* is isomorphically transformed to the factor group H_0/E_0^*. By the statement of (3.1.5), the class group Δ/E of K therefore has the factor group Δ/H^* (which is) isomorphic to the class group Δ_0/E_0 of K_0. Hence, since $h_0 = [\Delta_0 : E_0]$ is a divisor of $h = [\Delta : E]$, the desired result follows:

Theorem 3.10 *The relative class number h^* of K is a positive rational integer.*

Precisely it holds that h^* is the order of the subgroup H^*/E of the class group Δ/E of K. On account of the meaning of H^*, one has here the following arithmetic meaning of h^*:

Theorem 3.11 *The relative class number h^* of K is the order of the group H^* that consists of the classes whose relative norms fall into the principal class of K_0.*[10]

The essential kernel of our proofs of the results described in Theorems 3.10 and 3.11, as here briefly summarized once more, is the demonstration that every ordinal class of K_0 contains relative norms of classes of K. This fact follows from the matter that the congruence group H_0 of K_0 constructed by relative norms of ideals of K has first index 2 by defining modulo $\mathfrak{d}_0 p_\infty$, the conductor of K/K_0, while it has index 1 by defining modulo a proper divisor of this conductor (here divisor 1).[11] We have already given the proof more broadly and we also (here) continue the already-started investigation in order to gain at the same time a deep

---

[10]This meaning of h^* was expressed without proof by Kummer [18] for the special case of the p-th cyclotomic field P_p, p a prime number. Probably Kummer had deduced it by virtue of his tacit assumption, not admissible in general, on the condition that the classes of K_0 are embedded into the ones of K (See Sect. 1.6, footnote 23, p. 18–19) and he had thought that any detailed proof was not needed.

[11]Behind this proof method, there remains a general theorem that Kronecker [13] had already striven to prove but sought in an incorrect direction (see Sect. 1.6, footnote 23, p. 18–19). He asserted:

> Let K_0 be an arbitrary algebraic number field and K a relative abelian number field (extension) of K_0 such that the classification of congruence classes corresponding to K is distinct from that of the ordinal division classes of K_0. Then the class number h_0 of K_0 is a divisor of the class number h of K, and the quotient h^*, the relative class number of K/K_0, is the order of the group H^* of the classes of K whose relative norms fall into the principal class of K_0.

The proof arises exactly as in the special case we explore here, where it is shown that the *factor group* of the class group of K by the subgroup of H^* is transformed by the relative norm isomorphically to the class group of K_0. Kronecker attempted, against this, to show that a *subgroup* of the class group of K is isomorphic to the class group of K_0.

insight into the arithmetic structure of the relative quadratic field K/K_0, which is fundamental for our further research.

From (3.1.4) and (3.1.5) it follows moreover that

$$[E_0^* : N_0] = 2^\gamma, \quad [H_0 : N_0] = h_0 2^\gamma, \tag{3.1.6}$$

where, for abbreviation, we let $\gamma = \delta + q^* - 1$. By the homomorphism $1 + J$ the subgroup H^*/H of the genus group is, in addition, isomorphically transformed to the subgroup E_0^*/N_0 of the norm residue class group. Hence by (3.1.6) there arises a supplement to Theorems 3.10 and 3.11:

Theorem 3.12 *The relative class number of h^* of K is divisible by the genus factor 2^γ. Here*

$$\gamma = \delta + q^* - 1$$

where δ is the number of distinct prime divisors of K_0 appearing in the relative discriminant $\mathfrak{d}_0$ of K/K_0 and q^* is the number of independent units η_0^* of K_0 that are norm residues mod $\mathfrak{d}_0 \mathfrak{p}_\infty$ of K/K_0—and then also relative norms of numbers of K—with respect to the squares ε_0^2 of units ε_0 of K_0.

The quotient $h^*/2^\gamma$ is the order of classes in the principal genus H of K/K_0, while 2^γ is the order of the group H^*/H of genera of K/K_0 whose relative norms fall in the principal class of K_0.

By (3.1.6) the inequality $\gamma \geqq 0$ holds surely and hence

$$\delta + q^* \geqq 1, \tag{3.1.7}$$

that is, $\delta = 0$ and $q^* = 0$ cannot hold simultaneously. Now $\delta = 0$ means that $\mathfrak{d}_0 = 1$, and hence K/K_0 is unramified. In this case the norm residues mod $\mathfrak{d}_0 \mathfrak{p}_\infty$ are just totally positive numbers of K_0. Therefore inequality (3.1.7) says that in the case of K/K_0 being unramified every totally positive unit of K_0 is not always square in K_0, or in a different way says that K/K_0 must be ramified if every totally positive unit of K_0 is square. The last condition therefore is synonymous with (the condition) that $\Sigma_0 \not\equiv 0 \pmod 2$ holds for the regulatrix Σ_0 of K_0, i.e., with (the condition) that the signatures of fundamental units of K_0 are independent. Thus we have as a result from (3.1.7) the next fact, which is interesting for our research and also for itself.

Theorem 3.13 *If the signatures of fundamental units of K_0 are independent, then the quadratic extension K/K_0 is ramified.*

Particularly if K_0 is a real quadratic field, then the assumption means that the fundamental unit of K_0 has norm -1. Consequently, by this assumption there exists no imaginary absolute abelian unramified relative quadratic field K/K_0. Correspondingly the meaning of Theorem 3.13 lies in the general case.

We have so far discussed and made full use of only the part relevant to the homomorphism $1 + J$ of our schematic diagram in Fig. 3.1. This part was sufficient for the proof given earlier of the integrality of the relative class number h^* and its arithmetic meanings. For the sake of completeness and in regard to our subsequent research on the divisibility of the full class number h by 2, we will discuss later the part relevant to the homomorphism $1 - J$ of our schematic diagram in Fig. 3.1, which is, after all, as important as the already-discussed part for the accomplishment of the proof of the class-field-theoretic property (3.1.1) of K/K_0 and its results.

A is the group of ambiguous classes of K/K_0. By the homomorphism $1 - J$ the factor group Δ/A is transformed isomorphically to the group H/E of the classes in the principal genus.[12] From this isomorphism, as a supplement to the meaning of Theorem 3.12, a further arithmetic meaning of the quotient $\frac{h^*}{2^\gamma}$ arises as the index of the subgroup A of ambiguous classes in the full class group Δ in K.

The obvious thought to easily deduce the divisibility of h by h_0 and thereby the integrality of h^* from the existence and the meaning of the subgroup A, as Kummer apparently had in mind, does not immediately lead us to the goal. Indeed, the h_0 classes of K_0 provide, by the embedding in K, the h_0 formal ambiguous classes of K/K_0, but it is not said that these h_0 classes are different from each another; as we will see later, they could be reduced rather to a factor group of order $\frac{1}{2}h_0$.[13] For the exact accomplishment of this thought, one must provide evidence of $\gamma \geqq 0$, that is, of inequality (3.1.7) above, or a proof of Theorem 3.13. In the case where K/K_0 is ramified, one can merely deduce[14] the divisibility of h by h_0 and then the integrality of h^* immediately from the class-field-theoretic formula[15]

$$a = [A : E] = h_0 2^\gamma = h_0 2^{\delta + q^* - 1}$$

for the number a of ambiguous classes of K/K_0—or from the consideration that is equivalent to its proof—therefore indeed without acquiring a subgroup of the class group of K isomorphic to the class group of K_0 and hence (without) acquiring the arithmetic meaning of h^*. It is doubtful for me whether, in addition to the above-determined *factor group* Δ/H^* isomorphic to the class group of K_0, such a *subgroup* of Δ could also be given in a generally valid and systematic way, for example, as a definite subgroup A_0 of A of order h_0 and with index 2^γ.

The essential result relating to the homomorphism $1 - J$ is the so-called principal genus theorem $H = \Delta^{1-J}$, by which every class of the principal genus of K/K_0

[12] See Klassenkörperbericht [5], Part I, §**6**, Theorem (B'''').

[13] See Theorem 3.18 in Sect. 3.4. For the cyclotomic fields that only Kummer and Kronecker treated, this case did not occur, as we ascertain in Theorem 3.28 in Sect. 3.7. Consequently Kummer and Kronecker could succeed in the way given in the text. (See footnote 23, p. 18–19, in Sect. 1.6 above.)

[14] For the cyclotomic field $K = P_f$ one deduces immediately from our later Theorem 3.19 in Sect. 3.5 that K/K_0 is ramified or unramified according as f is a prime-power or a composite.

[15] See Klassenkörperbericht [5], Part Ia, §**13**, Theorem 13.

is expressed as a symbolic $(1 - J)$-th power of a class of K; conversely it is clear directly by definition that every such power belongs to the principal genus. For quadratic fields, from the principal genus theorem, there follows the famous fact that the exponent γ of the genus factor coincides with the 2-rank of the class group Δ, that is, the number of elements of the basis of Δ of 2-power order. We will generalize this fact to the present case of the relative quadratic field K/K_0.

For this purpose we start with the identity valid for every class C of K:

$$C^2 = C^{1-J} C^{1+J}. \tag{3.1.8}$$

If C runs over the class group Δ of K, then C^{1-J} runs over the principal genus $H = \Delta^{1-J}$ by the principal genus theorem, while C^{1+J} runs over the group $\Delta^* = \Delta^{1+J}$ of classes containing the relative norms of divisors of K. Since these norms lie in K_0 and since every class of K generated by a class of K_0 contains relative norms of divisors of K, the group Δ^* is just generated by the embedding of the class group Δ_0 of K_0 in the class group Δ of the extension K; as we have already remarked previously and we will prove in Theorem 3.18, Sect. 3.4, the group Δ_0 either survives or is reduced to a factor group of half order. Among the groups Δ, H, and Δ^*, in virtue of the identity (3.1.8), there exists the relation

$$\Delta^2 \Delta^* = H \Delta^*.$$

With this help the 2-rank r of Δ can be calculated as follows[16]:

$$\begin{aligned} 2^r &= [\Delta : \Delta^2] = [\Delta : \Delta^2 \Delta^*][\Delta^2 \Delta^* : \Delta^2] = [\Delta : H\Delta^*][\Delta^2 \Delta^* : \Delta^2] \\ &= \frac{[\Delta : H]}{[H\Delta^* : H]}[\Delta^2 \Delta^* : \Delta^2] = \frac{[\Delta : H]}{[\Delta^* : H \cap \Delta^*]}[\Delta^2 \Delta^* : \Delta^2]. \end{aligned}$$

Herein, first by the results regarding $1 + J$, there arises

$$[\Delta : H] = h_0 \, 2^{\gamma}.$$

For a class C^* of Δ^*, furthermore by $C^* = C^{*J}$, the relation of the definition of H, $C^{*1+J} = 1$, is tantamount to $C^{*2} = 1$; therefore, it follows that

$$[\Delta^* : H \cap \Delta^*] = \frac{h_0^*}{2^{r_0^*}},$$

[16] For this see the reduction principle in Klassenkörperbericht [5], Part Ia, §**1**, Lemma 1, or Hasse [7], §**14**, a), 3.

where h_0^* denotes the order of Δ^* and r_0^* the 2-rank of Δ^*.[17] Finally it holds that

$$[\Delta^2\Delta^* : \Delta^2] = 2^{s_0^*},$$

where s_0^* denotes the 2-rank of Δ^* in Δ, that is, the number of independent basis elements of Δ^* (it is sufficient that they are of 2-power order) with respect to squares (of elements) of Δ. By combining the above, there arises

$$2^r = \frac{h_0}{h_0^*}\, 2^{\gamma+r_0^*+s_0^*}.$$

By the subsequent result anticipated above on the embedding of Δ_0 in Δ, the group Δ therefore has 2-rank[18]

$$r = \gamma + r_0^* + s_0^* + \kappa, \tag{3.1.9}$$

where γ is the exponent of the genus factor of K/K_0 defined in Theorem 3.12, r_0^* and s_0^* have the meanings given above and $\kappa = 0$ or 1 according as the class group Δ_0 of K_0 survives or is reduced to a factor group of half order by being embedded in the class group Δ of K. Especially, if h_0 is odd, then one has $r_0^* = 0$, $s_0^* = 0$ and $\kappa = 0$ and hence (one has) the simple relation $r = \gamma$ as familiar in quadratic fields. Hereof we will make an interesting application in Sect. 3.19.

3.2 The Unit Index Q

Now we turn to the research of the unit index Q of K/K_0 appearing as the factor in the front of the relative class number formula $(*)$. Though the question is, as we will see immediately, the decision of the alternative $Q = 1$ or 2 for which one would believe that it is easily treated, this research will cover, in fact, a wide range (of

---

[17](Translator's remark) Though the group $H \cap \Delta^*$ is of 2-elementary, the 2-rank of this group is not always equal to r_0^*. Hence the equation $[\Delta^* : H \cap \Delta^*] = \frac{h_0^*}{2^{r_0^*}}$ should be read as $[\Delta^* : H \cap \Delta^*] \geqq \frac{h_0^*}{2^{r_0^*}}$. More precisely, see F. Lemmermeyer, On 2-class field towers of imaginary quadratic number fields, *J. Théor. Nombres Bordeaux* (2) **6** (1994), 261–272. The translator would like to thank Franz Lemmermeyer for informing him of this incorrectness.

[18](Translator's remark) The equality (3.1.9) does not hold in general. As remarked by Lemmermeyer, since $\frac{h_0^*}{2^{r_0^*}} \leqq [\Delta^* : H \cap \Delta^*] \leqq h_0^*$, it holds generally that

$$\gamma + s_0^* + \kappa \leqq r \leqq \gamma + r_0^* + s_0^* + \kappa.$$

For the field $K = \mathbf{Q}(\sqrt{-1}, \sqrt{5 \cdot 101})$, we actually have

$$r = 2,\ \ \gamma + r_0^* + s_0^* + \kappa = 2 + 1 + 0 + 0 = 3.$$

The value $r = 2$ is due to a table of Ken Yamamura.

subjects), and we will be able to state for the first time the results in Sect. 3.6 after accomplishing all the required interim research.

The unit index Q is, according to its definition given by (1.5.2), the index of the subgroup generated by the associated units $\zeta\varepsilon_0$ to

$$\text{units } \varepsilon_0 \text{ in } K_0$$

with

$$\text{roots of unity } \zeta \text{ in } K$$

in the group of all

$$\text{the units } \varepsilon \text{ in } K.$$

Making use of my short notation for the index of number groups,[19] this index is written in the form

$$Q = [\varepsilon : \zeta\varepsilon_0] \tag{3.2.1}$$

with reference to the notation just introduced above.

In advance we present the following remarks. The conjugate transformation in K is exchangeable with the transition to the square of the absolute value; this is because, if we let J be, as before, the generating automorphism of K/K_0 (transformation by complex-conjugate), then we have, since K is abelian, $\alpha^{S(1+J)} = \alpha^{(1+J)S}$, therefore $|\alpha^S|^2 = (|\alpha^2|)^S$ for all α of K and for all S of $\mathfrak{G}$. Hereby a unit ε of K with $|\varepsilon| = 1$ has the property $|\varepsilon^S| = 1$ for all S of $\mathfrak{G}$ and hence is a root of unity ζ in K.[20]

From these remarks there arises moreover by the isomorphism principle (the homomorphism theorem)[21] another expression for the index in question:

$$Q = [|\varepsilon| : |\varepsilon_0|], \tag{3.2.1$'$}$$

that is, Q is also the index of the group of absolute values of units of K_0 in that of K. Apart from the slightly simple formation of the definition of Q obtained here, it is more convenient for our research and also more natural for the point of arithmetic to take a basis on the first-mentioned expression (3.2.1) of Q.[22]

---

[19] See Klassenkörperberichit [5], Part Ia, §**1**.

[20] See, for example, Zahlberichit [10], §**21**, Theorem 48, or Hasse [7], §**28**, b), 1.

[21] See Kassenörperberichit [5], Part Ia, §**1**, Lemma 2, or Hasse [7], §3, p.18.

[22] An arithmetic equivalence to (3.2.1′) is the further expression (3.2.3b) derived below.

Like the research on the divisor class group in Sect. 3.1, we here consider the behavior of the group ε with the two subgroups ζ and ε_0 by the two homomorphisms $1-J$ and $1+J$.

By the homomorphism $1-J$ the group ε is mapped onto the subgroup

$$\varepsilon^{1-J} = \frac{\varepsilon}{\overline{\varepsilon}} = \zeta^*,$$

which is contained in the group ζ according to the remark mentioned above. Since $\zeta^{1-J} = \zeta^2$ and since the subgroup ε_0 in the group ε is characterized by $\varepsilon_0^{1-J} = 1$, there arises the relation that

$$\varepsilon^{1-J} = \zeta^2 \text{ is tantamount to } \varepsilon = \zeta\varepsilon_0. \tag{3.2.2a}$$

From the application of $1-J$ on (3.2.1), it holds therefore that

$$Q = [\zeta^* : \zeta^2] = 1 \text{ or } 2 \tag{3.2.3a}$$

by the isomorphic principle; the latter could occur because the group ζ is cyclic.

By the homomorphism $1+J$ the group ε is mapped onto the subgroup

$$\varepsilon^{1+J} = \varepsilon\overline{\varepsilon} = \mathbf{N}(\varepsilon),$$

where $\mathbf{N}$ denotes the relative norm map of K/K_0. This subgroup is contained in the group ε_0. Since $\varepsilon_0^{1+J} = \varepsilon_0^2$ and since, by the remark mentioned above, the subgroup ζ is characterized by $\zeta^{1+J} = 1$ in the group ε, there arises the relation that

$$\varepsilon^{1+J} = \varepsilon_0^2 \text{ is tantamount to } \varepsilon = \zeta\varepsilon_0. \tag{3.2.2b}$$

From the application of $1+J$ on (3.2.1), it holds therefore by the isomorphic principle that

$$Q = [\mathbf{N}(\varepsilon) : \varepsilon_0^2] = 2^q, \tag{3.2.3b}$$

where q has the meaning from the theorem on the unit principal genus[23] for the quadratic field K/K_0, namely, q denotes the number of independent units of K_0 of the form $\eta_0 = \mathbf{N}(\varepsilon)$ of K_0 with respect to the group ε_0^2.[24] This holds because for the index Q only the two values 1 and 2 are possible by (3.2.3a), and for the exponent q

[23]See Klassenkörperbericht [5], Part Ia, §**12**, Theorem 12.

[24]Take notice of the distinction between this number q and the number q^* that appears in (3.1.3) and (appears) subsequently as the number of independent units of K_0 of the form $\eta_0^* = \mathbf{N}(\theta)$ (θ not necessarily unit) with respect to the group ε_0^2. It holds that $0 \leqq q \leqq q^* \leqq p \leqq n_0$, where p is the number of independent totally positive units of K_0 with respect to the group ε_0^2 and n_0 denotes, as so far, the degree of K_0.

in (3.2.3b) only two values 0 and 1 are possible (here in the underlying special case K/K_0 of general class field theory).

Summing up relations (3.2.2a), (3.2.2b), (3.2.3a), and (3.2.3b), we obtain, as a first result on the unit index Q of K/K_0, the following.

Theorem 3.14 *It holds that either* $Q = 1$ *or* $Q = 2$.

If all the units ε *of* K *have the two equivalent properties*

$$\frac{\varepsilon}{\overline{\varepsilon}} = \zeta^2, \quad N(\varepsilon) = \varepsilon_0^2,$$

then $Q = 1$.

If, however, there exists a unit ε^* *of* K *with the two equivalent properties*

$$\frac{\varepsilon^*}{\overline{\varepsilon^*}} = \zeta^* \quad \text{is not square in } K, \tag{3.2.4a}$$

$$N(\varepsilon^*) = \varepsilon_0^* \quad \text{is not square in } K_0, \tag{3.2.4b}$$

then $Q = 2$.

When one multiplies the unit ε^* of K appearing in (3.2.4a) and (3.2.4b) by a root of unity ζ of K, then the root of unity ζ^* is multiplied by ζ^2. Hence one can normalize ε^* such that the root of unity ζ^* is of 2-power order, and actually this order is necessarily the highest one among such orders possible in K, that is, (the highest order is) the contribution 2^ω of the prime number 2 in the number w of roots of unity in K. We can therefore settle the condition (3.2.4a) in the normalized form

$$\varepsilon^* = \zeta_{2^\omega}\overline{\varepsilon^*}. \tag{3.2.4a$_0$}$$

The unit ε_0^* of K_0 appearing in (3.2.4b) remains unaffected in the normalization of ε^* and ζ^*.

In the following we will successively have to distinguish the situation into the two cases of $\omega = 1$ and $\omega \geqq 2$, and then of $w \not\equiv 0 \pmod 4$ and $w \equiv 0 \pmod 4$. In the first case, K contains only the square roots of unity ± 1 as roots of unity of 2-power order; in the second case, K contains in addition at least the fourth roots of unity $\pm\sqrt{-1}$. The distinction between the two cases plays a role over and over again for our present task, the determination of the unit index. It (the distinction) has the following meaning for the structure of K/K_0. In the case of $w \equiv 0 \pmod 4$, we have the special simple Kummer-generator $K = K_0(\sqrt{-1})$, by which K is expressed by the composite of K_0 and the imaginary quadratic field $P_4 = P(\sqrt{-1})$. Then the characters χ_1 of K/K_0 are derived from the characters χ_0 of K_0 in the form of $\chi_1 = \cup\chi_0$, where $\cup$ is the generating character of P_4 given by

$$\cup(x) = (-1)^\alpha \quad \text{for } x \equiv (-1)^\alpha \pmod 4. \tag{3.2.5}$$

This quadratic character $\cup$ appears over and over in the further process of our work. This is also simply defined as the (unique) character with conductor 4. In the case of $w \not\equiv 0 \pmod 4$, the Kummer-generators of K/K_0 and the relation between the characters χ_0 of K_0 and χ_1 of K/K_0 cannot be given in a same simple and generally valid way.

By Theorem 3.14 the alternative $Q = 1$ or 2 is attributed to the non-existence or to the existence, respectively, of a unit ε^* of K with properties (3.2.4a) and (3.2.4b). It is our task to attribute further this alternative to properties of characters of K. Indeed, this has not been completely succeeded, because, as we will see, there joins in the process a so-called non-abelian property, namely, the behavior of the divisor classes of K_0 by the embedding into that of K, which by Artin [2] depends essentially on a non-abelian extension of the Galois group $\mathfrak{G}$ of K,[25] while the characters χ of K can basically describe the abelian properties of K, as they are expressed in the group $\mathfrak{G}$ itself. First we attribute the non-existence or the existence of the unit ε^* in question to the type of Kummer-generators of the real absolute abelian relative quadratic field (extension) K_0'/K_0 (that is) connected to the extension K/K_0. Thereupon we characterize the type of Kummer-generators of K_0'/K_0 by the behavior of ramification of K_0'/K_0 except for the remaining problem of class-embedding. (This[26] is only unessentially different from the behavior of ramification of K/K_0 itself.) Finally we describe the behavior of ramification of K_0'/K_0 by the characters of K, while the class-embedding problem must remain untouched.

3.3 Criterion for $Q = 1$ or 2 by a Kummer-Generator

As in Sect. 3.2 let 2^ω be the contribution of 2 to the number w of roots of unity in K. Then P_{2^ω}, but not $P_{2^{\omega+1}}$, is contained in K. Since $P_{2^{\omega+1}}$ is absolute abelian and quadratic over P_{2^ω}, then the composite field

$$K' = K P_{2^{\omega+1}}$$

is absolute abelian and quadratic over K. This field K' is imaginary. Its maximal real subfield

$$K_0' = (K P_{2^{\omega+1}})_0$$

is quadratic over K_0. (See Fig. 3.2.) The relative quadratic field K_0'/K_0 defined above and connected to K/K_0 is the one for which we will attribute the problem of

---

[25]To this see the fundamental execution in my Klassenkörperbericht [5], Part II, §**27**, where I reported on the cited work of Artin. See the remark there at the end of No. 3 (in Part II, §**27**).

[26](Translator's remark) "This" means the behavior of ramification of K_0'/K_0.

Fig. 3.2 Relation among the fields

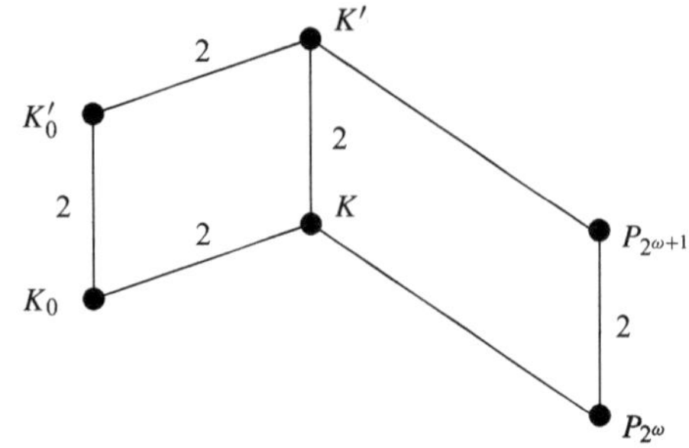

the existence of units treated in Theorem 3.14 to its Kummer-generators according to the statement at the end of Sect. 3.2.

In the case of $w \not\equiv 0 \pmod 4$, i.e., $\omega = 1$, there exists a simple relation between Kummer-generators of K/K_0 and of K_0'/K_0. Putting a Kummer-generator of K/K_0 as

$$K = K_0(\sqrt{-\mu_0}), \tag{3.3.1}$$

where μ_0 is a totally positive number of K_0, one has

$$K' = KP_4 = K(\sqrt{-1}) = K_0(\sqrt{-\mu_0}, \sqrt{-1}) = K_0(\sqrt{\mu_0}, \sqrt{-1});$$

therefore K_0'/K_0 has a Kummer-generator

$$K_0' = K_0(\sqrt{\mu_0}). \tag{3.3.1$'$}$$

In the case of $w \equiv 0 \pmod 4$, i.e., $\omega \geqq 2$, where K/K_0 has Kummer-generator $K = K_0(\sqrt{-1})$, the Kummer-generator of the extension K_0'/K_0 can be explicitly given. First $P_{2^{\omega+1}}/P_{2^{\omega+1},0}$ has a Kummer-generator

$$P_{2^{\omega+1}} = P_{2^{\omega+1},0}(\sqrt{-1}).$$

Since $\sqrt{-1}$ is contained in K, one has here

$$K' = KP_{2^{\omega+1}} = KP_{2^{\omega+1},0}$$

and hence

$$K_0' = K_0 P_{2^{\omega+1},0} \tag{3.3.2}$$

(different than the case of $w \not\equiv 0 \pmod 4$, where $P_{2^{\omega+1},0} = P_{4,0} = P$ is contained in K_0). Since $P_{2^\omega,0}$ is contained in K_0, the point here is only the assignment of the Kummer-generator of $P_{2^{\omega+1}}/P_{2^{\omega+1},0}$. (See Fig. 3.3.)

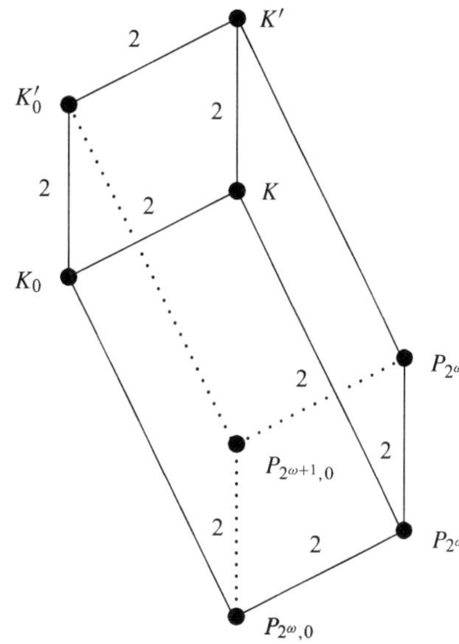

Fig. 3.3 ($\omega \geqq 2$) Relation among the fields

For this purpose we consider the number

$$\lambda_{2^\omega} = 1 + \zeta_{2^\omega} \quad \text{with} \quad \lambda_{2^\omega} = \zeta_{2^\omega}\,\overline{\lambda}_{2^\omega} \tag{3.3.3}$$

of P_{2^ω}. It represents the (unique) prime divisor of 2 in P_{2^ω}. Its relative norm to $P_{2^\omega,0}$,

$$\lambda_{2^\omega,0} = \mathbf{N}(\lambda_{2^\omega}) = \lambda_{2^\omega}\,\overline{\lambda_{2^\omega}} = (1+\zeta_{2^\omega})(1+\zeta_{2^\omega}^{-1}) = (\zeta_{2^{\omega+1}} + \zeta_{2^{\omega+1}}^{-1})^2, \tag{3.3.4}$$

represents the (unique) prime divisor of 2 in $P_{2^\omega,0}$. Between these two numbers there exists the divisor equation

$$\lambda_{2^\omega,0} \cong \lambda_{2^\omega}^2, \tag{3.3.5}$$

that is, both sides are different from each other only by a unit factor (of P_{2^ω}). The base (number) of the square in (3.3.4) lies in $P_{2^{\omega+1},0}$ but not in $P_{2^\omega,0}$ on account of the property of the prime divisor.

Therefore one has

$$P_{2^{\omega+1},0} = P_{2^\omega,0}(\sqrt{\lambda_{2^\omega,0}}), \tag{3.3.6}$$

and then also by (3.3.2)

$$K_0' = K_0(\sqrt{\lambda_{2^\omega,0}}). \tag{3.3.7}$$

For the lowest value $\omega = 2$ of this case, it holds that

$$\lambda_{2^2} = 1 + \zeta_{2^2} = 1 + \sqrt{-1}, \quad \text{then} \quad \lambda_{2^2,0} = \mathbf{N}(1 + \sqrt{-1}) = N(1 + \sqrt{-1}) = 2,$$

and hence

$$P_{2^3,0} = P_{2^2,0}(\sqrt{2}) = P(\sqrt{2}), \ K_0' = K_0(\sqrt{2}) \quad \text{for} \ \omega = 2.$$

Moreover, according to (3.3.4) it holds generally that $(\lambda_{2^\omega,0} - 2)^2 = \lambda_{2^{\omega-1},0}$, and hence

$$P_{2^4,0} = P_{2^3,0}\left(\sqrt{2+\sqrt{2}}\right), \qquad K_0' = K_0\left(\sqrt{2+\sqrt{2}}\right) \qquad \text{for} \ \omega = 3,$$

$$P_{2^5,0} = P_{2^4,0}\left(\sqrt{2+\sqrt{2+\sqrt{2}}}\right), \ K_0' = K_0\left(\sqrt{2+\sqrt{2+\sqrt{2}}}\right) \ \text{for} \ \omega = 4,$$

$$\ldots.$$

Now we connect the criterion for $Q = 2$ obtained in Theorem 3.14 with the Kummer-generator of K_0'/K_0. To this purpose we remark beforehand that the radicand[27] of the Kummer-generator of the relative quadratic field K_0'/K_0 is uniquely determined up to arbitrary quadratic factor of the basis field K_0.

Suppose first again that $w \not\equiv 0 \pmod 4$. If $Q = 2$, then by Theorem 3.14 there exists a unit ε^* of K with the property (3.2.4a$_0$), and then here $\varepsilon^* = -\overline{\varepsilon^*}$. Therefore $\varepsilon^{*2} = -\varepsilon_0^*$ is a unit of K_0, where ε_0^* is the unit defined by (3.2.4b). Hence it follows that

$$K = K_0(\sqrt{-\varepsilon_0^*}).$$

Then by (3.3.1) and (3.3.1′) it holds that

$$K_0' = K_0(\sqrt{\varepsilon_0^*}).$$

Conversely if a radicand of the Kummer-generator of K_0'/K_0 is selected as a unit ε_0^* of K_0, then $\varepsilon^* = \sqrt{-\varepsilon_0^*}$ is a unit of K so that $\varepsilon^* = -\overline{\varepsilon^*}$ holds and that $\mathbf{N}(\varepsilon^*) = \varepsilon_0^*$ is not square in K_0, which has therefore the properties (3.2.4a$_0$) and (3.2.4b). Thus in this case it holds that $Q = 2$.

Suppose second that $w \equiv 0 \pmod 4$. If $Q = 2$, then by Theorem 3.14 there exists a unit ε^* of K with the property (3.2.4a$_0$), and then $\varepsilon^* = \zeta_{2^\omega}\overline{\varepsilon^*}$. Since $\lambda_{2^\omega} = \zeta_{2^\omega}\overline{\lambda_{2^\omega}}$ also holds by (3.3.3), it follows that

$$\lambda_{2^\omega} = \varepsilon^* \gamma_0 \tag{3.3.8}$$

[27](Translator's remark) A radicand is the quantity under a radical sign. For example, 2 is the radicand of $\sqrt{2}$.

with some number γ_0 of K_0, and hence by (3.3.4) from the relative norm mapping there exists the relation

$$\lambda_{2^\omega,0} = \varepsilon_0^* \gamma_0^2, \tag{3.3.9}$$

where ε_0^* is defined by (3.2.4b). Therefore the Kummer-generator (3.3.7) of K_0'/K_0 is reduced to

$$K_0' = K_0(\sqrt{\varepsilon_0^*}).$$

Conversely if a radicand of the Kummer-generator of K_0'/K_0 can be selected as a unit ε_0^* of K_0, then one has Eq. (3.3.9) with a number γ_0 of K_0. By (3.3.4) and (3.3.5) one moreover has Eq. (3.3.8) with a unit ε^* of K for which $\mathbf{N}(\varepsilon^*) = \varepsilon_0^*$ is not square in K_0 and then the unit ε_0^* has the property (3.2.4b)—and then also (3.2.4a). Thus in this case it holds that $Q = 2$.

Consequently the following criterion for the alternative $Q = 1$ or 2 has been proved:

Theorem 3.15 *Let 2^ω be the exact 2-power in w and let $K' = KP_{2^{\omega+1}}$.*
Then, $Q = 2$ if and only if a Kummer-generator of K_0'/K_0 is of type

$$K_0' = K_0(\sqrt{\varepsilon_0^*}) \quad \textit{with a unit } \varepsilon_0^* \textit{ of } K_0.$$

Otherwise, $Q = 1$.

We remark moreover that for $\omega = 1$, in this criterion, by (3.3.1) and (3.3.1′) the field K/K_0 itself can also simply take the place of the relative quadratic field K_0'/K_0 derived from K/K_0, because in this case the sign of the radicand of the Kummer-generator is out of the picture.

3.4 Criterion for $Q = 1$ or 2 by Ramification and Class Problem

By virtue of the arithmetic theory of relative Kummer fields[28] the necessary and sufficient condition for $Q = 2$ obtained in Theorem 3.15, namely, the condition under which a radicand of the Kummer-generator of K_0'/K_0 can be selected from a unit of K_0, is satisfied at most when K_0'/K_0 has the following condition for ramification: the prime divisors in K_0 of every prime number $p \neq 2$ are unramified

[28] See Klassenkörperbericht [5], Part Ia, §**11**, Theorems 9 and 10.

in K_0'/K_0 and the prime divisors of $p = 2$ in K_0 are not *at most ramified*, that is, the ramification number v—the number of its ramification groups distinct from 1—does not have the possible theoretical maximal value $2e_0$, the double of the ramification index e_0 of prime number 2 in K_0.[29] A relative quadratic field is called *at most unessentially ramified* if it has this property of ramification, and otherwise called *essentially ramified*. Then $Q = 2$ can at most occur when K_0'/K_0 is at most unessentially ramified. Conversely, if K_0'/K_0 is at most unessentially ramified, then by the well-known theory, in any case (of $Q = 1$ or 2), the Kummer-generator of K_0'/K_0 is of the type

$$K_0' = K_0(\sqrt{\mu_0}) \quad \text{with } \mu_0 \cong \mathfrak{m}_0^2 \text{ in } K_0, \tag{3.4.1}$$

that is, its radicand μ_0 is the square of a divisor $\mathfrak{m}_0$ of K_0.[30] Then two cases are possible: either $\mathfrak{m}_0$ belongs to the principal class of K_0—then μ_0 can be reduced to a unit ε_0^* through the square of a number factor of K_0, and hence $Q = 2$ by Theorem 3.15—or $\mathfrak{m}_0$ belongs to a class $M_0 \neq 1$ of K_0—then μ_0 cannot be reduced to a unit, and hence $Q = 1$ by Theorem 3.15. We call K_0'/K_0 *of unit type* in the former case and *of class type* in the latter case.[31]

Thus by Theorem 3.15 the following criterion for the alternative $Q = 1$ or 2 has been obtained.

Theorem 3.16 *Let 2^ω be the exact 2-power in w and let $K' = KP_{2^{\omega+1}}$.*
If K_0'/K_0 is essentially ramified, then $Q = 1$.
If, however, K_0'/K_0 is at most unessentially ramified, then $Q = 2$ or 1 according as K_0'/K_0 is of unit type or of class type.

Also again in this criterion, as in Theorem 3.15, for $\omega = 1$ the field K/K_0 itself can simply take the place of K_0'/K_0. For $\omega \geqq 2$, note that the prime divisor $\lambda_{2^\omega,0}$ of 2 is actually at most ramified in $P_{2^{\omega+1},0}/P_{2^\omega,0}$, but the prime divisors of 2 are not necessarily at most ramified in K_0'/K_0; by the extension of the basis field $P_{2^\omega,0}$ to K_0, by which $P_{2^{\omega+1},0}$ is extended to K_0', the number $\lambda_{2^\omega,0}$ could be ramified into the

---

[29] So far as the prime divisors of 2 are actually ramified, either $1 \leqq v \leqq 2e_0 - 1$ holds and v is odd, or $v = 2e_0$ holds. The latter occurs only in the case of the highest (at most) ramification: it is also simply characterized by the case where v is even.

[30] (Translator's remark) By Theorems 9 and 10 in Klassenkörperbericht [5], Part Ia, if the present condition (3.4.1) holds for some divisor $\mathfrak{m}_0$ of K_0, then K_0'/K_0 is at most unessentially ramified.

[31] When we make use of both concepts (of ramifications), to abbreviate the notation, we will not always explicitly express but tacitly implicate the premise on which the concepts rely for its foundation that K_0'/K_0 is at most unessentially ramified.

prime divisors with relative indices divisible by 2.[32] The alternative of whether this is indeed the case or not is really the point for our criterion.[33]

By Theorem 3.16 the alternative $Q = 1$ or 2 traces back to the essential part of the alternative of whether K_0'/K_0 is essentially ramified or at most unessentially ramified: an alternative that can be determined by the characters of K by virtue of class field theory, on which we will elaborate in Sect. 3.5. If K_0'/K_0 is at most unessentially ramified, then there remains certainly the further alternative to determine whether K_0'/K_0 is of unit type or of class type.

The last case, namely, where K_0'/K_0 is of class type, in which $Q = 1$ as in the case of being essentially ramified, can be characterized by the state of the divisor classes of K_0 in the embedding into that of K. If this case occurs, then one has by (3.4.1) for $\omega = 1$[34]

$$\mathfrak{m}_0^2 \cong \mu_0 \cong -\mu_0 \quad \text{in } K_0, \text{ then } \mathfrak{m}_0 \cong \sqrt{-\mu_0} \quad \text{in } K,$$

and for $\omega \geqq 2$ with attention to (3.3.7) and (3.3.5)

$$\mathfrak{m}_0^2 \cong \mu_0 = \lambda_{2^\omega,0}\gamma_0^2 \quad \text{in } K_0, \text{ then } \mathfrak{m}_0 \cong \lambda_{2^\omega}\gamma_0 \quad \text{in } K.$$

In both cases the class $M_0 \neq 1$ of K_0 determined by $\mathfrak{m}_0$ falls in the principal class in K. Conversely, if a class $M_0 \neq 1$ of K_0 falls in the principal class of K and if $\mathfrak{m}_0$ is a divisor in M_0, then it holds that

$$\mathfrak{m}_0 \cong \mu \quad \text{in } K$$

and so $\frac{\mu}{\overline{\mu}} \cong 1$. Therefore $\frac{\mu}{\overline{\mu}} = \zeta$ is a root of unity in K by the remark mentioned in Sect. 3.2, and this root of unity ζ is not square in K, because, otherwise, μ could be reduced to a number of K_0 through a factor of a root of unity in K, and then $M_0 = 1$ could occur. Consequently one can normalize μ to

$$\mu = \zeta_{2^\omega}\overline{\mu}.$$

For $\omega = 1$ one has

$$\mathfrak{m}_0 \cong \mu = -\overline{\mu} \quad \text{in } K, \text{ then } \mathfrak{m}_0^2 \cong \mu^2 = -\mathbf{N}(\mu) = -\mu_0 \quad \text{in } K_0,$$

---

[32](Translator's remark) We actually have an example where in K_0'/K_0 the prime divisors of 2 are ramified but are not at most ramified: $K = P(\sqrt{-1}, \sqrt{p})$, p an odd prime with $p \equiv 3 \pmod 4$. Actually for this field we have $Q = 2$, as we will see in the end of Sect. 3.8, and K_0'/K_0 is of unit type.

[33]Remark moreover that in the case of $\omega \geqq 2$ the matter is only the prime divisors of 2 for the property of ramification, because the Kummer-generator $\lambda_{2^\omega,0}$ itself contains such divisors and does not contain the prime divisors of any other prime numbers.

[34](Translator's remark) See Sect. 3.3, (3.3.1) and (3.3.1′).

where μ_0 is a number of K_0 defined by the last equation; therefore, it follows that this μ_0 is eligible for the radicand appearing in the Kummer-generator of (3.3.1) and (3.3.1′), and hence K_0'/K_0 is (at most unessentially ramified and) of class type. For $\omega \geqq 2$ one has by (3.3.3) and (3.3.4) that

$$\mathfrak{m}_0 \cong \mu = \lambda_{2^\omega}\gamma_0 \quad \text{in } K, \text{ then } \mathfrak{m}_0^2 \cong \mathbf{N}(\mu) = \lambda_{2^\omega,0}\gamma_0^2 \quad \text{in } K_0$$

with γ_0 being a number of K_0; therefore, it follows again with regard to Kummer-generator (3.3.7) that K_0'/K_0 is of class type.

Thus we have proved, as a supplement of Theorem 3.16, the following:

Theorem 3.17 *For an imaginary abelian number field K, the quadratic extension K_0'/K_0 is (at most unessentially ramified and) of class type if and only if there exists a divisor class $M_0 \neq 1$ of K_0 that falls into the principal class of K.*

By the above-mentioned proof, when K_0'/K_0 is of class type, only one class of order 2 determined by the radical $\sqrt{\mu_0} \cong \mathfrak{m}_0$ of Kummer-generator (3.4.1) of K_0'/K_0 comes into consideration for M_0. Then the following arises moreover:

Theorem 3.18 *There exists at most one divisor class $M_0 \neq 1$ of K_0 that falls into the principal class of K.*

Hereby we have proved the fact anticipated in Sect. 3.1 that the class group of K_0 is, by the embedding into that of K, either preserved or reduced to the factor group of half order, namely, to the factor group (of the class group of K) by the subgroup generated by M_0 of order 2.

The knowledge obtained in Theorems 3.16, 3.17, and 3.18 suggests that we should introduce two new invariants in place of the unit index Q of K/K_0, namely, the *ramification index Q^* of K/K_0* given by

$Q^* = 1$ or 2 according as K_0'/K_0 is essentially ramified or at most unessentially ramified,

and the *class index k of K/K_0* given by

$k = 1$ or 2 according as the class group of K_0 is, by the embedding to that of K, preserved or reduced to the factor group of half order.[35]

The ramification index Q^* is really an invariant of the auxiliary relative field K_0'/K_0 corresponding to K/K_0, however it is also indirectly regarded as an invariant of K/K_0, as this fits well for the aim of the introduction of the notation of Q^*; we will actually describe it again by characters of K (but not of K_0').[36] The class index k indicates what fraction the order h_0 of the class group of K_0 is reduced to by the

[35](Translator's remark) Then it holds always that $Q^* = k\,Q$.

[36](Translator's remark) We will see in Theorem 3.22 a necessary and sufficient condition for $Q^* = 1$.

embedding into that of K. Set $k = 2^\kappa$, then $\kappa = 0$ or 1 is the contribution appearing in our 2-rank formula (3.1.9).The quotient

$$h_0^* = \frac{h_0}{k}$$

is the order of the group Δ^* of the classes of K generated by divisors of K_0, as it was introduced in the proof of (3.1.9).

By introducing this new construction, one can set the relative class number formula $(*)$ in a slightly altered form:

$$\frac{h}{h_0^*} = k\,h^* = Q^*\,w \prod_{\chi_1} \frac{1}{2f(\chi_1)} \sum_{x \bmod f(\chi_1)}^{+} (-\chi_1(x)x)\,, \tag{3.4.2}$$

which, in place of the relative class number $h^* = \frac{h}{h_0}$ itself, expresses the index $k\,h^* = \frac{h}{h_0^*}$ of the subgroup Δ^* in the full class group Δ of K. This altered relative class number formula (3.4.2) is considered to be a theoretically interesting sidepiece to the original relative class number formula $(*)$. For the aim to be pursued here, i.e., the calculation of the relative class number h^*, it (formula (3.4.2)) is however less suitable; the reason is that the factor k included here surely has also the relative-class-number-like meaning by itself, in so far as it deals with the relation between the classes of K_0 and that of K, but the significance of the relative class number formula is just to lead the calculation of the relative class number to the calculation of forms from some other calculations in the realm of elementary ideas as much as possible. Practically this appears in every given case—whether it is in the inspection of special kinds of fields or it is in individual numerical cases. For the determination of the alternative $Q = 1$ or 2, it is easier to attribute the definitions of unit type and of class type from the Kummer-generator by means of Theorem 3.16 than to take as a basis the criterion from Theorem 3.17, that is, to return to the definition of the class index k.

3.5 Description of Ramification by the Characters

For simplicity and also because it is interesting of itself in the area of our research, we first describe the state of ramification of K/K_0 by characters of K. Here we will not make use of the assumption that K is imaginary.[37] Therefore later we can transfer our results easily to the auxiliary relative field K_0'/K_0 corresponding to

[37](Translator's remark) We actually assume in this section that K is an abelian number field and that K/K_0 is a quadratic extension.

K/K_0 indicated in Sect. 3.3, as is required for the criterion of Theorem 3.16, where we trace back the characters of K_0' to the ones of K.

According to the decomposition law of class field theory, the state of ramification of a prime number p in the field K is determined by the character group X of K in the following way.[38] Let X_p be the subgroup of X of all the characters χ of K whose conductor $f(\chi)$ is not divisible of p. Then X_p is the character group of the inertia field K_p of p, and hence $[X : X_p] = [K : K_p]$ is the ramification index of p in K. One can also describe X_p as the subgroup of all the characters whose p-component is $\chi_p = 1$. The factor group X/X_p is represented by the differently appearing p-component χ_p of characters χ.[39] Then the number of these χ_p is the ramification index of p in K.[40]

For the prime divisors of p the inertia field in the relative field K/K_0 is the composite K_0K_p. Since K/K_0 has a degree of a prime number (namely degree 2), the ramification occurs exactly when $K_0K_p = K_0$, that is, exactly when K_p is contained in K_0. By class field theory one recognizes when there arises the case where the character group X_p of K_p is contained in the character group X_0 of K_0. Consequently the prime divisors of p are ramified in K/K_0 if and only if all the characters χ of K with conductor $f(\chi)$ not divisible by p belong to (the group of) the characters χ_0 of K_0. Therefore, expressed another way, the following has been proved:

Theorem 3.19 *The prime divisors of p are ramified in K/K_0 if and only if*

$$p \mid f(\chi_1) \quad \textit{for all characters } \chi_1 \textit{ of } K/K_0. \tag{3.5.1}$$

For the prime number $p = 2$ it still remains to investigate moreover when the highest ramification occurs in the case of ramification. This investigation takes some pains.

Now, assume therefore that the prime divisors of 2 are ramified in K/K_0. In this case X_2 is contained in X_0. Group X_0/X_2 is represented by the characters χ_0 of K_0 with different 2-component χ_{02}; the number of these χ_{02} is equal to the ramification index e_0 of 2 in K_0. According to the general ramification theory of relative cyclic fields of prime degree,[41] the ramification number v of the prime divisors of 2 in K/K_0 is given by the contribution of prime number 2 to the relative discriminant $\mathfrak{d}_0$ of K/K_0; namely, it is

$$\mathfrak{d}_{02} = \mathfrak{z}_0^{v+1},$$

[38] To this see the general detailed comments at the end of Sect. 1.1. Here we select slightly different notations in which the dependence of the prime number p emerges.

[39] (Translator's remark) This sentence means that the factor group X/X_p is represented by the characters χ with different p-component χ_p.

[40] (Translator's remark) See Washington, *Introduction to Cyclotomic Fields*, 2nd ed., Springer 1997, Theorem 3.5, p. 24.

[41] See Klassenkörperbericht [5], Part Ia, §**9**, Theorem 3_2.

where $\mathfrak{z}_0$ is the product of the distinct prime divisors of 2 in K_0.[42] By taking the (absolute) norm of K_0, it follows that

$$N_0(\mathfrak{d}_{02}) = 2^{(v+1)\frac{n_0}{e_0}}. \tag{3.5.2}$$

The quotient $\frac{n_0}{e_0}$ appearing in the exponent here is equal to the order of X_2 by virtue of the previous description of e_0. By the general casket formula[43] for the relative discriminant[44] it holds in the present case that

$$d_2 = d_{02}^2 N_0(\mathfrak{d}_{02})$$

for the contributions of prime number 2 to the discriminants of K and K_0. By the conductor-discriminant formula (1.3.2) it holds moreover that

$$d_2 = \prod_{\chi_0} f(\chi_{02}) \cdot \prod_{\chi_1} f(\chi_{12}),$$
$$d_{02} = \prod_{\chi_0} f(\chi_{02}).$$

Hence there arises

$$N_0(\mathfrak{d}_{02}) = \frac{\prod_{\chi_1} f(\chi_{12})}{\prod_{\chi_0} f(\chi_{02})}. \tag{3.5.3}$$

Let ψ be a fixed character of K/K_0 of which we will later make use according to circumstances; then the characters χ_0 and χ_1 are united by $\chi_1 = \psi\chi_0$, and hence $\chi_{12} = \psi_2\chi_{02}$. Hereafter it will be the point for the calculation of the ramification number v to determine the quotient of conductors

$$\frac{f(\chi_{12})}{f(\chi_{02})} = \frac{f(\psi_2\chi_{02})}{f(\chi_{02})}$$

by appropriate selection of ψ. Since the $\frac{n_0}{e_0}$ characters χ_0 (of K_0) in a class mod X_2 have the same 2-component χ_{02} in each class, it is enough to let (χ_0 with) χ_{02}

[42] In the following we express in such a case the 2-contributions and the 2-components by adding subscript 2 as index; remark here the rule $f_2(\chi) = f(\chi_2)$.

[43] (Translator's remark) The casket formula is $|D_K| = N_{k/P}(D_{K/k})|D_k|^{[K:k]}$ for a field extension K/k where D_K and D_k are the discriminants of K and k, respectively, and $D_{K/k}$ is the relative discriminant of K/k.

[44] See, for example, Zahlbericht [10], §**15**, Theorem 39, or Hasse [7], §**25**, e).

run over a system of representatives of X_0/X_2. By this meaning of χ_{02} therefore, according to (3.5.2) and (3.5.3), it holds exactly that[45]

$$2^{v+1} = \prod_{\chi_{02}} \frac{f(\psi_2\chi_{02})}{f(\chi_{02})}. \tag{3.5.4}$$

Now let $f_2 = 2^\rho$,[46] then $\rho \geqq 1$ by the assumption and moreover necessarily $\rho \geqq 2$, since 2^1 itself cannot be any contribution of the conductor. Corresponding to the basis expression

$$x \equiv (-1)^\alpha (1+2^2)^\beta \pmod{2^\rho} \quad (\alpha \mod 2,\ \beta \mod 2^{\rho-2}) \tag{3.5.5}$$

of the prime residue group mod 2^ρ, the following two characters $\cup$ and φ_ρ constitute a basis of the full character group mod 2^ρ:

$$\cup(x) = (-1)^\alpha, \quad \varphi_\rho(x) = \zeta^\beta_{2^{\rho-2}}. \tag{3.5.6}$$

The character $\cup$, which we already introduced in (3.2.5) has conductor $f(\cup) = 2^2$ and order 2. The character φ_ρ, which is determined up to the algebraic-conjugates, has conductor $f(\varphi_\rho) = 2^\rho$ and order $2^{\rho-2}$ for $\rho > 2$; and for $\rho = 2$ it is reduced to the principal character 1 and can remain out of consideration. Instead of φ_ρ, $\varphi'_\rho = \cup\varphi_\rho$ can be taken as basis elements. In the following we denote by $\hat{\varphi}_\rho$ a character appropriately selected from the two characters φ_ρ and φ'_ρ in each case.

Since X_0/X_2 has index 2 in X/X_2, exactly one of the following three cases arises for generators of the two factor groups[47]:

	(a)	(b)	(c)
X/X_2	$\cup, \hat{\varphi}_\rho$	$\cup, \hat{\varphi}_\rho$	$\hat{\varphi}_\rho$
X_0/X_2	$\cup^2 = 1, \hat{\varphi}_\rho$	$\cup, \hat{\varphi}_\rho^2$	$\hat{\varphi}_\rho^2$
	$\rho \geqq 2$; for $\rho = 2$, $\hat{\varphi}_\rho$ is omitted	$\rho \geqq 3$	$\rho \geqq 3$

This table means that in each case X/X_2 and X_0/X_2 are generated by characters of X and X_0 with given 2-components (in the table). For the character ψ, in each case,

[45](Translator's remark) In this product $\prod_{\chi_{02}}$ the characters χ_0 with 2-component χ_{02} run over a system of representatives χ_0 of the classes $\chi_0 X_2$ of X_0/X_2. Therefore the notation $\prod_{\chi_{02}}$ in (3.5.4) would be better than $\prod_{\chi_0}$ in the original book published in 1985. In the original book in 1952 it was printed as $\prod_{\chi_{02}}$.

[46](Translator's remark) f_2 is the 2-part of the conductor of f of K.

[47](Translator's remark) In the original book the condition "$\rho \geqq 2$" in case (a) was not printed. The translator added it for the convenience for the reader.

we can select such a character whose 2-component ψ_2 is assembled by generators of X/X_2 but not of X_0/X_2, because ψ then belongs to X, but not to X_0, as demanded. Now we treat successively the three cases mentioned above.

(a) Here one can take $\psi_2 = \cup$. Then one has

$$\chi_{02} = \hat{\varphi}_\rho^\beta,\ \psi_2\chi_{02} = \cup\hat{\varphi}_\rho^\beta \quad (\beta \bmod 2^{\rho-2}),$$

and hence

$$\frac{f(\psi_2\chi_{02})}{f(\chi_{02})} = \frac{f(\cup\hat{\varphi}_\rho^\beta)}{f(\hat{\varphi}_\rho^\beta)} = \left\{\begin{array}{l} \dfrac{2^2}{1} = 2^2 \quad \text{for } \beta \equiv 0 \pmod{2^{\rho-2}} \\[2ex] \dfrac{2^{\rho-\sigma}}{2^{\rho-\sigma}} = 1 \text{ for } (\beta, 2^{\rho-2}) = 2^\sigma \text{ with } 0 \leqq \sigma \leqq \rho - 3 \end{array}\right\}.$$

Hence from (3.5.4) one has

$$2^{v+1} = \prod_{\beta \bmod 2^{\rho-2}} \frac{f(\cup\hat{\varphi}_\rho^\beta)}{f(\hat{\varphi}_\rho^\beta)} = 2^2,$$

and therefore

$$v = 1.$$

On the other hand, in this case one has $e_0 = [X_0 : X_2] = 2^{\rho-2}$, and hence $2e_0 = 2^{\rho-1}$. Therefore the highest ramification does not arise; rather v has the possible smallest value.[48]

(b) Here one can take $\psi_2 = \hat{\varphi}_\rho$. Then one has

$$\chi_{02} = \cup^\alpha\hat{\varphi}_\rho^{2\beta},\ \psi_2\chi_{02} = \cup^\alpha\hat{\varphi}_\rho^{2\beta+1} \quad (\alpha \bmod 2,\ \ \beta \bmod 2^{\rho-3}),$$

and hence

$$\frac{f(\psi_2\chi_{02})}{f(\chi_{02})} = \frac{f(\cup^\alpha\hat{\varphi}_\rho^{2\beta+1})}{f(\hat{\varphi}_\rho^{2\beta})} = \left\{\begin{array}{ll} \dfrac{2^\rho}{1} = 2^\rho & \text{for } \alpha \equiv 0 \pmod 2, \\ & \quad \beta \equiv 0 \pmod{2^{\rho-3}} \\ \dfrac{2^\rho}{2^2} = 2^{\rho-2} & \text{for } \alpha \equiv 1 \pmod 2, \\ & \quad \beta \equiv 0 \pmod{2^{\rho-3}} \\ \dfrac{2^\rho}{2^{\rho-\sigma-1}} = 2^{\sigma+1} & \text{for } \alpha \text{ arbitrary, } (\beta, 2^{\rho-3}) = 2^\sigma \\ & \text{with } 0 \leqq \sigma \leqq \rho - 4 \end{array}\right\}.$$

[48] For completeness we also give the value $2e_0$ in this and the other two cases, though one already knows easily the existence of the highest ramification ($v = 2e_0$), which is interesting to us, by knowing whether v is even or not; see footnote 29, p. 82, in Sect. 3.4.

The number of the pairs of residue classes α mod 2 and β mod $2^{\rho-3}$ for which the last relation (condition) with fixed σ is fulfilled is $2 \cdot 2^{\rho-4-\sigma} = 2^{\rho-3-\sigma}$. Consequently there arises from (3.5.4) with easy calculation in the exponent (partial summation)

$$2^{v+1} = \prod_{\substack{\alpha \bmod 2 \\ \beta \bmod 2^{\rho-3}}} \frac{f(\mho^{\alpha}\hat{\varphi}_{\rho}^{2\beta+1})}{f(\mho^{\alpha}\hat{\varphi}_{\rho}^{2\beta})} = 2^{\rho} \cdot 2^{\rho-2} \cdot 2^{\sum_{\sigma=0}^{\rho-4}(\sigma+1)2^{\rho-3-\sigma}} = 2^{2^{\rho-1}},$$

and therefore

$$v = 2^{\rho-1} - 1.$$

On the other hand, in this case one has $e_0 = [X_0 : X_2] = 2 \cdot 2^{\rho-3} = 2^{\rho-2}$, and hence $2e_0 = 2^{\rho-1}$. Therefore the highest ramification does not arise; rather v has the largest possible odd value.

(c) Here one can again take $\psi_2 = \hat{\varphi}_{\rho}$. Then one has

$$\chi_{02} = \hat{\varphi}_{\rho}^{2\beta}, \quad \psi_2\chi_{02} = \hat{\varphi}_{\rho}^{2\beta+1} \quad (\beta \bmod 2^{\rho-3}),$$

and hence

$$\frac{f(\psi_2\chi_{02})}{f(\chi_{02})} = \frac{f(\hat{\varphi}_{\rho}^{2\beta+1})}{f(\hat{\varphi}_{\rho}^{2\beta})} = \left\{ \begin{array}{ll} \dfrac{2^{\rho}}{1} = 2^{\rho} & \text{for } \beta \equiv 0 \pmod{2^{\rho-3}} \\[2ex] \dfrac{2^{\rho}}{2^{\rho-\sigma-1}} = 2^{\sigma+1} & \text{for } (\beta, 2^{\rho-3}) = 2^{\sigma} \\ & \quad \text{with } 0 \leq \sigma \leq \rho-4 \end{array} \right\}.$$

The number of the residue classes β mod $2^{\rho-3}$ for which the last relation (condition) with fixed σ is fulfilled is $2^{\rho-4-\sigma}$. Consequently there arises from (3.5.4) with easy calculation in the exponent

$$2^{v+1} = \prod_{\beta \bmod 2^{\rho-3}} \frac{f(\hat{\varphi}_{\rho}^{2\beta+1})}{f(\hat{\varphi}_{\rho}^{2\beta})} = 2^{\rho} \cdot 2^{\sum_{\sigma=0}^{\rho-4}(\sigma+1)2^{\rho-4-\sigma}} = 2^{2^{\rho-2}+1},$$

and therefore

$$v = 2^{\rho-2}.$$

On the other hand, in this case one has $e_0 = [X_0 : X_2] = 2^{\rho-3}$ and hence $2e_0 = 2^{\rho-2}$. Therefore the highest ramification does arise.

Thus the ramification number v has been determined in all cases. The highest ramification arises only in case (c). This case is now, as opposed to the other two cases (a) and (b), characterized by the fact that the character $\mho$ does not appear as 2-component χ_2 of any character χ of K, or therefore that every character χ of K has the conductor $f(\chi)$ that is either not divisible by 2 or divisible by at least 2^3. By adding particularly the condition of Theorem 3.19 for the existence of ramification, there arises a supplement to Theorem 3.19:

Theorem 3.20 *The prime divisors of 2 are at most ramified in K/K_0 if and only if*

$$2^3 \mid f(\chi_1) \quad \textit{for all characters } \chi_1 \textit{ of } K/K_0 \tag{3.5.7}$$

and

$$2 \nmid f(\chi_0) \;\; \textit{or} \;\; 2^3 \mid f(\chi_0) \quad \textit{for all characters } \chi_0 \textit{ of } K_0. \tag{3.5.8}$$

The results obtained in Theorems 3.19 and 3.20, as mentioned already in the beginning of this section, hold also for K_0'/K_0. In this case the state of ramification in K_0'/K_0 is described by the characters of K_0'/K_0 and of K_0. For our application we moreover have to trace back the characters of K_0'/K_0 to the ones of K/K_0.

Since, by the meaning of 2^ω as the contribution of prime number 2 to the number w of the roots of unity in K, P_{2^ω} is actually contained in K but $P_{2^{\omega+1}}$ is not, the $2n_0$ characters $\psi\chi_0$, $\psi\chi_1$ of the extension $K' = KP_{2^{\omega+1}}$ over K are, first of all, added to the $2n_0$ characters χ_0, χ_1 of K where ψ is a fixed character of $P_{2^{\omega+1}}/P_{2^\omega}$. Then among the $4n_0$ characters χ' of K', the $2n_0$ characters of K_0' are characterized by $\chi'(-1) = 1$.

In the case of $\omega = 1$ it holds then that $\psi = \mho$. Because $\mho(-1) = -1$, then $\chi_0, \mho\chi_1$ are the $2n_0$ characters of K_0', and hence $\mho\chi_1$ are the n_0 characters of K_0'/K_0. Accordingly, in the transition of K/K_0 to K_0'/K_0, the characters χ_1 are to be replaced by the $\mho\chi_1$ in Theorems 3.19 and 3.20. In this case condition (3.5.1) in question for $p \neq 2$ and (the condition) (3.5.7) do not change according to $f(\mho) = 2^2$. This is also clear from the beginning, because in this case the states of ramification in K/K_0 and in K_0'/K_0 are actually the same on the basis of the relations (3.3.1) and (3.3.1$'$) of the Kummer-generators.

In the case of $\omega \geqq 2$, $\psi = \varphi_{\omega+1}$ can be selected. Because $\varphi_{\omega+1}(-1) = 1$, then χ_0 and $\varphi_{\omega+1}\chi_0$ are the $2n_0$ characters of K_0', and hence $\varphi_{\omega+1}\chi_0$ are the n_0 characters of K_0'/K_0. Accordingly, in the transition from K/K_0 to K_0'/K_0, the characters χ_1 are to be replaced by the $\varphi_{\omega+1}\chi_0$ in Theorem 3.20; in this case, condition (3.5.1) for $p \neq 2$ of Theorem 3.19 does not come into consideration, because here on the basis of the Kummer-generator (3.3.7) only the divisors of 2 can be ramified. By replacing χ_1 by $\varphi_{\omega+1}\chi_0$ in condition (3.5.7) that only comes into consideration, one has the condition

$$2^3 \mid f(\varphi_{\omega+1}\chi_0) \quad \text{for all characters } \chi_0 \text{ of } K_0.$$

This means that no character χ_0 has 2-component of the form $\mho^\alpha \varphi_{\omega+1}^{-1}$. Since the 2-components of χ_0 form a group, this is synonymous with saying that no character

χ_0 has 2-component of the form $\mathsf{U}^\alpha \varphi_\rho$ with $\rho \geqq \omega + 1$, or therefore, by virtue of $f(\mathsf{U}^\alpha \varphi_\rho) = 2^\rho$, it is synonymous with saying that

$$2^{\omega+1} \nmid f(\chi_0) \quad \text{for all characters } \chi_0 \text{ of } K_0.$$

The combination of this condition and (3.5.8), instead of (3.5.7) and (3.5.8), produces the sole condition

$$\left\{ \begin{array}{l} \text{either } 2 \nmid f(\chi_0), \\ \\ \quad \text{or } 2^3 \mid f(\chi_0) \text{ and } 2^{\omega+1} \nmid f(\chi_0) \quad \text{for all characters } \chi_0 \text{ of } K_0 \end{array} \right\}, \tag{3.5.9}$$

which expresses, in other words, that the conductor $f(\chi_0)$ of any character χ_0 of K_0 must always be either not divisible by 2 or divisible by at least 2^3 and at most 2^ω.

Thus the following theorem has been proved:

Theorem 3.21 *Let 2^ω be the exact 2-power in w and let $K' = K P_{2^{\omega+1}}$.*

In the case of $\omega = 1$ the prime divisors of a prime number $p \neq 2$ are ramified in K_0'/K_0 if and only if condition (3.5.1) *of Theorem 3.19 is fulfilled, and the prime divisors of prime number 2 are at most ramified in K_0'/K_0 if and only if conditions* (3.5.7) *and* (3.5.8) *of Theorem 3.20 are fulfilled.*

In the case of $\omega \geqq 2$ the prime divisors of $p \neq 2$ are unramified in K_0'/K_0, and the prime divisors of prime number 2 are at most ramified in K_0'/K_0 if and only if condition (3.5.9) *is fulfilled.*

3.6 Criteria for $Q = 1$ or 2 by Characters and Class Problem

By virtue of Theorem 3.21, first of all, the state of ramification for every *fixed* prime number p in the auxiliary relative field K_0'/K_0 is described by characters of K. Summarizing now the conditions established there for *all* primes p, we obtain criteria for deciding whether K_0'/K_0 itself is essentially ramified or at most unessentially ramified, and hence by Theorem 3.16 we obtain criteria for the alternative $Q = 1$ of 2. These criteria read as follows:

Theorem 3.22 *Let 2^ω be (the 2-power) exactly contained in the number w of the roots of unity in K and let $K' = K P_{2^{\omega+1}}$.*

(a) $\underline{w \not\equiv 0 \pmod 4 \quad (\omega = 1).}$

It holds that $\underline{Q = 1}$ *if either there exists a prime number $p \neq 2$ with*

$$p \mid f(\chi_1) \quad \textit{for all characters } \chi_1 \textit{ of } K/K_0, \tag{3.6.1}$$

or for the prime number $p = 2$ the (following) two conditions are fulfilled:

$$2^3 \mid f(\chi_1) \quad \textit{for all characters } \chi_1 \textit{ of } K/K_0 \tag{3.6.2}$$

and

$$\textit{either } 2 \not| f(\chi_0) \textit{ or } 2^3 \mid f(\chi_0) \quad \textit{for all characters } \chi_0 \textit{ of } K_0. \tag{3.6.3}$$

In the other case, $Q = 2$ or 1 according as K_0'/K_0 (or—as is the same—K/K_0) is of unit type or of class type.[49]

(b) $w \equiv 0 \pmod 4 \quad (\omega \geqq 2)$.

It holds that $Q = 1$ if for the prime number $p = 2$ the (following) condition is fulfilled[50]*:*

$$\textit{either } 2 \not| f(\chi_0) \textit{ or } 2^3 \mid f(\chi_0) \textit{ and } 2^{\omega+1} \not| f(\chi_0) \textit{ for all characters } \chi_0 \textit{ of } K_0. \tag{3.6.4}$$

In the other case, $Q = 2$ or 1 according as K_0'/K_0 is of unit type or of class type.[51]

By the conceptional construction introduced at the end of Sect. 3.4, the assertions $Q = 1$ and $Q = 2$ or 1 can be also expressed as the forms $Q^* = 1, k = 1$ and $Q^* = 2, k = 1$ or 2.

In this theorem, which describes an example of not entirely easily visible but formally logical relations, the example being pleasant and interesting to logicians and also arising from "practice," we have deviated from the German language rule of style and punctuation—used by non-mathematicians—for the benefit of adaption to the writing of logical concepts more than usual, and (we have) emphasized moreover by the intervening spaces the logical particles "and" and "or" as well as "exist" and "all." It is decisively important to use accurately these particles for the application of the criteria in the given cases and also when one has to install explicitly the assumption lying in "the other cases."

While Theorem 3.22 permits us in some circumstance to read $Q = 1$ in given cases only by characters of K, its application to the certification of $Q = 2$ needs in every case the more difficult distinction between unit type and class type (or the class embedding problem).

[49] According to the stipulation in footnote 31, p. 82, in Sect. 3.4, this assertion implicates that under the assumption given by "the other case" the fields K_0'/K_0 and K/K_0 are at most unessentially ramified.

[50] (Translator's remark) Condition (3.6.4) also says that "either $2 \not| f(\chi_0)$ or $2^3 \mid f(\chi_0)$" and "$2^{\omega+1} \not| f(\chi_0)$" for all characters χ_0 of K_0.

[51] This assertion also implies that K_0'/K_0 is at most unessentially ramified.

The road to the results pronounced in Theorem 3.22 was not simple. It may be disproportionate that we have to be engaged in such extensive research in order to control merely the seemingly insignificant factor $Q = 1$ or 2 appearing in the relative class number formula (∗). However experience really teaches us that in algebraic number theory non-trivial statements on units—and here the question is such a matter—can always be obtained only with great pains.

3.7 Types of Fields with $Q = 1$ and Types of Fields with $Q = 2$

First we can easily establish two types of imaginary abelian number fields K with $Q = 1$ only by the conditions of the characters of K in Theorem 3.22.

Theorem 3.23 *For an imaginary abelian number field K with prime-power conductor $f = p^\rho$, we have $Q = 1$.*

Proof If $p \neq 2$, then $w \not\equiv 0 \pmod 4$. Since also $p | f(\chi)$ for all characters $\chi \neq 1$, (3.6.1) is fulfilled for the prime number p appearing in f.

If $p = 2$ and $w \not\equiv 0 \pmod 4$, then U does not appear among the characters χ. Hence (3.6.2) and (3.6.3) are fulfilled.

If $p = 2$ and $w \equiv 0 \pmod 4$, then U actually appears among the χ, but not among the χ_0. Assume that the character φ_σ with $\sigma \geqq \omega + 1$ appears among the χ_0, then the whole character group mod 2^σ generated by U and φ_σ would be contained in X and hence P_{2^σ} would be contained in K, which contradicts the meaning of 2^ω as the contribution of 2 in w. Thus (3.6.4) is fulfilled. □

Theorem 3.24 *For an imaginary cyclic field K we have $Q = 1$.*

Proof Let

$$\chi = \prod_{p|f} \chi_p \tag{3.7.1}$$

be a generating character of K in the form of decomposition into components. Since K is to be imaginary, it holds that $\chi(-1) = -1$. The characters χ_0, χ_1 are powers of χ:

$$\chi_0 = \chi^g = \prod_{p|f} \chi_p^g, \quad \chi_1 = \chi^u = \prod_{p|f} \chi_p^u \tag{3.7.2}$$

with even g and odd u. Since K is to be cyclic, whereas P_{2^ω} is not cyclic for $\omega \geqq 3$, it holds that either $\omega = 1$ or $\omega = 2$.

First suppose that $\omega = 1$. If the component χ_p (of χ) for some prime divisor $p \neq 2$ of f has even order, then for odd u the powers χ_p^u are not the principal character, i.e., $\chi_p^u \neq 1$, and so $p \mid f(\chi^u)$. Hence (3.6.1) is fulfilled for this prime p. However, if for all prime divisors $p \neq 2$ of f the χ_p have odd order u_p, then, by letting u_0 be the least common multiple of u_p, the character $\chi^{u_0} = \chi_2^{u_0}$ appears in X (and χ^{u_0} is distinct from the principal character 1 because $\chi(-1) = -1$). Since χ_2 is of 2-power order, χ_2 itself also appears in X and in fact (it appears) in the χ_1, so that $\chi_2 = \cup\varphi_\rho$ holds necessarily with $\rho \geqq 3$ because $\omega = 1$. Hence $\chi_2^g = \varphi_\rho^g \neq \cup$ for all even g, and $\chi_2^u = \cup\varphi_\rho^u \neq 1, \cup$ for all odd u. Thus (3.6.3) and (3.6.2) are fulfilled.

Further suppose that $\omega = 2$. Since $\cup$ appears in X in this case, and in fact among the χ_1, the character $\chi_2 = \cup$ appears necessarily, so that all the $\chi_0 = \chi^g = \prod_{p|f,\, p\neq 2} \chi_p^g$ have conductor $f(\chi_0)$ not divisible by 2. Thus (3.6.4) is fulfilled. □

A broader type of imaginary abelian number fields K with $Q = 1$ arises from the original criterion in Theorem 3.14 connecting to the definition of Q.

Theorem 3.25 *If K is an imaginary abelian number field such that for the maximal real subfield K_0 the signatures of a system of fundamental units are independent, then $Q = 1$.*

Proof The assumption here (in the theorem) can also be stated that in K_0 every totally positive unit is square. Then there exists no unit ε_0^* in K_0 with the property (3.2.4b), because such a unit should be totally positive as a relative norm from the imaginary field K. Thus it holds that $Q = 1$. □

Indeed, the existence of the fields of the type in Theorem 3.25 cannot be established by means of characters of K as in Theorems 3.23 and 3.24. For the special case where K is an imaginary biquadratic field and so K_0 is a real quadratic field, the type of fields in Theorem 3.25 exists exactly when the fundamental unit of K_0 has norm -1.

Now we will show further that there exist certain types of imaginary abelian number fields K for which $Q = 2$ holds. In this case, according to Theorem 3.23, we assume here from the beginning that the conductor of K is composite.

We start with the fact that under this assumption the number

$$\eta = \prod_{\substack{a \bmod f \\ a \text{ in } H}} (1 - \zeta_f^a) \tag{3.7.3}$$

is a unit of K,[52] and we investigate under what assumption on K this unit η has the property that the corresponding root of unity

$$\zeta = \frac{\eta}{\overline{\eta}} \quad \text{is not square in } K, \tag{3.7.4}$$

[52]This unit η is connected with the cyclotomic unit of K_0 introduced in Sect. 2.4. If m has the subsequent meaning given (defined below) in this book, then $(-1)^m \eta\overline{\eta} = (-1)^m \mathbf{N}(\eta) = \lambda_0^2$ is the square of λ_0 (in Sect. 2.4 it is denoted by λ and belongs to K_0 or is quadratic over K_0), from which the cyclotomic unit of K_0 is derived by the formation of conjugate-quotient by (2.4.4).

so that the condition (3.2.4a) is fulfilled. Then $Q = 2$ by Theorem 3.14.

By virtue of $1 - \zeta_f^a = -\zeta_f^a(1 - \zeta_f^{-a})$ it holds that

$$\zeta = (-1)^m \zeta_f^A,$$

where

$$m = \sum_{\substack{a \bmod f \\ a \text{ in } H}} 1 = \frac{\varphi(f)}{n}$$

is the number of the residue classes a mod f in H and

$$A \equiv \sum_{\substack{a \bmod f \\ a \text{ in } H}} a \pmod{f}$$

is their sum.[53] As so far, let 2^ω be the contribution of the prime number 2 to w and 2^ρ the contribution to f. If f is odd, then $\rho = 0$ and so ζ is not square in K if and only if

$$m \not\equiv 0 \pmod{2}. \tag{3.7.5$_0$}$$

If f is even, then $\rho \geqq 2$ and so $A \equiv 0 \pmod{2^{\rho-\omega}}$ because ζ_f^A is a root of unity in K. And hence ζ is not square in K if and only if in the case of $\omega = 1$,

$$m + \frac{A}{2^{\rho-1}} \not\equiv 0 \pmod{2} \tag{3.7.5$_1$}$$

and in the case of $\omega \geqq 2$,

$$\frac{A}{2^{\rho-\omega}} \not\equiv 0 \pmod{2}. \tag{3.7.5$_2$}$$

Hence the question has turned out to determine further the residue class A mod $2^{\rho-\omega+1}$ for even f.

We determine A mod 2^ρ even more accurately. For this purpose let m_0 be the number of the residue classes a_0 mod f from H with $a_0 \equiv 1 \pmod{2^\rho}$, and let b run over a system of representatives of distinct residue classes mod 2^ρ provided by the classes from H. Then it holds that

$$A \equiv m_0 B \pmod{2^\rho},$$

[53](Translator's remark) The number A is defined by this congruence.

with[54]

$$B \equiv \sum_{\substack{b \bmod 2^\rho \\ b \text{ in } H}} b \pmod{2^\rho}.$$

The group of residue classes b mod 2^ρ is the narrowest upper-group of H with 2-power conductor; the order is $\frac{m}{m_0}$. Let Ψ be the subgroup of all characters in X with 2-power conductor, then the residue classes b mod 2^ρ are characterized by $\psi(b) = 1$ for all ψ in Ψ.

1. Assume first that $\omega = 1$. Then Ψ is generated neither by $\cup$ alone nor by a pair of $\cup$ and φ_ρ ($\rho \geqq 3$), and hence Ψ is necessarily cyclic and generated by one of the following characters:

$$\text{(a)}\ \varphi_\sigma\ (2 \leqq \sigma \leqq \rho), \qquad \text{(b)}\ \cup\varphi_\sigma\ (3 \leqq \sigma \leqq \rho)$$

In case (a) b runs over all the residue classes b mod 2^ρ with $b \equiv \pm 1 \pmod{2^\sigma}$. Since these pairs of the opposite $\pm b$ mod 2^ρ come together, the congruence

$$B \equiv 0 \pmod{2^\rho}$$

holds and then also

$$A \equiv 0 \pmod{2^\rho}.$$

Hence

$$m + \frac{A}{2^{\rho-1}} \equiv m \pmod 2. \qquad (3.7.6_1\text{a})$$

In case (b) b runs over all the residue classes b mod 2^ρ with $b \equiv 1, -1 + 2^{\sigma-1} \pmod{2^\sigma}$. Since these $2^{\rho-\sigma}$ pairs come each together with sum $2^{\sigma-1}$ mod 2^ρ, the congruence

$$B \equiv 2^{\rho-1} \pmod{2^\rho}$$

holds and hence

$$A \equiv m_0 2^{\rho-1} \pmod{2^\rho}.$$

---

[54](Translator's remark) The number B is defined by this congruence.

Therefore

$$m + \frac{A}{2^{\rho-1}} \equiv m + m_0 \equiv m_0 \pmod 2, \tag{3.7.6$_1$b}$$

where the last congruence holds, because in this case $m = m_0 2^{\rho-\sigma+1} \equiv 0 \pmod 2$.[55]

2. Assume further that $\omega \geqq 2$. Then Ψ contains all the characters mod 2^ω, but by the existence of $\mho$ there are no more characters φ_σ or $\mho\varphi_\sigma$ $(\sigma > \omega)$. Hence b runs over all the residue classes b mod 2^ρ with $b \equiv 1 \pmod{2^\omega}$, and so

$$B \equiv \sum_{c \bmod 2^{\rho-\omega}} (1 + c\,2^\omega) \equiv 2^{\rho-\omega} + 2^{\rho-1}(1 + 2^{\rho-\omega}) \pmod{2^\rho},$$

and then

$$\begin{aligned} A &\equiv m_0(2^{\rho-\omega} + 2^{\rho-1}(1 + 2^{\rho-\omega})) \pmod{2^\rho} \\ &\equiv m_0 2^{\rho-\omega} \pmod{2^{\rho-\omega+1}}. \end{aligned}$$

Consequently

$$\frac{A}{2^{\rho-\omega}} \equiv m_0 \pmod 2. \tag{3.7.6$_2$}$$

By (3.7.5$_0$) and (3.7.5$_1$), (3.7.5$_2$) together with (3.7.6$_1$a), (3.7.6$_1$b), and (3.7.6$_2$) we have exactly determined when the root of unity ζ in (3.7.4) is not square and consequently $Q = 2$. As the condition (of Q being equal to 2) it has been shown that m, respectively m_0, are odd according as the assumptions made for (3.7.5$_0$) and (3.7.6$_1$a), respectively (3.7.6$_1$b) and (3.7.6$_2$), are presented. The cases where m comes to the point, contrary to the cases where m_0 comes to the point, are characterized by the following: in the first cases, there exist at most characters χ_0 of K_0 with 2-power conductor,[56] while in the last cases there exists at least a (one) character χ_1 of K/K_0 with 2-power conductor. Thus we have, as result of our research, the following:

Theorem 3.26 *Let K be an imaginary abelian number field of degree n and with composite conductor $f = 2^\rho f_0$ $(\rho \geqq 0,\ f_0$ odd).*

---

[55] Also in case (a) the congruence $m = m_0 2^{\rho-\sigma+1} \equiv 0 \pmod 2$ holds. For a clear formulation of the last result, it is however appropriate to take there the congruence value m mod 2.

[56] (Translator's remark) This means that there exists no odd character χ_1 of K/K_0 with 2-power conductor in the first case. The first case where m comes to the point is the one with odd f, or with even f and $\omega = 1$ as in case (a). The last case where m_0 comes to the point is the one with even f and $\omega = 1$ as in case (b) or with $\omega \geqq 2$.

If there exist no characters χ_1 of K/K_0 with 2-power conductor, then $Q = 2$ surely when[57] *the number $m = \frac{\varphi(f)}{n}$ of the residue classes a mod f in H is odd.*

If there exists a character χ_1 of K/K_0 with 2-power conductor, then $Q = 2$ surely when the number m_0 (appearing in m) of the residue classes a_0 mod f in H with $a_0 \equiv 1 \pmod{2^\rho}$ is odd.

The assumptions of this theorem are specially fulfilled for the cyclotomic fields $K = P_f$ with composite conductor f, because for such fields the congruence group H consists of one residue class $a \equiv 1 \mod f$. Since the cyclotomic fields $K = P_{p^\rho}$ with prime-power conductor p^ρ fall within Theorem 3.23, the following holds accordingly:

Theorem 3.27 *For the cyclotomic field $K = P_f$, $Q = 1$ or 2 according as f is a prime power or a composite.*

Regarding the fields K with $Q = 2$ exhibited in Theorem 3.26, we note the following. Except for the unit η of (3.7.3) and its conjugates as well as the units derived from products of their powers and combination of roots of unity in K (cyclotomic units), one does not know further units of K in general form. Therefore one cannot exhibit broader types of imaginary abelian number fields K with $Q = 2$ except the ones in Theorem 3.26, because for the unit derived from η, as is easily seen, the corresponding root of unity is a power of the root of unity ζ of (3.7.4), so that no extension of the results in Theorem 3.26 arises from these units.

For the fields K in Theorem 3.26, it is determined along with $Q = 2$ by Theorem 3.16 that K_0'/K_0 is at most unessentially ramified and of unit type, and then by Theorem 3.17 the class-embedding problem is solved because only the principal class of K_0 falls into the principal class of K, i.e., the class index $k = 1$. The latter ($k = 1$) holds also for the fields in Theorems 3.23 and 3.24, because for such fields (the fact that) $Q = 1$ has been shown in the way that the ramification index Q^* has been determined as $Q^* = 1$, namely, that the fulfillment of the character condition for K_0'/K_0 being essentially ramified has been proved. Since the fields in Theorem 3.27 fall into the fields in Theorems 3.23 and 3.26, we can therefore ascertain the following:

Theorem 3.28 *For the imaginary abelian number fields K in Theorems 3.23, 3.24, 3.26, and 3.27, the class index k is equal to 1, that is, only the principal class of K_0 falls into the principal class of K.*

This does not hold anymore for the fields K in Theorem 3.25. For them K/K_0 is actually ramified by Theorem 3.13, and then in the case of $w \not\equiv 0 \pmod 4$, K_0'/K_0 is also ramified,[58] however it is not said that K_0'/K_0 is moreover essentially

[57](Translator's remark) Here "surely when" does not mean "if and only if." It means "if" with emphasis. We usually use "exactly if" as "if and only if."

[58](Translator's remark) This statement is not correct. We have a counterexample: let $K = P(\sqrt{-2}, \sqrt{-p})$, p an odd prime with $p \equiv 5 \pmod 8$. Then the fundamental unit ε of $K_0 = P(\sqrt{2p})$ has norm -1 and $K_0'/K_0 = P(\sqrt{2}, \sqrt{p})/P(\sqrt{2p})$ is unramified.

ramified. In the case of $w \equiv 0 \pmod 4$, it is not even said that K_0'/K_0 is ramified in general,[59] which would also be necessary to exclude the fields K of class type for K_0'/K_0—the fields K of unit type have already been excluded in the proof of Theorem 3.25.

For the cyclotomic fields the result of Theorem 3.28 has already been proved by Kronecker [12] in an essentially same way.

To the decision of the alternative $Q = 1$ or 2, the following fact could eventually also be convenient in certain cases.

Theorem 3.29 *Let $\tilde{K}$ be an imaginary subfield of K, then the unit index $\tilde{Q}$ of $\tilde{K}/\tilde{K}_0$ is a divisor of the unit index Q of K/K_0, and then $\tilde{Q} = 1$ for $Q = 1$ and $Q = 2$ for $\tilde{Q} = 2$.*[60]

Proof It is sufficient to prove the last relation. Now the fields $\tilde{K}'$ and $\tilde{K}_0'$ obtained by $\tilde{K}$ are subfields of the auxiliary fields K' and K_0' obtained by K, respectively. Hence a Kummer-generator of $\tilde{K}_0'/\tilde{K}_0$ is also one such of K_0'/K_0.[61] Then if $\tilde{Q} = 2$, that is, $\tilde{K}_0'/\tilde{K}_0$ is of unit type, then the same holds also for K_0'/K_0, that is, $Q = 2$. □

Finally we here further pay attention to the sufficient criterion for $Q = 2$ deduced from Theorems 3.26 and 3.29 that will be used in the proof of Theorem 3.35 below.[62]

3.8 Imaginary Bicyclic Biquadratic Fields

For an explanation of the results on the unit index Q of K/K_0, we consider imaginary bicyclic biquadratic fields.

They are the fields

$$K = K_1K_2 = P(\sqrt{-f_1}, \sqrt{-f_2}), \tag{3.8.1}$$

---

[59](Translator's remark) Indeed for example, for the field $K = P(\sqrt{-1}, \sqrt{-2p})$, p an odd prime with $p \equiv 5 \pmod 8$, the extension K_0'/K_0 is unramified.

[60](Translator's remark) This theorem is not correct. In fact, we have a counterexample: $K = P(\sqrt{-1}, \sqrt{2}, \sqrt{17})$, $\tilde{K} = P(\sqrt{-1}, \sqrt{34})$. This counterexample was known at least in the 1970s. See Martinet's Preface in the original book published in 1985 or M. Hirabayashi and K. Yoshino, Remarks on unit indices of imaginary abelian number fields, *Manuscripta Math.* **60** (1988), 423–436. Martinet gave a reformed theorem (see p. xxi):

Theorem 3.29′ *Let $\tilde{K}$ be an imaginary subfield of K. Let $\mu_{\tilde{K}}$ and μ_K be the groups of roots of unity in $\tilde{K}$ and K, respectively. If the index of $\mu_{\tilde{K}}$ in μ_K is odd, then the unit index $\tilde{Q}$ of $\tilde{K}/\tilde{K}_0$ divides the unit index Q of K/K_0.*

In the original book Hasse demonstrated two results by using Theorem 3.29: one is the statement in Sect. 3.15 (p. 136) that is written in italics, and the other is the former part of Theorem 3.41, both of which are revised in this translation.

[61](Translator's remark) The incorrectness of this theorem is due to this argument.

[62](Translator's remark) We will also reform this proof in footnotes 98 and 99 there.

where

$$K_1 = P(\sqrt{-f_1}), \quad K_2 = P(\sqrt{-f_2}) \tag{3.8.2}$$

are two distinct imaginary quadratic fields with discriminants $-f_1$, $-f_2$ (with conductors f_1, f_2). The maximal real subfield of K is the real quadratic field

$$K_0 = P(\sqrt{f_0}) \tag{3.8.3}$$

whose discriminant (conductor) f_0 is determined by

$$f_0 \underset{2}{=} f_1 f_2 \tag{3.8.4}$$

(equality up to a rational square factor) and by the well-known divisibility and the congruence conditions for the discriminants of quadratic fields.[63]

For these fields there exist the relative class number formulas closely related to the product formula of class numbers that was proved in the special case of $f_1 = 4$ of so-called Dirichlet biquadratic fields

$$K = K_0 P_4 = P(\sqrt{f_0}, \sqrt{-1}) \tag{3.8.5}$$

by Dirichlet [4][64] and in the general case first by Bachmann [3][65] and then by Amberg [1] as well as by Herglotz [8].

Let h, h_0, h_1, h_2 be the class numbers of K, K_0, K_1, K_2 and w, w_1, w_2 the numbers of roots of unity in K, K_1, K_2 (for K_0 the number of roots of unity is $w_0 = 2$). Further let ε, ε_0 always be the fundamental units of K, K_0, where we suppose that the unique normalization $\varepsilon_0 > 1$ is chosen for ε_0 as usual and we would demand that $|\varepsilon| > 1$ for ε in every case. Then the regulators of K, K_0 are $R = 2\log|\varepsilon|$, $R_0 = \log\varepsilon_0$ and the unit index of K/K_0 is given as

$$Q = \frac{2R_0}{R} = \frac{\log\varepsilon_0}{\log|\varepsilon|}$$

---

[63] For the discriminant d of a quadratic field, as for every algebraic number field, it holds first that $d \neq 1$ and $d \equiv 0$ or $1 \pmod 4$. Further d contains prime numbers $p \neq 2$ at most with first power and $p = 2$ at most with third power, and hence d is composed of the contributions of powers p^1 $(p \neq 2)$ and 2^2 or 2^3. Conversely every rational integer d with this property is the discriminant of a quadratic field, and exactly one quadratic field belongs to it, namely, the field $K = P(\sqrt{d})$ with conductor $f = |d|$ and with the generating character $\chi(x) = \left(\frac{d}{x}\right)$ (Kronecker's symbol).

[64] See to this also Zahlbericht [10], §**87**, as well as an arithmetic proof of Hilbert [9].

[65] Bachmann's proof is not based upon today's concept of algebraic integers. By letting Δ_0, Δ_1, and Δ_2 be the square-free kernels of f_0, f_1, and f_2, respectively, he instead considers as algebraic integers only the rational integral functions (forms) of $\sqrt{\Delta_0}$, $\sqrt{-\Delta_1}$ and $\sqrt{-\Delta_2}$ with coefficients of rational integers.

by virtue of (1.5.2). From the product formulas (1.2.1) for the zeta functions of $K,\ K_0,\ K_1,\ K_2$, one obtains the product formula

$$\frac{\zeta_K(s)}{\zeta(s)} = \frac{\zeta_{K_0}(s)}{\zeta(s)} \cdot \frac{\zeta_{K_1}(s)}{\zeta(s)} \cdot \frac{\zeta_{K_2}(s)}{\zeta(s)}$$

by the elimination of L-functions. According to the residue formulas (1.2.2) for $K,\ K_0,\ K_1,\ K_2$, it follows further from the product formula above that

$$\frac{(2\pi)^2 hR}{w\sqrt{d}} = \frac{2h_0R_0}{\sqrt{f_0}} \cdot \frac{2\pi h_1}{w_1\sqrt{f_1}} \cdot \frac{2\pi h_2}{w_2\sqrt{f_2}}.$$

By the conductor product formula (1.3.2), it holds now that

$$d = f_0\, f_1\, f_2.$$

As one can easily confirm, it holds further that

$$w = \frac{1}{2}\, w_1\, w_2,$$

except for the special field $K = P(\sqrt{-1}, \sqrt{-2}) = P_{23}$ with $f_1 = 2^2,\ f_2 = 2^3$ for which $w_1 = 4,\ w_2 = 2,\ w = 8$ and hence $w = w_1w_2$. Except for this special case there arises the product formula for the class numbers

$$h = \frac{1}{2}\, Q\, h_0\, h_1\, h_2. \tag{3.8.6}$$

By this formula the relative class number h^* of K/K_0 for the considered field K is expressed in the form[66]

$$h^* = \frac{1}{2}\, Q\, h_1\, h_2. \tag{3.8.7}$$

[66]For the real bicyclic biquadratic field $K = P(\sqrt{f_1}, \sqrt{f_2})$, one obtains the product formula

$$hR = h_0R_0 \cdot h_1R_1 \cdot h_2R_2$$

in the entirely corresponding way, and hence (one obtains) the formula for the product of class numbers

$$h = \frac{1}{4}\, Q\, h_0\, h_1\, h_2 \quad \text{with} \quad Q = \frac{R(\varepsilon_0, \varepsilon_1, \varepsilon_2)}{R},$$

which one also deduces easily by the application of result (2.10.3) of our second way of transformation on K and $K_0,\ K_1,\ K_2$. The quotient Q appearing here, analogous to our unit index, indicates how much higher the system of fundamental units $\varepsilon_0,\ \varepsilon_1,\ \varepsilon_2$ of quadratic subfields $K_0,\ K_1,\ K_2$ is than that of fundamental units of K. Then there arises the task to generalize this *unit index Q of K* to arbitrary non-cyclic fields and to develop it to a theory analogous to that of Sects. 3.4–3.7. For the fields considered here it is essentially only known by Amberg [1] that in every case $Q = 1,\ 2$ or 4.

In the mentioned special field $K = P(\sqrt{-1}, \sqrt{-2}) = P_{2^3}$, instead of the formulas above, it holds that $h = Q\, h_0\, h_1\, h_2$ and $h^* = Q\, h_1\, h_2$. In this case, as determined in Theorem 3.27, $Q = 1$ and hence $h = h_0 h_1 h_2$ and $h^* = h_1 h_2$. Since here $h_0,\ h_1,\ h_2 = 1$ are known, it also follows that $h = 1$ and $h^* = 1$. For simplicity we exclude in the following consideration this exceptional case, which is no longer interesting.

The various forms in which formula (3.8.6) for the product of class numbers appears in the literature are distinguished from our point of view by how exactly and in what way the unit index $Q = 1$ or 2 of K/K_0 is determined. In the following we will summarize all the statements on Q that arise by the general theory of the unit index developed in Sects. 3.2–3.7 for the special fields considered here.

The two cases for w to be distinguished by the general theory are determined here in the following way:

Case I. $w \not\equiv 0 \pmod 4$, when $f_1 \neq 4,\ f_2 \neq 4$.
Case II. $w \equiv 0 \pmod 4$, when, e.g., $f_1 = 4$.

In Case I, as up to the present, for the description of K we will turn from the three invariants $f_0,\ f_1,\ f_2$ of K connected by relation (3.8.4) to the last two invariants $f_1,\ f_2$ corresponding to the imaginary subfields $K_1,\ K_2$. Case II is the special one of Dirichlet biquadratic fields (3.8.5) already emphasized above; in this case, we describe K by the remaining unique independent invariant f_0, which corresponds to the real quadratic subfield K_0. By our assumption $K \neq P_{2^3}$, here $f_0 \neq 2^3$, and for the contribution 2^ω of 2 to w, it holds that $\omega = 2$. Instead of $f_0,\ f_1,\ f_2$ we use their quadratic kernels $\Delta_0,\ \Delta_1,\ \Delta_2$ for the description of K in both cases, which will be simpler.

According to the definition, $Q = 1$ if $\varepsilon_0 = \varepsilon$ holds by a suitable normalization of ε; and $Q = 2$ if $\varepsilon_0 = \zeta\varepsilon^2$ with a root of unity ζ in K. In the latter case ζ is not square in K, since otherwise the imaginary field K would be generated by adjoining the real number $\sqrt{\varepsilon_0}$ of K_0. Hence ζ can be normalized to $\zeta = \zeta_{2^\omega}$. Then, according to appropriate normalization of ε, one has by the definition of the unit index

$$\left\{\begin{array}{l} Q = 1, \text{ if } \varepsilon_0 = \varepsilon \\ \\ Q = 2, \text{ if } \left\{\begin{array}{ll} \varepsilon_0 = -\varepsilon^2 & \text{in Case I} \\ \\ \varepsilon_0 = \sqrt{-1}\,\varepsilon^2 & \text{in Case II} \end{array}\right\} \end{array}\right\}. \tag{3.8.8}$$

From Theorem 3.25, as already remarked there, it is deduced that

$$Q = 1, \quad \text{if} \quad N_0(\varepsilon_0) = -1. \tag{3.8.9}$$

In the following we treat the two Cases I and II separately:

Case I. $K = P(\sqrt{-\Delta_1}, \sqrt{-\Delta_2})$ **with** $\Delta_1 \neq 1,\ \Delta_2 \neq 1$

The conditions of characters in Theorem 3.22 imply that

$$Q = 1, \quad \text{if} \left\{ \begin{array}{l} \Delta_1 \text{ and } \Delta_2 \text{ have in common a prime divisor } p \neq 2 \\ \text{or the prime divisor 2, and in the latter case it} \\ \text{holds that } \Delta_0 \equiv 1 \pmod 4 \end{array} \right\}. \tag{3.8.10$_{\mathrm{I}}$}$$

If these conditions are not fulfilled, that is, if

$$\left\{ \begin{array}{l} \Delta_1 \text{ and } \Delta_2 \text{ do not have in common any prime divisor } p \neq 2, \\ \text{and in the case where they have in common the prime divisor 2} \\ \text{it holds that } \Delta_0 \equiv -1 \pmod 4 \end{array} \right\}, \tag{$\overline{3.8.10_{\mathrm{I}}}$}$$

then the auxiliary relative quadratic field K_0'/K_0, which is given by

$$K_0' = K_0(\sqrt{\Delta_1}) = K_0(\sqrt{\Delta_2}),$$

is at most unessentially ramified, and by Theorem 3.22 (or already by Theorem 3.16) $Q = 2$ or 1 according as it is of unit type or of class type. By definition K_0'/K_0 is of unit type exactly if [67]

$$\Delta_1 = \eta_0 \, \gamma_0^2$$

holds with a unit η_0 and a number γ_0 (an integer itself) of K_0. Therefore, by taking the norm, it follows that

$$\pm \, \Delta_1 = N_0(\gamma_0) \quad \text{with an integer } \gamma_0 \text{ of } K_0. \tag{3.8.11$_{\mathrm{I}}$}$$

One obtains at the same time that $N_0(\eta_0) = 1$ holds, as this must arise from (3.8.9) in this case. Conversely, if (3.8.11$_{\mathrm{I}}$) is fulfilled along with ($\overline{3.8.10_{\mathrm{I}}}$) (in this situation the integrality of γ_0 must be demanded emphatically), then K_0'/K_0 is of unit type. Namely, since under the assumption ($\overline{3.8.10_{\mathrm{I}}}$) K_0'/K_0 is at most unessentially ramified, one has $\Delta_1 \cong \mathfrak{d}_{10}^2$ with a (integral) divisor $\mathfrak{d}_{10}$ of K_0. By ($\overline{3.8.10_{\mathrm{I}}}$) this divisor is composed of only ramified prime divisors of K_0. Hence from (3.8.11$_{\mathrm{I}}$) it follows that $N_0\left(\frac{\gamma_0}{\mathfrak{d}_{10}}\right) = 1$, and then, as is well known, (it holds that) $\frac{\gamma_0}{\mathfrak{d}_{10}} = \frac{\mathfrak{a}_0}{\mathfrak{a}_0'}$ with a pair of conjugate integral divisors $\mathfrak{a}_0, \mathfrak{a}_0'$ of K_0 that can be supposed to be coprime each other without restriction and to be coprime to the ramified prime divisors of K_0. Since γ_0 and $\mathfrak{d}_{10}$ are integral, this holds only for $\mathfrak{a}_0, \mathfrak{a}_0' = 1$. Consequently it

[67](Translator's remark) Here "exactly if" means "if and only if."

follows that $\mathfrak{d}_{10} \cong \gamma_0$, and then, in fact, $\Delta_1 \cong \gamma_0^2$, that is, $\Delta_1 = \eta_0 \gamma_0^2$ with a unit η_0 of K_0. Thus the following has been proved:

$$\text{If } \overline{(3.8.10_{\rm I})} \text{ is fulfilled, then } \left\{ \begin{array}{ll} Q = 2, & \text{when } (3.8.11_{\rm I}) \text{ is fulfilled} \\ \\ Q = 1, & \text{when } (3.8.11_{\rm I}) \text{ is not fulfilled} \end{array} \right\}. \tag{3.8.12$_{\rm I}$}$$

The condition (3.8.11$_{\rm I}$) is expressed in the rational form by the solvability of

$$\frac{x^2 - \Delta_0 y^2}{4} = \pm 1$$

in the rational integers or, as easily seen, also of

$$\frac{\Delta_1 x_1^2 - \Delta_2 x_2^2}{4} = \pm 1;$$

in the latter form, the result of Amberg [1] is expressed by a minor modification feasible in consideration of (3.8.9).

For the solvability of (3.8.11$_{\rm I}$), necessary conditions can be given by the genus characters of K_0 in a well-known way, namely, by means of the quadratic norm residue symbols in the form

$$\left(\frac{\pm f_1,\ f_0}{p}\right) = \left(\frac{\pm f_1,\ \mp f_2}{p}\right) = 1 \quad \text{for all prime divisors } p \neq 2 \text{ of } f_0, \tag{3.8.13$_{\rm I}$}$$

or also by means of the quadratic residue symbols in the form

$$\left\{ \begin{array}{l} \left(\dfrac{\pm f_1}{p_2}\right) = 1 \text{ for all prime divisors } p_2 \neq 2 \text{ of } f_2 \\ \\ \left(\dfrac{\mp f_2}{p_1}\right) = 1 \text{ for all prime divisors } p_1 \neq 2 \text{ of } f_1 \end{array} \right\} \tag{3.8.13$'_{\rm I}$}$$

with two fixed opposite signs of f_1 and f_2 to each other (independently to the prime divisors $p,\ p_1,\ p_2$). If these conditions are fulfilled, then (3.8.11$_{\rm I}$) is in every case solvable for a number γ_0 of K_0, γ_0 being an integer or not. The remaining integrality condition in (3.8.11$_{\rm I}$) cannot be comprehensible in general by the congruence conditions. This is the non-abelian kernel of the class-embedding problem of K/K_0, which is here reduced to the problem of whether the divisor $\mathfrak{d}_{10}$ defined above belongs to the principal class of K_0 or not. Only when the principal genus of K_0 is composed of odd number of classes, that is, when the 2-class group of K_0 is of type $(2, 2, \ldots, 2)$, then the genus character conditions ($3.8.13_{\mathrm{I}}$) and ($3.8.13'_{\mathrm{I}}$) are also sufficient for the solvability of ($3.8.11_{\mathrm{I}}$) in rational integers. This is because, since $\mathfrak{d}_{10}^2 \cong \Delta_1$ is contained in the principal class, then—and also only then—the belonging of $\mathfrak{d}_{10}$ to the principal class is deduced from the belonging of $\mathfrak{d}_{10}$ to the principal genus. This special condition is particularly satisfied if K_0 has only a unique genus, that is, if the class number of K_0 is odd.

Case II. $K = P(\sqrt{\Delta_0}, \sqrt{-1})$ **with** $\Delta_0 \neq 2$

The conditions of characters in Theorem 3.22 imply that

$$Q = 1, \quad \text{if} \quad \Delta_0 \equiv 1 \pmod 4). \tag{$3.8.10_{\mathrm{II}}$}$$

The auxiliary relative quadratic field K'_0/K_0 is here given by

$$K'_0 = K_0(\sqrt{2}).$$

If $\Delta_0 \not\equiv 1 \pmod 4$, then it is at most unessentially ramified and hence it is exactly of unit type provided that

$$\pm 2 = N_0(\gamma_0) \quad \text{with an integer } \gamma_0 \text{ of } K_0 \tag{$3.8.11_{\mathrm{II}}$}$$

holds.[68] One can prove this (equation) entirely correspondingly to the above by replacing Δ_1 with 2. Thus one has the following:

$$\text{If} \quad \Delta_0 \not\equiv 1 \pmod 4, \ \text{then} \ \begin{Bmatrix} Q = 2, & \text{when } (3.8.11_{\mathrm{II}}) \text{ is fulfilled} \\ \\ Q = 1, & \text{when } (3.8.11_{\mathrm{II}}) \text{ is not fulfilled} \end{Bmatrix}. \tag{$3.8.12_{\mathrm{II}}$}$$

The condition ($3.8.11_{\mathrm{II}}$) is expressed in the rational form by the solvability of

$$\frac{x^2 - \Delta_0 y^2}{4} = \pm 2$$

in the rational integers. For this solvability it is necessary that the genus character conditions hold:

$$\left(\frac{\pm 2,\ f_0}{p}\right) = 1 \quad \text{for all prime divisors } p \neq 2 \text{ of } f_0, \tag{$3.8.13_{\mathrm{II}}$}$$

or also

$$\left(\frac{\pm 2}{p}\right) = 1 \quad \text{for all prime divisors } p \neq 2 \text{ of } f_0, \tag{$3.8.13'_{\mathrm{II}}$}$$

---

68 (Translator's remark) "$\pm 2 = N(\gamma_0)$" in the original book should be "$\pm 2 = N_0(\gamma_0)$."

with one fixed signs of 2 (independent of the prime divisors p). Only when the principal genus of K_0 has an odd number of classes, then particularly when K_0 has one genus, these conditions are also sufficient for the solvability of (3.8.11$_{\mathrm{II}}$) in rational integers.

As the first real quadratic field with the principal genus of even number of classes, one finds in the table of Sommer [23] the field

$$K_0 = P(\sqrt{82}),$$

and, in fact, for this field the prime divisors of 2 lie in the principal genus; however, (they lie) not in the principal class, but (they lie) in the class of order 2 of the cyclic class group of order 4. Therefore the two fields

$$K = P(\sqrt{-2}, \sqrt{-41}) \text{ (Case I) and } K = P(\sqrt{-1}, \sqrt{-82}) \text{ (Case II)}$$

present examples that the field $K_0' = K_0(\sqrt{2})$ is of class type and $Q = 1$, $Q^* = 2$ and $k = 2$.[69]

Note, moreover, that Hilbert [9][70] has stated the criterion for $Q = 1$ or 2 to Case II of Dirichlet biquadratic fields, the only case treated by him, in the following different form:

$$Q = 1 \text{ or } 2 \text{ according as } N_0(\varepsilon) = \pm 1 \text{ or } \pm\sqrt{-1}, \tag{3.8.14$_{\mathrm{II}}$}$$

where N_0 denotes the relative norm of $K = K_0(\sqrt{-1})$ with respect to $P(\sqrt{-1})$. The validity of (3.8.14$_{\mathrm{II}}$) arises immediately from condition (3.8.8) of the definition with attention to (3.8.9).

We still have not yet applied the sufficient criterion for $Q = 1$ of Theorem 3.23 and the sufficient criterion for $Q = 2$ of Theorem 3.26. The conductor f of K appearing in these criteria is the least common multiple of conductors f_0, f_1, and f_2 of K_0, K_1, and K_2, respectively. Within Theorem 3.23 there falls only the above-excluded field

$$K = P(\sqrt{-1}, \sqrt{2}) = P_{2^3},$$

for which $Q = 1$ is obtained by departing from (not using) the rule (3.8.12$_{\mathrm{II}}$), and therefore its role as the exception becomes hereby clear. Under both cases of

[69](Translator's remark) Recall that the relation $Q^* = kQ$ holds for any imaginary abelian field K, which is given in footnote 35, p. 84, in Sect. 3.4.

[70]See also Zahlbericht [10], §**87**.

Theorem 3.26, as easily assured, there fall only the two series of fields[71]

$$\left.\begin{cases} K = P(\sqrt{-q_1}, \sqrt{-q_2}) \text{ with prime numbers } q_1, q_2 \equiv -1 \pmod 4 \\ K = P(\sqrt{-1}, \sqrt{-q}),\ P(\sqrt{-1}, \sqrt{-2q}),\ P(\sqrt{-2}, \sqrt{-q}), \\ \qquad P(\sqrt{-2}, \sqrt{-2q}) \text{ with a prime number } q \equiv -1 \pmod 4 \end{cases}\right\} . \tag{3.8.15}$$

These are exactly the imaginary bicyclic biquadratic fields K for which the maximal real subfields K_0 have indeed composite conductor f_0 but further have one genus, so that for these (fields K_0) the necessary conditions ($3.8.13'_{\mathrm{I}}$) and ($3.8.13'_{\mathrm{II}}$) are trivially fulfilled and are at the same time sufficient for $Q = 2$. Then the criterion of Theorem 3.26 affords here, exceeding these trivial cases, no contribution to the decision on the class embedding problem.

3.9 Preparation for Direct Proof of the Integrality

Now we turn to the task of making clear the integrality of the relative class number h^* of K/K_0 directly from expression $(*)$. The reshaping of the contributions of individual characters χ_1 of K/K_0 demanded for this expression will at the same time provide useful forms of these contributions for the actual calculation of h^* in given cases.

To begin with, we give to expression $(*)$ a form from which its arithmetic structure emerges clearly. For this purpose we collect the characters $\chi = \chi_1$ of K/K_0 (as with the characters χ_0 of K_0 in Sect. 2.7) into Frobenius classes. Since the characters χ_0 of K_0 will not appear anymore, we drop the index (subscript) 1 from the characters χ_1 of K/K_0. In the following let ψ run always over a system of representatives of the Frobenius classes of χ, and let n_ψ denote the order of ψ; moreover, for short, let P_ψ be the field P_{n_ψ} to which ψ belongs, and N_ψ the norm map of this field. The formations n_ψ, P_ψ, and N_ψ as well as the conductor $f(\psi)$ are invariants of the classes $\chi = \psi^\mu$ where μ runs over a system of prime residues mod n_ψ. Moreover, give attention to the fact that the required property $\psi(-1) = -1$ is particular to the class of ψ, because by its existence the order n_ψ is even, so then all the μ are odd. The arithmetical meaning of the collection in a class lies in the fact that the characters of a class are composed of a full system of algebraic-conjugate characters, so that the product over such characters is merely the norm N_ψ of factors

[71] (Translator's remark) We will use this fact in the proof of Theorem 3.40.

(corresponding to ψ, for example). Hence the relative class number formula (∗) can be expressed in the form

$$h^* = Q\,w \prod_{\psi} N_{\psi}(\Theta(\psi)), \tag{3.9.1}$$

where the sums

$$\Theta(\psi) = \frac{1}{2f(\psi)} \sum_{x \bmod f(\psi)}^{+} (-\psi(x)x) \tag{3.9.2}$$

are numbers of the cyclotomic fields P_{ψ} with the apparent denominator $2f(\psi)$.

Moreover, these numbers $\Theta(\psi)$ can be decomposed additionally into two sums of numbers of P_{ψ} with the apparent denominators 2 and $f(\psi)$, respectively, by making use of the anti-symmetry $\psi(x) \to -\psi(x)$ for $x \to f(\psi) - x$ of the value-partition of ψ in the system of the least positive prime residues $x \bmod f(\psi)$. By means of the collection of the two anti-symmetry summands, one obtains indeed

$$\Theta(\psi) = \frac{1}{2}A(\psi) + \frac{1}{f(\psi)}B(\psi) \tag{3.9.3}$$

with

$$A(\psi) = \sum_{\pm x \bmod f(\psi)}^{+} \psi(x), \quad B(\psi) = \sum_{\pm x \bmod f(\psi)}^{+} (-\psi(x)x), \tag{3.9.4}$$

where $\sum^{+}_{\pm x \bmod f(\psi)}$, as previously defined in Sect. 1.4, denotes the summation over the half system of the least positive prime residues $x \bmod f(\psi)$ (the x prime to $f(\psi)$ with $0 < x < \frac{1}{2}f(\psi)$).

In the special case $f(\psi) = 2^{\rho}$ ($\rho \geqq 3$), the value-partition of ψ in the least positive prime residues $x \bmod 2^{\rho}$ has still a further anti-symmetry that leads to a further reduction of the sums $\Theta(\psi)$. The above-mentioned anti-symmetry originated from the always-existing solution $u \equiv -1 \pmod{f(\psi)}$ of the congruence $u^2 \equiv 1 \pmod{f(\psi)}$ in connection with the required property $\psi(-1) = -1$. For $f(\psi) = 2^{\rho}$ ($\rho \geqq 3$), the congruence $u^2 \equiv 1 \pmod{f(\psi)}$ has then the further solution $u \equiv 1 + 2^{\rho-1} \pmod{2^{\rho}}$, which generates along with the first-mentioned the full solutions of type (2, 2); and since ψ must have conductor $f(\psi) = 2^{\rho}$, it also holds necessarily that $\psi(1 + 2^{\rho-1}) = -1$. From this there arises the further anti-symmetry $\psi(x) \to -\psi(x)$ for $x \to x + 2^{\rho-1}$, which states along with the first-mentioned that, corresponding to the quadruple

$$x,\ 2^{\rho-1} - x,\ 2^{\rho-1} + x,\ 2^{\rho} - x,$$

there exist correlatively the character values

$$\psi(x),\ \psi(x),\ -\psi(x),\ -\psi(x).$$

If x runs over the half system of the least positive prime residues mod $2^{\rho-1}$, then these quadruples run just over the system of the least positive prime residues mod 2^{ρ} in its partition of the four quantities by the consecutive quarter system

$$0 \ldots 1\cdot 2^{\rho-2} \ldots 2\cdot 2^{\rho-2} \ldots 3\cdot 2^{\rho-2} \ldots 4\cdot 2^{\rho-2}.$$

If x runs over an arbitrary half system of the prime residues mod $2^{\rho-1}$, then the quadruples run over a corresponding system of the full prime residues mod 2^{ρ}. Therefore for the sum $\Theta(\psi)$ there arises the further reduction

$$\Theta(\psi) = \frac{1}{2^{\rho+1}} \sum_{\pm x \bmod 2^{\rho-1}}^{+} (-\psi(x)(x + (2^{\rho-1} - x) - (2^{\rho-1} + x) - (2^{\rho} - x))),$$

and hence simply

$$\Theta(\psi) = \frac{1}{2} \sum_{\pm x \bmod 2^{\rho-1}}^{+} \psi(x) \qquad \text{for } f(\psi) = 2^{\rho} \ (\rho \geqq 3). \tag{3.9.5}$$

Thus the apparent denominator $2f(\psi) = 2^{\rho+1}$ in (3.9.2) and (3.9.3) is reduced to 2^{1}.

The sums $A(\psi)$ and $B(\psi)$ in (3.9.4) are not purely algebraic constructions in so far as, in them, the fixed half system of prime residues mod $f(\psi)$ for the summation is required to be characterized by a quantity instruction that eventually goes back to the analytic origin for determination of the class number.[72] However, in the research of the integrality pursued here, one can now take no account of this quantity instruction. Namely, with transition to an arbitrary half system of the prime residues modulo $f(\psi)$ (with omission of the index (superscript) $+$ in the symbol of the sums), the congruences

$$A(\psi) \equiv \sum_{\pm x \bmod f(\psi)} \psi(x) \pmod 2, \quad B(\psi) \equiv \sum_{\pm x \bmod f(\psi)} (-\psi(x)x) \pmod{f(\psi)} \tag{3.9.4$'$}$$

hold obviously. For the determination of the actual denominator of $\Theta(\psi)$ and of the precise additive congruence value of $\Theta(\psi)$ mod$^{+}1$,[73] one may therefore replace the

[72] See the deduction of formula (1.4.1b).

[73] To this see Klassenkörperbericht [5], Part Ia, §**3**, and Hasse [7], §**24**, c).
(Translator's remark) The congruence $a \equiv b \pmod{{}^{+}1}$ means that $a = b + g$ with some integer g.

sums of expression (3.9.4) in (3.9.3) by the sums of (3.9.4′), which are congruent to them. We will use this replacement only in the case where the two apparent denominators 2 and $f(\psi)$ are relatively prime. In this case the actual denominator interesting to us is equal to the product of the actual denominators of $\frac{1}{2}A(\psi)$ and $\frac{1}{f(\psi)}B(\psi)$. Then these last denominators can be determined, as mentioned above, by taking the congruence values (3.9.4′) instead of the precise values (3.9.4). However, for the actual calculation of the relative class number, the congruence values (3.9.4′) are not sufficient; on the contrary, one has to determine the precise summation values (3.9.4) in every case.

3.10 The Characters with Composite Conductor

We start with an examination of the contributions[74] $N_\psi(\Theta(\psi))$ to the relative class number formulas (3.9.1) and (3.9.2) that arise from characters ψ with composite conductor.

To a proper decomposition $f = f_1 f_2$ into two relatively prime factors f_1, f_2, there corresponds a proper decomposition $\psi = \psi_1\psi_2$ into two characters ψ_1, ψ_2 such that $f(\psi_1) = f_1$, $f(\psi_2) = f_2$. Here the characters ψ_1, ψ_2 do not necessarily belong to the character group X of K. By $\psi(-1) = -1$ one may put, for example,

$$\psi_1(-1) = 1, \quad \psi_2(-1) = -1.$$

Now let x_1 and x_2 run over the systems of the least positive prime residues mod f_1 and mod f_2, respectively, and put

$$\frac{x}{f} = \frac{x_1}{f_1} + \frac{x_2}{f_2}; \tag{3.10.1}$$

then x runs over a system of the prime residues mod f, and the inequality $0 < x < 2f$ always holds for these numbers x. If one restricts x_1 and x_2 to the half systems of the least positive prime residues mod f_1 and mod f_2 and (if one) adds to them the complements $f_1 - x_1$ and $f_2 - x_2$, respectively, then for the corresponding system of the prime residues x mod f, there appears a decomposition into four quarter systems whose reductions to the least positive residues x_0, x_0', x_0'' and x_0''' mod f,

[74](Translator's remark) In the following, as with other cases, we call the norm $N_\psi(\Theta(\psi))$ the *contribution* (of ψ to the relative class number formula). For the values of $N_\psi(\Theta(\psi))$, see Table 3.1 in the Appendix.

respectively, by virtue of subtraction of the multiple $\varepsilon(x_1, x_2)f$ with $\varepsilon(x_1, x_2) = 0$ or 1: indeed (the decomposition occurs) by the scheme

$$\frac{x_0}{f} = \frac{x_1}{f_1} + \frac{x_2}{f_2} - 0,$$

$$\frac{x_0'}{f} = \frac{f_1 - x_1}{f_1} + \frac{x_2}{f_2} - \varepsilon(x_1, x_2),$$

$$\frac{x_0''}{f} = \frac{x_1}{f_1} + \frac{f_2 - x_2}{f_2} - (1 - \varepsilon(x_1, x_2)),$$

$$\frac{x_0'''}{f} = \frac{f_1 - x_1}{f_1} + \frac{f_2 - x_2}{f_2} - 1,$$

where[75]

$$\varepsilon(x_1, x_2) = 0 \text{ or } 1 \quad \text{according as } \frac{x_1}{f_1} > \frac{x_2}{f_2} \text{ or } \frac{x_1}{f_1} < \frac{x_2}{f_2}. \tag{3.10.2}$$

By

$$x \equiv f_2 x_1 \pmod{f_1}, \quad x \equiv f_1 x_2 \pmod{f_2},$$

one now has

$$\psi_1(x) = \psi_1(f_2)\psi_1(x_1), \quad \psi_2(x) = \psi_2(f_1)\psi_2(x_2).$$

By executing here the above-defined partition (of the prime residue system) into the system of quarters, it arises that, corresponding to the quadruple

$$x_0,\ x_0',\ x_0'',\ x_0''',$$

[75] (Translator's remark) By corresponding x_0, x_0', x_0'' and x_0''' to the points (x_1, x_2), $(f_1 - x_1, x_2)$, $(x_1, f_2 - x_2)$, and $(f_1 - x_1, f_2 - x_2)$ in the (x_1, x_2) plane, respectively, they lie as follows:

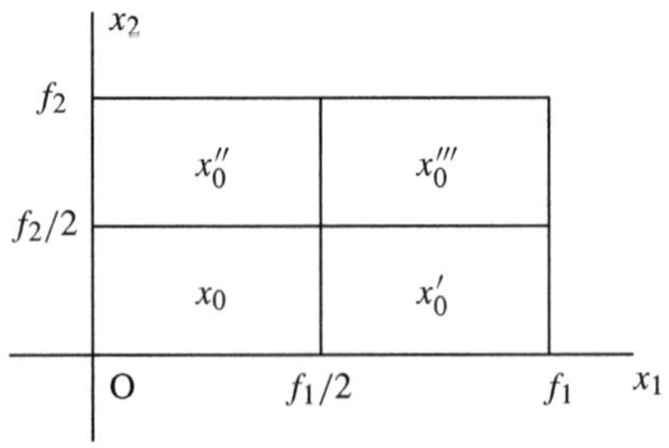

there exist correlatively the values of characters $\psi = \psi_1\psi_2$ with the relation

$$\psi(x_0),\ \psi(x_0),\ -\psi(x_0),\ -\psi(x_0),$$

which is entirely analogous to the partition of the values of ψ in the special case $f(\psi) = 2^\rho$ ($\rho \geqq 3$) considered in Sect. 3.9 on the system of quarters there, and it arises only that the system of quarters here are produced not in numerical order but at random. Therefore, as in Sect. 3.9, for the sum $\Theta(\psi)$ it holds here from (3.9.2) that the reduction is

$$\begin{aligned}\Theta(\psi) &= \frac{1}{2f}\sum_{x_0}(-\psi(x_0)(x_0 + x_0' - x_0'' - x_0''')) \\ &= \psi_1(f_2)\psi_2(f_1) \\ &\quad\times \sum_{\pm x_1 \bmod f_1}^{+} \sum_{\pm x_2 \bmod f_2}^{+} \left(-\psi_1(x_1)\psi_2(x_2)\left(\frac{2x_2}{f_2} - \varepsilon(x_1, x_2)\right)\right).\end{aligned}$$

By $\psi_1(-1) = 1$ and $\psi_1 \neq 1$, it holds here that

$$\sum_{\pm x_1 \bmod f_1} \psi_1(x_1) = \frac{1}{2}\sum_{x_1 \bmod f_1} \psi_1(x_1) = 0.$$

Hence the double-sum corresponding to the first summand in the parentheses has value 0, so that only the double-sum corresponding to the second summand in the parentheses remains. Consequently from the meaning in (3.10.2) of $\varepsilon(x_1, x_2)$, there arises the simple formula[76]:

$$\Theta(\psi) = \psi_1(f_2)\psi_2(f_1) \sum_{\substack{\pm x_1, \pm x_2 \bmod f_1, f_2 \\ \frac{x_1}{f_1} < \frac{x_2}{f_2}}}^{+} \psi_1(x_1)\psi_2(x_2) \qquad (3.10.3)$$

for $\psi = \psi_1\psi_2$ with relatively prime $f(\psi_1) = f_1 \neq 1$, $f(\psi_2) = f_2$ and $\psi_1(-1) = 1$, $\psi_2(-1) = -1$. Here the anti-symmetry of setting $\psi_1(-1) = 1$ and $\psi_2(-1) = -1$ corresponds to the anti-symmetry of the summation condition $\frac{x_1}{f_1} < \frac{x_2}{f_2}$. It does not especially need to be required that $f_2 \neq 1$ also holds, because it follows from the requirement that $\psi_2(-1) = -1$.

[76](Translator's remark) In summation (3.10.3) the pairs $(x_1,\ x_2)$ must satisfy the conditions

$$1 \leqq x_1 < \frac{f_1}{2},\quad 1 \leqq x_2 < \frac{f_2}{2} \quad\text{and}\quad \frac{x_1}{f_1} < \frac{x_2}{f_2}.$$

By means of formula (3.10.3) the integrality of $\Theta(\psi)$ has particularly been established in the considered case, and hence the following theorem has been proved:

Theorem 3.30 *If ψ is a character of K/K_0 with composite conductor $f(\psi)$, then $N_\psi(\Theta(\psi))$ is a rational integer.*

One should also have been able to prove this fact (Theorem 3.30) in a slightly simpler way through mere determination of congruence values of $\Theta(\psi)$ $\bmod^{+}1$ by composing a half system of the prime residues mod $f(\psi)$ of such a (half) system mod $f(\psi_1)$ and a full system of the prime residues mod $f(\psi_2)$ by virtue of (3.10.1) and then by determining the congruence values of the sums $A(\psi)$ mod 2 and $B(\psi)$ mod $f(\psi)$ by virtue of (3.9.4). Indeed, through (3.9.4′) on account of this decomposition (3.10.1) of a half system of prime residues x mod $f(\psi)$, one can read immediately the congruences $A(\psi) \equiv 0 \pmod 2$ and $B(\psi) \equiv 0 \pmod{f(\psi)}$, and hence by (3.9.3) one obtains the congruence $\Theta(\psi) \equiv 0 \ (\bmod^{+}1)$, that is, the integrality of $\Theta(\psi)$. If one takes again the reduction to the least positive residues by the above-mentioned decomposition (3.10.1) of the half system of prime residues mod $f(\psi)$, then for the precise value of $\Theta(\psi)$, in the place of (3.10.3) one has the following formula constructed similarly and valid under the same assumption[77]:

$$\Theta(\psi) = \psi_1(f_2)\psi_2(f_1) \sum_{\substack{\pm x_1, x_2 \bmod f_1, f_2 \\ \frac{x_1}{f_1}+\frac{x_2}{f_2}>1}} \psi_1(x_1)\psi_2(x_2) \qquad (3.10.3')$$

For the special case of the cyclotomic fields $K = P_f$ with composite conductor f, Kummer [19][78] has, essentially in this way, expressed (calculated) the contributions of characters with composite conductor to the relative class number formula, (which have) nevertheless two deflections.[79] He once summarized over the system of full prime residues and then also took as a basis the complete prime-power decomposition of f.

[77] (Translator's remark) In the summation the pairs (x_1, x_2) must satisfy the conditions

$$1 \leqq x_1 < \frac{f_1}{2}, \quad 1 \leqq x_2 < f_2 \quad \text{and} \quad \frac{x_1}{f_1} + \frac{x_2}{f_2} > 1.$$

Since $\sum_{\pm x_1 \bmod f_1} \psi_1(x_1) = 0$, the sum in (3.10.3′) takes the *same* value as that in the Martinet's Correction of the original book published in 1985 whose conditions are

$$1 \leqq x_1 < \frac{f_1}{2}, \quad 1 \leqq x_2 < \frac{f_2}{2} \quad \text{and} \quad \frac{x_1}{f_1} + \frac{x_2}{f_2} > 1.$$

[78] (Translator's remark) "Kummer [20]" in the original book should be replaced by "Kummer [19]" as in footnote 165, p. 188, in the Appendix.

[79] (Translator's remark) See footnotes 165, 166 and 167, p. 188, in the Appendix.

We have so far put at the forefront the slightly laborious formula (3.10.3) to be proved and (have) completely carried out its proof, because formula (3.10.3) has the advantage over (3.10.3′) of symmetry on the theoretical standpoint and also, for the actual calculation of relative class numbers, the advantage that in the sum the two summation quantities x_1 and x_2 are restricted to the *half system* of positive prime residues. Hereby in numerical cases the labor of calculation is saved. It is obvious that one has the considerable saving of the calculation work that formula (3.10.3) affords to the contrary to the original formulas (3.9.3) and (3.9.4); while one needs to arrange for (3.10.3) the value partitions of components ψ_1 and ψ_2 in the half systems of positive prime residues mod $f(\psi_1)$ and mod $f(\psi_2)$, one had to arrange for (3.9.3) and (3.9.4) the value partition of ψ itself in the half system of the least positive residues (which is) broader several times (as much as the former one).

By way of explanation we here deal with an example. Let $\psi_{105} = \psi_{21}\psi_5$ be the product of the quadratic character $\psi_{21} = \psi_3\psi_7$ with conductor 21 and one of the two biquadratic characters ψ_5 with conductor 5.[80] It holds that $\psi_{21}(-1) = \psi_3(-1)\psi_7(-1) = (-1)\cdot(-1) = 1$ and $\psi_5(-1) = -1$. The value partition of ψ_{21} and ψ_5 in the half system of the least positive prime residues x mod 21 and y mod 5 as well as the solutions of the inequality $\frac{x}{21} < \frac{y}{5}$, that is, $5x < 21y$, arise schematically as follows:

x	1	2	4	5	8	10
$\psi_3(x)$	1	−1	1	−1	−1	1
$\psi_7(x)$	1	1	1	−1	1	−1
$\psi_{21}(x)$	1	−1	1	1	−1	−1

Scheme 1a.

y	1	2
$\psi_5(y)$	1	i

Scheme 1b.

		x	1	2	4	5	8	10
		$5x$	5	10	20	25	40	50
y	$21y$							
1	21		1	1	1	0	0	0
2	42		1	1	1	1	1	0

Scheme 2.

[80](Translator's remark) The two odd characters with conductor 5 are χ_5 and its conjugate $\overline{\chi}_5$. Hence, $\psi_5 = \chi_5$ or $\psi_5 = \overline{\chi}_5$.

Here in the last scheme the values of $\varepsilon(x_1, x_2) = 0$ or 1 introduced in (3.9.2) are tabulated. Hence by (3.9.3) it holds that

$$\Theta(\psi_{105}) = \psi_{21}(5)\psi_5(21)$$
$$\times ((1 + \psi_{21}(2) + \psi_{21}(4)) + (1 + \psi_{21}(2) + \psi_{21}(4) + \psi_{21}(5) + \psi_{21}(8))\psi_5(2))$$
$$= 1 + i,$$

and hence

$$N(\Theta(\psi_{105})) = 2.$$

By the calculation in accordance with (3.9.3) and (3.9.4), one has to arrange, by simple selection of the solutions of the inequalities, the value partition of $\psi_{105} = \psi_3\psi_7\psi_5$ in the 24-term half system of the least positive prime residues mod 105. Certainly this is more laborious. Moreover, from the calculation of $\Theta(\psi_{105}) = \frac{1}{2}A(\psi_{105}) + \frac{1}{105}B(\psi_{105})$, there appear considerably larger numbers in the terms of the sum $B(\psi_{105})$.

3.11 Supplement of Gauss' Lemma

In order to draw a much more important conclusion beyond Theorem 3.30 from the formula (3.10.3), we prove first a lemma that belongs to the range of ideas regarding Gauss' Lemma on the theory of quadratic residue. To present a complete number-theoretic result, we express this lemma slightly more broadly than what it will actually be required later.

The question is (to evaluate) the number of solutions of the inequality $\frac{x_1}{f_1} < \frac{x_2}{f_2}$ with x_1 and x_2 in the half system of the least positive prime residues mod f_1 and mod f_2, respectively, appearing in Sect. 3.10, and actually (we treat this problem) under the special assumption that one of the two numbers f_1, f_2 is a prime number.[81]

Lemma *For a natural number $m \neq 1, 2$ and for a prime number $p \neq 2$ not appearing in m, let N be the number of solutions of $\frac{x}{m} < \frac{y}{p}$ with x and y in the half systems of the least positive prime residues* mod m *and* mod p*, respectively.*[82]

---

[81] *Addendum in proof* (1951): I am indebted to the friendly information of G. Beyer for the remark that this lemma follows from a theorem in Winogradow [26] by slight transformation.

[82] (Translator's remark) Here we treat the case where

$$0 < x < \frac{m}{2} \text{ with } (x, m) = 1, \quad y = 1, 2, \ldots, \frac{p-1}{2}.$$

If m is a composite, then N is even.
If $m = q^\rho$ is a power of prime q, then $(-1)^N = \left(\frac{q}{p}\right)$.

For our application we will need only the case where m is composite and moreover under the restriction that m (as the conductor of a character) is not of the form $2m_0$ with odd m_0. Under this assumption, in the following proofs of the five separated individual cases, three slightly complicated cases will not be used.[83]

Proof The integers x and y must always run over the half systems of the least positive prime residues mod m and mod p, respectively, and hence

$$0 < x < \frac{m}{2} \text{ with } (x, m) = 1, \quad y = 1, 2, \ldots, \frac{p-1}{2}.$$

The N solutions x, y to be considered are characterized by

$$px - my < 0.$$

For every such solution it holds that

$$0 < my - px < \frac{p}{2}m.$$

If one decomposes the interval $0 \ldots \frac{p}{2}m$ into the $\frac{p+1}{2}$ subintervals

$$\begin{array}{ccccc} 0 \ldots \frac{1}{2}m & \ldots \frac{3}{2}m & \ldots\ldots & \frac{p-2}{2}m & \ldots \frac{p}{2}m \\ 0 & 1 & \ldots\ldots & & \frac{p-1}{2} \end{array}$$

whose 0-th (sub-interval) has length $\frac{m}{2}$, while the first, ... , the $\frac{p-1}{2}$-th (sub-intervals) have length m, then for the number N there arises the decomposition

$$N = N_0 + N_1 + \cdots + N_{\frac{p-1}{2}},$$

where N_k denotes the number of the solutions x, y with $my - px$ being in the k-th interval.

An equation of the form

$$px = \pm x' + mz$$

[83](Translator's remark) See the application after Theorem 3.31.

with a rational integer z, where x' belongs again to the system of x,[84] and hence every reduction of multiple px of x (belongs) to its absolutely least residue $\pm x'$ mod m, we call, in short, a *reduction* and we name $\pm x'$ its *residue* and z its *multiplicator*. The last one (the multiplicator), as one immediately see, belongs to the system $z = 0,\ 1,\ \ldots,\ \frac{p-1}{2}$. Let R_z be the number of reductions with fixed multiplicator z, and moreover U the number of reductions with odd multiplicator.

First N_0 is the number of solutions $x,\ y$ of

$$-\frac{m}{2} < px - my < 0.$$

These solutions correspond uniquely to the reductions with negative residue $-x'$ because the multiplicator z is not equal to 0 by this last inequality. Further, for $k = 1,\ 2,\ \ldots,\ \frac{p-1}{2}$, N_k is the number of solutions $x,\ y$ of

$$-\frac{m}{2} - km < px - my < \frac{m}{2} - km.$$

Writing this (inequality) in the form

$$-\frac{m}{2} < px - m(y-k) < \frac{m}{2},$$

one sees that these solutions correspond uniquely to the reductions such that its multiplicator belongs to the system $z = 0,\ 1,\ \ldots,\ \frac{p-1}{2} - k$. One obtains these last reductions from the set of all $\frac{\varphi(m)}{2}$ reductions from which one omits the reductions with $z = \frac{p-1}{2} - (k-1), \ldots, \frac{p-1}{2}$. Consequently one has

$$N_k = \frac{\varphi(m)}{2} - \left(R_{\frac{p-1}{2}-(k-1)} + \cdots + R_{\frac{p-1}{2}}\right) \quad \left(k = 1,\ 2,\ \ldots,\ \frac{p-1}{2}\right).$$

Therefore, it follows that

$$\begin{aligned} N_1 + \cdots + N_{\frac{p-1}{2}} &= \frac{p-1}{2}\frac{\varphi(m)}{2} - \left(R_1 + 2R_2 + \cdots + \frac{p-1}{2}R_{\frac{p-1}{2}}\right) \\ &\equiv \frac{p-1}{2}\frac{\varphi(m)}{2} - \sum_u R_u \pmod 2, \end{aligned}$$

where u runs over the odd integers of the series $1,\ \ldots,\ \frac{p-1}{2}$, and hence

$$N_1 + \cdots + N_{\frac{p-1}{2}} \equiv \frac{p-1}{2}\frac{\varphi(m)}{2} + U \pmod 2.$$

[84](Translator's remark) This means that $1 \leqq x' < \frac{m}{2},\ (x',\ m) = 1$.

Thus for the number N in question there arises the congruence

$$N \equiv \frac{p-1}{2}\frac{\varphi(m)}{2} + N_0 + U \pmod 2, \tag{3.11.1}$$

where N_0 is the number of reductions (of px) with negative residue and U the number of reductions with odd multiplicator.

For the rest of this proof, we have to break down the proof into some cases according as m is odd or even and composite or prime-power.

First let m be *an odd integer*. The multiplication of all reductions (of px) provides by a well-known method the congruence

$$p^{\frac{1}{2}\varphi(m)} \equiv (-1)^{N_0} \pmod m, \tag{3.11.2$_1$}$$

while by $\sum_x z \equiv U \pmod 2$ the addition of all reductions for odd m provides the congruence

$$U \equiv 0 \pmod 2.$$

If, now, m is *composite*, then one has by Fermat's Little Theorem

$$p^{\frac{1}{2}\varphi(m)} \equiv 1 \pmod m,$$

and hence one has by (3.11.2$_1$)

$$N_0 \equiv 0 \pmod 2;$$

moreover, one has

$$\frac{\varphi(m)}{2} \equiv 0 \pmod 2.$$

Thus by (3.11.1) there arises

$$N \equiv 0 \pmod 2. \tag{3.11.3$_1$a}$$

If, however, m is *a prime-power*, $m = q^\rho$, then one has by Euler's criterion

$$p^{\frac{1}{2}\varphi(m)} \equiv \left(\frac{p}{q}\right) \pmod m,$$

and hence one has by (3.11.2$_1$)

$$(-1)^{N_0} = \left(\frac{p}{q}\right) \text{ (Gauss' Lemma);}$$

moreover, one has

$$\frac{\varphi(m)}{2} \equiv \frac{q-1}{2} \pmod 2.$$

Thus by (3.11.1) and by the quadratic reciprocity law, there arises

$$(-1)^N = (-1)^{\frac{p-1}{2}\frac{q-1}{2}} \left(\frac{p}{q}\right) = \left(\frac{q}{p}\right). \tag{3.11.3$_1$b}$$

Second let m be *an even integer*. Then the x are odd, and the reduction equation draws by itself the congruence

$$px \equiv \pm x'(1+m)^z \pmod{2m}.$$

The multiplication of all reductions then provides here the stronger congruence

$$p^{\frac{1}{2}\varphi(m)} \equiv (-1)^{N_0}(1+m)^U \pmod{2m}. \tag{3.11.2$_2$}$$

If, now, m is *composite*, but first not of the form $m = 2q^\rho$ (with prime number $q \neq 2$), then one has by Fermat's Little Theorem with a slight supplement regarding the 2-contribution to m,

$$p^{\frac{1}{2}\varphi(m)} \equiv 1 \pmod{2m},$$

and hence one has by (3.11.2$_2$)

$$N_0,\ U \equiv 0 \pmod 2;$$

moreover, one has

$$\frac{\varphi(m)}{2} \equiv 0 \pmod 2.$$

Thus by (3.11.1) there arises

$$N \equiv 0 \pmod 2. \tag{3.11.3$_2$a}$$

If *further* $m = 2q^\rho$, then by $p \equiv (-1)^{\frac{p-1}{2}} \pmod 4$, one has

$$p^{\frac{1}{2}\varphi(m)} \equiv (-1)^{\frac{p-1}{2}\frac{q-1}{2}} \pmod 4,$$

and hence one has by (3.11.2$_2$)

$$N_0 + U \equiv \frac{p-1}{2}\frac{q-1}{2} \pmod 2;$$

moreover, one has

$$\frac{\varphi(m)}{2} \equiv \frac{q-1}{2} \pmod 2.$$

Thus by (3.11.1) there arises again

$$N \equiv 0 \pmod 2. \tag{3.11.3$_2$a$'$}$$

If, finally, m is *a prime-power*, then here $m = 2^\rho$ $(\rho \geqq 2)$, and by setting $\left(\frac{2}{p}\right) = (-1)^\delta$, one has, by virtue of the basis expression of p mod $2^{\rho+1}$,

$$p^{\frac{1}{2}\varphi(m)} \equiv (-1)^{\frac{p-1}{2}\frac{\varphi(m)}{2}}(1+m)^\delta \pmod{2m},$$

and hence one has by (3.11.2$_2$)

$$N_0 \equiv \frac{p-1}{2}\frac{\varphi(m)}{2}, \quad U \equiv \delta \pmod 2.$$

Thus by (3.11.1) there arises

$$(-1)^N = \left(\frac{2}{p}\right). \tag{3.11.3$_2$b}$$

By the establishments (3.11.3$_1$a), (3.11.3$_1$b), (3.11.3$_2$a), (3.11.3$_2$a$'$), and (3.11.3$_2$b), the lemma has been proved in every of the five separated cases. □

3.12 The Characters of 2-Power Order and with Composite Conductor

Here we obtain by the lemma proved in Sect. 3.11 the already-announced result that surpasses Theorem 3.30 from formula (3.10.3). It is relevant to the divisibility of $N_\psi(\Theta(\psi))$ by 2 for characters ψ of 2-power order n_ψ and with composite conductor $f(\psi)$. Just by this result we will succeed in Sect. 3.15 in eliminating the still-unsettled possibility of Kummer [20], i.e., the power 2^{r-2} in the denominator of h^*.[85]

[85] See footnote 22, p. 18, in Sect. 1.6.

If $n_\psi = 2^\mu$, then $\psi(x) \equiv 1 \pmod{\mathfrak{z}}$ holds for all the values of characters ψ, where $\mathfrak{z}$ is the unique prime divisor of 2 of degree 1 in the field $P_\psi = P_{2^\mu}$. If, moreover, $f(\psi)$ is composite, then from (3.10.3) it follows that the congruence

$$\Theta(\psi) \equiv N(f_1, f_2) \pmod{\mathfrak{z}}$$

holds, where $f = f_1 f_2$ is some proper decomposition of $f = f(\psi)$ and $N(f_1, f_2)$ denotes the number of solutions of $\frac{x_1}{f_1} < \frac{x_2}{f_2}$ with x_1 and x_2 being in the half systems of the least positive prime residues mod f_1 and mod f_2, respectively. By taking the norm, there arises moreover

$$N_\psi(\Theta(\psi)) \equiv N(f_1, f_2) \pmod{2}. \tag{3.12.1}$$

Here it does not matter whether the order of (writing) $f_1 = f(\psi_1)$ and $f_2 = f(\psi_2)$, with $\psi_1(-1) = 1$ and $\psi_2(-1) = -1$ assigned by virtue of (3.10.3), is selected or the reverse order (is selected), because, in fact, the double sum running over the complete half system of prime residues has value 0 and then the double sum with the restrictive inequality $\frac{x_1}{f_1} < \frac{x_2}{f_2}$ alters only the sign by changing the order of f_1 and f_2, which is insignificant for the congruence mod 2. One adds (here the assumption) that the composite conductor $f(\psi)$ contains at least a prime number $p \neq 2$ and that, since the order n_ψ is a power of 2, this (prime number p) appears in $f(\psi)$ only to the first power,[86] then here one can apply the decomposition $f = f_1 f_2$ underlying in (3.10.3) in the form $f = mp$ with a prime number $p \neq 2$. Then, according to the lemma in Sect. 3.11, the conclusion follows from (3.12.1):

Theorem 3.31 *Let ψ be a character of K/K_0 whose order n_ψ is 2-power and whose conductor $f(\psi)$ is composite.*

If at least three distinct prime numbers appear in $f(\psi)$, then

$$N_\psi(\Theta(\psi)) \equiv 0 \pmod{2}.$$

If, however, only two distinct prime numbers p, q appear in $f(\psi)$ in which, say, $p \neq 2$, then

$$N_\psi(\Theta(\psi)) \equiv 0 \text{ or } 1 \pmod{2} \quad \text{according as} \quad \left(\frac{q}{p}\right) = 1 \text{ or } -1.$$

For our application we will need only the first case where at least three distinct prime numbers appear in $f(\psi)$ with the restriction that there exists a proper decomposition $\psi = \psi_1\psi_2\psi_3$ with pairwise relatively prime conductors $f(\psi_1) = f_1$, $f(\psi_2) = f_2$ and $f(\psi_3) = f_3$, where all the values $\psi_1(-1)$, $\psi_2(-1)$ and $\psi_3(-1)$

[86](Translator's remark) This means that $p \mid f(\psi)$ and $p^2 \nmid f(\psi)$.

are equal to -1.[87] One could try to prove the assertion $N_\psi(\Theta(\psi)) \equiv 0 \pmod 2$ in this case in a different—probably simpler—way in which, namely, one applies from the beginning the method used for the deduction of formula (3.10.3) to a three-term decomposition $\psi = \psi_1\psi_2\psi_3$. This method is quite feasible. In this case under the above-mentioned restriction for the decomposition, one obtains the formula[88]

$$\Theta(\psi) = -\psi_1(f_2 f_3)\psi_2(f_1 f_3)\psi_3(f_1 f_2) \sum_{(x_1,\, x_2,\, x_3)\,\text{in}\,\mathfrak{T}} \psi_1(x_1)\psi_2(x_2)\psi_3(x_3), \tag{3.12.2}$$

where the summation region $\mathfrak{T}$ is given by the following four inequalities for $\xi_1 = \frac{x_1}{f_1}$, $\xi_2 = \frac{x_2}{f_2}$, $\xi_3 = \frac{x_3}{f_3}$:

$$\xi_1 + \xi_2 + \xi_3 < 1,$$

$$-\xi_1 + \xi_2 + \xi_3 > 0,$$

$$\xi_1 - \xi_2 + \xi_3 > 0,$$

$$\xi_1 + \xi_2 - \xi_3 > 0,$$

[87](Translator's remark) We will also use the latter case of Theorem 3.31 in later applications. See Theorem 3.39.

[88](Translator's remark) Under this restriction we also have a formula analogous to the one in Sect. 3.10:

$$\Theta(\psi) = \psi_1(f_2 f_3)\psi_2(f_3 f_1)\psi_3(f_1 f_2)$$
$$\times \sum_{\pm x_1 \bmod f_1}^{+} \sum_{\pm x_2 \bmod f_2}^{+} \sum_{\pm x_3 \bmod f_3}^{+} \psi_1(x_1)\psi_2(x_2)\psi_3(x_3)(2 + \varepsilon - \eta_1 - \eta_2 - \eta_3),$$

where

$$\varepsilon = \varepsilon(x_1,\, x_2,\, x_3) = \begin{cases} 0, & \text{if } \frac{x_1}{f_1} + \frac{x_2}{f_2} + \frac{x_3}{f_3} < 1, \\ 1, & \text{if } \frac{x_1}{f_1} + \frac{x_2}{f_2} + \frac{x_3}{f_3} > 1, \end{cases}$$
$$\eta_1 = \eta_1(x_1,\, x_2,\, x_3) = \begin{cases} 0, & \text{if } -\frac{x_1}{f_1} + \frac{x_2}{f_2} + \frac{x_3}{f_3} < 0, \\ 1, & \text{if } -\frac{x_1}{f_1} + \frac{x_2}{f_2} + \frac{x_3}{f_3} > 0, \end{cases}$$
$$\eta_2 = \eta_2(x_1,\, x_2,\, x_3) = \begin{cases} 0, & \text{if } \frac{x_1}{f_1} - \frac{x_2}{f_2} + \frac{x_3}{f_3} < 0, \\ 1, & \text{if } \frac{x_1}{f_1} - \frac{x_2}{f_2} + \frac{x_3}{f_3} > 0, \end{cases}$$
$$\eta_3 = \eta_3(x_1,\, x_2,\, x_3) = \begin{cases} 0, & \text{if } \frac{x_1}{f_1} + \frac{x_2}{f_2} - \frac{x_3}{f_3} < 0, \\ 1, & \text{if } \frac{x_1}{f_1} + \frac{x_2}{f_2} - \frac{x_3}{f_3} > 0. \end{cases}$$

or, in short,[89]

$$2\,\mathrm{Max}\{\xi_i\} < \sum_{i=1}^{3} \xi_i < 1. \tag{3.12.3}$$

These inequalities define in the space of the orthogonal coordinate system with coordinates (x_1, x_2, x_3) the interior of the tetrahedron $\mathfrak{T}$ whose vertices are the origin and the three non-adjacent and non-opposite vertex points of the rectangular parallelepiped with surfaces parallel to the axes in the first octant with sides of length $\frac{f_1}{2}, \frac{f_2}{2}, \frac{f_3}{2}$. Since $\mathfrak{T}$ is contained in this rectangle, the restriction of x_1, x_2, x_3 to the half systems of the least positive prime residues modulo f_1, f_2, f_3 does not need especially to be indicated here. The summation is taken over all the lattices in the interior of $\mathfrak{T}$, in which only the coordinates x_1, x_2, x_3 (that are) relatively prime to f_1, f_2, f_3, respectively, and (that are) distinct from the origin make contributions to the sum.

In the case of $n_\psi = 2^\mu$, it follows from (3.12.2) that the congruence analogous to (3.12.2) is

$$N_\psi(\Theta(\psi)) \equiv N(f_1, f_2, f_3) \pmod 2, \tag{3.12.4}$$

where $N(f_1, f_2, f_3)$ denotes the number of the lattices in $\mathfrak{T}$ with coordinates x_1, x_2, x_3 relatively prime to f_1, f_2, f_3, respectively. For the proof of a partial assertion of Theorem 3.31 (which is) necessary for our later application, one would have to show that this number is even. For this I have not yet succeeded in giving a direct proof similar to that of the lemma in Sect. 3.11. The fact that $N(f_1, f_2, f_3)$ is even emerges really from congruence (3.12.4) in comparison with the result in Theorem 3.31.[90]

Formula (3.12.2) can be used with advantage for numerical calculation of the contributions $N_\psi(\Theta(\psi))$ to the relative class number in the case of more-multiple composite conductor $f(\psi)$ to make further the reduction already achieved by (3.10.3) to small values. Indeed in the example of $\psi_{105} = \psi_3\psi_7\psi_5$ treated at the end of Sect. 3.10, the conditions of restricting $\psi_1(-1), \psi_2(-1), \psi_3(-1) = -1$ are satisfied. For the lattice points (x_1, x_2, x_3) to be considered (Scheme 1), the (double figures) decimal fraction expansions of ξ_1, ξ_2, ξ_3 (Scheme 2) show that exactly the four lattice points indicated by * satisfy inequality (3.12.3), that is, (they) lie in $\mathfrak{T}$.

---

89 (Translator's remark) In the original this is printed as

$$2\,\mathrm{Max}\,\xi_i < \sum \xi_i < 1 \quad (i = 1, 2, 3).$$

90 Here the above-mentioned restriction on the values $\psi_1(-1), \psi_2(-1), \psi_3(-1)$ must be placed. But numerical examples seem to confirm that $N(f_1, f_2, f_3)$ is also even without this restriction and then (the evenness holds) for all pairwise relatively prime numbers $f_1, f_2, f_3 \neq 1, 2$. For this, see the problem provided by me [6].

3	7	5	3	7	5	Σ	𝔗
1	1	1	<u>0.33...</u>	0.14...	0.20...	0.67...	*
1	1	2	0.33...	0.14...	<u>0.40...</u>	0.87...	*
1	2	1	<u>0.33...</u>	0.28...	0.20...	0.81...	*
1	2	2	0.33...	0.28...	<u>0.40...</u>	1.01...	
1	3	1	0.33...	<u>0.42...</u>	0.20...	0.95...	*
1	3	2	0.33...	<u>0.42...</u>	0.40...	1.15...	

Scheme 1. Scheme 2 (Max. underlined.).

Thus there arises

$$\Theta(\psi_{105}) = -\psi_3(35)\psi_7(15)\psi_5(21)(1 + \psi_5(2) + \psi_7(2) + \psi_7(3)) = 1 + i,$$

and hence

$$N(\Theta(\psi_{105})) = 2,$$

as already found in Sect. 3.10.

3.13 The Characters of Odd Prime-Power Conductor

We consider further the contributions $N_\psi(\Theta(\psi))$ to the relative class number formulas (3.9.1) and (3.9.2) that come from characters ψ with odd prime-power conductor $f(\psi) = p^\rho$ $(p \neq 2)$. To determine the exact denominators for them, we calculate by (3.9.4′) the congruence values $A(\psi)$ mod 2 and $B(\psi)$ mod p^ρ, and hence we obtain a statement of the congruence value $\Theta(\psi)$ mod$^+1$ by (3.9.3) and by the remark at the end of Sect. 3.9.

Let g be a primitive root mod p with normalization $g^{p-1} \equiv 1 \pmod{p^\rho}$. Then the group of prime residues mod p^ρ has the basis expression[91]

$$x \equiv g^\alpha(1+p)^\beta \pmod{p^\rho} \quad (\alpha \bmod p-1, \;\; \beta \bmod p^{\rho-1}). \tag{3.13.1}$$

Correspondingly a character ψ with conductor $f(\psi) = p^\rho$ is given by

$$\psi(x) = \zeta^{-\alpha} Z^{-\beta}, \tag{3.13.2}$$

[91] (Translator's remark) The original expression (3.13.1) does not have "(mod p^ρ)."

where ζ is a primitive m-th root of unity with $m|(p-1)$ and Z a primitive $p^{\rho-1}$-th root of unity.[92] Then the order of ψ is $n_\psi = mp^{\rho-1}$. By $\psi(-1) = -1$, m is a divisor of $p-1$ such that $\frac{p-1}{m}$ is odd and that therefore the highest power of 2 dividing $p-1$ divides also m, because the congruence $-1 \equiv g^{\frac{p-1}{2}} \pmod{p^\rho}$ holds and hence $\psi(-1) = \zeta^{-\frac{p-1}{2}} = (-1)^{\frac{p-1}{m}}$. The field P_ψ is composed of the relatively prime cyclotomic fields P_m and $P_{p^{\rho-1}}$ of degree $\varphi(m)$ and $\varphi(p^{\rho-1})$, respectively.

One obtains a half system of the prime residues mod p^ρ if in the basis expression (3.13.1) one restricts to α mod $\frac{p-1}{2}$ while β mod $p^{\rho-1}$ remains. Hence the congruence values of sums $A(\psi)$ mod 2 and $B(\psi)$ mod p^ρ can be calculated by (3.9.4′) as follows:

Calculation of $A(\psi)$ mod 2

By (3.13.2) it holds that

$$A(\psi) \equiv \sum_{\pm x \bmod p^\rho} \psi(x) \equiv \sum_{\alpha \bmod \frac{p-1}{2}} \zeta^{-\alpha} \sum_{\beta \bmod p^{\rho-1}} Z^{-\beta} \pmod{2}. \tag{3.13.3}$$

Herein

$$\sum_{\alpha \bmod \frac{p-1}{2}}^{+} \zeta^{-\alpha} = \frac{\zeta^{-\frac{p-1}{2}} - 1}{\zeta^{-1} - 1} = -\zeta \frac{2}{1-\zeta}. \tag{3.13.4}$$

If now $m \neq 2^\mu$, then m contains, other than the contribution 2^μ, still an odd component (divisor) $\neq 1$ and so $1-\zeta$ is a unit, and then it follows from (3.13.4) that

$$\sum_{\alpha \bmod \frac{p-1}{2}} \zeta^{-\alpha} \equiv 0 \pmod{2} \quad \text{for } m \neq 2^\mu. \tag{3.13.4a}$$

If, however, $m = 2^\mu$, then $1-\zeta$ represents the unique prime divisor of 2 of degree 1 in the field $P_m = P_{2^\mu}$, and hence it follows from (3.13.4) that

$$\sum_{\alpha \bmod \frac{p-1}{2}}{}' \zeta^{-\alpha} \equiv 0 \left(\bmod \frac{2}{\mathfrak{l}}\right), \quad \text{but not} \pmod{2} \quad \text{for } m = 2^\mu. \tag{3.13.4b}$$

[92] The minus sign in the exponent will prove itself to be convenient for the later calculation. See to this the remark in Sect. 1.5, footnote 17, p. 15.

Furthermore

$$\sum_{\beta \bmod p^{\rho-1}} Z^{-\beta} = \left\{ \begin{array}{l} 1 \text{ for } \rho = 1 \\ \\ 0 \text{ for } \rho \geqq 2 \end{array} \right\}. \tag{3.13.5}$$

Bringing (3.13.4a), (3.13.4b), and (3.13.5) into (3.13.3), there arises

$$A(\psi) \equiv 0 \left\{ \begin{array}{ll} (\text{mod } 2) & \text{for } n_\psi \neq 2^\mu \\ \\ \left(\text{mod } \frac{2}{3}\right), \text{ but not (mod 2)} & \text{for } n_\psi = 2^\mu \end{array} \right\}. \tag{3.13.6}$$

Notice here that $\rho = 1$ holds necessarily for $n_\psi = 2^\mu$. From (3.13.6) it follows, by dividing by 2, that

$$\frac{1}{2}A(\psi) \equiv 0 \left\{ \begin{array}{ll} (\text{mod}^+ 1) & \text{for } n_\psi \neq 2^\mu \\ \\ \left(\text{mod}^+ \frac{1}{3}\right), \text{ but not } (\text{mod}^+ 1) & \text{for } n_\psi = 2^\mu \end{array} \right\}. \tag{3.13.7}$$

Calculation of $B(\psi)$ mod p^ρ

By (3.13.2) it holds that

$$B(\psi) \equiv \sum_{\pm x \bmod p^\rho} (-\psi(x)x) \equiv - \sum_{\alpha \bmod \frac{p-1}{2}} \zeta^{-\alpha} g^\alpha \sum_{\beta \bmod p^{\rho-1}} Z^{-\beta}(1+p)^\beta \pmod{p^\rho}. \tag{3.13.8}$$

Herein[93]

$$- \sum_{\alpha \bmod \frac{p-1}{2}}^{+} \zeta^{-\alpha} g^\alpha = - \frac{(\zeta^{-1}g)^{\frac{p-1}{2}} - 1}{\zeta^{-1}g - 1} = \zeta \, \frac{g^{\frac{p-1}{2}} + 1}{g - \zeta} = \zeta \prod_u (g - \zeta^u), \tag{3.13.9}$$

[93](Translator's remark) The last equality of (3.13.9) holds surely for $m = p - 1$ but not in general. For $p = 7$ and $m = 2$, this equality does not hold. As we will see just below, we will use this equality only in the case of $m = p - 1$.

where u runs over the odd residues $\not\equiv 1 \pmod{p-1}$. If, now, $m \neq p-1$, then $g-\zeta$ is relatively prime to p, while $g^{\frac{p-1}{2}}+1 \equiv 0 \pmod{p^\rho}$ holds. Consequently it follows from (3.13.9) that

$$-\sum_{\alpha \bmod \frac{p-1}{2}} \zeta^{-\alpha} g^{\alpha} \equiv 0 \pmod{p^\rho} \quad \text{for } m \neq p-1. \tag{3.13.9a}$$

If, however, $m = p-1$, then $g-\zeta$ is divisible by the ρ-th power of $\mathfrak{p}$, $\mathfrak{p}$ being one of the $\varphi(p-1)$ prime divisors of p of degree 1 in the field $P_m = P_{p-1}$, but $g-\zeta$ is relatively prime to $\varphi(p-1)-1$ conjugate prime divisors $\mathfrak{p}', \mathfrak{p}'', \ldots \neq \mathfrak{p}$. As $g-\zeta$ corresponds to $\mathfrak{p}$, the $\varphi(p-1)-1$ linear factors $g-\zeta^u$, with u being relatively prime to $p-1$ and $u \not\equiv 1 \pmod{p-1}$, correspond to conjugate prime divisors $\mathfrak{p}', \mathfrak{p}'', \ldots \neq \mathfrak{p}$, respectively, while the remaining linear factors $g-\zeta^u$ (with odd u not relatively prime to $p-1$) are relatively prime to p. Consequently it follows from (3.13.9) that

$$-\sum_{\alpha \bmod \frac{p-1}{2}} \zeta^{-\alpha} g^{\alpha} \equiv 0 \pmod{\mathfrak{p}'^{\rho}\mathfrak{p}''^{\rho}\ldots} \text{ but relatively prime to } \mathfrak{p} \quad \text{for } m = p-1. \tag{3.13.9b}$$

In the special case of $p=3$, it holds necessarily by $\psi(-1) = -1$ that $m = p-1 = 2$. Then $P_{p-1} = P$ and hence $\mathfrak{p} = p$ is the unique prime divisor of p; therefore (3.13.9b) states that

$$-\sum_{\alpha \bmod \frac{p-1}{2}} \zeta^{-\alpha} g^{\alpha} \quad \text{is relatively prime to } p \qquad \text{for } p = 3. \tag{3.13.9b$'$}$$

This is clear without calculation, because the sum can be reduced to a unique term 1.

For $p \neq 3$, on the contrary, p has more prime divisors $\mathfrak{p}, \mathfrak{p}', \ldots$ in P_{p-1}, and the additional statement obtained in (3.13.9b) that the considered sum is relatively prime to $\mathfrak{p}$ will be useless to us, because from the statement it cannot be concluded that the norm of the sum on the left-hand side is divisible only by the norm of $\mathfrak{p}'^{\rho}\mathfrak{p}''^{\rho}\ldots$ and then only by the power $p^{\rho(\varphi(p-1)-1)}$.

Furthermore for $\rho \geq 2$ one has

$$\sum_{\beta \bmod p^{\rho-1}}^{+} Z^{-\beta}(1+p)^{\beta} = \frac{(Z^{-1}(1+p))^{p^{\rho-1}}-1}{Z^{-1}(1+p)-1} = Z\frac{(1+p)^{p^{\rho-1}}-1}{1+p-Z}. \tag{3.13.10}$$

The numerator of the last expression (fraction) is divisible exactly by p^ρ and the denominator exactly by the unique prime divisor $\mathfrak{P}$ of p in the field $P_{p^{\rho-1}}$ generated by $1 - Z$ of degree 1. Hence it follows from (3.13.10) that

$$\sum_{\beta \bmod p^{\rho-1}} Z^{-\beta}(1+p)^\beta \equiv 0 \left(\bmod \frac{p^\rho}{\mathfrak{P}}\right), \quad \text{but not } (\bmod\ p^\rho), \qquad (3.13.11)$$

first for $\rho \geqq 2$ and also trivially for $\rho = 1$ where the sum (in (3.13.11)) can be reduced to the unique term 1 and $\mathfrak{P} = p$.

By bringing (3.13.9a), (3.13.9b), (3.13.9b′), and (3.13.11) into (3.13.8), there arises

$$B(\psi) \equiv 0 \left\{ \begin{array}{ll} (\bmod\ p^\rho) & \text{for } n_\psi \neq \varphi(p^\rho) \\ \left(\bmod \frac{p^\rho}{\mathfrak{P}^*}\right), \text{ but for } p = 3 \text{ not } (\bmod\ p^\rho) & \text{for } n_\psi = \varphi(p^\rho) \end{array} \right\}, \qquad (3.13.12)$$

where $\mathfrak{P}^*$ is the unique prime divisor of $\mathfrak{p}$ in the (composite) field $P_{p-1}P_{p^{\rho-1}}$ of degree 1.[94] From (3.13.12) it follows by dividing by p^ρ that

$$\frac{1}{p^\rho} B(\psi) \equiv 0 \left\{ \begin{array}{ll} (\bmod^+ 1) & \text{for } n_\psi \neq \varphi(p^\rho) \\ \left(\bmod^+ \frac{1}{\mathfrak{P}^*}\right), \text{ but for } p = 3 \text{ not } (\bmod^+ 1) & \text{for } n_\psi = \varphi(p^\rho) \end{array} \right\}. \qquad (3.13.13)$$

Through (3.9.3) the results (3.13.7) and (3.13.13) provide, by addition, a statement on the congruence value $\Theta(\psi) \bmod^+ 1$ and hence a statement on the denominator of $N_\psi(\Theta(\psi))$. Since in each case at most *one* prime divisor $\mathfrak{z}$ of 2 and *one* prime divisor $\mathfrak{P}^*$ of p appear in this denominator, one obtains by taking the norm the following facts on the denominator of $N_\psi(\Theta(\psi))$:

Theorem 3.32 *Let ψ be a character of K/K_0 with odd prime-power conductor $f(\psi) = p^\rho$ $(p \neq 2)$. Then $N_\psi(\Theta(\psi))$ is a rational number with denominator at most $2p$.*

---

[94] Notice that one could not be allowed to conclude $B(\psi) \equiv 0 \ (\bmod\ p^\rho \frac{p^\rho}{\mathfrak{P}})$ or $(\bmod\ \mathfrak{p}'^\rho \mathfrak{p}''^\rho \ldots \frac{p^\rho}{\mathfrak{P}})$ (according as $m \neq p-1$ or $m = p-1$) merely through the product process of modules. Namely, although the product of evaluated sums (3.13.9a), (3.13.9b), and (3.13.11) is only the congruence value $B(\psi) \bmod p^\rho$, one has rather to take the greatest common divisor of p^ρ and the divisor modules generated by the product process; this (the greatest common divisor) has, corresponding to both cases (3.13.9a) and (3.13.9b), the values p^ρ or $\frac{p^\rho}{\mathfrak{P}^*}$ in the text (according as $m \neq p-1$ or $m = p-1$).

The prime factor 2 of the denominator really appears exactly when (if and only if) ψ has 2-power order, $n_\psi = 2^\mu$; in this case, it holds that $\rho = 1$ and that 2^μ is the highest 2-power appearing in $p - 1$.

The prime factor p of the denominator at most appears (only) when ψ has order $n_\psi = \varphi(p^\rho)$; for $p = 3$ the factor really appears.

3.14 The Characters of 2-Power Conductor

We finally consider the contributions $N_\psi(\Theta(\psi))$ to the relative class number formulas (3.9.1) and (3.9.2) that come from characters ψ with 2-power conductor $f(\psi) = 2^\rho$. They are the characters introduced by (3.5.5) and (3.5.6) so far as $\psi(-1) = -1$, that is, $\psi = \cup\varphi_\rho \ \ (\rho \geqq 2)$.

For $\rho = 2$, therefore, $\psi = \cup$, so $P_\psi = P$, and one has by (3.9.3) and (3.9.4) directly

$$\Theta(\psi) = \frac{1}{2} - \frac{1}{2^2} = \frac{1}{2^2} \qquad \text{for } \psi = \cup, \tag{3.14.1}$$

therefore also

$$N_\psi(\Theta(\psi)) = \frac{1}{2^2} \qquad \text{for } \psi = \cup. \tag{3.14.2}$$

For $\rho \geqq 3$, by corresponding to the basis expression

$$x \equiv (-1)^\alpha (1 + 2^2)^\beta \pmod{2^\rho} \qquad (\alpha \mod 2, \ \ \beta \mod 2^{\rho-2}) \tag{3.14.3}$$

of the group of the prime residues mod 2^ρ, the character $\psi = \cup\varphi_\rho$ is given by

$$\psi(x) = (-1)^\alpha Z^\beta, \tag{3.14.4}$$

where Z is a $2^{\rho-2}$-th primitive root of unity. The order of ψ is $n_\psi = 2^{\rho-2}$. The field P_ψ is the cyclotomic field $P_{2^{\rho-2}}$ of degree $\varphi(2^{\rho-2}) = 2^{\rho-3}$.

In (3.9.5) it has already been determined that

$$\Theta(\psi) = \frac{1}{2} \sum_{\pm x \bmod 2^{\rho-1}}^{+} \psi(x) \qquad \text{for } \psi = \cup\varphi_\rho \ \ (\rho \geqq 3). \tag{3.14.5}$$

The reduction of the full system of the prime residues mod 2^ρ to the quarter system of least positive residues, namely, to the half system of the least positive residues mod $2^{\rho-1}$, did mean algebraically the reduction of the group of the prime residues mod 2^ρ to the factor group by the subgroup of solutions of $u^2 \equiv 1 \pmod{2^\rho}$, a subgroup of type (2, 2). The latter reduction can also be executed in the way that, corresponding to the basis expression

$$u \equiv (-1)^{\mu}((1+2^2)^{2^{\rho-3}})^{\nu} \pmod{2^\rho} \qquad (\mu,\ \nu \bmod 2)$$

of these solutions, one restricts the exponent region to $\alpha \bmod 1$, $\beta \bmod 2^{\rho-3}$ in the basis expression (3.14.3). On account of $\psi(u) = \pm 1 \equiv 1 \pmod 2$, it holds by (3.14.4) that the congruence

$$\Theta(\psi) \equiv \frac{1}{2} \sum_{\beta \bmod 2^{\rho-3}} Z^\beta \ (\mathrm{mod}^+ 1) \qquad \text{for } \psi = \cup \varphi_\rho \ (\rho \geqq 3)$$

holds.

Herein

$$\sum_{\beta \bmod 2^{\rho-3}}^{+} Z^\beta = \frac{Z^{2^{\rho-3}} - 1}{Z-1} = \frac{2}{1-Z}.$$

Thus it follows that

$$\Theta(\psi) \equiv 0 \ \left(\mathrm{mod}^+ \frac{1}{\mathfrak{z}}\right), \ \text{but not } (\mathrm{mod}^+ 1) \qquad \text{for } \psi = \cup \varphi_\rho \ \ (\rho \geqq 3), \tag{3.14.6}$$

where $\mathfrak{z}$ is the unique prime division of 2 of degree 1 in the field $P_{2^{\rho-2}}$ generated by $1-Z$. By taking the norm, there arises from (3.14.6)

$$N_\psi(\Theta(\psi)) \equiv 0 \ \left(\mathrm{mod}^+ \frac{1}{2}\right) \ \text{but not } (\mathrm{mod}^+ 1) \qquad \text{for } \psi = \cup \varphi_\rho \ \ (\rho \geqq 3). \tag{3.14.7}$$

By (3.14.2) and (3.14.7) the actual denominator of $\Theta(\psi)$ in this considered case has been exactly determined:

Theorem 3.33 *If $\psi = \cup \psi_\rho$ is a character of K/K_0 with conductor $f(\psi) = 2^\rho$, then $N_\psi(\Theta(\psi))$ is a rational number whose denominator is exactly 2^2 for $\rho = 2$ and exactly 2 for $\rho \geqq 3$.*

We, moreover, give another proof of fact (3.14.6) and hence of (3.14.7) and Theorem 3.33, which offers us a formula that will be important later. For this we denote by $s_{2^\rho}(x) = (-1)^{\delta_{2^\rho}(x)}$ the sign of the absolutely least residue of x mod 2^ρ. The notation of the exponent $\delta_{2^\rho}(x)$ has already been introduced generally in

Sect. 2.11 for formula (2.11.5).[95] While the matter is that these exponents are only mod 2, we will normalize them by fixing $\delta_{2^\rho}(x) = 0$ or 1; then it holds that

$$\delta_{2^\rho}(x) = \frac{1 - s_{2^\rho}(x)}{2}. \tag{3.14.8}$$

The function $s_{2^\rho}(x)$ of the prime residue classes $x \bmod 2^\rho$ has the two same anti-symmetries $s_{2^\rho}(x) \to -s_{2^\rho}(x)$ for $x \to -x$ and for $x \to x + 2^{\rho-1}$ as those of $\psi(x)$ that were determined in Sect. 3.9, so that the product $\psi(x)s_{2^\rho}(x)$ is invariant by the above transformations. Consequently formula (3.14.5) can be described in the invariable form

$$\Theta(\psi) = \frac{1}{2} \sum_{\pm x \bmod 2^{\rho-1}} \psi(x)s_{2^\rho}(x) \qquad \text{for } \psi = \cup\varphi_\rho \quad (\rho \geqq 3), \tag{3.14.5'}$$

in which the summation can be extended here over an *arbitrary* half system of prime residues mod $2^{\rho-1}$.

Moreover, let a fixed residue class $z \bmod 2^\rho$ be selected such that $\psi(z)$ is a primitive $2^{\rho-2}$-th root of unity, say, the basis class $z \equiv 1 + 2^2 \pmod{2^\rho}$. Then by the multiplication of $1 - \overline{\psi}(z) \cong \mathfrak{z}$, there arises from formula (3.14.5′)

$$\begin{aligned}(1 - \overline{\psi}(z))\Theta(\psi) &= \frac{1}{2} \sum_{\pm x \bmod 2^{\rho-1}} \psi(x)s_{2^\rho}(x) - \frac{1}{2} \sum_{\pm x \bmod 2^{\rho-1}} \psi(z^{-1}x)s_{2^\rho}(x) \\ &= \frac{1}{2} \sum_{\pm x \bmod 2^{\rho-1}} \psi(x)s_{2^\rho}(x) - \frac{1}{2} \sum_{\pm x \bmod 2^{\rho-1}} \psi(x)s_{2^\rho}(zx) \\ &= \sum_{\pm x \bmod 2^{\rho-1}} \psi(x)\frac{s_{2^\rho}(x) - s_{2^\rho}(zx)}{2},\end{aligned}$$

and therefore from (3.14.8)

$$(1 - \overline{\psi}(z))\Theta(\psi) = - \sum_{\pm x \bmod 2^{\rho-1}} \psi(x)(\delta_{2^\rho}(x) - \delta_{2^\rho}(zx)) \quad \text{for } \psi = \cup\varphi_\rho \quad (\rho \geqq 3). \tag{3.14.9}$$

Here by definition it holds that $\delta_{2^\rho}(x) = 0$ or 1 according as the absolutely least residue of $x \bmod 2^\rho$ is positive or negative.

---

[95] (Translator's remark) As shown in Sect. 2.11, we could define $\delta_{2f}(x)$ as normalized by

$$\delta_{2f}(x) = \begin{cases} 1 \text{ if } -f < x < 0, \\ 0 \text{ if } 0 < x < f. \end{cases}$$

The right-hand side of formula (3.14.9) is of the type of the linear factor of the regulatrix in (2.11.5). We will have to apply it in Sect. 3.20, and in fact this relation to (2.11.5) will then play an important role. Also for us, formula (3.14.9) will be useful in Sect. 3.16 in the numerical calculation of the relative class number of imaginary abelian number fields with 2-power conductor.

By virtue of formula (3.14.9), first of all, it becomes clear that $\mathfrak{Z}\Theta(\psi)$ is integral. One also easily deduces from this formula that $\mathfrak{Z}\Theta(\psi)$ is relatively prime to 2. In fact, selecting the half system of the exponential prime residues $x \equiv z^{\nu}$ $(\bmod\ 2^{\rho-1})$ $(\nu \mod 2^{\rho-2})$ with $z \equiv 1+2^2 \pmod{2^\rho}$ and noticing that $\psi(x) \equiv 1$ $(\bmod\ \mathfrak{Z})$, one has from (3.14.9)

$$\begin{aligned}(1-\overline{\psi}(z))\Theta(\psi) &\equiv -\sum_{\pm\nu \bmod 2^{\rho-3}} (\delta_{2^\rho}(z^{\nu}) - \delta_{2^\rho}(z^{\nu+1})) \\ &\equiv -\delta_{2^\rho}(1) + \delta_{2^\rho}(z^{2^{\rho-3}}) \pmod{\mathfrak{Z}},\end{aligned}$$

and owing to $z^{2^{\rho-3}} \equiv 1 + 2^{\rho-1} \pmod{2^\rho}$, one obtains

$$(1-\overline{\psi}(z))\Theta(\psi) \equiv 1 \pmod{\mathfrak{Z}}. \tag{3.14.10}$$

Thus the congruence (3.14.6) has actually been recognized as a result of (3.14.9).

3.15 Direct Proof of the Integrality

Now we have come to the application of the results on the numbers $N_\psi(\Theta(\psi))$ obtained in Theorems 3.30–3.33, from Sects. 3.10–3.14, to a direct proof of the integrality of the relative class number h^* of K/K_0.

By (3.9.1) the formula of h^* was

$$h^* = Qw \prod_{\psi} N_\psi(\Theta(\psi)), \tag{3.15.1}$$

where ψ runs over a system of representatives of classes of algebraic-conjugate characters of K/K_0. In Theorems 3.30, 3.32, and 3.33 it has been established that, in the contributions $N_\psi(\Theta(\psi))$ of these representatives ψ to the relative class number h^*, only the following three types of denominator factors appear:

(a) When $f(\psi) = p^\rho$ $(p \neq 2)$ and $n_\psi = \varphi(p^\rho)$, the prime denominator factor p appears occasionally for $p \neq 3$ and really for $p = 3$.
(b) When $f(\psi) = 2^\rho$, i.e., $\psi = \cup\varphi_\rho$, the square prime denominator factor 2^2 really appears for $\rho = 2$ and the prime denominator factor 2 really appears for $\rho \geqq 3$.
(c) When $f(\psi) = p \neq 2$ and $n_\psi = 2^\mu$, the prime denominator factor 2 really appears.

In each of these cases the character ψ is uniquely determined by $f(\psi)$ up to algebraic-conjugate, so that among the representatives for a given prime-power $f(\psi) = p^\rho$ $(p \neq 2)$, $f(\psi) = 2^\rho$ or $f(\psi) = p \neq 2$, there appears, respectively, at most one character of type (a), (b), or (c).

First we show that the prime denominator factors of the two types (a) and (b) are reduced by the number w of roots of unity in K appearing as a factor in (3.15.1).

Proof of the Reduction in Case of (a) [96]For $p \neq 2$ a character ψ with conductor $f(\psi) = p^\rho$ and of degree $n_\psi = \varphi(p^\rho)$ generates the group of characters of the cyclotomic field P_{p^ρ}. Its existence therefore means that P_{p^ρ} is contained in K. If p^ω divides exactly w, then there appear exactly ω characters of this type among the representatives ψ, namely, the characters ψ with conductor $f(\psi) = p^\rho$, $\rho = 1, 2, \ldots, \omega$. As the prime denominator factors p really correspond to them, these factors are reduced by the contribution p^ω to w. □

Proof of the Reduction in Case of (b) For $p = 2$ the character $\cup\varphi_2 = \cup$ generates the group of characters of the cyclotomic field P_{2^2}, and a character $\cup\varphi_\rho$ $(\rho \geqq 3)$ together with $\cup$ generates the group of characters of the cyclotomic field P_{2^ρ}. Then the existence of $\cup\varphi_{2^2} = \cup$ means that P_{2^2} is contained in K, and the existence of $\cup\varphi_\rho$ $(\rho \geqq 3)$ together with $\cup$ means that even P_{2^ρ} is contained in K. If 2^ω divides exactly w, then in the case of $\omega \geqq 2$ there appear exactly $\omega - 1$ characters among the representatives ψ of type of $\cup\varphi_\rho$, namely, with $\rho = 2, \ldots, \omega$. Then, corresponding to them the prime denominator factor 2^2 for $\rho = 2$ and 2 for $\rho \geqq 3$ are reduced by the contribution 2^ω to w. In the case of $\omega = 1$ there *does not* appear $\cup\varphi_2 = \cup$ in the (representatives) ψ, and at most *one* $\cup\varphi_\rho$ $(\rho \geqq 3)$ can appear (in the ψ) because from the existence of $\cup\varphi_\sigma$ $(\sigma > \rho)$ it could follow that $(\cup\varphi_\sigma)^{2^{\sigma-\rho}} = \varphi_\rho$ and hence $\cup$ could also appear. Therefore, at most one prime denominator factor 2 corresponding to $\cup\varphi_\rho$ is again reduced by the contribution 2^1 to w. □

Beyond these proofs (a) and (b) there arise additional special results that the prime denominator factor of type (a) for $p = 3$ and the prime denominator factor $p = 2$ of type (b) appear not only *occasionally* but *really*. We will explore these results in Sect. 3.16.

We call the prime denominator factor of type (c) to still be investigated and also the corresponding character ψ with $f(\psi) = p \neq 2$ and of $n_\psi = 2^\mu$ *irregular*. For the treatment (of the irregular denominators and characters) it is appropriate to include also in this concept the following two prime denominator factors 2 of type (b) and their corresponding characters:

(c_1) for $w \equiv 0 \pmod 4$, the actually appearing character $\cup$ with one of the corresponding two prime denominator factors 2,
(c_2) for $w \not\equiv 0 \pmod 4$, the occasionally appearing character $\cup\varphi_\rho$ $(\rho \geqq 3)$ with the corresponding prime denominator factor 2.

[96](Translator's remark) For the reader's convenience we give here this caption.

The other characters and (also the corresponding) prime denominator factors are to be called *regular*. This definition of irregularity or regularity depends generally only on the character itself and not on the field K; only for the character $\cup\varphi_\rho$ $(\rho \geqq 3)$ (however), it depends on K, namely, on the alternative of whether $w \not\equiv 0 \pmod 4$ or $w \equiv 0 \pmod 4$.

Thus we can state the results of the above proofs (of the reduction in cases) (a) and (b) as follows:

Theorem 3.34 *In the relative class number formula*

$$h^* = 2\,Q\,\frac{w}{2}\prod_{\psi} N_{\psi}(\Theta(\psi)), \tag{3.15.1'}$$

the regular prime denominator factors p of contributions $N_\psi(\Theta(\psi))$ are reduced by the p-contributions to $\frac{w}{2}$.

More precisely, if one expresses formula (3.15.1′) *in the form*

$$h^* = 2\,Q\prod_{\psi} h^*_{\psi}, \tag{3.15.2}$$

with

$$h^*_\psi = \left\{\begin{array}{l} pN_\psi(\Theta(\psi)) \text{ for } f(\psi) = p^\rho \ (p \neq 2) \text{ and } n_\psi = \varphi(p^\rho) \\ \\ 2N_\psi(\Theta(\psi)) \text{ for } f(\psi) = 2^\rho \text{ in the case of } w \equiv 0 \pmod 4 \\ \\ N_\psi(\Theta(\psi)) \quad \text{otherwise} \end{array}\right\}, \tag{3.15.3}$$

then the contributions h^*_ψ are rational integers for regular characters ψ and rational numbers with denominator exactly 2 *for irregular characters ψ.*

Denote by r the number of irregular characters ψ. Hence $2^{r-1}h^*$ is an integer and so is $2^{r-2}h^*$ for $Q = 2$. For the cyclotomic field P_f with composite conductor f, there exists exactly one irregular representative ψ for every prime number $p|f$, and by Theorem 3.27 it holds that $Q = 2$ for the field; consequently, we have the result of Kummer [20] introduced in Sect. 1.6.

Proof of the Integrality of h^* [97]Now we are in the position to deal with the still remaining irregular prime denominator factors 2 in the relative class number formula (3.15.2). Since in (3.15.2) there remains the factor $2Q = 2$ or 2^2, a complete reduction of these prime denominator factors can be achieved, in general, only in

[97](Translator's remark) From now on, by showing the reduction in case of (c), we prove Theorem 3.35, which shows that h^* is an integer.

the way that certain contributions h^*_ψ are divisible by 2. In fact, in Theorem 3.31 we have obtained a result of this type. By its help we show here that sufficiently many contributions h^*_ψ divisible by 2 are available for reducing all the irregular denominator factors 2 with the aid of the factor $2Q$.

Let $\psi_1, \ldots, \psi_r$ be the r irregular characters of representatives ψ of characters of K/K_0. Then also all the products χ of odd number of characters from $\psi_1, \ldots, \psi_r$ different from each other appear among the characters of K/K_0, because by the odd number of the terms it holds that $\chi(-1) = -1$. These products χ are not algebraic-conjugate with each other, so that we can assume without generality that such characters appear among the representatives ψ. For any such product ψ of at least three terms, the contribution $h^*_\psi = N_\psi(\psi)$ defined by (3.15.3) is divisible by 2 on account of Theorem 3.31, because the order n_ψ of ψ is a power of 2 since this is the case for $\psi_1, \ldots, \psi_r$ and (because) the conductor $f(\psi)$ contains at least three different prime numbers since $\psi_1, \ldots, \psi_r$ have pairwise relatively prime conductors.

It is sufficient for our proof to pick up the products of three characters from $\psi_1, \ldots, \psi_r$. In order to show that the contributions h^*_ψ corresponding to $\binom{r}{3}$ products ψ reduce the r irregular prime denominator factors 2 by the aid of the factor $2Q = 2^{1+q}$, we have to confirm only the existence of the inequality

$$1 + q + \binom{r}{3} \geqq r$$

for all $r \geqq 0$. For every $r > 3$ this inequality is really fulfilled and (it holds) in fact independently of the value $q = 0$ or 1—likewise also in the trivial cases of $r = 0$ and $r = 1$. However, for $r = 2$ and $r = 3$ the value $q = 1$, i.e., $Q = 2$, must be ascertained for the proof. In what follows we show, in connection with the criteria in Sect. 3.7, the verification of the following slightly more general fact:

If the irregular characters among the representatives ψ of characters of K/K_0 are $r \geqq 2$ in number, then $Q = 2$.

For the remaining proof we depend on the sufficient criterion for $Q = 2$ in Theorem 3.26 and also on the combination with Theorem 3.29.[98] By Theorem 3.29 it is sufficient to show that for any pair of different irregular representatives ψ_1, ψ_2 the corresponding imaginary field $\tilde{K}$—with ψ_1, ψ_2 as generators—has property $\tilde{Q} = 2$. Since, in every case, $\tilde{K}$ has composite conductor $\tilde{f} = f(\psi_1)f(\psi_2)$, Theorem 3.26 is applicable to $\tilde{K}$. According as which type of (c), (c_1), or (c_2),

[98](Translator's remark) As pointed out in footnote 60, p. 100, in Sect. 3.7 regarding Theorem 3.29, we need to prove the statement just above (written in italics) without using Theorem 3.29, for which it is sufficient to reform the following part (c″) as shown by the translator's remark there.

the pair of irregular characters $\psi_1,\ \psi_2$ takes, we have here to distinguish (the pair) into the following two cases:

$$(\mathrm{c}')\ f(\psi_1) = p_1 \neq 2, \quad f(\psi_2) = p_2 \neq 2,$$

$$(\mathrm{c}'')\ f(\psi_1) = p_1 \neq 2, \quad f(\psi_2) = 2^{\rho}.$$

(c′) Since there exist no characters with 2-power conductor among the ones of $\tilde{K}/\tilde{K}_0$, the first case of Theorem 3.26 occurs. The field $\tilde{K}$ has degree $\tilde{n} = n_{\psi_1} n_{\psi_2} = 2^{\mu_1} 2^{\mu_2}$, where $2^{\mu_1},\ 2^{\mu_2}$ are the highest powers appearing in $p_1 - 1$, $p_2 - 1$. Hence the number in question,

$$\tilde{m} = \frac{\varphi(\tilde{f})}{\tilde{n}} = \frac{(p_1 - 1)(p_2 - 1)}{2^{\mu_1} 2^{\mu_2}},$$

becomes odd here.

(c″) [99]Since there exists a character with 2-power conductor among the ones of $\tilde{K}/\tilde{K}_0$, the second case of Theorem 3.26 occurs. The residue classes $\tilde{a}_0$ mod $\tilde{f}$ in $\tilde{H}$ with $\tilde{a}_0 \equiv 1 \pmod{2^{\rho}}$ are expressed by direct combination of the $\frac{\varphi(p_1)}{n_{\psi_1}} = \frac{p_1 - 1}{2^{\mu_1}}$ residue classes $\tilde{a}_1$ mod p_1 with $\psi_1(\tilde{a}_1) = 1$ and of the one class $\tilde{a}_2 \equiv 1 \pmod{2^{\rho}}$. Hence again the number in question, $\tilde{m} = \frac{p_1 - 1}{2^{\mu_1}}$, becomes odd here.

In the two cases (c′) and (c″), it follows therefore from Theorem 3.26 that $\tilde{Q} = 2$. Thus, by the description above our direct proof of the integrality of the relative class number h^* has been completed. More precisely the following theorem has been proved. □

Theorem 3.35 *In the relative class number formula* (3.15.2) *the remaining irregular prime denominator factors* 2 *are reduced by the factor* $2Q$ *and by the contributions* h^*_{ψ} *of* $\binom{r}{3}$ *products* ψ *of three different characters from the irregular representatives* $\psi_1, \ldots, \psi_r$.

By Theorems 3.34 and 3.35, exceeding the direct proof of the integrality, we have entered fairly deeply into the arithmetic structure of the relative class number, where we have moreover determined the integrality of individual contributions h^*_{ψ} up to occasional denominator 2 and pursued exactly the reduction of the denominators 2 remaining there by the other contributions. Surely no arithmetic meaning of

[99](Translator's remark) When $w \not\equiv 0 \pmod 4$, w the number of roots of unity in K, we have $Q = 2$ by this argument (c″) and by Theorem 3.29′ in footnote 60, p. 100, regarding Theorem 3.29. However when $w \equiv 0 \pmod 4$, to use Theorem 3.29′ we need to take as the field $\tilde{K}$ the composite field of the w-th cyclotomic field and the field corresponding to the cyclic group $\langle \psi_1 \rangle$. Then the field $\tilde{K}$ has unit index 2 by Theorem 3.26 (not by the argument in (c″)) and then we have $Q = 2$ by Theorem 3.29′.

contributions h^*_ψ appears there. For the cyclotomic field P_p with prime conductor p, Kummer [18] attempted to give such a meaning. However, his construction needs at least critical re-examination because he treated logarithms of ideal prime factors without detailed explanation.

In an explanation of Theorem 3.35 we consider some examples of imaginary abelian number fields K with $r = 3$, i.e., with three irregular characters being not algebraic-conjugate with each other. The series of irregular characters, arranged by increasing conductor, begins with[100]

$$\psi_3,\ \psi_4 = \cup,\ \psi_5,\ \psi_7,\ \psi_8 = \cup\varphi_3 \text{ (if } \cup \text{ does not appear)},\ \psi_{11},\ \psi_{13},\ldots.$$

Here the subscript of ψ always indicates the conductor $f(\psi)$; then for every prime number $p \neq 2$, here ψ_p denotes the character ψ, uniquely determined up to algebraic-conjugate, with conductor p and of degree the highest power 2^μ appearing in $p-1$. Then $\psi_3,\ \psi_7$ are the quadratic characters of conductors 3, 7; further ψ_5 is one of the conjugate biquadratic character of conductor 5, ... To the triples

$$\psi_7,\ \psi_3,\ \psi_5;$$

$$\psi_4,\ \psi_3,\ \psi_5;$$

$$\psi_8,\ \psi_3,\ \psi_5,$$

as generating characters, there correspond the imaginary abelian number fields

$$K_{105} \text{ with odd conductor } f = 105 = 7\cdot 3\cdot 5 \text{ and } w = 30 \not\equiv 0 \pmod 4,$$

$$K_{60} \text{ with even conductor } f = 60 = 4\cdot 3\cdot 5 \text{ and } w = 60 \equiv 0 \pmod 4,$$

$$K_{120} \text{ with even conductor } f = 120 = 8\cdot 3\cdot 5 \text{ and } w = 30 \not\equiv 0 \pmod 4,$$

whose Galois groups are always of type $(2,\ 2,\ 4)$. The field K_{105} originates from the cyclotomic field $P_{105} = P_7P_3P_5$, where one replaces the field P_7 by the quadratic field $P_7^{(2)} = P(\sqrt{-7})$ of three-term period[101]; the field K_{60} originates from the cyclotomic field $P_{60} = P_4P_3P_5$ itself; the field K_{120} originates from the cyclotomic field $P_{120} = P_8P_3P_5$, where one replaces the field $P_8 = P(\sqrt{-1},\ \sqrt{2})$ by the quadratic field $P_8^{(2)} = P(\sqrt{-2})$.

A full system of representatives of non-algebraic-conjugate characters ψ of K/K_0 are always generated by the addition of the product ψ_f with conductor f of the three generating irregular characters; pay attention here to the fact that

[100](Translator's remark) The condition "if $\cup$ does not appear" is necessary for the definition of irregular character.

[101](Translator's remark) The three-term periods are $\eta_0 = \frac{-1+\sqrt{-7}}{2}$, $\eta_1 = \frac{-1-\sqrt{-7}}{2}$.

$\psi(-1) = -1$ must hold. Now one finds immediately the values of h^*_ψ, by the definition (3.15.3) of h^*_ψ, for ψ_3, ψ_5, ψ_7 by the basic formula (3.9.2) and for ψ_4, ψ_8 by the formulas (3.14.1) and (3.9.5):

$$h^*_{\psi_3} = 3N(\Theta(\psi_3)) = 3\Theta(\psi_3) = 3 \cdot \frac{-1+2}{2 \cdot 3} = 3 \cdot \frac{1}{2 \cdot 3} = \frac{1}{2},$$

$$h^*_{\psi_4} = h^*_{\mathsf{U}} = 2N(\Theta(\mathsf{U})) = 2\Theta(\mathsf{U}) = 2 \cdot \frac{1}{2^2} = \frac{1}{2},$$

$$h^*_{\psi_5} = 5N(\Theta(\psi_5)) = 5N\left(\frac{-1-2i+3i+4}{2 \cdot 5}\right) = 5N\left(\frac{3+i}{2 \cdot 5}\right) = 5 \cdot \frac{10}{2^2 \cdot 5^2} = \frac{1}{2},$$

$$h^*_{\psi_7} = N(\Theta(\psi_7)) = \Theta(\psi_7) = \frac{-1-2+3-4+5+6}{2 \cdot 7} = \frac{7}{2 \cdot 7} = \frac{1}{2},$$

$$h^*_{\psi_8} = h^*_{\mathsf{U}\varphi_3} = N(\Theta(\mathsf{U}\varphi_3)) = \Theta(\mathsf{U}\varphi_3) = \frac{1}{2}.$$

As already determined in the example at the end of Sects. 3.10 and 3.12, it holds further that

$$h^*_{\psi_{105}} = N(\Theta(\psi_{105})) = N(1+i) = 2.$$

By corresponding short calculations[102] one similarly finds

$$h^*_{\psi_{60}} = N(\Theta(\psi_{60})) = N(-i(1+i)) = 2,$$

$$h^*_{\psi_{120}} = N(\Theta(\psi_{120})) = N(-(1+i)) = 2.$$

[102]For example, when one proceeds by the method of Sect. 3.10, one only needs to fill the three Schemes 1a, 1b, and 2 there, as broadly as they correspond to the solutions of the inequality in (3.10.3). Therefore it is practical that one starts with Scheme 2 in Sect. 3.10; in the cases above of $\psi_{60} = \psi_{12}\psi_5$ and $\psi_{120} = \psi_{24}\psi_5$, the following extracts are sufficient for them. (Translator's remark: Actually, in the first and second tables, the numbers 1 and 0 mean that the inequalities $5x < 12y$ and $5x < 24y$ hold or not, respectively.)

		x	1	5	......	29
		$5x$	5	25	......	145
y	$12y$					
1	12		1	0	......	0
2	24		1	0	......	0

($5x < 12y$)

		x	1	5	7	11	......	59
		$5x$	5	25	35	55	......	295
y	$24y$							
1	24		1	0	0	0	......	0
2	48		1	1	1	0	......	0

($5x < 12y$)

Thus the relative class numbers of K/K_0 for the three fields $K = K_{105},\ K_{60},\ K_{120}$ have been calculated by (3.15.2) to be

$$h^* = 2 \cdot 2 \cdot \frac{1}{2} \cdot \frac{1}{2} \cdot \frac{1}{2} \cdot 2 = 1.$$

The factor $2Q = 2 \cdot 2$ standing at the front of the formula reduces only two of the three irregular prime denominator factors 2, while the third denominator factor is reduced by the last contribution $h^*_{\psi_f} = 2$ of the product ψ_f of the three generating irregular characters.

3.16 Theorem of Weber and Its Supplement

Now we have come to the supplemental results announced already in Sect. 3.15 (obtained) from the two proofs of (a) and (b) there for the special cases of $p = 3$ and 2. As already mentioned, these results arise from the facts that the prime denominator factors $p = 3$ and 2 in (a) and (b) in Sect. 3.15, respectively, actually appear. Therefore for these reductions, as the above-mentioned proofs show, the whole contributions of $p = 3$ and 2 to w are actually used, except the contribution of 2 to w in the case where $w \not\equiv 0 \pmod 4$ and no characters $\cup\varphi_p$ with 2-power conductor appear among the representatives ψ. Hence when, except for the characters of types (a) and (b) in Sect. 3.15 treated in the proofs, there exist no more such characters of K/K_0, that is, when the conductor f of K is a power of $p = 3$ or 2, it can be deduced from (3.15.1) that h^* is not divisible by $p = 3$ or 2, respectively; for $p = 2$ pay attention to the fact that $Q = 1$ by Theorem 3.23 and that the above-mentioned exceptional case surely never exists.

Corresponding to Schemes 1a and 1b in Sect. 3.10, the following extracts are sufficient:

x	1	5	 29
$\psi_4(x)$	1	1	....
$\psi_3(x)$	1	-1	....
$\psi_{12}(x)$	1	-1	....

x	1	5	7	 59
$\psi_8(x)$	1	-1	-1	....
$\psi_3(x)$	1	-1	1	....
$\psi_{24}(x)$	1	1	-1	....

y	1	2
$\psi_5(y)$	1	i

Hence one can read by virtue of formula (3.10.3)

$$\Theta(\psi_{60}) = \psi_{12}(5)\psi_5(12)(\psi_{12}(1)\psi_5(1) + \psi_{12}(1)\psi_5(2)) = (-1) \cdot i \cdot (1+i) = -i(1+i),$$

$$\Theta(\psi_{120}) = \psi_{24}(5)\psi_5(24)(\psi_{24}(1)\psi_5(1) + \psi_{24}(1)\psi_5(2) + \psi_{24}(5)\psi_5(2) + \psi_{24}(7)\psi_5(2))$$

$$= 1 \cdot (-1) \cdot (1 + i + i - i) = -(1+i).$$

Thus the values inserted above have been calculated.

Thus we have proved the following two facts:

Theorem 3.36 *For the imaginary abelian number fields K with 2-power conductor $f = 2^\rho$, the relative class number h^* of K is not divisible by 2.*

Theorem 3.37 *For the imaginary abelian number field K with 3-power conductor $f = 3^\rho$, the relative class number h^* of K is not divisible by 3.*

For $p = 2$, as imaginary abelian number fields with 2-power conductor $f = 2^\rho$, there exist only one for $\rho = 2$ and a pair for $\rho \geqq 3$, namely, for every $\rho \geqq 2$ there exists the cyclotomic field $K = P_{2^\rho}$ whose character group is generated by $\cup$ and $\cup\varphi_\rho$, and for $\rho \geqq 3$, in addition, the cyclic subfield $K = P'_{2^\rho}$ contained in the field P_{2^ρ} whose character group is generated only by $\cup\varphi_\rho$. (See Fig. 3.4.) The latter field is given by

$$P'_{2^\rho} = P_{2^{\rho-1},0}(\sqrt{-\lambda_{2^{\rho-1},0}}), \tag{3.16.1}$$

where $\lambda_{2^{\rho-1},0}$ is the number introduced in (3.3.3) and (3.3.4),

$$\lambda_{2^2,0} = 2, \quad \lambda_{2^3,0} = 2 + \sqrt{2}, \quad \lambda_{2^4,0} = 2 + \sqrt{2 + \sqrt{2}}, \ \ldots;$$

therefore

$$P'_{2^3} = P(\sqrt{-2}), \ P'_{2^4} = P(\sqrt{-(2+\sqrt{2})}), \ P'_{2^5} = P(\sqrt{-(2+\sqrt{2+\sqrt{2}})}), \ \ldots.$$

For $p = 3$, since here the group of the prime residue classes mod 3^ρ is cyclic of order $\varphi(3^\rho) = 2 \cdot 3^{\rho-1}$, as imaginary abelian number fields K with conductor

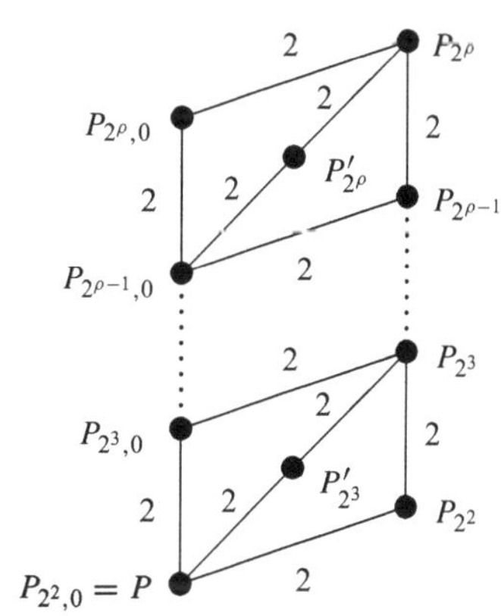

Fig. 3.4 Rclation among thc fields with 2-power conductor

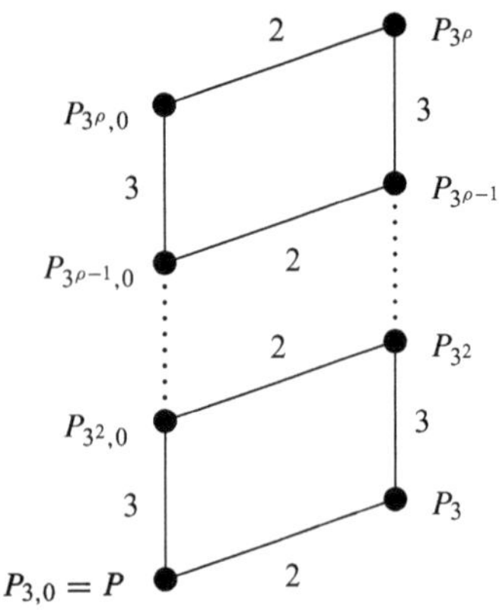

Fig. 3.5 Relation among the fields with 3-power conductor

$f = 3^\rho$, for every $\rho \geqq 3$ there exists only one such field,[103] namely, the cyclotomic field $K = P_{3^\rho}$. (See Fig. 3.5.)

Therefore, by expressing Theorems 3.36 and 3.37 in such a way that a class of fields in them (theorems) emerges, the theorems can be restated as follows:

Theorem 3.36′ *For the cyclotomic* $K = P_{2^\rho}$ $(\rho \geqq 2)$ *and its imaginary subfield* $K = P'_{2^\rho}$ $(\rho \geqq 3)$, *the relative class number* h^* *of* K/K_0 *is not divisible by* 2.

Theorem 3.37′ *For the cyclotomic* $K = P_{3^\rho}$ $(\rho \geqq 1)$, *the relative class number* h^* *of* K/K_0 *is not divisible by* 3.

Theorem 3.36′ was proved by Weber [24, 25], in the same way as here but certainly there only for the cyclotomic field P_{2^ρ} itself and not for the imaginary subfield P'_{2^ρ}, as this could have arisen immediately. By adding (to Theorem 3.36′) Weber's result stated in Theorem 2.6, Sect. 2.6, and proved there in a new way that for the real subfield $K_0 = P_{2^\rho,0}$ the class number h_0 is also not divisible by 2, the following theorem arises:

Theorem 3.38 *For an imaginary abelian number field* K *with* 2*-power conductor* $f = 2^\rho$, *the full class number* h *is not divisible by* 2.

By being expressed by a class of emerging fields, this becomes the main theorem of Weber[104]:

Theorem 3.38′ *For the cyclotomic field* P_{2^ρ} $(\rho \geqq 2)$, *its real subfield* $P_{2^\rho,0}$ $(\rho \geqq 2)$ *and its imaginary subfield* P'_{2^ρ} $(\rho \geqq 3)$, *the full class numbers* h *are not divisible by* 2.

As mentioned above, the field P'_{2^ρ} was not cited by Weber [24, 25], though this could have occurred immediately by the proof of Weber.

[103](Translator's remark) This "for every $\rho \geqq 3$" printed in the original book could be replaced by "for every $\rho \geqq 1$."

[104]To this see the statement in footnote 4, p. 25, in Sect. 2.1.

So far as I know, Theorem 3.37′ is a supplement of Weber's Theorem 3.36′, which has not been remarked so far. The question of whether the class number h_0 of $K_0 = P_{3^\rho,0}$ is not divisible by 3 analogously to Weber's Theorem 2.6, and then of whether the supplement for $p = 3$ also holds analogously to Theorem 3.38 and to Weber's main theorem, Theorem 3.38′, I have to leave open for the time being.

In connection with the preceding theorems, I enter here more precisely into the actual calculation of the relative class numbers $h^*_{2^\rho}$ of $P_{2^\rho}/P_{2^\rho,0}$ and $h^*_{3^\rho}$ of $P_{3^\rho}/P_{3^\rho,0}$, namely, I develop a *systematic* method of numerical calculation of these relative class numbers. This method may serve as a model of what I strive for, the calculation of the class number of general abelian number fields according to the statement in my Preface.

Calculation of the Relative Class Numbers $h^*_{2^\rho}$ *of* $P_{2^\rho}/P_{2^\rho,0}$

Between the relative class numbers $h^*_{2^\rho}$ of $P_{2^\rho}/P_{2^\rho,0}$ and $h^{*\,\prime}_{2^\rho}$ of $P'_{2^\rho}/P_{2^{\rho-1},0}$, there exists a system of recurrence relations given by Weber [25]. By the relative class number formula (3.15.1) and since here $Q = 1$ by Theorem 3.23, one has on the one hand

$$h^*_{2^\rho} = 2^\rho \prod_{\sigma=2}^{\rho} N_{\varphi_\sigma}(\Theta(\cup\varphi_\sigma)) \qquad (\rho \geqq 2) \tag{3.16.2}$$

and on the other hand

$$h^{*\,\prime}_{2^\rho} = 2N_{\varphi_\rho}(\Theta(\cup\varphi_\rho)) \qquad (\rho \geqq 3). \tag{3.16.3}$$

As is well known and determined by (3.14.1), it holds that

$$h^*_{2^2} = 2^2\Theta(\cup) = 1. \tag{3.16.4}$$

By (3.16.2), (3.16.3), and (3.16.4), there arises a system of recurrence relations

$$h^*_{2^\rho} = h^{*\,\prime}_{2^3} \cdots h^{*\,\prime}_{2^\rho} \qquad (\rho \geqq 3). \tag{3.16.5}$$

According to (3.16.5) the factors $h^{*\,\prime}_{2^\rho}$ are to be regarded as arithmetic construction bricks of $h^*_{2^\rho}$.

For the numerical calculation of $h^{*\,\prime}_{2^\rho}$ we start, as did Weber [25], from formula (3.14.9) for the expression $(1 - \overline{\varphi}_\rho(z))\Theta(\cup\varphi_\rho)$. With regard to (3.16.3) it yields by the absolute norm an expression of $h^{*\,\prime}_{2^\rho}$, namely,

$$h^{*\,\prime}_{2^\rho} = N_{\varphi_\rho}\left(-\sum_{\pm x \bmod 2^{\rho-1}} \cup(x)\varphi_\rho(x)(\delta_{2^\rho}(x) - \delta_{2^\rho}(zx))\right) \qquad (\rho \geqq 3). \tag{3.16.6}$$

Selecting again in (3.16.6), as at the end of Sect. 3.14, the half system of exponential prime residues

$$x \equiv z^{\nu} \pmod{2^{\rho-1}} \quad (\nu \bmod 2^{\rho-3}) \quad \text{with } z \equiv 1+2^2 \equiv 5 \pmod{2^{\rho}},$$

one obtains Weber's formula

$$h_{2^{\rho}}^{*\,\prime} = N_{2^{\rho-2}}\left(\sum_{\nu \bmod 2^{\rho-3}} e_{2^{\rho}}^{(\nu)}\zeta_{2^{\rho-2}}^{\nu}\right) \quad \text{with } e_{2^{\rho}}^{(\nu)} = \delta_{2^{\rho}}(5^{\nu+1}) - \delta_{2^{\rho}}(5^{\nu}) \ (\rho \geqq 3), \tag{3.16.6'}$$

where $N_{2^{\rho-2}}$ denotes the absolute norm of the cyclotomic field $P_{2^{\rho-2}}$.

The coefficient system $e_{2^{\rho}}^{(\nu)}$ introduced in (3.16.6′) is connected with the sign-alternation in the sequence of the absolutely least residues of powers $5^{\nu} \bmod 2^{\rho}$. Namely, this means that $e_{2^{\rho}}^{(\nu)} = 0, 1, -1$ according as, in this sequence by the transformation from 5^{ν} to $5^{\nu+1}$, there exists no sign-alternation, sign-alternation from + to − or sign-alternation from − to +, respectively. Hence for the calculation of the coefficients $e_{2^{\rho}}^{(\nu)}$, the following scheme (of the absolutely least residues of 5^{ν} mod 2^{ρ}) arises:

5^0	5^1	5^2	5^3	5^4	5^5	5^6	5^7	5^8	5^9	5^{10}	5^{11}	5^{12}	5^{13}	5^{14}	5^{15}	5^{16}	…	
1	−3																	mod 2^3
1	5	−7																mod 2^4
1	5	−7	−3	−15														mod 2^5
1	5	25	−3	−15	−11	9	−19	−31										mod 2^6
1	5	25	−3	−15	53	9	45	−31	−27	−7	−35	−47	21	−23	13	−63		mod 2^7
…																		…

By (3.16.6′) one reads out the following values from the sign-alternation in the rows of this scheme:

$$\left.\begin{aligned}
h_{2^3}^{*\,\prime} &= N_{2^1}(\zeta_{2^1}^{0}) \\
h_{2^4}^{*\,\prime} &= N_{2^2}(\zeta_{2^2}^{1}) \\
h_{2^5}^{*\,\prime} &= N_{2^3}(\zeta_{2^3}^{1}) \\
h_{2^6}^{*\,\prime} &= N_{2^4}(\zeta_{2^4}^{2} - \zeta_{2^4}^{5} + \zeta_{2^4}^{6}) \\
h_{2^7}^{*\,\prime} &= N_{2^5}(\zeta_{2^5}^{2} - \zeta_{2^5}^{4} + \zeta_{2^5}^{7} - \zeta_{2^5}^{12} + \zeta_{2^5}^{13} - \zeta_{2^5}^{14} + \zeta_{2^5}^{15}) \\
&\cdots\cdots\cdots\cdots
\end{aligned}\right\} \tag{3.16.7}$$

Now we have to develop further a method of calculation of the norms in (3.16.7). To do this one can actually treat them recurrently by virtue of the famous rule

$$N_{2^{\rho+1}}(\xi_{2^{\rho+1}}) = N_{2^\rho}(\mathbf{N}_{2^{\rho+1}/2^\rho}(\xi_{2^{\rho+1}})) \quad \text{for } \xi_{2^{\rho+1}} \text{ of } P_{2^{\rho+1}}, \tag{3.16.8}$$

where $\mathbf{N}_{2^{\rho+1}/2^\rho}$ denotes the relative norm of $P_{2^{\rho+1}}/P_{2^\rho}$. Corresponding to the basic irreducible equations

$$\zeta_{2^{\rho+1}}^{2^\rho} = -1 \quad \text{for } P_{2^{\rho+1}}, \tag{3.16.9a}$$

$$\zeta_{2^{\rho+1}}^{2} = \zeta_{2^\rho} \quad \text{for } P_{2^{\rho+1}}/P_{2^\rho}, \tag{3.16.9b}$$

there exists for a number $\xi_{2^{\rho+1}}$ of $P_{2^{\rho+1}}$ the unique basis expression

$$\xi_{2^{\rho+1}} = a_0 + a_1\zeta_{2^{\rho+1}} + \cdots + a_{2^\rho-1}\zeta_{2^{\rho+1}}^{2^\rho-1} = [a_0,\, a_1,\, \ldots,\, a_{2^\rho-1}]$$

$$\text{with rational numbers } a_0,\, a_1,\, \ldots,\, a_{2^\rho-1}, \tag{3.16.10a}$$

$$\xi_{2^{\rho+1}} = \alpha_{2^\rho} + \beta_{2^\rho}\zeta_{2^{\rho+1}} \quad \text{with } \alpha_{2^\rho},\, \beta_{2^\rho} \text{ of } P_{2^\rho}, \tag{3.16.10b}$$

where in the first equation an abridged way of writing (of numerical expression) appropriate for numerical calculation is introduced. Noting the basic equation (3.16.9b) in (3.16.10a) and accordingly summing up the terms of even and odd exponents of $\zeta_{2^{\rho+1}}$,[105], respectively, one obtains the coefficients in (3.16.10b); they (the coefficients) then arise immediately in the basis expression (3.16.10a) (for P_{2^ρ}), namely,

$$\alpha_{2^\rho} = [a_0,\, a_2,\, \ldots,\, a_{2^\rho-2}], \qquad \beta_{2^\rho} = [a_1,\, a_3,\, \ldots,\, a_{2^\rho-1}]. \tag{3.16.11}$$

Hence by the basic equation (3.16.9b) it holds that

$$\mathbf{N}_{2^{\rho+1}/2^\rho}(\alpha_{2^\rho} + \beta_{2^\rho}\zeta_{2^{\rho+1}}) = \alpha_{2^\rho}^2 - \zeta_{2^\rho}\beta_{2^\rho}^2. \tag{3.16.12}$$

The expression appearing here on the right-hand side is transformed in accordance with (3.16.11) by making use of the basic equation (3.16.9a) (for P_{2^ρ}) into the basis expressions (3.16.10a) (for P_{2^ρ}) and (3.16.10b) (for $P_{2^\rho}/P_{2^{\rho-1}}$). Let the result of this reduction be[106]

$$\alpha_{2^\rho}^2 - \zeta_{2^\rho}\beta_{2^\rho}^2 = \alpha_{2^{\rho-1}} + \beta_{2^{\rho-1}}\zeta_{2^\rho}. \tag{3.16.13}$$

Therefore, by (3.16.8), (3.16.12), and (3.16.13), there arises

$$N_{2^{\rho+1}}(\alpha_{2^\rho} + \beta_{2^\rho}\zeta_{2^{\rho+1}}) = N_{2^\rho}(\alpha_{2^{\rho-1}} + \beta_{2^{\rho-1}}\zeta_{2^\rho}). \tag{3.16.14}$$

---

[105] (Translator's remark) "$\zeta_{2^{\rho+2}}$" in the original book should be "$\zeta_{2^{\rho+1}}$."

[106] (Translator's remark) We define $\alpha_{2^{\rho-1}}$ and $\beta_{2^{\rho-1}}$ by this Eq. (3.16.13).

By repeating the application of the recurrence formula (3.16.14), where the reduction is always to be executed from (3.16.12) to (3.16.13),[107] it follows finally that

$$N_{2^{\rho+1}}(\alpha_{2^\rho} + \beta_{2^\rho}\zeta_{2^{\rho+1}}) = N_{2^2}(\alpha_{2^1} + \beta_{2^1}\zeta_{2^2}) = \alpha_{2^1}^2 + \beta_{2^1}^2, \tag{3.16.15}$$

where α_{2^1} and β_{2^1} lie in $P_{2^1} = P$.[108] Then, by (3.16.15) the norm calculation has been achieved.

As for the norms in (3.16.7), the first three trivially have value 1. For the other two norms given in (3.16.7), the developed recursion process is taken as follows (by omitting intermediate calculations in the last two norms):

$$\begin{aligned}
h_{2^6}^{*\,\prime} &= N_{2^4}([0,\ 0,\ 1,\ 0,\ 0,\ -1,\ 1,\ 0]) = N_{2^4}([0,\ 1,\ 0,\ 1] + [0,\ 0,\ -1,\ 0]\zeta_{2^4}) \\
&= N_{2^3}([0,\ 1,\ 0,\ 1]^2 - \zeta_{2^3}[0,\ 0,\ -1,\ 0]^2) = N_{2^3}([-2,\ 1,\ 0,\ 0]) \\
&\qquad\qquad = N_{2^3}([-2,\ 0] + [1,\ 0]\zeta_{2^3}) \\
&= N_{2^2}([-2,\ 0]^2 - \zeta_{2^2}[1,\ 0]^2) = N_{2^2}([4,\ -1]) = N_{2^2}(4 - \zeta_{2^2}) \\
&= 4^2 + 1^2 = 17,
\end{aligned}$$

---

[107] One can execute the operations of square-forming and multiplication with ζ_{2^ρ} by a simple scheme, which reads, for example for $\rho = 3$,

	a_0	a_1	a_2	a_3
a_0	a_0^2	a_0a_1	a_0a_2	a_0a_3
a_1	$-a_1a_3$	a_1a_0	a_1^2	a_1a_2
a_2	$-a_2^2$	$-a_2a_3$	a_2a_0	a_2a_1
a_3	$-a_3a_1$	$-a_3a_2$	$-a_3^2$	a_3a_0
	*	*	*	*

$$\zeta_{2^3}[a_0,\ a_1,\ a_2,\ a_3] = [-a_3,\ a_0,\ a_1,\ a_2].$$

In the place of the stars, as by ordinary multiplication of numbers, the sums of the columns are to be registered.

(Translator's remark) In the table, for example, the second row shows that, letting $\zeta = \zeta_{2^3}$,

$$a_1\zeta_{2^3}[a_0,\ a_1,\ a_2,\ a_3] = a_1\zeta(a_0 + a_1\zeta + a_2\zeta^2 + a_3\zeta^3) = -a_1a_3 + a_1a_0\zeta + a_1^2\zeta^2 + a_1a_2\zeta^3.$$

Let A_0, A_1, A_2, and A_3 be the sums of the first, second, third, and fourth columns, respectively. Then we have

$$(a_0 + a_1\zeta + a_2\zeta^2 + a_3\zeta^3)^2 = A_0 + A_1\zeta + A_2\zeta^2 + A_3\zeta^3.$$

[108] (Translator's remark) The norm N_{2^1} in (3.16.15) in the original book should be N_{2^2}. The norm $N_{2^{\rho-1}}$ in (3.16.15) in the original book published in 1952 should be $N_{2^{\rho+1}}$.

$$h_{2^7}^{*\,\prime} = N_{2^5}([0,\ 1,\ -1,\ 0,\ 0,\ 0,\ -1,\ -1] + [0,\ 0,\ 0,\ 1,\ 0,\ 0,\ 1,\ 1]\zeta_{2^5})$$

$$= N_{2^4}([0,\ 3,\ 0,\ 1] + [-2,\ 0,\ -1,\ -2]\zeta_{2^4})$$

$$= N_{2^3}([2,\ 12] + [-3,\ 0]\zeta_{2^3})$$

$$= N_{2^2}(-140 + 39\zeta_{2^2})$$

$$= 140^2 + 39^2 = 21{,}121.$$

For this number calculation there exists a practical test that for the number $h_{2^\rho}^{*\,\prime}$ (relatively prime to 2), due to the norm expression (3.16.6′), the congruence

$$h_{2^\rho}^{*\,\prime} \equiv 1 \pmod{2^{\rho-2}}$$

must hold by the property of the class field of $P_{\rho-2}$.[109]

By the calculation mentioned above and by (3.16.5), the first five relative class numbers $h_{2^\rho}^{*\,\prime}$ and $h_{2^\rho}^{*}$ have the following values:

$$h_{2^3}^{*\,\prime} = 1, \qquad h_{2^3}^{*} = 1,$$

$$h_{2^4}^{*\,\prime} = 1, \qquad h_{2^4}^{*} = 1,$$

$$h_{2^5}^{*\,\prime} = 1, \qquad h_{2^5}^{*} = 1,$$

$$h_{2^6}^{*\,\prime} = 17 \text{ (prime number)}, \qquad h_{2^6}^{*} = 17,$$

$$h_{2^7}^{*\,\prime} = 21{,}121 \text{ (prime number)}, \ h_{2^7}^{*} = 17 \cdot 12{,}121.$$

These values have already been calculated by Weber [25]. However he did not indicate any systematic procedure for the calculation of norms, and he restricted to the mere announcement of the result for the last value. I have moreover calculated by the systematic process above the following value:

$$h_{2^8}^{*\,\prime} = 29{,}102{,}880{,}226{,}241.$$

[109]During the execution of the calculation, moreover, a consecutive check is desirable. One obtains such a check, for example, by carrying out at the same time the calculation mod p, where p is an appropriate prime number. Particularly, the prime numbers p with $p \equiv 1 \pmod{2^{\rho-2}}$ (for example, such smallest) are suitable, because for them (the primes p) the existent $2^{\rho-2}$-th roots of unity are congruent to rational integers. Then one works favorably by tables of indices of residue system mod p for a primitive root (mod p) as a role of the basis, for example, as they are in *Canon Arithmeticus* of Jacobi [11].

I have not yet checked whether this value is a prime or not.[110] The whole mechanical process of calculation, in which very large numbers appear at the final stage, needed only about 2 h without help of a table of squares, for which (the calculation by) using a table of squares could have been saved.

Since $h_{23}^{*}{}'$, $h_{24}^{*}{}'$, $h_{25}^{*}{}' = 1$ and since $h_{26}^{*}{}', \ldots \equiv 1 \pmod{2^4}$ hold, we can moreover determine by (3.16.5) that

$$h_{2^\rho}^{*} \equiv 1 \pmod{2^4} \qquad (\rho \geqq 3).$$

Calculation of the Relative Class Numbers $h_{3^\rho}^{*}$ of $P_{3^\rho}/P_{3^\rho,0}$

Here there exists no analogue to the subfields P_{2^ρ}'. However the relative class numbers $h_{3^\rho}^{*}$ are also constructed here—and the correspondence holds completely in general—by simpler arithmetic construction bricks, namely, by the contributions h_{ψ}^{*} of representatives ψ of characters of $P_{3^\rho}/P_{3^\rho,0}$.

By corresponding to the basis expression[111]

$$x \equiv (-1)^{\alpha}(1+3)^{\beta} \pmod{3^\rho} \qquad (\alpha \bmod 2,\ \beta \bmod 3^{\rho-1})$$

of the group of the prime residues mod 3^ρ, the group of characters of P_{3^ρ} is generated by the two characters

$$\omega(x) = (-1)^{\alpha}, \quad \psi_\rho(x) = \zeta_{3^{\rho-1}}^{\beta},$$

which are regarded as analogues of the generating characters $\cup$, ψ_ρ of P_{2^ρ}.[112] The characters $\omega\psi_\sigma$ ($\sigma = 1, \ldots, \rho$) construct a system of non-algebraic-conjugate characters ψ of $P_{3^\rho}/P_{3^\rho,0}$. By the relative class number formula (3.15.2) and since here $Q = 1$ again by Theorem 3.23, one has therefore, analogously to (3.16.2),

$$h_{3^\rho}^{*} = 2\prod_{\sigma=1}^{\rho} h_{\omega\psi_\sigma}^{*} \qquad (\rho \geqq 1), \tag{3.16.16}$$

where by (3.15.3) it holds that, analogously to (3.16.3),

$$h_{\omega\psi_\rho}^{*} = 3N_{\psi_\rho}(\Theta(\omega\psi_\rho)) \qquad (\rho \geqq 1). \tag{3.16.17}$$

---

[110](Translator's remark) As remarked by Martinet in the second edition, this value is a prime.

[111](Translator's remark) In the original book this "(mod 3^ρ)" is missing.

[112](Translator's remark) The letter υ in the original book should be $\cup$.

For $\rho = 1$ one has $\psi_\rho = 1$, and as is well known, by (3.9.2) it holds that, analogously to (3.16.4),

$$h_{3^1}^* = 2h_\omega^* = 2 \cdot 3 \cdot \Theta(\omega) = 2 \cdot 3 \cdot \frac{-1+2}{2 \cdot 3} = 1. \tag{3.16.18}$$

Thus, by (3.16.16), (3.16.17), and (3.16.18), there arises the system of recurrence relations analogous to (3.16.5)

$$h_{3^\rho}^* = h_{\omega\psi_2}^* \dots h_{\omega\psi_\rho}^* \qquad (\rho \geqq 2). \tag{3.16.19}$$

Since the characters $\omega\psi_\rho$ are regular for $\rho \geqq 2$, the arithmetic construction bricks $h_{\omega\psi_\rho}^*$ of $h_{3^\rho}^*$ appearing in (3.16.19) are integral and then natural numbers.

For the numerical calculation of $h_{\omega\psi_\rho}^*$ we first develop an analogue to Weber's formula (3.16.6′) taken above as a starting point. This (Weber's formula (3.16.6′)) traces back to formula (3.14.9), which itself was obtained by formula (3.14.5) (or (3.9.5)). Therefore we must first present the deduction of an analogue of the latter-mentioned formula.

For this purpose, analogously to the anti-symmetry in the proof of (3.9.5), we depend here on the legality of the value distribution of characters $\omega\psi_\rho$ in the system of the least positive residues mod 3^ρ, and in fact here we have to take as a basis the classification of this residue system in the three thirds system[113]

$$0 \dots 1 \cdot 3^{\rho-1} \dots 2 \cdot 3^{\rho-1} \dots 3 \cdot 3^{\rho-1},$$

which is related to the cycle of three terms of solutions of the congruence $u^3 \equiv 1 \pmod{3^\rho}$ that is generated by $u \equiv 1 + 3^{\rho-1} \pmod{3^\rho}$.[114] For $u \equiv 1 + 3^{\rho-1} \pmod{3^\rho}$, it holds that

$$\omega(u) = 1, \quad \psi_\rho(u) = \varepsilon;$$

here the primitive third root of unity $\zeta_{3^{\rho-1}}^{3^{\rho-2}} = \zeta_3 = \varepsilon$ $(= \frac{-1+\sqrt{-3}}{2})$ is fixed (adopted) for a clearly arranged form of later formulas. Consequently for the value distributions of ω and ψ_ρ in the above-mentioned thirds system, one immediately deduces the rules

$$\omega(x + 3^{\rho-1}) = \omega(x), \quad \psi_\rho(x + 3^{\rho-1}) = \varepsilon^{\omega(x)}\psi_\rho(x). \tag{3.16.20}$$

---

[113](Translator's remark) Let thirds system consist of, for example, classes $a \bmod 3^\rho$ with $0 \leqq a \leqq 3^{\rho-1}$, $3^{\rho-1}+1 \leqq a \leqq 2 \cdot 3^{\rho-1}$ or $2 \cdot 3^{\rho-1}+1 \leqq a \leqq 3^\rho$.

[114](Translator's remark) The solutions of the congruence $u^3 \equiv 1 \pmod{3^\rho}$ are

$$u \equiv 1,\ 1+3^{\rho-1},\ 1+2 \cdot 3^{\rho-1} \pmod{3^\rho}.$$

Moreover, on account of

$$\omega(-x) = -\,\omega(x), \quad \psi_\rho(-x) = \psi_\rho(x), \tag{3.16.21}$$

there arise further the rules

$$\omega(3^{\rho-1} - x) = -\,\omega(x), \quad \psi_\rho(3^{\rho-1} - x) = \varepsilon^{-\omega(x)}\psi_\rho(x), \tag{3.16.22}$$

which correspond to the classification of the three thirds system (the system of residues mod $3^{\rho-1}$) in each two half system mod $3^{\rho-1}$. On the basis of the legalities (3.16.20) and (3.16.22) of the value distributions of ω and ψ_ρ, for the underlying number $\Theta(\omega\psi_\rho)$ in (3.16.17) that originated from the definition formula (3.9.2), one obtains the following transformation, by which the apparent denominators are reduced except for the remaining denominator $\sqrt{-3}$:

$$\begin{aligned}
\Theta(\omega\psi_\rho) &= -\frac{1}{2\cdot 3^\rho} \sum_{x \bmod 3^\rho}{}^{+} \omega(x)\psi_\rho(x)x \\
&= -\frac{1}{2\cdot 3^\rho} \sum_{x \bmod 3^{\rho-1}}{}^{+} \omega(x)\psi_\rho(x) \\
&\qquad \times \left(x + \varepsilon^{\omega(x)}(x + 3^{\rho-1}) + \varepsilon^{2\omega(x)}(x + 2\cdot 3^{\rho-1})\right) \\
&= -\frac{1}{2\cdot 3} \sum_{x \bmod 3^{\rho-1}}{}^{+} \omega(x)\psi_\rho(x)(\varepsilon^{\omega(x)} + 2\varepsilon^{2\omega(x)}) \\
&= -\frac{1}{2\cdot 3} \sum_{\pm x \bmod 3^{\rho-1}}{}^{+} \omega(x)\psi_\rho(x) \\
&\qquad \times \left((\varepsilon^{\omega(x)} + 2\varepsilon^{2\omega(x)}) - \varepsilon^{-\omega(x)}(\varepsilon^{-\omega(x)} + 2\varepsilon^{-2\omega(x)})\right) \\
&= -\frac{1}{3} \sum_{\pm x \bmod 3^{\rho-1}}{}^{+} \omega(x)\psi_\rho(x)(\varepsilon^{2\omega(x)} - 1) \\
&= -\frac{1}{3} \sum_{\pm x \bmod 3^{\rho-1}}{}^{+} \varepsilon^{\omega(x)}\psi_\rho(x)\cdot\omega(x)(\varepsilon^{\omega(x)} - \varepsilon^{-\omega(x)});
\end{aligned}$$

since $\omega(x)(\varepsilon^{\omega(x)} - \varepsilon^{-\omega(x)}) = \varepsilon - \varepsilon^{-1} = \sqrt{-3}$ holds, analogously to (3.9.5) (or (3.14.5)), one has finally

$$\Theta(\omega\psi_\rho) = \frac{1}{\sqrt{-3}} \sum_{\pm x \bmod 3^{\rho-1}}{}^{+} \varepsilon^{\omega(x)}\psi_\rho(x) \qquad (\rho \geqq 2). \tag{3.16.23}$$

In order to obtain moreover an analogue to the invariant way of writing (3.14.5′), we introduce functions of the prime residues mod 3^ρ analogous to functions $s_{2^\rho}(x)$, $\delta_{2^\rho}(x)$ there. To do this, denote by $r_{3^\rho}(x)$ the least positive residue of x mod 3^ρ. Then we set

$$\delta_{3^\rho}(x) = \left\{ \begin{array}{rl} -1 \text{ for } & 0 < r_{3^\rho}(x) < 3^{\rho-1} \\ 0 \text{ for } & 3^{\rho-1} < r_{3^\rho}(x) < 2\cdot 3^{\rho-1} \\ 1 \text{ for } & 2\cdot 3^{\rho-1} < r_{3^\rho}(x) < 3\cdot 3^{\rho-1} \end{array} \right\}$$

and by this let

$$s_{3^\rho}(x) = \varepsilon^{-\delta_{3^\rho}(x)\omega(x)}.$$

With regard to the first formulas of (3.16.20) and (3.16.21), it holds that

$$\delta_{3^\rho}(x+3^{\rho-1}) \equiv \delta_{3^\delta}(x) + 1 \pmod 3, \qquad s_{3^\rho}(x+3^{\rho-1}) = \varepsilon^{-\omega(x)} s_{3^\rho}(x),$$

$$\delta_{3^\rho}(-x) = -\delta_{3^\delta}(x), \qquad s_{3^\rho}(-x) = s_{3^\rho}(x),$$

so that with regard to the second formulas of (3.16.20) and (3.16.21), the function $\psi_\rho(x)s_{3^\rho}(x)$ is invariant under the substitutions $x \to x + 3^{\rho-1}$ and $x \to -x$. For the x in the half system of the least positive residues mod $3^{\rho-1}$, it holds that $\delta_{3^\rho}(x) = -1$ and $s_{3^\rho}(x) = \varepsilon^{\omega(x)}$. Consequently, for formula (3.16.23) there arises the invariant way of writing

$$\Theta(\omega\psi_\rho) = \frac{1}{\sqrt{-3}} \sum_{\pm x \bmod 3^{\rho-1}} \psi_\rho(x) s_{3^\rho}(x) \tag{3.16.23′}$$

analogous to (3.14.5′), where the summation can be taken over an *arbitrary* half system of prime residues mod $3^{\rho-1}$.

In order to obtain moreover an analogue to (3.14.9) from this (3.16.23′), we select a fixed residue class z mod 3^ρ such that $\psi_\rho(z)$ is a primitive $3^{\rho-1}$-th root of unity, for example, $z \equiv 1+3 \bmod 3^\rho$. By multiplying formula (3.16.23′) by $1-\overline{\psi}_\rho(z)$, there arises

$$(1-\overline{\psi}_\rho(z))\Theta(\omega\psi_\rho) = \sum_{\pm x \bmod 3^{\rho-1}} \psi_\rho(x) \frac{s_{3^\rho}(x) - s_{3^\rho}(zx)}{\sqrt{-3}} \quad (\rho \geqq 2), \tag{3.16.24}$$

and hence, by taking the norm of (3.16.24), there arises with regard to (3.16.17)

$$h^*_{\omega\psi_\rho} = N_{\psi_\rho}\left(\sum_{\pm x \bmod 3^{\rho-1}} \psi_\rho(x)\frac{s_{3^\rho}(x) - s_{3^\rho}(zx)}{\sqrt{-3}}\right) \quad (\rho \geqq 2). \tag{3.16.25}$$

Since the primitive third roots of unity are congruent mod $\sqrt{-3}$ to each other, the apparent denominators in (3.16.24) and (3.16.25) are reduced. The explicit expression of coefficients of $\psi_\rho(x)$ in rational integral form is slightly more laborious than that of (3.14.9). Namely, one has here no explicit expression of normalized exponents $\mu = 0, \pm 1$ of an exponent ε^μ by this power analogous to (3.14.8). However, in this respect the following relation holds:

$$\varepsilon^\mu = \frac{2 - 3\mu^2 + \mu\sqrt{-3}}{2} \qquad \text{for } \mu = 0, \pm 1.$$

From this it follows that the relation

$$\frac{\varepsilon^{-\mu} - \varepsilon^{-\mu'}}{\sqrt{-3}} = \frac{-3(\mu^2 - \mu'^2) - (\mu - \mu')\sqrt{-3}}{2\sqrt{-3}} = (\mu - \mu')\frac{-1 + (\mu + \mu')\sqrt{-3}}{2}$$

$$= \left\{\begin{matrix} -\mu & \text{for } \mu' = -\mu \\ (\mu - \mu')\varepsilon^{\mu+\mu'} & \text{for } \mu' \neq -\mu \end{matrix}\right\}$$

is valid for $\mu, \mu' = 0, \pm 1$.[115] Hence, for $\mu = \delta_{3^\rho}(x)\omega(x)$, $\mu' = \delta_{3^\rho}(zx)\omega(zx) = \delta_{3^\rho}(zx)\omega(x)$, there arises

$$\frac{s_{3^\rho}(x) - s_{3^\rho}(zx)}{\sqrt{-3}}$$

$$= \left\{\begin{matrix} -\delta_{3^\rho}(x)\omega(x) & \text{for } \delta_{3^\rho}(zx) = -\delta_{3^\rho}(x) \\ (\delta_{3^\rho}(x) - \delta_{3^\rho}(zx))\omega(x)\,\varepsilon^{(\delta_{3^\rho}(x)+\delta_{3^\rho}(zx))\omega(x)} & \text{for } \delta_{3^\rho}(zx) \neq -\delta_{3^\rho}(x) \end{matrix}\right\}. \tag{3.16.26}$$

Substituting these values for the coefficients in the right-hand sides of (3.16.24) and (3.16.25), one obtains analogues to (3.14.9) and to formula (3.16.6) above.

115(Translator's remark) "$\mu' \neq -\mu$" in the relation above in the original book should be "$\mu' \neq -\mu$" without parenthesis.

Finally, in order to obtain an analogue to Weber's formula (3.16.6′), we select the half system of exponential prime residues

$$x \equiv z^{\nu} \pmod{3^{\rho}} \quad (\nu \bmod 3^{\rho-2}) \quad \text{with } z \equiv 1+3 \equiv 4 \pmod{3^{\rho}}.$$

Thus there arises by (3.16.25) the formula analogous to (3.16.6′)

$$h^*_{\omega\psi_\rho} = N_{3^{\rho-1}}\left(\sum_{\nu \bmod 3^{\rho-2}} e_{3^\rho}^{(\nu)} \zeta_{3^{\rho-1}}^{\nu}\right) \quad (\rho \geqq 2), \tag{3.16.27}$$

where $N_{3^{\rho-1}}$ denotes the norm of the cyclotomic field $P_{3^{\rho-1}}$ and the coefficients are, according to (3.16.26), given by

$$e_{3^\rho}^{(\nu)} = \frac{s_{3^\rho}(4^\nu) - s_{3^\rho}(4^{\nu+1})}{\sqrt{-3}} = \left\{\begin{array}{ll} -\delta^{(\nu)} & \text{for } \delta^{(\nu+1)} = -\delta^{(\nu)} \\ (\delta^{(\nu)} - \delta^{(\nu+1)})\varepsilon^{\delta^{(\nu)}+\delta^{(\nu+1)}} & \text{for } \delta^{(\nu+1)} \neq -\delta^{(\nu)} \end{array}\right\}, \tag{3.16.28}$$

with $\delta^{(\nu)} = \delta_{3^\rho}(4^\nu)$ for short.

Corresponding completely to the situation in Sect. 3.14 that a further proof of the fact (3.14.6) or (3.14.10) has arisen from (3.14.9), which is equivalent to Weber's Theorem 3.36′ by formulas (3.14.3) and (3.14.5) above, a further proof of Theorem 3.37′ arises here from (3.16.27). Namely, denote by $\mathfrak{t}$ the unique prime divisor of 3 in the field $P_{3^{\rho-1}}$, then $\zeta_{3^{\rho-1}}^{\nu} \equiv 1 \pmod{\mathfrak{t}}$ for all integers ν, and one has

$$\sum_{\nu \bmod 3^{\rho-2}} e_{3^\rho}^{(\nu)} \zeta_{3^{\rho-1}}^{\nu} \equiv \sum_{\nu \bmod 3^{\rho-2}} e_{3^\rho}^{(\nu)} \pmod{\mathfrak{t}}.$$

Then by (3.16.28) it holds that

$$\begin{aligned}\sum_{\nu \bmod 3^{\rho-2}} e_{3^\rho}^{(\nu)} &= \sum_{\nu \bmod 3^{\rho-2}} \frac{s_{3^\rho}(4^\nu) - s_{3^\rho}(4^{\nu+1})}{\sqrt{-3}} = \frac{s_{3^\rho}(1) - s_{3^\rho}(1+3^{\rho-1})}{\sqrt{-3}} \\ &= \frac{\varepsilon - 1}{\sqrt{-3}} = -\varepsilon^{-1},\end{aligned}$$

so that by (3.16.27) the $h^*_{\omega\psi_\rho}$ are, in fact, prime to 3 and so are the $h^*_{3^\rho}$ by (3.16.19). Therefore, by virtue of the class field property of $P_{3^{\rho-1}}$, it follows from the norm expression (3.16.27) that the congruence[116]

$$h^*_{\omega\psi_\rho} \equiv 1 \pmod{3^{\rho-1}}$$

[116](Translator's remark) This congruence holds for $\rho \geqq 2$.

holds more precisely and hence by (3.16.19) in any case ($\rho \geqq 1$)

$$h_{3^\rho}^* \equiv 1 \pmod 3, \tag{3.16.29}$$

which we will later sharpen a little more.

The coefficients $e_{3^\rho}^{(\nu)}$ appearing in (3.16.27) and given in (3.16.28) are 0 or sixth roots of unity $\pm 1, \pm\varepsilon, \pm\varepsilon^2$. They are connected with the alternation of the thirds system to which the least positive residues mod 3^ρ of power series 4^ν belong. If the least positive residues 4^ν mod 3^ρ and $4^{\nu+1}$ mod 3^ρ lie in the same system ($\delta^{(\nu+1)} = \delta^{(\nu)}$), then $\varepsilon_{3^\rho}^{(\nu)} = 0$. On the other hand, if an alternation of the thirds system occurs by the transformation from 4^ν to $4^{\nu+1}$ ($\delta^{(\nu+1)} \neq \delta^{(\nu)}$), then $e_{3^\rho}^{(\nu)}$ is one of the sixth roots of unity and in fact $e_{3^\rho}^{(\nu)}$ corresponds to

$$\left\{\begin{array}{l} \left\{\begin{array}{llll} \text{cycle of alternations } (\delta^{(\nu)}), \delta^{(\nu+1)}) & (+1,\ 0), & (0,\ -1), & (-1,\ +1) \\ \text{cycle of roots of unity} \quad e_{3^\rho}^{(\nu)} & \varepsilon, & \varepsilon^2, & 1 \end{array}\right\}, \\ \left\{\begin{array}{llll} \text{cycle of alternations } (\delta^{(\nu)}), \delta^{(\nu+1)}) & (-1,\ 0), & (0,\ +1), & (+1,\ -1) \\ \text{cycle of roots of unity} \quad e_{3^\rho}^{(\nu)} & -\varepsilon^2, & -\varepsilon, & -1 \end{array}\right\}. \end{array}\right\} \tag{3.16.30}$$

Moreover, the thirds system, which corresponds to the definition of $\delta_{3^\rho}(x)$ above, are characterized by the values $-1,\ 0,\ +1$. (See Fig. 3.6.)

Hence, for the calculation of the coefficients $e_{3^\rho}^{(\nu)}$, the following scheme (of $\delta^{(\nu)} = \delta_{3^\rho}(4^\nu)$) arises:

4^0	4^1	4^2	4^3	4^4	4^5	4^6	4^7	4^8	4^9	......	
1	4										mod 3^2
−1	0										
1	4	16	10								mod 3^3
−1	−1	0	0								
1	4	16	64	13	52	46	22	7	28		mod 3^4
−1	−1	−1	+1	−1	0	0	−1	−1	0		
....											....

Fig. 3.6 The values of $\delta_{3^\delta}(x)$

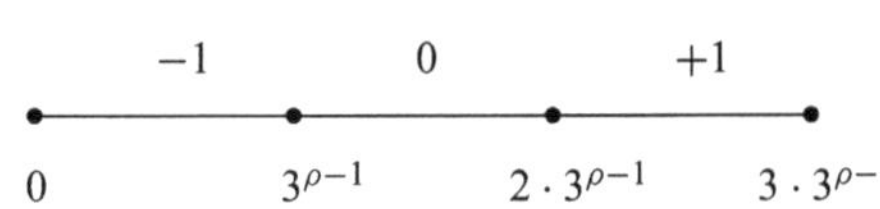

By (3.16.27) and (3.16.30) one reads (the following values) from the alternation of the thirds system in the rows of this scheme:

$$\left.\begin{aligned} h^*_{\omega\psi_2} &= N_{31}(-\varepsilon^2\zeta_3^0) \\ h^*_{\omega\psi_3} &= N_{32}(-\varepsilon^2\zeta_{32}^1) \\ h^*_{\omega\psi_4} &= N_{33}(\zeta_{33}^2 - \zeta_{33}^3 - \varepsilon^2\zeta_{33}^4 + \varepsilon^2\zeta_{33}^6 - \varepsilon^2\zeta_{33}^8) \\ &\dots\dots\dots\dots\dots\dots\dots . \end{aligned}\right\} \qquad (3.16.31)$$

For the calculation of the norms in (3.16.31), one can develop again a recurrence procedure, which is analogous to the formulas given extensively in (3.16.8)–(3.16.15) above, only that one calculates practically (the norms) in the field $P(\varepsilon) = P_3$ of the coefficients $e_{3^\rho}^{(\nu)} = 0, \pm 1, \pm\varepsilon, \varepsilon^2$ instead of in the rational number field P. Then one has the basic equations

$$\zeta_{3^{\rho+1}}^{3^\rho} = \varepsilon \quad \text{for } P_{3^{\rho+1}}/P(\varepsilon), \qquad \zeta_{3^{\rho+1}}^3 = \zeta_{3^\rho} \quad \text{for } P_{3^{\rho+1}}/P_{3^\rho}$$

and the basis expressions

$$\xi_{3^{\rho+1}} = a_0 + a_1\zeta_{3^{\rho+1}} + \dots + a_{3^\rho-1}\zeta_{3^{\rho+1}}^{3^\rho-1} = [a_0, a_1, \dots, a_{3^\rho-1}]$$
$$\text{with } a_0, a_1, \dots, a_{3^\rho-1} \text{ in } P(\varepsilon),$$

$$\xi_{3^{\rho+1}} = \alpha_{3^\rho} + \beta_{3^\rho}\zeta_{3^{\rho+1}} + \gamma_{3^\rho}\zeta_{3^{\rho+1}}^2 \quad \text{with} \left\{\begin{aligned} \alpha_{3^\rho} &= [a_0, a_3, \dots, a_{3^\rho-3}] \\ \beta_{3^\rho} &= [a_1, a_4, \dots, a_{3^\rho-2}] \\ \gamma_{3^\rho} &= [a_2, a_5, \dots, a_{3^\rho-1}] \end{aligned}\right\} \text{ in } P_{3^\rho}.$$

For the relative norm one has here

$$N_{3^{\rho+1}/3^\rho}(\xi_{3^{\rho+1}}) = \alpha_{3^\rho}^3 + \zeta_{3^\rho}\beta_{3^\rho}^3 + \zeta_{3^\rho}^2\gamma_{3^\rho}^3 - 3\zeta_{3^\rho}\alpha_{3^\rho}\beta_{3^\rho}\gamma_{3^\rho}.$$

Consequently there arises the recurrence formula

$$N_{3^{\rho+1}}(\alpha_{3^\rho}+\beta_{3^\rho}\zeta_{3^{\rho+1}}+\gamma_{3^\rho}\zeta_{3^{\rho+1}}^2)=N_{3^\rho}(\alpha_{3^\rho}^3+\zeta_{3^\rho}\beta_{3^\rho}^3+\zeta_{3^\rho}^2\gamma_{3^\rho}^3-3\zeta_{3^\rho}\alpha_{3^\rho}\beta_{3^\rho}\gamma_{3^\rho})$$

and hence by the running of the recurrence (there arises) the final formula[117]

$$N_{3^{\rho+1}}(\alpha_{3^\rho}+\beta_{3^\rho}\zeta_{3^{\rho+1}}+\gamma_{3^\rho}\zeta_{3^{\rho+1}}^2)=N_{3^1}(\alpha_{3^1}^3+\zeta_{3^1}\beta_{3^1}^3+\zeta_{3^1}^2\gamma_{3^1}^3-3\zeta_{3^1}\alpha_{3^1}\beta_{3^1}\gamma_{3^1}),$$

in which the norm of $P_{3^1}=P(\varepsilon)$ can be calculated in a well-known way.

For the third norm given in (3.16.31)—the first two norms have trivial value 1—this recurrence procedure without showing the reduction calculation[118] is given as follows:

$$\begin{aligned}
h^*_{\omega\psi_4} &= N_{3^3}([0,\ 0,\ 1,\ -1,\ -\varepsilon^2,\ 0,\ \varepsilon^2,\ 0,\ -\varepsilon^2]) \\
&= N_{3^3}([0,\ -1,\ \varepsilon^2]+[0,\ -\varepsilon^2,\ 0]\zeta_{3^3}+[1,\ 0,\ -\varepsilon^2]\zeta_{3^3}^2) \\
&= N_{3^2}([0,\ -1,\ \varepsilon^2]^3+\zeta_{3^2}[0,\ -\varepsilon^2,\ 0]^3+\zeta_{3^2}^2[1,\ 0,\ -\varepsilon^2]^3 \\
&\qquad -3\zeta_{3^2}[0,\ -1,\ \varepsilon^2][0,\ -\varepsilon^2,\ 0][1,\ 0,\ -\varepsilon^2]) \\
&= N_{3^2}((-\varepsilon-2\varepsilon^2)+(-\varepsilon+3\varepsilon^2)\zeta_{3^2}+(-\varepsilon-2\varepsilon^2)\zeta_{3^2}^2) \\
&= N_3((-\varepsilon-2\varepsilon^2)^3+\varepsilon(-\varepsilon+3\varepsilon^2)^3+\varepsilon^2(-\varepsilon-2\varepsilon^2)^3 \\
&\qquad -3\varepsilon(-\varepsilon-2\varepsilon^2)(-\varepsilon+3\varepsilon^2)(-\varepsilon-2\varepsilon^2)) \\
&= N_3(41\varepsilon+57\varepsilon^2)=N_3(-49-8\sqrt{-3})=49^2+3\cdot 8^2=2593.
\end{aligned}$$

Hereby and by (3.16.19) the first three contributions $h^*_{\omega\psi_\rho}$ and the relative class numbers $h^*_{3^\rho}$ $(\rho \geqq 2)$ have the following values:

$$h^*_{\omega\psi_2}=1, \qquad h^*_{3^2}=1,$$

$$h^*_{\omega\psi_3}=1, \qquad h^*_{3^3}=1,$$

$$h^*_{\omega\psi_4}=2593 \text{ (prime number)}, \ h^*_{3^4}=2593.$$

[117](Translator's remark) "$\cdots-\zeta_{3^1}\alpha_{3^1}\beta_{3^1}\gamma_{3^1}$" in the original book should be "$\cdots-3\zeta_{3^1}\alpha_{3^1}\beta_{3^1}\gamma_{3^1}$," which Martinet pointed out in his Preface in the second edition.

[118]For this reduction one can develop a completely corresponding schematic procedure, as it has been suggested previously in footnote 107, p. 146.

These values have already been found in the table of the relative class number h_f^* of $P_f/P_{f,0}$ for all $f \leqq 100$ calculated by Kummer [19]. I have moreover calculated by the above-mentioned procedure the next value:

$$h_{\omega\psi_5}^* = 5{,}764{,}966{,}319{,}758{,}245{,}087{,}799.$$

I have not yet checked whether this value is also a prime number or not.[119] Since $h_{\omega\psi_2}^* = 1$, $h_{\omega\psi_3}^* = 1$, $h_{\omega\psi_4}^* \equiv 1 \pmod{3^4}$ (instead of only mod 3^3),[120] congruence (3.16.29) above is sharpened to

$$h_{3^\rho}^* \equiv 1 \pmod{3^4} \qquad (\rho \geqq 2). \tag{3.16.29'}$$

Remarks on the General Case $f(\psi) = p^\rho$ ($p \neq 2$, $\rho \geqq 2$)

In connection with these special investigations we will further make a remark on a generalization of formula (3.16.23) derived in the preceding special case $f(\psi) = 3^\rho$ ($\rho \geqq 2$) to the case of arbitrary higher power $f(\psi) = p^\rho$ ($p \neq 2$, $\rho \geqq 2$). For this purpose we settle in formula (3.9.2) the system of the least positive prime residues mod p^ρ in the form $x + yp^{\rho-1}$ such that x runs over the system of the least positive prime residues mod $p^{\rho-1}$ and y over the system of the full least positive residues mod p. In accordance with the decomposition

$$x + yp^{\rho-1} \equiv x(1 + yx^{-1}p^{\rho-1}) \equiv x(1+p)^{yx^{-1}p^{\rho-2}} \pmod{p^\rho},$$

where x^{-1} is to be understood as a representative of the inverse class of x mod p, one has then the rule

$$\psi(x + yp^{\rho-1}) = \psi(x)\zeta^{yx^{-1}}, \tag{3.16.32}$$

where

$$\zeta = \psi(1 + p^{\rho-1}) = \psi(1+p)^{p^{\rho-2}}$$

[119](Translator's remark) As remarked by Martinet, this number is a composite:

$$5{,}764{,}966{,}319{,}758{,}245{,}087{,}799 = 6{,}252{,}002{,}011 \cdot 922{,}099{,}242{,}709.$$

[120](Translator's remark) We have shown in this section that $h_{\omega\psi_\rho}^* \equiv 1 \pmod{3^{\rho-1}}$ for every $\rho \geqq 2$.

is a primitive p-th root of unity. Consequently there arises first the transformation

$$\begin{aligned}\Theta(\psi) &= -\frac{1}{2p^{\rho}} \sum_{x \bmod p^{\rho}}^{+} \psi(x)x \\ &= -\frac{1}{2p^{\rho}} \sum_{x \bmod p^{\rho-1}}^{+} \sum_{y \bmod p}^{+} \psi(x+yp^{\rho-1})(x+yp^{\rho-1}) \\ &= -\frac{1}{2p^{\rho}} \sum_{x \bmod p^{\rho-1}}^{+} \psi(x) \sum_{y \bmod p}^{+} (x+yp^{\rho-1})\zeta^{yx^{-1}} \\ &= -\frac{1}{2p} \sum_{x \bmod p^{\rho-1}}^{+} \psi(x) \sum_{y \bmod p}^{+} y\zeta^{yx^{-1}} \\ &= \frac{1}{2} \sum_{x \bmod p^{\rho-1}}^{+} \frac{\psi(x)}{1-\zeta^{x^{-1}}};\end{aligned}$$

the last equation is obtained immediately by attention to the fact confirmed by multiplication that

$$(\zeta + 2\zeta^2 + \cdots + (p-1)\zeta^{p-1})(\zeta - 1) = p$$

for every primitive p-th root of unity ζ. In the formula obtained here we moreover reduce the summation over the *half* system of the least positive residues x mod $p^{\rho-1}$, in which we apply the rule

$$\psi(p^{\rho-1} - x) = -\psi(x)\zeta^{-x^{-1}}$$

obtained by (3.16.32) and by $\psi(-1) = -1$. Since $1 - \zeta^{x^{-1}}$ has the same behavior as $\psi(x)$ by the substitution $x \to p^{\rho-1} - x$, the expression $\psi(x)/(1 - \zeta^{x^{-1}})$ is invariant under this substitution. So there arises the further transformation

$$\Theta(\psi) = \sum_{\pm x \bmod p^{\rho-1}}' \frac{\psi(x)}{1-\zeta^{x^{-1}}} \quad \text{with } \zeta = \psi(1+p^{\rho-1}). \tag{3.16.33}$$

This formula can also be written in the form

$$\Theta(\psi) = \frac{1}{1-\zeta} \sum_{\pm x \bmod p^{\rho-1}}^{+} \psi(x)\eta^{(x^{-1})}, \tag{3.16.33'}$$

where (the set of units)

$$\eta^{(x^{-1})} = \frac{1-\zeta}{1-\zeta^{x^{-1}}} = 1 + \zeta^{x^{-1}} + \cdots + \zeta^{(x-1)x^{-1}}$$

is the famous system of units of the cyclotomic field P_p (this system is also describable easily by cyclotomic units in Sect. 2.4).

In the special case of $p = 3$, the obtained formula (3.16.33′) turns out to be the previous formula (3.16.23) when one pays attention to

$$\frac{1}{1-\varepsilon} = \frac{\varepsilon}{\varepsilon - \varepsilon^{-1}} = \frac{\varepsilon}{\sqrt{-3}}$$

and $\eta^{(1)} = 1$ and $\eta^{(2)} = 1 + \varepsilon^2 = -\varepsilon$ are valid for the primitive cube root of unity $\zeta = \varepsilon$.

The initial apparent denominator $2p^\rho$ of (3.9.2) is reduced to be the remaining denominator $1-\zeta$ in (3.16.33) and (3.16.33′), and then to the unique prime divisor of p in the cyclotomic field P_p. This is not yet the real denominator; by Theorem 3.32, this is rather 1 or one of the $\psi(p-1)$ prime divisors of p in the cyclotomic field $P_{p-1}P_{p^{\rho-1}}$ (the latter case appears at most in the special case where the order of ψ is $n_\psi = \varphi(p^\rho)$—then $1-\zeta$ represents the $p^{\rho-2}$-th power of the relative norm of this prime divisor for $P_{p-1}P_{p^{\rho-1}}/P_{p^{\rho-1}}$). Hereafter, further transformations of (3.16.33) and (3.16.33′) by which this denominator law could be clear would have to be possible. In the special case of $p = 3$, such a transformation above has already been realized in connection with (3.16.23) and also with result (3.16.24)—or by taking the norms in (3.16.27) and (3.16.28). I have not yet been able to carry out this generalization to arbitrary odd prime number p. Nevertheless, result (3.16.33′) presents surely also an essential facilitation for the calculation of the relative class number contributions $N_\psi(\Theta(\psi))$ of characters ψ with higher odd prime-power conductor.

Therefore, by way of example,[121] one finds easily for the essentially unique character ψ_2 of $P_{5^2}/P_{5^2,0}$

$$\Theta(\psi_2) = \frac{1 + \psi_2(2)(1+\zeta^3)}{1-\zeta} \qquad \text{with} \quad \zeta = \psi_2(1+5).$$

By virtue of $2^4 = 16 = 1+3\cdot 5 \equiv (1+5)^3 \pmod{5^2}$, it holds here that $\psi_2(2)^4 = \zeta^3$, therefore $\psi_2(2) = i\zeta^2$ and hence

$$\Theta(\psi_2) = \frac{1 + i\zeta^2(1+\zeta^3)}{1-\zeta}.$$

[121](Translator's remark) As Martinet pointed out, in the following three equations the powers ζ^2 in $(1+\zeta^2)$ in the original book should be replaced by ζ^3 and the third equation should be corrected as printed here.

From this equation the norm map of $P_{4\cdot 5}/P_5$ produces

$$\mathbf{N}(\Theta(\psi_2)) = \frac{1+\zeta^4(1+\zeta^3)^2}{(1-\zeta)^2} = \frac{2+2\zeta^2+\zeta^4}{(1-\zeta)^2} = \frac{1+\zeta^2}{1-\zeta};$$

here the reduction of $1-\zeta$ meets our above-mentioned general knowledge on the still-necessarily-existing denominator reduction in formula (3.16.33′). Then the norm map of P_5 produces moreover

$$N_{\psi_2}(\Theta(\psi_2)) = N(\mathbf{N}(\Theta(\psi_2))) = \frac{1}{5}.$$

For the essentially unique character ψ_1 of $P_5/P_{5,0}$, one has directly from (3.9.2)

$$\Theta(\psi_1) = -\frac{1+2i-3i-4}{2\cdot 5} = \frac{3+i}{2\cdot 5} = \frac{1}{3-i},$$

and hence

$$N_{\psi_1}(\Theta(\psi_1)) = \frac{1}{2\cdot 5}.$$

By (3.9.1) and by giving attention to Theorem 3.23, there arise therefore the values for the relative class numbers h_5^* and $h_{5^2}^*$ of $P_5/P_{5,0}$ and $P_{5^2}/P_{5^2,0}$:

$$h_5^* = 2\cdot 5\cdot N_{\psi_1}(\Theta(\psi_1)) = 1,$$

$$h_{5^2}^* = 2\cdot 5^2\cdot N_{\psi_1}(\Theta(\psi_1))\cdot N_{\psi_2}(\Theta(\psi_2)) = 1.$$

I have further calculated the next value by the above-mentioned procedure:

$$h_{5^3}^* = 2\cdot 5^3\cdot N_{\psi_1}(\Theta(\psi_1))\cdot N_{\psi_2}(\Theta(\psi_2))\cdot N_{\psi_3}(\Theta(\psi_3)) = 57{,}708{,}445{,}601.$$

It is left open whether this value is a prime number or not.[122]

[122](Translator's remark) As remarked by Martinet in the second edition, this number is a composite:

$$h_{5^3}^* = 57{,}708{,}445{,}601 = 2801\cdot 20{,}602{,}801.$$

3.17 Remarks on the Genus Factor

In the direct proof of the integrality in Sect. 3.15, the divisibility of certain contributions $h^*_\psi = N_\psi(\Theta(\psi))$ by 2 established in Theorem 3.31 has not yet been fully utilized in the proof of Theorem 3.35—except for the products of three terms of irregular characters $\psi_1, \ldots, \psi_r$, the products of five terms, seven terms, ..., which indeed also provide contributions h^*_ψ divisible by 2. Moreover one can further turn to the algebraic-conjugate characters in the factors of these products[123]; then indeed one has to select again a system of representatives of non-algebraic-conjugate characters from all the products appearing in this way. In conclusion, one need not restrict oneself to the irregular characters $\psi_1, \ldots, \psi_r$ and to the characters of K/K_0 produced (by such characters) as above, but one can take into consideration all characters ψ of K/K_0 of 2-power order n_ψ and with the composite conductor $f(\psi)$ only in the case where $f(\psi)$ contains only two distinct prime divisors $p \neq 2$, q with the restriction $\left(\frac{q}{p}\right) = 1$ being moreover required. For all these characters ψ the contribution h^*_ψ is indeed divisible by 2 by Theorem 3.31 and by (3.15.3); furthermore, Theorem 3.31 states that in the last-mentioned case with $\left(\frac{q}{p}\right) = -1$, the contribution h^*_ψ is not divisible by 2. So one obtains beyond the integrality of h^* the divisibility of h^* by a certain power 2^ν ($\nu \geqq 0$).

Now one could think that just in this way the genus factor 2^γ ($\gamma \geqq 0$) of h^* determined by class field theory in Theorem 3.12 arises in the direct way. However, this is *not* always the case. Rather $\nu < \gamma$ or $\nu > \gamma$ can also occur. In the first case Theorem 3.31 does not yet provide the full genus factor, while in the second case Theorem 3.31 provides the statement beyond the result of the genus theory, namely, that the number $\frac{h^*}{2^\gamma}$ of the classes in the principal genus of K/K_0 is moreover divisible by $2^{\nu-\gamma}$.

We will explain this fact by an example of an imaginary quadratic field $K = P(\sqrt{-f})$. Here, on the one hand the number q^* appearing in Theorem 3.12 is $q^* = 0$, and then in a well-known way[124]

$$\gamma = \delta - 1,$$

where δ is the number of distinct prime divisors of the conductor f. On the other hand, one has here by formulas (3.15.2) and (3.15.1′)

$$h = h^* = 2h^*_\psi = 2\,\frac{w}{2}\,\Theta(\psi),$$

[123](Translator's remark) For example, as representatives we can take $\psi_1\psi_2\psi_3$ as well as $\psi_1\psi_2\psi_3^\mu$ with $(\mu, n_{\psi_3}) = 1$.

[124](Translator's remark) We have defined γ by $\gamma = \delta + q^* - 1$ in Sect. 3.1.

where ψ is the generating character of K. If $\gamma = 0$, then f is a power of a prime (a prime $p \equiv -1 \pmod 4$, or 2^2 or 2^3), and so ψ is irregular. Hence by Theorem 3.34 h_ψ^* has denominator 2. Therefore it holds that $\nu = 0$ and then $\nu = \gamma$. Moreover it follows from the well-known fact obtained by the genus theory that in this case h is not divisible by 2. If $\gamma \geqq 1$, then f is composite, and so ψ is regular. Hence by Theorem 3.34 h_ψ^* is integral. If much more $\gamma \geqq 2$, then f contains at least three distinct prime divisors. Therefore it holds that $\nu = 2$, and then $\nu = \gamma$ or $\nu < \gamma$ according as $\gamma = 2$ or $\gamma > 2$. But, if $\gamma = 1$, then f contains only two distinct primes $p \neq 2$, q, and so $\nu = 2$ or 1. Therefore $\nu > \gamma$ or $\nu = \gamma$ according as $\left(\frac{q}{p}\right) = 1$ or -1; in the latter case, h_ψ^* is not divisible by 2, and then h is divisible exactly by $2^1 = 2$. Thus we have the following statement beyond the result of the genus theory:

Theorem 3.39 *If $K = \mathbf{Q}(\sqrt{-f})$ is an imaginary quadratic field whose conductor f contains exactly two prime divisors $p \neq 2$, q, then the number $\frac{h}{2}$ of the classes in the principal genus of K is divisible by 2 or not according as $\left(\frac{q}{p}\right) = 1$ or -1.*

According to the above, Theorem 3.31 is, in general, not sufficient to prove the divisibility of h^* by the genus factor 2^γ directly from the relative class number formula (3.15.2). One needs rather a generalization in the direction that under appropriate conditions of ψ the divisibility of $N_\psi(\Theta(\psi))$ by a higher power of 2 could be determined. Such a kind of generalization of Theorem 3.31 has not been successful for me at all, even in the case of imaginary quadratic fields. Since the divisor 2^ν of h^* obtained by the above-mentioned method of Theorem 3.31 expresses nothing definitive, I abandon here its exact description and formulation for a corresponding generalization of Theorem 3.35 to an arbitrary imaginary abelian number field.

3.18 Divisibility by the Relative Class Number of a Subfield

We ask whether for an imaginary subfield $\tilde{K}$ of K the relative class number $\tilde{h}^*$ of $\tilde{K}/\tilde{K}_0$ is a divisor of the relative class number h^* of K/K_0.

For an investigation of this question, we rely on the form of the relative class number formula given by (3.15.2), namely,

$$h^* = 2Q \prod_\psi h_\psi^*, \tag{3.18.1}$$

where the contribution h^*_ψ is defined by (3.15.3) for every representative ψ of characters of K/K_0. One can also express the definition of contribution h^*_ψ as follows:

$$h^*_\psi = \left\{ \begin{array}{ll} pN_\psi(\Theta(\psi)) & \text{if } \psi \text{ is a generating character } \psi \text{ of } P_{p^\rho}\ (p \neq 2); \\ & \text{or if } w \equiv 0 \pmod 4 \text{ and } \psi \text{ together with } \cup \\ & \text{is a generating character of } P_{2^\rho}\ (p = 2), \\ N_\psi(\Theta(\psi)) & \text{otherwise} \end{array} \right\}. \tag{3.18.2}$$

Therefore the contribution h^*_ψ depends—completely corresponding to the regularity or irregularity of ψ as in Sect. 3.15—generally only on the character ψ and not on the field K; (however) only for the special characters $\psi = \cup\varphi_\rho$ $(\rho \geqq 3)$ it is also a matter of the field K, namely, a matter of whether the field K has the property $w \equiv 0 \pmod 4$ or not. As we have shown in the proof of (b) in Sect. 3.15, among the characters of K/K_0 for $w \not\equiv 0 \pmod 4$ there appears at most *one* character of type $\cup\varphi_\rho$ $(\rho \geqq 3)$ for which the contribution $h^*_\psi = N_\psi(\Theta(\psi))$ is to be assigned *without* 2, differently from the general rule.

Dividing the relative class number formula (3.18.1) for K/K_0 by the corresponding formula for $\tilde{K}/\tilde{K}_0$, one obtains accordingly[125]

$$\frac{h^*}{\tilde{h}^*} = 2^\tau \frac{Q}{\tilde{Q}} \prod_{\psi'} h^*_{\psi'} \qquad \text{with } \tau = 0 \text{ or } 1, \tag{3.18.3}$$

where ψ' runs over a system of non-algebraic-conjugate characters of K/K_0 that are not characters of $\tilde{K}/\tilde{K}_0$. Here by (3.18.2) in general we have $\tau = 0$; (however) $\tau = 1$ occurs only when $w \equiv 0 \pmod 4$, $\tilde{w} \not\equiv 0 \pmod 4$, and among the characters of $\tilde{K}/\tilde{K}_0$ there appears one character of type $\cup\psi_\rho$ $(\rho \geqq 3)$, that is, one odd character of 2-power conductor.

In order to obtain examples where $\tilde{h}^*$ is not a divisor of h^*, we consider the imaginary bicyclic biquadratic fields treated in Sect. 3.8. As in (3.8.1)–(3.8.4), let

$$K = K_1K_2 = P(\sqrt{-f_1}, \sqrt{-f_2}),$$

$$K_1 = P(\sqrt{-f_1}), \quad K_2 = P(\sqrt{-f_2})$$

$$K_0 = P(\sqrt{f_0}) \quad \text{with } f_0 \underset{2}{=} f_1 f_2,$$

[125]In the following the quantities regarding $\tilde{K}$ are indicated by tilde.

and let h_1, h_2 be the class numbers of K_1, K_2, respectively. If one takes, say, $\tilde{K} = K_2$ as an imaginary subfield, then $\tilde{h}^* = h_2$, and the quotient in question is expressed by (3.8.7) in the form

$$\frac{h^*}{\tilde{h}^*} = \frac{h^*}{h_2} = \frac{1}{2} Q h_1, \tag{3.18.4}$$

provided that the special case $K = P_{2^3}$ excluded in Sect. 3.8 is not taken. Then, by appropriate selections of f_1 and f_2, one can obtain the case where not only the quotient (of h^* by $\tilde{h}^*$) but also the quotient of the full class number h of K by $\tilde{h} = h_2$ of $\tilde{K} = K_2$, which is expressed by (3.8.6) in the form

$$\frac{h}{\tilde{h}} = \frac{h}{h_2} = \frac{1}{2} Q h_0 h_1, \tag{3.18.5}$$

become fractions. We will determine *all* the fields of considered type (i.e., imaginary bicyclic biquadratic fields) with this property.

Quotient (3.18.5) and then also quotient (3.18.4) are fractions with denominator 2 exactly when simultaneously

$$h_0 \text{ is odd}, \quad h_1 \text{ is odd and } \quad Q = 1.$$

By the genus theory of quadratic fields, the class number h_0 is odd exactly when K_0 is of genus one, that is, when f_0 is of one of two types:

$$f_0 = p \quad \text{or} \quad f_0 = q_1 q_2 \tag{3.18.6}$$

with a prime number $p \equiv 1 \pmod 4$, which also may be replaced by 2^3, or with two distinct prime numbers q_1, q_2, each of which also may be replaced by 2^2 or 2^3.[126] In the same way h_1 is odd exactly when f_1 is of type

$$f_1 = q \tag{3.18.7}$$

[126](Translator's remark) See, for example, H. Hasse, *Number Theory*, reprint of the 1980 edition translated by H.G. Zimmer, Springer 2002, pp. 589–590. A quadratic field $P(\sqrt{d})$ has odd class number if and only if d is one of the following:

$$\begin{cases} d = -1,\ -2,\ -q & \text{for a prime } q \text{ with } q \equiv -1 \pmod 4, \\ d = 2,\ p & \text{for a prime } p \text{ with } p \equiv 1 \pmod 4, \\ d = q,\ 2q,\ q_1 q_2 & \text{for primes } q,\ q_1,\ q_2 \text{ with } q \equiv q_1 \equiv q_2 \equiv -1 \pmod 4. \end{cases}$$

with odd prime $q \equiv -1 \pmod 4$, which also can be replaced by 2^2 or 2^3.[127] If we suppose that the two conditions (3.18.6) and (3.18.7) are satisfied, then we have

$$f_2 = pq \quad \text{or} \quad f_2 = q_1q_2q, \tag{3.18.8}$$

provided that only in the latter case q is different from q_1, q_2. This last assumption is necessary for our purpose—and actually with regard to the side case in the sense that the simultaneous replacements (q_1 or $q_2 \to 2^2$ or 2^3; $q \to 2^2$ or 2^3) are excluded—to fulfill the mentioned condition $Q = 1$. Namely, if q coincides with q_1 or q_2—with respect to the side case in the above-mentioned sense—then the cases of $Q = 2$ arise directly, which are determined by (3.8.15). In the first case (of (3.18.8)) the simultaneous replacement ($p \to 2^3$, $q \to 2^2$ or 2^3), which leads to the exceptional case $K = P_{2^3}$, is to be excluded. Apart from these cases the remaining condition $Q = 1$ is indeed fulfilled by (3.8.10$_{\mathrm{I}}$) or (3.8.10$_{\mathrm{II}}$), so that the fields K with the demanded property have actually been obtained by (3.18.6), (3.18.7), and (3.18.8). Thus we can establish the following:

Theorem 3.40 [128]*For the imaginary bicyclic biquadratic fields of two types*

$$K = P(\sqrt{-q}, \sqrt{-pq}) \quad \left\{ \begin{array}{l} p \text{ a prime number} \equiv 1 \pmod 4 \text{ or } \to 2^3 \\ q \text{ a prime number} \equiv -1 \pmod 4 \text{ or } \to 2^2 \text{ or } 2^3 \\ \text{not simultaneously } p \to 2^3,\ q \to 2^2 \text{ or } 2^3 \end{array} \right\}$$

---

[127]We indicate this replacement of the type of symbols $p \equiv 1 \pmod 4$ and $q \equiv -1 \pmod 4$ by the notation ($p \to 2^3$) and ($q \to 2^2$ or 2^3), respectively. Some statements and formulas in the text are then modified slightly depending on the meaning; for example, in (3.18.6) by the simultaneous replacement ($q_1 \to 2^2$, $q_2 \to 2^3$) the conductor $f_0 = 2^3$ is meaningfully to be understood, because just by the liberation of square factor 2^2 the formal product $2^2 \cdot 2^3$ becomes the conductor of a real quadratic field.

[128](Translator's remark) For the reader's convenience we make here the theorem more readable.

Theorem 3.40 (Again) *Among the imaginary bicyclic biquadratic fields K with class number h and their imaginary subfields $\tilde{K}$ with class number $\tilde{h}$, only for the following two types of fields the quotient $h/\tilde{h}$ is not integral; in these cases, the quotient $h/\tilde{h}$ has denominator 2:*

$$K = \boldsymbol{Q}(\sqrt{-q}, \sqrt{-pq}) \quad \text{and} \quad \tilde{K} = \boldsymbol{Q}(\sqrt{-pq}),$$

where p and q are prime numbers with $p \equiv 1 \pmod 4$ and $q \equiv -1 \pmod 4$, or p is replaced by 2^3 and q also by 2^2 or 2^3, but not simultaneously:

$$K = \boldsymbol{Q}(\sqrt{-q}, \sqrt{-2q}), \qquad \tilde{K} = \boldsymbol{Q}(\sqrt{-2q});$$

$$K = \boldsymbol{Q}(\sqrt{-1}, \sqrt{-p}), \qquad \tilde{K} = \boldsymbol{Q}(\sqrt{-p});$$

$$K = \boldsymbol{Q}(\sqrt{-2}, \sqrt{-2p}), \qquad \tilde{K} = \boldsymbol{Q}(\sqrt{-2p});$$

$$K = \boldsymbol{Q}(\sqrt{-q}, \sqrt{-q_1q_2q}) \qquad \tilde{K} = \boldsymbol{Q}(\sqrt{-q_1q_2q}),$$

and

$$K = P(\sqrt{-q}, \sqrt{-q_1q_2q}) \left\{ \begin{array}{l} q_1, q_2, q \text{ different prime numbers} \equiv -1 \pmod 4 \\ \quad \text{or} \to 2^2 \text{ or } 2^3 \\ \text{not simultaneously } q_1 \text{ or } q_2 \to 2^2 \text{ or } 2^3, \\ \quad q \to 2^2 \text{ or } 2^3 \end{array} \right\}$$

with the imaginary subfields

$$\tilde{K} = P(\sqrt{-pq}) \quad \text{and} \quad \tilde{K} = P(\sqrt{-q_1q_2q}),$$

respectively, and only for these two types of fields, the class number h of K is not divisible by the class number $\tilde{h}$ of $\tilde{K}$; rather the quotient $\frac{h}{\tilde{h}}$ has denominator 2.

The quotient $\frac{Q}{\tilde{Q}}$ is integral by Theorem 3.29[129]; it has value 1, 2. By Theorem 3.34 the contributions $h^*_{\psi'}$ are integral for regular characters ψ', while they have denominator 2 for irregular characters ψ'. Let $\tilde{\psi}_1, \ldots, \tilde{\psi}_{\tilde{r}}$ be the representatives of characters $\tilde{\psi}$ of $\tilde{K}/\tilde{K}_0$ that are irregular with respect to K, and $\psi'_1, \ldots, \psi'_{r'}$ the representatives ψ' of irregular characters of K/K_0 that are not characters of $\tilde{K}/\tilde{K}_0$. Here one can *not* conclude from (3.18.3), as in Sect. 3.15 from (3.15.2), that the r' irregular denominator factors 2 are reduced by the existing factor $2^\tau \frac{Q}{\tilde{Q}}$ and by the contributions of $\binom{r'}{3}$ three-term products of $\psi'_1, \ldots, \psi'_{r'}$; this is because all these three-term products do not necessarily appear among the characters ψ', but they may even appear among the characters $\tilde{\psi}$. Surely among the characters ψ' there appear the $\binom{\tilde{r}}{2} r'$ three-term products from two of $\tilde{\psi}_1, \ldots, \tilde{\psi}_{\tilde{r}}$ and one of $\psi'_1, \ldots, \psi'_{r'}$; hence if $\tilde{r} \geqq 2$, then all the r' irregular denominator factors are

where q_1, q_2, q are distinct prime numbers with $q_1 \equiv q_2 \equiv q \equiv -1 \pmod 4$; or q_1, q_2, q are replaced by 2^2 or 2^3, but not simultaneously "q_1 or q_2 is replaced by 2^2 or 2^3" and "q by 2^2 or 2^3":

$$K = Q(\sqrt{-1}, \sqrt{-q_1q_2}), \qquad \tilde{K} = Q(\sqrt{-q_1q_2});$$

$$K = Q(\sqrt{-2}, \sqrt{-2q_1q_2}), \qquad \tilde{K} = Q(\sqrt{-2q_1q_2});$$

$$K = Q(\sqrt{-q}, \sqrt{-q_1q}), \qquad \tilde{K} = Q(\sqrt{-q_1q});$$

$$K = Q(\sqrt{-q}, \sqrt{-2q_1q}), \qquad \tilde{K} = Q(\sqrt{-2q_1q});$$

$$K = Q(\sqrt{-q}, \sqrt{-2q}), \qquad \tilde{K} = Q(\sqrt{-2q}).$$

[129](Translator's remark) As we have already remarked, Theorem 3.29 is not correct and so in the following demonstration of Theorem 3.41 we must consider also the case of $\frac{Q}{\tilde{Q}} = \frac{1}{2}$. Actually we need to reform Theorem 3.41 as shown in footnote 133 below.

reduced.[130] As we will show later in examples, for $\tilde{r} = 0, 1$, all the r' irregular denominator factors are not required to be reduced and the quotient is not necessarily integral.[131] We abandon the problem (of determination) of the exact upper bound of the denominator of the quotient[132] and only state the following result of our investigation:

Theorem 3.41 [133]*If $\tilde{K}$ is an imaginary subfield of K, then the quotient $h^*/\tilde{h}^*$ of the relative class number h^* of K/K_0 by $\tilde{h}^*$ of $\tilde{K}/\tilde{K}_0$ has denominator at most $2^{r'}$, where r' is the number of irregular characters of K/K_0 that really are not characters of $\tilde{K}$.*

Especially, if there exist at least two characters of $\tilde{K}/\tilde{K}_0$ that are irregular with respect to K, then $\tilde{h}^*$ is a divisor of h^*.[134]

3.19 Imaginary Abelian Number Fields with Odd Class Number

For the cyclotomic field $K = P_p$ with prime conductor $p \neq 2$, Kummer deduced from his class number formula two interesting theorems on the divisibility of the full class number $h = h_0 h^*$ by prime numbers.

The first theorem of Kummer [15, 17] is relevant to the divisibility of the class number h of P_p by the prime number p. He states that h_0 can be divisible by p only

---

[130](Translator's remark) As we have seen in the italicized, conditional statement in Sect. 3.15 (p. 136), if K has at least two representatives of irregular characters, then $Q = 2$. Hence the latter part of Theorem 3.41 is valid by this argument.

[131](Translator's remark) As for $\tilde{r} = 1$, see the examples in footnote 133 below.

[132](Translator's remark) K. Horie has completely solved this problem: On the ratio between relative class numbers, *Math. Z.* **65** (1989), 505–521.

[133](Translator's remark) The former part of this theorem is incorrect. In fact, we have counterexamples:

$$K = P(\sqrt{-1}, \sqrt{2}, \sqrt{pq}) \quad \text{and} \quad \tilde{K} = P(\sqrt{-1}, \sqrt{2pq})$$

for $(p, q) = (17, 3), (41, 3), (137, 3), (17, 11), (41, 11), \ldots$. For these fields we have $r' = 0$ and

$$2^{\tau} = 2^0 = 1, \quad \frac{Q}{\tilde{Q}} = \frac{1}{2} \quad \text{and} \quad \prod_{\psi'} h^*_{\psi'} \text{ is odd.}$$

Therefore in the former part we should replace "at most $2^{r'}$" by "at most $2^{r'+1}$." For more precision, see M. Hirabayashi, K. Yoshino, Remarks on unit indices of imaginary abelian number fields II, *Manuscripta Math.* **64** (1989), 235–489.

[134](Translator's remark) As shown in a previous footnote, under the assumption we have $Q = 2$ and hence $2^{\tau}\frac{Q}{\tilde{Q}}$ is an integer. In the final sentence in Theorem 3.41 "$\tilde{K}/K_0$" in the original book should be "$\tilde{K}/\tilde{K}_0$."

when h^* is also divisible by p, and he gives moreover as a necessary and sufficient condition—and hence also for the divisibility of h by p—that the prime number p appears in the numerator of one of the Bernoulli numbers $B_2, B_4, \ldots, B_{p-3}$.[135] Kummer [16] derived this theorem with regard to his proof of Fermat's Last Theorem for the regular primes p (not appearing in the class number h of P_p) as a criterion with rational numbers for the regularity of p. Due to this significance he is well known, and this proof is written in detail in Hilbert's Zahlbericht.[136] Accordingly, I will not enter further into his result here.

The second theorem of Kummer [21], which is referenced only briefly in Hilbert's Zahlbericht [10], is less well known. It is relevant to the divisibility of the class number h of P_p by the prime number 2 and, correspondingly to the first (theorem), states that h_0 can be divisible by 2 only when h^* is also divisible by 2, so that h is divisible by 2 if and only if h^* is divisible by 2. This is also a criterion essentially with rational numbers for the divisibility of h by 2, though different from the previous case of divisibility by p; for the divisibility of h^* by 2, no explicit reduction of the divisibility property with rational numbers has been given. The determination on the divisibility of h^* by 2 can indeed be decided in every given case by the relative class number formula or by the formulas and results of its individual factors obtained in Sects. 3.9–3.14, whereas the determination on the divisibility of h_0 by 2, and hence the determination on the divisibility of h by 2, would require difficult calculation of fundamental units of K_0 according to the class number formula.

There arises the question of whether there exists a corresponding relation between the divisibilities of h^* and of h_0 by 2 also for an arbitrary abelian number field K, that is, whether the non-divisibility of h_0 by 2 always follows also from the non-divisibility of h^* by 2. This question is to be answered in the negative, which we will see from examples of imaginary bicyclic biquadratic fields. However we will be able to deduce certain general necessary conditions for the non-divisibility of h^* by 2 from the genus theory of K/K_0 developed in Sect. 3.1 that refers itself to the arithmetic structure of K and to the units of K_0, and we will show that these necessary conditions together with the non-divisibility of h_0 by 2 (especially to be required) are also sufficient for the non-divisibility of h^* by 2 and hence of h

135(Translator's remark) It would be preferable that we read "$B_2, B_4, \ldots, B_{p-3}$" for "$B_2, B_4, \ldots, B_{\frac{p-3}{2}}$" in the original book, because we "usually" define the Bernoulli number B_n by

$$\frac{x}{e^x - 1} = B_0 + \frac{B_1}{1!}x + \frac{B_2}{2!}x^2 + \cdots.$$

Kummer defined B_n by

$$\frac{1}{e^x - 1} = \frac{1}{x} - \frac{1}{2} + \frac{B_1}{2!}x - \frac{B_2}{4!}x^3 + \frac{B_3}{6!}x^5 - \cdots.$$

136See Zahlbericht [10], §§**137–139**.

by 2. This result, in fact, is less interesting from the point of view of numeration because the non-divisibility of h_0 by 2 is more difficult to establish than the non-divisibility of h^* by 2. But our result is theoretically meaningful because it attributes the arithmetic facts on the over-field K/K_0 to the arithmetic facts on the under-field K_0 of K. After the conceptional (theoretical) methods of the genus theory have been exhausted without having led to a generalization of Kummer's theory, we will show in Sect. 3.20 in a numerical way, namely, in comparison with our formulas for h^* and for h_0, that the theorem of Kummer on the divisibility by 2 is always generalized to an imaginary cyclic field. The implementation of the research on the divisibility of h by 2 outlined above is very well suited to the conclusion of this work, because almost all the knowledge obtained in the process of the implementation is required or always referred.

We start with an application of the genus theory developed in Sect. 3.1 to the present problem setting. By Theorem 3.12 the number h^* is divisible by the genus factor 2^γ, whose exponent is $\gamma = \delta + q^* - 1 \geqq 0$. Hence, as a necessary condition for "h^* being odd," there arises the existence of the relation $\delta + q^* = 1$ or therefore one of the following two pairs of relations:

$$\delta = 0,\ q^* = 1 \quad \text{or} \quad \delta = 1,\ q^* = 0.$$

Here δ is the number of distinct prime divisors of K_0 in the relative discriminant $\mathfrak{d}_0$ of K/K_0 and q^* the number of independent units ε_0 of K_0 that are norm residues mod $\mathfrak{d}_0 p_\infty$ of K/K_0 with respect to squares (of units) of K_0. Then, by Theorem 3.19 there appear in $\mathfrak{d}_0$ the prime divisors of all and only the prime numbers p that arise in the conductors $f(\chi_1)$ of all characters χ_1 of K/K_0. The requirement, $\delta = 0$ or 1, therefore means that there exists no or only one such prime number p, and in the latter case this prime number p is not decomposed (only is inert and/or is ramified) in K_0, i.e., that $\chi_0(p) \neq 1$ holds for all the characters $\chi_0 \neq 1$ of K_0. By the product formula for the norm residue symbol,[137]

$$\prod_{\mathfrak{p}_0} \left(\frac{\varepsilon_0, K/K_0}{\mathfrak{p}_0} \right) = 1,$$

where $\mathfrak{p}_0$ runs over all the prime divisors and infinite places (divisors) of K_0, and moreover, under the assumption that $\delta = 1$, the condition of the norm residue for the unique prime divisor $\mathfrak{p}_0$ of $\mathfrak{d}_0$ is a result of the conditions of the norm residues for the infinite places comprising p_∞. The last conditions demand just that ε_0 is totally positive. For $\delta = 0$ or $\delta = 1$, therefore, the quantity q^* is the number of independent totally positive units of K_0 with respect to squares (of units) of K_0 or—which is the same as—the number of independent relations among the signatures of a system of

[137]For this see Klassenkörperbericht [5], Part II, §**6**, (6) and (5′) as well as §**7** and also Chapter III.

fundamental units of K_0, and therefore the condition $q^* = 0$ or 1 means that there exists no or only one such relation.[138]

The derived conditions mentioned above are necessary for "h^* being odd" and at the same time sufficient for $\gamma = 0$. If one adds the condition "h_0 being odd," then by (3.1.9), as already remarked subsequently there, the full class number h of K has 2-rank $r = \gamma$; then it follows that $r = 0$, i.e., h is odd and therefore h^* is also odd.

Thus we have proved the following:

Theorem 3.42 *Let K be an arbitrary imaginary abelian number field. For the relative class number h^* of K/K_0 being odd, it is necessary that the following three conditions are satisfied:*

1. *There exists (a) no prime number or (b) only one prime number p whose prime divisors of K_0 ramify in K/K_0, i.e., p appears in $f(\chi_1)$ for all characters χ_1 of K/K_0.*
2. *In case (b), the prime number p does not split in K_0/P, that is, $\chi_0(p) \neq 1$ for all characters $\chi_0 \neq 1$ of K_0.*
3. *In case (b), the signatures of a system of fundamental units of K_0 are independent; in case (a), among these signatures there exists exactly one relation.*

If these three conditions are satisfied simultaneously, then K/K_0 is of genus one.[139]

If, moreover,

4. *the class number h_0 of K_0 is odd,*

then h^* is also odd.

Hence conditions 1–4 are necessary and sufficient for the oddness of the full class number $h = h_0 h^*$ of K.

This theorem contains the establishment that "h^* being odd" follows from "h_0 being odd" under the assured additional conditions, while Kummer's theorem just conversely states that in the special case of $K = P_p$ "h_0 being odd" follows from "h^* being odd." Now in the proof of Theorem 3.42 only the assumption "h^* being odd" has been used insofar as it is relevant to the genus factor 2^γ, which must then have exponent $\gamma = 0$. By fully utilizing the assumption "h^* being odd," from which the complementary factor $\frac{h^*}{2^\gamma}$ must also be odd, one could think generally to conclude "h_0 being odd." However, so far as I see, the genus theory gives no means for this conclusion from the number $\frac{h^*}{2^\gamma}$ of classes in the principal genus of K/K_0 to the class number h_0 of K_0.

---

[138] For $\delta = 0$, as seen immediately, the product formula for the norm residue symbol provides the relation $N_0(\varepsilon_0) = 1$ for every unit ε_0 of K_0. Hence one sees that for $\delta = 0$ there exists *at least* one relation among the signatures of a system of fundamental units of K_0, as already determined in another way in Theorem 3.13. Therefore, for $\delta = 0$ and $q^* = 1$ there exists *exactly* one such relation.

[139] (Translator's remark) This means that the genus factor $2^\gamma = 1$, i.e., $\gamma = 0$.

Indeed, as we will show here in examples of imaginary bicyclic biquadratic fields, there can really exist (fields with) odd h^* and even h_0.[140] As in (3.8.1)–(3.8.4), let

$$K = K_1 K_2 = P(\sqrt{-f_1}, \sqrt{-f_2}),$$

$$K_1 = P(\sqrt{-f_1}), \quad K_2 = P(\sqrt{-f_2}),$$

$$K_0 = P(\sqrt{f_0}) \quad \text{with} \quad f_0 \underset{2}{=} f_1 f_2,$$

and let h_1, h_2 be the class numbers of K_1, K_2, respectively, and ε_0 a fundamental unit of K_0.

Making use of Theorem 3.42, we first determine all the fields K of this type with odd class number h. Conditions 1, 2, and 3 say here that one of the following two cases exists:

(a) f_1, f_2 have no common prime divisor; then $N_0(\varepsilon_0) = 1$.
(b) f_1, f_2 have exactly one common prime divisor q; then $\left(\frac{f_0}{q}\right) = -1$ or $q \mid f_0$ (the latter case occurs only for $q = 2$ in question); then $N_0(\varepsilon_0) = -1$.

Condition 4 says that, as already determined in (3.18.6), one of the following two cases exists:

(c) $f_0 = p$; then $N_0(\varepsilon_0) = -1$.
(d) $f_0 = q_1 q_2$; then $N_0(\varepsilon_0) = 1$.

Here again the establishment treated in (3.18.6) and (3.18.7) on the replacements ($p \to 2^3$) and ($q \to 2^2$ or 2^3) should be valid.[141] If h is odd, then h^* is odd and h_0 odd, and hence by Theorem 3.42 both "(a) or (b)" and "(c) or (d)" are satisfied; then there exists either the case of pair "(a) and (d)" or the case of pair "(b) and (c)." Conversely, by Theorem 3.42, in each case of these two pairs the class number h is odd. In the case of the pair "(a) and (d)" it follows that

$$f_1 = q_1, \quad f_2 = q_2.$$

In the case of the pair "(b) and (c)" it follows, say, that

$$f_1 = q, \quad f_2 = pq,$$

where q is also again submitted to the establishment treated in (3.18.6) and (3.18.7) —in case (b), first of all, this was not yet the case. The restriction $\left(\frac{p}{q}\right) = -1$ assigned by (b) in the principal case (the case where p and q are odd prime numbers)

[140](Translator's remark) In Theorem 3.44 we will give such fields.

[141]See also the remark there in Sect. 3.18, footnote 127, p. 165.

is accordingly to be replaced by the restriction $\left(\frac{2}{q}\right) = -1$ or $\left(\frac{p}{2}\right) = -1$ in the side cases ($p \to 2^3$) and ($q \to 2^2$ or 2^3), unless the special case ($p \to 2^3, q \to 2^2$ or 2^3), the special case $K = P_{2^3}$ of Sect. 3.8, is under consideration, and in the (special) case there exists no such restriction because the condition $2|f_0$ is satisfied of itself by (b). In the principal case with $\left(\frac{p}{q}\right) = -1$, $\left(\frac{q}{p}\right) = -1$ also holds by the law for quadratic residues, and then p and q are *of mutual non-residue*. The same form of expression is also applied to the above-mentioned restrictions in the side cases; then the special case $K = P_{2^3}$ is excluded from the case of the pair "(b) and (c)," which means, with regard to its existence (the existence of P_{2^3}), no restriction (is given) in the case of the pair "(a) and (d)."

Thus, as a result, we can determine the following:

Theorem 3.43 [142]*The imaginary bicyclic biquadratic fields K with odd class numbers are*

$$K = P(\sqrt{-q_1},\ \sqrt{-q_2})\ (q_1,\ q_2 \textit{ are distinct prime numbers } \equiv -1 \pmod 4 \textit{ or } \to 2^2 \textit{ or } 2^3),$$

$$K = P(\sqrt{-q},\ \sqrt{-pq}) \begin{Bmatrix} p \textit{ is a prime number } \equiv 1 \pmod 4 \textit{ or } \to 2^3 \\ q \textit{ is a prime number } \equiv -1 \pmod 4 \textit{ or } \to 2^2 \textit{ or } 2^3 \\ p,\ q \textit{ are of mutual non-residue} \end{Bmatrix}.$$

[142](Translator's remark) For the reader's convenience we make here the theorem more readable.

Theorem 3.43 (Again) *All the imaginary bicyclic biquadratic fields K with odd class number are given in the following. Let p, q_1, q_2, q be odd prime numbers. Then*

$$K = P_{2^3} = P(\sqrt{-1},\ \sqrt{-2}),$$

$$K = P(\sqrt{-q_1},\ \sqrt{-q_2}) \quad \textit{with } q_1 \equiv q_2 \equiv -1 \pmod 4,$$

$$K = P(\sqrt{-1},\ \sqrt{-q}) \quad \textit{with } q \equiv -1 \pmod 4,$$

$$K = P(\sqrt{-2},\ \sqrt{-q}) \quad \textit{with } q \equiv -1 \pmod 4,$$

$$K = P(\sqrt{-q},\ \sqrt{-pq}) \quad \textit{with } p \equiv 1 \pmod 4,\ q \equiv -1 \pmod 4,\ \left(\frac{p}{q}\right) = -1,$$

$$K = P(\sqrt{-q},\ \sqrt{-2q}) \quad \textit{with } q \equiv 3 \pmod 8,$$

$$K = P(\sqrt{-1},\ \sqrt{-p}) \quad \textit{with } p \equiv 5 \pmod 8,$$

$$K = P(\sqrt{-2},\ \sqrt{-2p}) \quad \textit{with } p \equiv 5 \pmod 8.$$

From now on we construct some examples of the treated type of fields (imaginary bicyclic biquadratic fields) for which h^* is indeed odd but h_0 is even. By (3.8.7) it holds that

$$h^* = \frac{1}{2}\, Q\, h_1\, h_2,$$

provided that the special field $K = P_{23}$, which is not taken into consideration here, is excluded. Hence it is necessary and sufficient for "h^* being odd" that either

(a′) h_1 is odd, h_2 is odd and $Q = 2$,

or, say,

(b′) h_1 is odd, h_2 is divisible exactly by 2^1 and $Q = 1$.

The requirement of the class numbers in (a′) means by the genus theory that f_1, f_2 are prime-powers. Then, since $f_0 \underset{2}{=} f_1 f_2$ has only two distinct prime divisors, at least one of which is $\equiv -1 \pmod 4$, (it holds that) h_0 is odd,[143] so that in this case no examples of the desired type have resulted. The requirement of the class numbers in (b′) means that by the genus theory and by Theorem 3.39 f_1 is a prime-power and f_2 has exactly two different prime divisors that are of mutual non-residue. If one excludes by virtue of Theorem 3.43 the combination of $f_1,\ f_2$ for which the class number h_0 is odd, then there remain the following cases with h_0 being even:

$$f_1 = q,\quad f_2 = pq',\quad f_0 = pqq' \quad \left\{ \begin{array}{l} q,\ q' \text{ different primes} \\ p,\ q' \text{ mutual non-residue} \end{array} \right\};$$

here one excludes the side cases in which "p and q" or "p and q'" are replaced by 2-powers simultaneously. In order that h^* becomes odd, the remaining demand that $Q = 1$ contained in (b′) is further to be fulfilled. For this purpose it is sufficient (not to be necessary) to prescribe the non-residue for $p,\ q$, because the necessary character conditions ($3.8.13'_{\mathrm{I}}$) or ($3.8.13'_{\mathrm{II}}$) for $Q = 2$ are then damaged. Consequently, if one gives first, say, p and then selects two different primes q, q' (both of which are) mutual non-residue with p—by this non-residue specification the forbidden side cases are previously excluded at the same time—so one obtains examples of the demanded type of fields. Thus we can state a result:

[143](Translator's remark) See the footnote 126, p. 164, where all the quadratic fields with odd class number are given.

Theorem 3.44 [144]*For the imaginary bicyclic biquadratic fields (among others)*

$$K = P(\sqrt{-q}, \sqrt{-pq'}) \left\{ \begin{array}{l} p \;\; a\ prime\ number\ \equiv 1 \pmod 4\ or\ \to 2^3 \\ q,\ q'\ \ different\ prime\ numbers\ \equiv -1 \pmod 4 \\ \qquad\qquad or \to 2^2\ or\ 2^3 \\ p, q\ \ and\ \ p, q'\ \ mutual\ non\text{-}residue \end{array} \right\}$$

with

$$K_0 = P(\sqrt{pqq'}),$$

the relative class number h^* of K/K_0 is odd, but the class number h_0 of K_0 is even.

The simplest example of fields of this type is

$$K = P(\sqrt{-1}, \sqrt{-10}) \quad \text{with} \quad K_0 = P(\sqrt{10}).$$

This corresponds to the case where $p = 5,\ q \to 2^2,\ q' \to 2^3$. The mutual non-residue specification is fulfilled by $\left(\frac{5}{2}\right) = \left(\frac{2}{5}\right) = -1$. In this example, in fact, it

---

[144](Translator's remark) For the reader's convenience we make here the theorem more readable.

Theorem 3.44 (Again) *Let p, q, q' be odd prime numbers. For the following imaginary biquadratic bicyclic fields K, the relative class number h^* of K is odd, but the class number h_0 of the maximal real subfield K_0 of K is even:*

$$K = P(\sqrt{-q}, \sqrt{-pq'}) \text{ with } p \equiv 1 \pmod 4,\ q \equiv q' \equiv -1 \pmod 4,\ q \neq q',\ \left(\frac{q}{p}\right) = \left(\frac{q'}{p}\right) = -1,$$

$$K = P(\sqrt{-q}, \sqrt{-2q'}) \text{ with } q \equiv q' \equiv 3 \pmod 8,\ q \neq q',$$

$$K = P(\sqrt{-1}, \sqrt{-pq'}) \text{ with } p \equiv 5 \pmod 8,\ q' \equiv -1 \pmod 4,\ \left(\frac{q'}{p}\right) = -1,$$

$$K = P(\sqrt{-q}, \sqrt{-p}) \quad \text{with } p \equiv 5 \pmod 8,\ q \equiv -1 \pmod 4,\ \left(\frac{q}{p}\right) = -1,$$

$$K = P(\sqrt{-2}, \sqrt{-pq'}) \text{ with } p \equiv 5 \pmod 8,\ q' \equiv -1 \pmod 4,\ \left(\frac{q'}{p}\right) = -1,$$

$$K = P(\sqrt{-q}, \sqrt{-2p}) \text{ with } p \equiv 5 \pmod 8,\ q \equiv -1 \pmod 4,\ \left(\frac{q}{p}\right) = -1,$$

$$K = P(\sqrt{-1}, \sqrt{-2p}) \text{ with } p \equiv 5 \pmod 8,$$

$$K = P(\sqrt{-2}, \sqrt{-p}) \quad \text{with } p \equiv 5 \pmod 8.$$

holds that[145]

$$h_1 = 1,\ h_2 = 2,\ h^* = 1,\ \text{but}\ h_0 = 2.$$

3.20 Imaginary Cyclic Fields with Odd Class Number

Since we have determined by Theorem 3.44 that the theorem of Kummer that is valid for $K = P_p$ $(p \neq 2)$—"h_0 being odd" follows from "h^* being odd"—is not true for *every* imaginary abelian number field K, we now turn ourselves to the demonstration that this theorem is always generalized to an imaginary *cyclic* field. Our proof covers Kummer's special case $K = P_p$ $(p \neq 2)$.

In the following let K be an imaginary cyclic field of degree n and χ a generating character of K. Let, moreover, Λ be the imaginary subfield of K whose degree is the highest power 2^ν adhering to $n = 2^\nu u_0$ $(\nu \geqq 1,\ u_0$ odd); then $\psi = \chi^{u_0}$ is a generating character of Λ. The subfield Λ is also characterized as the imaginary subfield of K of 2-power degree; in fact, it is imaginary because it is not contained in the maximal real subfield K_0 of degree $n_0 = 2^{\nu-1}u_0$ (or, because $\psi(-1) = -1$ holds), and any other subfield of K of 2-power degree is real because it is contained in the maximal real subfield Λ_0 of Λ of degree $2^{\nu-1}$. Therefore, the generating character of Λ is, unique up to algebraic-conjugate, characterized as a character of K/K_0 of 2-power order.

In the present cyclic case, condition 1 of Theorem 3.42 states that the subfield Λ, or the character ψ, has conductor of prime-power, $f(\psi) = p^\rho$. In fact, as in (3.7.1) and (3.7.2), let

$$\chi = \prod_{p|f} \chi_p, \quad \psi = \prod_{p|f} \chi_p^{u_0}, \quad \chi_1 = \prod_{p|f} \chi_p^u$$

be the component decompositions of the characters χ, $\psi = \chi^{u_0}$ and $\chi_1 = \chi^u$ (u odd) of K/K_0. That one prime number p appears in all the conductors $f(\chi_1)$ is necessary and sufficient for $\chi_p^u \neq 1$ to hold for all odd integers u, i.e., that its corresponding component χ_p has even order. Hence condition 1 of Theorem 3.42 states here that χ has at most one and then exactly one component of even order; the latter actually occurs because χ is of even order n. Since along with ψ the components $\psi_p = \chi_p^{u_0}$ are also of 2-power order, a component character χ_p is of even order if and only if the corresponding $\psi_p \neq 1$ holds. Consequently condition 1 of Theorem 3.42 is synonymous with the condition that for exactly one component $\psi_p \neq 1$ holds; that is, ψ has prime-power conductor $f(\psi) = p^\rho$.

145(Translator's remark) In the original book published in 1952 "$h_2 = 1$" is printed. But, in the book in 1985 it is changed to "$h_2 = 2$," which is correct.

With regard to condition 2 of Theorem 3.42, which actually occurs here by the demonstration above because case (b) there (in Theorem 3.42) really exists, we note that if condition 1 there (in Theorem 3.42) is satisfied, the prime number p is always non-decomposed in the subfield Λ and also surely in the subfield Λ_0 because the prime p is totally ramified in Λ by the demonstration above. Condition 2 of Theorem 3.42 demands further that the unique prime divisor of p in Λ_0 is non-decomposed (only is inert or is ramified) in K_0/Λ_0, i.e., that for the powers $\chi^g \neq 1$ (g even) as long as their conductors $f(\chi^g)$ are prime to p (this case at most occurs for $g \equiv 0 \pmod{2^\nu}$), $\chi^g(p) \neq 1$ always hold.

Since case (b) of Theorem 3.42 really exists there, condition 3 states here that the signatures of a system of fundamental units of K_0 are independent.

Thus the sharpening of Theorem 3.42 to be proved reads as follows:

Theorem 3.45 *Let K be an imaginary cyclic field. Then, the relative class number h^* of K/K_0 is odd if and only if the following four conditions are satisfied:*

1. *The imaginary subfield Λ of K of 2-power degree has prime-power conductor, or a character ψ of K/K_0 of 2-power order (uniquely determined up to algebraic-conjugates) has prime-power conductor $f(\psi) = p^\rho$.*
2. *The prime number p does not split in K_0, that is, for the powers $\chi^g \neq 1$ (g even) as long as its conductors $f(\chi^g)$ are prime to p (which at most occurs for $g \equiv 0 \pmod{2^\nu}$), $\chi^g(p) \neq 1$ holds.*
3. *The signatures of a system of fundamental units of K_0 are independent.*
4. *The class number h_0 of K_0 is odd.*

Therefore, especially, the full class number $h = h_0 h^*$ of K is odd if and only if h^* is odd.[146]

In Kummer's special case $K = P_p$ ($p \neq 2$), conditions 1 and 2 are satisfied themselves, and Theorem 3.45, beyond Kummer's result, states that, in addition to condition 4 of the class number, condition 3 of units is also necessary for "h^* being odd" and that these two conditions together with conditions 1 and 2 are also sufficient for the oddness of h^*.[147] The same holds more generally for the cyclotomic fields $K = P_{p^\rho}$ ($p \neq 2$), too, which are also cyclic. For the cyclotomic fields P_{2^ρ}, or rather for their imaginary subfields $K = P'_{2^\rho}$ given by (3.16.1) (only the latter are cyclic), Theorem 3.45 produces less than Theorems 2.6 and 3.36′ proved by Weber in Sects. 2.6 and 3.16, by which for these fields the class number h_0 and the relative class number h^* are *actually* odd and the signatures of a system of

[146](Translator's remark) As will be explained later, in order to prove Theorem 3.45 it suffices only to show that $2 \nmid h_0$ follows from $2 \nmid h^*$, which is proved by the congruence

$$h^* \equiv h_0 \sum\nolimits_0 \pmod 2,$$

where $\sum_0$ is the regulatrix of K_0.

[147]Though these additional statements, in fact, are not found to be expressed in Kummer [21], one can easily deduce them from the proof there.

fundamental units of $K_0 = P_{2^{\rho-1},0}$ are independent; in the proof of Theorem 3.45, after all, we will have to make use of Theorem 3.36′ and also of the way of the proof of Theorem 2.6. In the special case of imaginary quadratic fields K, conditions 2, 3, and 4 are trivially satisfied for $K_0 = P$, while condition 1 here states that K itself has prime conductor; in this case, Theorem 3.45 simply provides the well-known criterion from the genus theory for the non-divisibility of $h = h^*$ by 2. Moreover, our proof for this case becomes trivial because the class number formula for K_0 to which we have to relate the relative class number formula for K/K_0 becomes trivial; but we need not exclude the case of imaginary quadratic fields K.

Theorem 3.45 surpasses Theorem 3.42 formally only in (the assertion) that here, beside conditions 1, 2, and 3, condition 4 is also necessary for "h^* being odd." Therefore only this assertion must be proved now. However, by our proof in the present cyclic case, we will also be able to establish easily the other assertions, whose correctness has already been proved on the basis of general Theorem 3.42, and in fact this time (we will be able to establish our assertions) in a numerical way by comparison of the formulas for h^* and for h_0, the process being contrary to the conceptional proof of Theorem 3.42 that is rooted in the genus theory.

Proof of Theorem 3.45 We proceed in the following way. First we show that condition 1 is necessary for "h^* being odd." Second, under the assumption of this fulfillment (of condition 1), we deduce the congruence

$$h^* \equiv \prod_{\chi_0 \neq 1} (1 - \chi_0(p)) \cdot \prod_{\chi_0 \neq 1} \sum_{\pm x \bmod f(\chi_0)} \chi_0(x) \left(\delta_{2f(\chi_0)}(x) - \delta_{2f(\chi_0)}(zx)\right) \pmod 2, \tag{3.20.1}$$

where p is the prime number in condition 1 and moreover z is a representative (normalized to be odd)[148] of a generating class mod H corresponding to the field K, and $\delta_{2f(\chi_0)}(x)$, as already appeared in (2.11.5) and (3.14.9), is the exponent of the signature of the absolutely least residue of x mod $2f(\chi_0)$. Then $\delta_{2f(\chi_0)}(x) \equiv 0$ or $1 \pmod 2$ according as this residue is positive or negative. By (2.3.1) the first product in the right-hand side in (3.20.1) has value

$$\prod_{\chi_0 \neq 1} (1 - \chi_0(p)) = \left\{ \begin{array}{ll} 0, & \text{if } p \text{ is decomposed in } K_0 \\ \\ n_p, & \text{if } p \text{ is non-decomposed in } K_0 \end{array} \right\}, \tag{3.20.2}$$

where in the latter case n_p is the degree of the unique prime divisor of p in K_0; since, by the fulfillment of condition 1, the prime number p is, as already remarked, totally ramified in Λ and also in Λ_0 and since, by definition, K/Λ and also K_0/Λ_0 have odd degree u_0, the degree n_p is surely odd. If h^* is odd, then by Eqs. (3.20.1) and (3.20.2) it is necessary that p is non-decomposed in K_0. Consequently, beside

[148] See footnote 12, p. 32, in Sect. 2.4.

condition 1, also condition 2 has been proved to be necessary for "h^* being odd," and moreover it has been established that by the fulfillment of the two necessary conditions the number h^* is congruent mod 2 to the second product in the right-hand side in (3.20.1). The latter product is the expression founded in (2.11.5) for the regulatrix $\sum(\eta_{n'_0}^{Z^{-\nu'_0}})$ of a system of cyclotomic units of subfields of K_0 appearing in Theorem 2.9.[149] By virtue of the fulfillment of the necessary conditions 1 and 2, it follows further from (3.20.1) that the congruence[150]

$$h^* \equiv \sum(\eta_{n'_0}^{Z^{-\nu'_0}}) \pmod 2$$

holds. Hence, from Theorem 2.9 and congruence (2.11.3), there arises the congruence

$$h^* \equiv h_0 \sum\nolimits_0 \pmod 2,$$

where $\sum_0$ is the regulatrix of K_0.[151] From this congruence one deduces that—under the assumption of conditions 1 and 2—conditions 3 and 4 are necessary for "h^* being odd," that is, the assertion of Theorem 3.45 to be proved.

Accordingly, it still remains to show that condition 1 is necessary for "h^* being odd" and that congruence (3.20.1) holds under its assumption (condition 1). Now we turn to the two proofs (of these assertions). The first proof is easily settled, while the second, which is regarded as the main part of our proof, requires some pains.

From the assumption that K is cyclic, it follows first that either $w \not\equiv 0 \pmod 4$ or $w = 4$, and further from Theorem 3.24 (it follows) that $Q = 1$. Hence the relative class number formula (3.9.1), as well as its alteration given by (3.15.2), states in the present cyclic case that

$$h^* = 2\frac{w}{2}\prod_{\chi_1}\Theta(\chi_1) = 2\frac{w}{2}\prod_{\psi_1}N_{\psi_1}(\Theta(\psi_1)) = 2\prod_{\psi_1}h^*_{\psi_1} \tag{3.20.3}$$

$$\text{with} \quad \frac{w}{2} \equiv 1 \pmod 2 \quad \text{or} \quad \frac{w}{2} = 2.$$

The contributions $h^*_{\psi'_1}$ of representatives ψ_1 to h^* have by Theorem 3.34 denominator 2 or are integral according as ψ_1 is irregular or regular in the sense there, that is, according as ψ_1 has prime-power conductor and 2-power order or

---

[149](Translator's remark) "$\sum(\eta_{n_0}^{Z^{-\nu'_0}})$" in the original should be read "$\sum(\eta_{n'_0}^{Z^{-\nu'_0}})$."

[150](Translator's remark) In the summation n'_0 runs over the set of the divisors n'_0 of $n_0 = [K_0 : P]$ and $\nu'_0 = 0, 1 \ldots, \varphi(n'_0) - 1$. To the definition of $\sum(\eta_{n'_0}^{Z^{-\nu'_0}})$, see (2.6.1).

[151]This (way of conclusion) is essentially the same way that leads to the proof of Theorem 2.5 in Sect. 2.6 and then of Weber's Theorem 2.6.

not.[152] If h^* is odd, then by (3.20.3) at least one representative ψ_1 has prime-power conductor and 2-power order. Since, as already emphasized, the generating character ψ of the imaginary subfield Λ of 2-power degree defined in the beginning is the unique character of K/K_0 of 2-power order up to algebraic-conjugate (characters), for the oddness of h^* this character ψ has consequently a prime-power conductor $f(\psi) = p^\rho$. Thus condition 1 has been proved to be necessary for "h^* being odd." In the following, assume that it (condition 1) is satisfied. Therefore we have further to prove the existence of congruence (3.20.1).

To this main part of our proof we make beforehand the following remarks. The characters χ_0 of K_0 and χ_1 of K/K_0 are expressed as powers of the generating character χ of K with even g and odd u mod n:

$$\chi_0 = \chi^g, \quad \chi_1 = \chi^u.$$

They are paired by the generating character $\psi = \chi^{u_0}$ of Λ in the form

$$\chi_1 = \psi\chi_0.$$

Here χ_0 is always a power of χ_1, because the odd-component of the order of χ_0 is the same as that of χ_1, while the 2-component of the order of $\chi_0 = \chi^g$ is an (even proper) divisor of that of $\chi_1 = \chi^u$. Consequently $f(\chi_0)$ is always a divisor of $f(\chi_1)$. We will have to separate the conductors $f(\psi)$ into two cases according as the prime number p appearing in $f(\psi) = p^\rho$ is not equal to 2 or equal to 2. For $p \neq 2$, $\rho = 1$ holds; for $p = 2$, $\rho \geqq 2$ always comes into question and then $\psi = \cup\varphi_\rho$.

Now, let z be a representative of a generating class mod H, so that $\chi(z)$ is therefore a primitive n-th root of unity. Then it holds that

$$\prod_{\chi_1}(1 - \overline{\chi}_1(z)) = \prod_{u \bmod n}(1 - \overline{\chi}(z)^u) = \left.\frac{t^n - 1}{t^{\frac{n}{2}} - 1}\right|_{t=1} = 2.$$

By inserting this expression of a product into the factor 2 at the beginning of expressions in (3.20.3), one has

$$h^* = \frac{w}{2}\prod_{\chi_1}(1 - \overline{\chi}_1(z))\Theta(\chi_1). \tag{3.20.4}$$

[152]For cyclic fields the definition of irregular and regular characters can be expressed in this simple way because there exist no cases where a character ψ_1 with prime-power conductor and of 2-power order is regular, that is, there exist no exceptional cases of $\psi_1 = \cup\psi_\rho$ $(\rho \geqq 3)$ for $w \equiv 0 \pmod 4$.

To take the product of $\Theta(\chi_1)$ and $1-\overline{\chi}_1(z)$, we use instead of the previous way of writing (3.9.2), namely,

$$\Theta(\chi_1) = \frac{1}{2f(\chi_1)} \sum_{x \bmod f(\chi_1)}^{+} (-\chi_1(x)x), \tag{3.20.5}$$

where x runs over the system of the *least positive* prime residues mod $f(\chi_1)$, the invariant way of writing

$$\Theta(\chi_1) = \frac{1}{2f(\chi_1)} \sum_{x \bmod f(\chi_1)} (-\chi_1(x)\, r_{f(\chi_1)}(x)), \tag{3.20.6}$$

where $r_{f(\chi_1)}(x)$ denotes the least positive residue x mod $f(\chi_1)$ and x can run over an *arbitrary* system of the prime residues mod $f(\chi_1)$.[153] Analogously to (3.14.9), it follows from (3.20.6) that

$$(1-\overline{\chi}_1(z))\,\Theta(\chi_1) = -\frac{1}{2f(\chi_1)} \sum_{x \bmod f(\chi_1)} \chi_1(x)\,(r_{f(\chi_1)}(x) - r_{f(\chi_1)}(zx)).$$

Because $r_{f(\chi_1)}(-x) = f(\chi_1) - r_{f(\chi_1)}(x)$, now by the transformation $x \to -x$ the difference $r_{f(\chi_1)}(x) - r_{f(\chi_1)}(zx)$ turns to the opposite value just like $\chi_1(x)$. Therefore the summation can be reduced to a *half system* of prime residues mod $f(\chi_1)$. As the result of our multiplication, it follows therefore that

$$(1-\overline{\chi}_1(z))\,\Theta(\chi_1) = -\frac{1}{f(\chi_1)} \sum_{\pm x \bmod f(\chi_1)} \chi_1(x)\,(r_{f(\chi_1)}(x) - r_{f(\chi_1)}(zx)). \tag{3.20.7}$$

Hereby the apparent denominator 2 appearing in (3.20.5) is removed,[154] and in regard to the summation region of the individual contributions of sums, the actual assimilation of the formula for h^* to that for h_0 is achieved; in the latter (formulas), the contributions of sums are also indeed extended over a *half system* of prime residues mod $f(\chi_0)$.

Now, in order to determine the congruence value of h^* mod 2 by the product formula (3.20.4), we introduce the above-mentioned formation $\chi_1 = \psi\chi_0$ in the individual contributions of sums (3.20.7) and determine its congruence value mod

[153] We have already proceeded in the special cases (3.14.5′) and (3.16.23′) completely correspondingly, where we had to execute such a multiplication of the sum under discussion.

[154] Kummer [14, 17, 20] has already applied this method, which is restricted in the present cyclic case, to remove the apparent denominator factor 2 of the relative class number formulas (3.9.1) and (3.9.2) in his proof of the integrality of h^* for the field $K = P_{p^\rho}$, and certainly also correspondingly (to remove) the apparent denominator factor p.

$\mathfrak{z}$, where $\mathfrak{z}$ is the unique prime divisor of 2 of degree 1 in the field $P_\psi = P_{2^\nu}$ of the values of ψ. Since $\psi(x) \equiv 1 \pmod{\mathfrak{z}}$ always holds, the factors $\chi_1(x)$ are then reduced to the $\chi_0(x)$. For the proof of congruence (3.20.1) it is the point to trace back from the $f(\chi_1)$ to the $f(\chi_0)$ (contained in $f(\chi_1)$ as the divisor by the above) in the summation $\sum_{\pm x \bmod f(\chi_1)}$ and in the factors $r_{f(\chi_1)}(x) - r_{f(\chi_1)}(zx)$. Hereby we have to separate the two already-mentioned cases $f(\psi) = p \neq 2$ and $f(\psi) = 2^\rho$.

I. $f(\psi) = p \neq 2$

Since (ψ is of the form) $\psi = \chi^{u_0}$ with odd u_0 and since in this case ψ has 2-component 1, χ also has 2-component 1. Then the same hold for all $\chi_1 = \chi^u$ and $\chi_0 = \chi^g$. Hence here all $f(\chi_1)$ and $f(\chi_0)$ are odd. Moreover, $\frac{w}{2}$ is odd.

For the special contribution (3.20.7) of $\chi_1 = \psi$ to (3.20.4), on account of $1 - \overline{\psi}(z) \cong \mathfrak{z}$, it holds that

$$(1 - \overline{\psi}(z))\,\Theta(\psi) \equiv 1 \pmod{\mathfrak{z}} \tag{3.20.8}$$

by Theorem 3.32 or rather by its proof.[155] One can also confirm easily this fact once again by determining the congruence values mod $\mathfrak{z}$ of sum (3.20.7) for this special case. Namely, first it follows from (3.20.7) that the congruence

$$(1 - \overline{\psi}(z))\,\Theta(\psi) \equiv \sum_{\pm x \bmod p} (r_p(x) - r_p(zx)) \pmod{\mathfrak{z}}$$

holds. Selecting for x the half system of the least positive prime residues mod p and applying the reduction

$$zx \equiv (-1)^{\delta_p(zx)} x' \pmod{p}$$

to the absolutely least residue mod p, where x' again runs over the half system of the least positive prime residues mod p, one has $r_p(x) = x$ and $r_p(zx) = x'$ or $p - x'$ according as $\delta_p(zx) \equiv 0$ or $1 \pmod{2}$, and therefore

$$r_p(x) - r_p(zx) \equiv x - x' + \delta_p(zx) \pmod{2}.$$

Consequently further

$$(1 - \overline{\psi}(z))\,\Theta(\psi) \equiv \sum_{\pm x \bmod p}^{+} \delta_p(zx) \pmod{\mathfrak{z}}.$$

[155] Namely, (one has congruence (3.20.8)) by congruence (3.13.6) obtained there in connection with (3.9.3); one can also directly conclude (3.20.8) backward from the fact (congruence) $2N_\psi(\Theta(\psi)) \equiv 1 \pmod{2}$ obtained by Theorem 3.32, because $\mathfrak{z}$ is the unique prime divisor of 2 in P_ψ and has degree 1.

By Gauss' Lemma it holds in general that

$$\sum_{\pm x \bmod p}{}^{+}\delta_p(ax) \equiv 0 \text{ or } 1 \pmod 2 \quad \text{according as } \left(\frac{a}{p}\right) = 1 \text{ or } -1.$$

Since the quadratic character $\left(\frac{x}{p}\right) = \psi(x)^{2^{\nu-1}}$ holds here, one has $\left(\frac{z}{p}\right) = \psi(z)^{2^{\nu-1}} = -1$, and hence $\sum_{\pm x \bmod p}{}^{+}\delta_p(zx) \equiv 1 \pmod 2$, which yields (3.20.8).

For the other contribution (3.20.7) of $\chi_1 \neq \psi$ in (3.20.4), introducing the formation $\chi_1 = \psi\chi_0$, for which $\chi_1(x) \equiv \chi_0(x) \pmod{\mathfrak{z}}$ holds, and taking into consideration that $f(\chi_1)$ and $\frac{w}{2}$ are odd in the present case and also that congruence (3.20.8) holds for $\chi_1 = \psi$, one obtains the congruence

$$h^* \equiv \prod_{\chi_0 \neq 1} \sum_{\pm x \bmod f(\psi\chi_0)} \chi_0(x)\,(r_{f(\psi\chi_0)}(x) - r_{f(\psi\chi_0)}(zx)) \pmod 2. \qquad (3.20.9)$$

In fact, at first this congruence holds as mod $\mathfrak{z}$, then by the rationality of both sides it holds also for mod 2.

Now, in the individual contributions of sums in (3.20.9) we reduce the sum $\sum_{\pm x \bmod f(\psi\chi_0)}$ to a sum $\sum_{\pm x \bmod f(\chi_0)}$. It holds that $f(\psi\chi_0) = pf(\chi_0)$ or $f(\chi_0)$ according as $p \nmid f(\chi_0)$ or $p \mid f(\chi_0)$. To accomplish the reduction we apply the general summation formula (2.2.3), and actually we apply it as a congruence mod 2. In fact, the sums of the contributions to be considered are

$$S_f(\chi_0) = \sum_{\pm x \bmod f} \chi_0(x) A_f(x)$$

of type there (in Sect. 2.2) with $f = f(\psi\chi_0)$ and the expression

$$A_f(x) = r_f(x) - r_f(zx)$$

depending only on the residue x mod f.[156] This expression satisfies the assumption (2.2.2), which we need here to confirm only congruence mod 2 and only for the one prime number $p \neq 2$ appearing in the reduction. Namely, it holds obviously that[157]

$$r_{f_0 p}(x_0 + y f_0) = r_{f_0}(x_0) + r_p(y) f_0,$$

$$r_{f_0 p}(x_0' p) = p\, r_{f_0}(x_0') \equiv r_{f_0}(x_0') \pmod 2,$$

[156](Translator's remark) "$A_f(\chi_0)$" in the original book should be "$A_f(x)$."

[157](Translator's remark) In the following equation we need the restriction $0 \leqq x_0 < f_0$.

and then on the one hand[158]

$$\sum_{y \bmod p} A_{f_0 p}(x_0 + y f_0) = \sum_{y \bmod p} (A_{f_0}(x_0) + A_p(y) f_0) = p A_{f_0}(x_0) + f_0 \sum_{y \bmod p} A_p(y)$$
$$\equiv A_{f_0}(x_0) + \alpha \pmod 2,$$

with the quantity α independent of x_0, and on the other hand

$$A_{f_0 p}(x_0' p) = p A_{f_0}(x_0') \equiv A_{f_0}(x_0') \pmod 2.$$

Therefore the summation formula (2.2.3) as a congruence mod 2 is actually applicable. It yields here

$$S_{f(\psi \chi_0)}(\chi_0) \equiv (1 - \chi_0(p)) S_{f(\chi_0)}(\chi_0) \pmod 2.$$

As a result of our reduction we therefore obtain from (3.20.9) the congruence

$$h^* \equiv \prod_{\chi_0 \neq 1} (1 - \chi_0(p)) \cdot \prod_{\chi_0 \neq 1} \sum_{\pm x \bmod f(\chi_0)} \chi_0(x) \left(r_{f(\chi_0)}(x) - r_{f(\chi_0)}(zx) \right) \pmod 2. \tag{3.20.10}$$

The obtained congruence (3.20.10) is different from congruence (3.20.1) to be proved only because in the congruence the functions $r_{f(\chi_0)}(x)$ take the places of the functions $\delta_{2f(\chi_0)}(x)$. Therefore our proof will be achieved if we show that the relation

$$r_{f_0}(x) - r_{f_0}(zx) \equiv \delta_{2f_0}(x) - \delta_{2f_0}(zx) \pmod 2 \tag{3.20.11}$$

holds for every $f(\chi_0) = f_0$. On account of the meaning of $\delta_{2f_0}(x)$ as sign-exponent of the absolutely least residue $r^*_{2f_0}(x)$ of x mod $2f_0$, one has here

$$r_{f_0}(x) = \left\{ \begin{array}{ll} r^*_{2f_0}(x) & \text{for } \delta_{2f_0}(x) \equiv 0 \pmod 2 \\ r^*_{2f_0}(x) + f_0 & \text{for } \delta_{2f_0}(x) \equiv 1 \pmod 2 \end{array} \right\}.$$

For the time being, one normalizes x to be odd in the half system of the prime residues x mod f_0, so that the $r^*_{2f_0}(x)$ are odd and then it follows that

$$r_{f_0}(x) \equiv 1 + \delta_{2f_0}(x) \pmod 2.$$

158(Translator's remark) In the following summations $\sum_{y \bmod p}$ the number y runs over $0, 1, 2, \ldots, p-1$. In the left-hand side the summation "$\sum_{y \bmod p} A_{f_0 p}(x_0 + yp)$" in the original book should be "$\sum_{y \bmod p} A_{f_0 p}(x_0 + y f_0)$."

Since the representative z is also normalized to be odd according to Sect. 2.4,[159] one has correspondingly

$$r_{f_0}(zx) \equiv 1 + \delta_{2f_0}(zx) \pmod 2.$$

Combining the two congruences above, one obtains indeed the relation (3.20.11) (and this congruence (3.20.11) is independent of the normalization to oddness of the prime residues x mod f_0 of the half system used in the proof, because by the transformations $x \to x + f_0$ and $x \to -x$, the expressions of the right- and left-hand sides are invariant).[160] By the statements obtained so far this establishment has completed the proof of Theorem 3.45 in the case of $f(\psi) = p \neq 2$.

II. $f(\psi) = 2^\rho$

By virtue of the form $\psi = \chi^{u_0}$ with odd u_0, the 2-components of χ and those of all odd characters $\chi_1 = \chi^u$ are in this case algebraic-conjugate to $\psi = \cup\varphi_\rho$. Hence all $f(\chi_1)$ are here divisible exactly by 2^ρ. On the contrary, for the $\chi_0 = \chi^g$, only powers 2^σ with $0 \leqq \sigma \leqq \rho - 1$ appear in the conductor $f(\chi_0)$. Moreover, here $\frac{w}{2} = 2$ or $\frac{w}{2} \equiv 1 \pmod 2$ according as $\rho = 2$ or $\rho \geqq 3$.

For the special contributions (3.20.7) to (3.20.4) of characters $\chi_1 = \psi^u$, which are algebraic-conjugate to ψ, it holds by (3.14.1), (3.14.6), or (3.14.10) that

$$2(1 - \overline{\psi}(z))\Theta(\psi) = 1 \qquad \text{for } \rho = 2 \text{ (then } \psi = \cup), \tag{3.20.12a}$$

$$(1 - \overline{\psi}^u(z))\Theta(\psi^u) \equiv 1 \pmod{\mathfrak{z}} \qquad \text{for } \rho \geqq 3 \text{ (then } \psi = \cup\varphi_\rho \neq \cup). \tag{3.20.12b}$$

For the treatment of the other contributions (3.20.7) to (3.20.4) of characters $\chi_1 \neq \psi^u$ not algebraic-conjugate to ψ, we write for short $f(\chi_1) = f = 2^\rho f_1$, where then $f_1 \neq 1$ and f_1 is odd. Let

$$x \equiv x_0 f_1 + x_1 2^\rho \pmod f, \tag{3.20.13}$$

then x runs over a half system of the prime residues mod f if x_1 runs over a half system of the prime residues modulo f_1 and x_0 runs over a complete system of the prime residues mod 2^ρ. As in (3.10.1), correspondingly it holds here that

$$\frac{r_f(x)}{f} = \frac{r_{2^\rho}(x_0)}{2^\rho} + \frac{r_{f_1}(x_1)}{f_1} - \varepsilon(x_0, x_1), \tag{3.20.14}$$

[159](Translator's remark) See footnotes 11 and 12 in Sect. 2.4 on p.31 and p. 32, respectively.

[160](Translator's remark) These parentheses are printed in the original.

with

$$\varepsilon(x_0, x_1) = 0 \text{ or } 1 \text{ according as } \frac{r_{2^\rho}(x_0)}{2^\rho} + \frac{r_{f_1}(x_1)}{f_1} < 1 \text{ or } > 1. \tag{3.20.15}$$

Consequently, executing the component decomposition of χ_1 corresponding to the splitting $f = 2^\rho f_1$ in sum (3.20.7) by the scheme in Sect. 3.10, there arises here by (3.20.14), analogously to relation (3.10.3), the relation

$$(1 - \overline{\chi}_1(z))\Theta(\chi_1) = \sum_{\pm x \bmod f} \chi_1(x)(\varepsilon(x_0, x_1) - \varepsilon(zx_0, zx_1)), \tag{3.20.16}$$

which is free from the apparent denominator $f(\chi_1)$, where the classes $x_0 \bmod 2^\rho$ and $x_1 \bmod f_1$ are associated with the classes $x \bmod f$ by (3.20.13). Now, by the definition in (3.20.15) of $\varepsilon(x_0, x_1)$, it holds that

$$\varepsilon(x_0, x_1) \equiv \delta_{2f}(r_{2^\rho}(x_0) f_1 + r_{f_1}(x_1) 2^\rho) \pmod 2$$

and correspondingly also

$$\begin{aligned} &\varepsilon(x_0, x_1) - \varepsilon(zx_0, zx_1) \\ &\quad \equiv \delta_{2f}(r_{2^\rho}(x_0) f_1 + r_{f_1}(x_1) 2^\rho) - \delta_{2f}(r_{2^\rho}(zx_0) f_1 + r_{f_1}(zx_1) 2^\rho) \pmod 2. \end{aligned}$$

On account of the invariance of $\delta_{2f}(x) - \delta_{2f}(zx) \bmod 2$ by the transformation $x \to x + f$, it follows more simply from the last relation (obtained) by (3.20.13) that

$$\varepsilon(x_0, x_1) - \varepsilon(zx_0, zx_1) \equiv \delta_{2f}(x) - \delta_{2f}(zx) \pmod 2.$$

Hereby there arises by (3.20.16) the congruence

$$(1 - \overline{\chi}_1(z))\Theta(\chi_1) \equiv \sum_{\pm x \bmod f(\chi_1)} \chi_1(x)(\delta_{2f(\chi_1)}(x) - \delta_{2f(\chi_1)}(zx)) \pmod 2. \tag{3.20.17}$$

This congruence proved initially for the $\chi_1 \neq \psi^u$ holds also for the $\chi_1 = \psi^u$ as a congruence mod$\mathfrak{z}$ instead of mod2 except for the trivial case $\psi = \cup$ from (3.20.12a). In fact, for $\psi = \cup\varphi_\rho$ ($\rho \geqq 3$) if one replaces the values of characters $\chi_1 = \psi^u = \cup\varphi_\rho^u$ with $f(\chi_1) = 2^\rho$ in the right-hand side of (3.20.17) by the values of characters $\chi_1' = \cup\varphi_{\rho+1}^u$ with $f(\chi_1') = 2^{\rho+1}$, then there arises the expression in the right-hand side of (3.14.9) (up to the sign and the replacement of ψ by ψ^u, which is not essential in the congruence). By (3.14.9) and (3.14.10) this last expression is $\equiv 1 \pmod{\mathfrak{z}'}$, where $\mathfrak{z}'$ is the prime divisor of 2 in $P_{2^{\rho}-1}$.[161] Because of $\chi_1(x) \equiv \chi_1'(x) \pmod{\mathfrak{z}'}$, the right-hand side of expression (3.20.17) is $\equiv 1 \pmod{\mathfrak{z}}$. Since, as already shown in (3.20.12b), the left-hand side of expression (3.20.17) is $\equiv 1 \pmod{\mathfrak{z}}$, and (3.20.17) holds as a congruence mod $\mathfrak{z}$ also for $\chi_1 = \psi^u$, provided $\psi \neq \mathsf{U}$. Thus we can state that (3.20.17) holds surely for all odd characters $\chi_1 \neq \psi$.[162]

Inserting the value of congruence (3.20.17) in (3.20.4) for the contribution (3.20.7) of $\chi_1 \neq \psi$ and then (inserting) $\chi_1(x) \equiv \chi_0(x) \pmod{\mathfrak{z}}$ for the formation $\chi_1 = \psi\chi_0$, and considering congruence (3.20.12a) and $\frac{w}{2} = 2$ or congruence (3.20.12b) and $\frac{w}{2} \equiv 1 \pmod 2$ for the contribution of $\chi_1 = \psi$, one obtains the congruence

$$h^* \equiv \prod_{\chi_0 \neq 1} \sum_{\pm x \bmod f(\psi\chi_0)} \chi_0(x)(\delta_{2f(\psi\chi_0)}(x) - \delta_{2f(\psi\chi_0)}(zx)) \pmod 2. \qquad (3.20.18)$$

Indeed, (it holds) first again only as congruence mod $\mathfrak{z}$ and then by the rationality of both sides also for mod 2.

Now, in the individual contributions of sums to (3.20.18), we again reduce the sum $\sum_{\pm x \bmod f(\psi\chi_0)}$ to a sum $\sum_{\pm x \bmod f(\chi_0)}$. Here it holds that $f(\psi\chi_0) = 2^{\rho} f_1$ (f_1 odd) and $f(\chi_0) = 2^{\sigma} f_1$ $(0 \leqq \sigma \leqq \rho - 1)$.[163] In order to execute the reduction, we apply, as before, the general summation formula (2.2.3) as a congruence mod 2. The contributions of the sums to be considered are, in fact,

$$S_f(\chi_0) = \sum_{\pm x \bmod f} \chi_0(x) A_f(x)$$

of type there (in Sect. 2.2) with $f = f(\psi\chi_0)$ and the expression

$$A_f(x) = \delta_{2f}(x) - \delta_{2f}(zx)$$

[161] (Translator's remark) The field "$P_{2^{\rho}+1}$" in the original book should be "$P_{2^{\rho}-1}$."

[162] We relate to this conclusion the remark connected with Theorem 3.45 that in our proof we have to make use of Weber's Theorem 3.36′ on the relative class number $h_{2^{\rho}}^{*}{}'$ of $P_{2^{\rho}}'$. In fact, either one of the two facts (3.14.6) and (3.14.10) used in the proof of (3.20.17) is equivalent to the fact that

$$h_{2^{\rho}}^{*}{}' = N_{\psi}((1 - \overline{\psi}(z))\Theta(\psi))|_{\psi = \mathsf{U}\varphi_{\rho}}$$

is odd.

(Translator's remark) In the equation just above "$|_{\psi=\varphi_{\rho}}$" in the original book should be "$|_{\psi=\mathsf{U}\varphi_{\rho}}$."

[163] (Translator's remark) "$f(\chi_0) = 2^{\rho-\sigma} f_1$" in the original book seems to be unsuitable because of the assumption $2^{\sigma} \| f(\chi_0)$ defined at the beginning of this case II; $f(\chi_0) = 2^{\sigma} f_1$ is preferable to $f(\chi_0) = 2^{\rho-\sigma} f_1$.

depending only on the residue class x mod f.[164] Then we have to show that this expression satisfies the assumption (2.2.2) as a congruence mod 2 for the one prime number $p = 2$ appearing in the reduction. This reduction is slightly more awkward than the previous one. As one can verify on the basis of the definition of $\delta_{2f}(x)$, it holds for $f = 2f_0$ that

$$\delta_{2f}(x_0) + \delta_{2f}(x_0 + f_0) \equiv \delta_{2f_0}(x_0) \pmod 2.$$

Then it follows that

$$\delta_{2f}(zx_0) + \delta_{2f}(z(x_0 + f_0)) \equiv \left\{ \begin{array}{ll} \delta_{2f_0}(zx_0) & \pmod 2 \text{ for } z \equiv 1 \pmod 4 \\ \delta_{2f_0}(zx_0) + 1 \pmod 2 & \text{for } z \equiv -1 \pmod 4 \end{array} \right\};$$

since in the first case the congruence $zf_0 \equiv f_0 \pmod{2f}$ holds, and hence $\delta_{2f}(z(x_0 + f_0)) \equiv \delta_{2f}(zx_0 + f_0) \pmod 2$, while in the second case the congruences $zf_0 \equiv -f_0 \equiv f_0 - f \pmod{2f}$ hold, and hence $\delta_{2f}(z(x_0 + f_0)) \equiv \delta_{2f}(zx_0 + f_0 - f) \equiv \delta_{2f}(zx_0 + f_0) + 1 \pmod 2$. Therefore it holds that

$$\sum_{y \bmod 2} A_{2f_0}(x_0 + yf_0) \equiv A_{f_0}(x_0) + \alpha \pmod 2$$

with a quantity α independent of x_0; thus $\alpha \equiv 0$ or $1 \pmod 2$ according as $z \equiv 1$ or $1 \pmod 4$. Hcrcby thc first assumption of (2.2.2) is satisfied. The second assumption is similarly satisfied, because for $f = 2f_0$ it holds clearly that

$$\delta_{2f}(2x_0') \equiv \delta_{2f_0}(x_0') \pmod 2,$$

and then correspondingly also

$$A_{2f_0}(2x_0') = A_{f_0}(x_0') \pmod 2.$$

Consequently the summation formula (2.2.3) is applicable as a congruence mod 2. It provides here

$$S_{f(\psi\chi_0)}(\chi_0) \equiv (1 - \chi_0(2)) \cdot S_{f(\chi_0)}(\chi_0) \pmod 2.$$

As a result of our reduction we therefore obtain from (3.20.18) the congruence

$$h^* \equiv \prod_{\chi_0 \neq 1} (1 - \chi_0(2)) \cdot \prod_{\chi_0 \neq 1} \sum_{\pm x \bmod f(\chi_0)} \chi_0(x)(\delta_{2f(\chi_0)}(x) - \delta_{2f(\chi_0)}(zx)) \pmod 2. \tag{3.20.19}$$

[164](Translator's remark) In this summation "$A_f(\chi_0)$" in the original book should be "$A_f(x)$."

Being deduced more simply than before, the obtained congruence (3.20.19) surely coincides here with congruence (3.20.1) to be proved. By the statements obtained so far, this establishment has completed the proof of Theorem 3.45 also in the case of $f(\psi) = 2^\rho$. □

Appendix: Tables of Relative Class Numbers

We give as an appendix a table of the contributions $N_\psi(\Theta(\psi))$ to the relative class number for all the characters ψ with $\psi(-1) = -1$ and with conductor $f \leqq 100$, which are calculated by the formulas in Sects. 3.10–3.16, as well as a table of the relative class numbers h^* of K/K_0 for all the abelian number fields K with conductor $f \leqq 100$. The latter table arises from the first table by formula (3.9.1), where the unit indices of K/K_0 are furthermore determined by the methods in Sects. 3.2–3.7, which we have also adopted in the table.

For the cyclotomic fields $K = P_f$ Kummer [17, 19][165] has already made tables of the latter kind.[166] Although the individual contributions, in fact, were to be calculated first, one could not draw these contributions from Kummer's tables anymore, so that for the transition to the subfields of cyclotomic fields, new calculations of contributions in our first table were required. Hereby the famous table of Reuschle [22] could be used effectively.[167]

Table of Contributions to Relative Class Number

Column 1 gives the conductors $f \leqq 100$ together with their decompositions into prime-powers $f = \prod_p p^\rho$. Here, among all the natural numbers $\leqq 100$, only 1 and numbers $\equiv 2 \pmod 4$ are deleted.

[165](Translator's remark) In the original book Hasse quoted Kummer [17, 20]. But this number [20] as well as Kummer [20] in footnote 167 below should be replaced by Kummer [19] as on this page.

[166]By comparison of Kummer's tables and our following tables, we have to remark that Kummer's first class number factor, as elaborated in Sect. 1.6, footnote 22, p. 18, coincides with our relative class number only for prime-power conductor $f = p^\rho$, while it is just $\frac{1}{2}h^*$ for composite conductor f. Moreover, give attention to the following errors in the Kummer's tables:

1. In Kummer [17] $7 \cdot 7 \cdot 29 \cdot 3851$ is printed incorrectly for $\lambda = 71$; This number is reported by Kummer [19] to be $7 \cdot 7 \cdot 79{,}241$.
2. In Kummer [19] $\frac{1}{2} \cdot 3 \cdot 67^2$ is printed incorrectly for $n = 92$; Kummer has later reported this number to be $\frac{1}{2} \cdot 3 \cdot 67$ by a remark of Reuschle.
3. In Kummer [19] 2^3 is printed incorrectly for $n = 68$; our calculation established the value of this number to be 2^2 in the Kummer normalization.

[167]Compare it with the remark given in my Preface on these tables.

Column 2 gives a system of representatives ψ of classes of algebraic-conjugate characters with $f(\psi) = f$ and $\psi(-1) = -1$. With only one exception of the least composite conductor $f = 12 = 2^2 \cdot 3$, there exists at least one such a class for every conductor f.

The representatives ψ are given by combination of basic characters that correspond to the prime-powers p^ρ constituting the conductor f. The affixed subscript always denotes here (differently than in the process of our foregoing research) the *conductor* of the basic character. As the basic characters the following are adopted:

For every $p \neq 2$

a character χ_p of order $p-1$ and with $\chi_p(-1) = -1$

and when $\rho \geqq 2$

a character ψ_{p^ρ} of order $p^{\rho-1}$ and with $\psi_{p^\rho}(-1) = 1$; then $\psi_{p^\rho}^p = \psi_{p^{\rho-1}}$

For the prime number $p = 2$

a character χ_{2^2} of order 2 and with $\chi_{2^2}(-1) = -1$

and when $\rho \geqq 3$

a character ψ_{2^ρ} of order $2^{\rho-2}$ and with $\psi_{2^\rho}(-1) = 1$; then $\psi_{2^\rho}^2 = \psi_{2^{\rho-1}}$.

In regard to the basic characters χ_p, ψ_{p^ρ} with respect to $p \neq 2$, see (3.13.1) and (3.13.2). The basic characters χ_{2^2}, ψ_{2^ρ} with respect to $p = 2$ were denoted by $\cup$, φ_ρ for short in our research because of their often appearances, and analogously to these, the basic characters χ_3, ψ_{3^ρ} with respect to $p = 3$ were denoted by ω, ψ_ρ in Sect. 3.16; here we also use the above-mentioned systematic expressions for these special basic characters.

We have attached the values of basic characters to a special *auxiliary table*, in fact, first because such a table is very desirable completely in general for when one would make numerical examples of the objects treated in future work, and second because (we have) particularly the following reason regarding the main table. We have fixed the basic characters only up to selection in their algebraic-conjugates. This, as a result, does not imply that a character ψ is determined in general only once by the indication of combination of basic characters up to selection among their algebraic-conjugates. Namely, if among the powers of basic characters constituting a character ψ there exist two powers whose orders have a common divisor $\neq 1, 2$ and if one replaces one of these two components by an algebraic-conjugate (for example, it suffices to replace the component by its complex-conjugate—denoted by bar—in the frame of our table), while one does not replace other powers by corresponding algebraic-conjugates, then there arises a character ψ' *not* algebraic-conjugate to ψ. In such cases where they appear for certain composite conductors f—in the first 100 conductors, indeed only for $f = 63, 80, 85, 95$—for the unique determination of the (Frobenius) classes of ψ and ψ', the coupling of the selections from the algebraic-conjugates for both of the components in question must be undertaken. This occurs here by setting that the values of all basic characters are constructed by the auxiliary table with a fixed system of roots of unity of prime-power order. Normalization[168] of this system among algebraic-conjugates, for example, by the analytic determination of roots of unity in as Sect. 1.3, is not required here, since the determination of character ψ among its algebraic-conjugates does not matter.

As in Reuschle [22] one can also fix uniquely the (Frobenius) class of character ψ by the indication of the least positive residues a mod f from the corresponding congruence group H_ψ with $\psi(a) = 1$. The list of these residue systems can easily be realized by the auxiliary table.

Column 3 gives the order n_ψ of ψ and simultaneously (gives) the index of H_ψ in the form of a prime decomposition.

Column 4 gives the degree $\varphi(n_\psi)$ of the cyclotomic field P_ψ generated by the values of ψ; then it is the number of algebraic-conjugate characters that constitute the class of ψ. As will be elaborated in the explanation of Table 3.2, this degree is important for the aim of the control there.

Column 5 gives the contributions $N_\psi(\Theta(\psi))$ of ψ to the relative class number h^* of K/K_0 for all such abelian number fields K that the ψ appear as characters of K/K_0; see to this formulas (3.9.1) and (3.9.2).

In the representation of these contributions (in Column 5), the prime factors of the denominators are listed first in the form of a fraction, which at most exist by Theorems 3.32 and 3.33 for prime-power conductor. Then the subsequent numerator, which expresses the essential part of the contribution, is decomposed into its prime-power factors. That one of the denominator factors, appearing at most once by the theory, is really reduced is expressed also by its appearance in the numerator factors.[169] It is interesting, in this respect, that there appears the different behavior of three famous Kummer's primes $p = 37,\ 59,\ 67$ in the first 100 (conductors) for which the relative class number h^* of $P_p/P_{p,0}$ is divisible by p. For the first two primes the denominator appears already in the contribution $N_{\chi_p}(\Theta(\chi_p))$; for the last one, it is just seen that one of the joining additional contributions $N_{\chi_p^d}(\Theta(\chi_p^d))$ (d an odd divisor of $p-1$) is divisible by p, so that in every case the joining factor $w = 2p$ divisible by p provides the factor p in h^*.

The table is divided into two parts. The first part contains the prime-power conductors $f = p^\rho$; these are arranged for clarity by the prime numbers p (not by

[168] For the χ_p the normalization in the auxiliary table is selected corresponding to *Canon Arithmeticus* of Jacobi [11], where the tabulated indices are composed of the exponents of primitive $(p-1)$-th roots of unity, and these roots of unity, corresponding to the prime decomposition of $p-1$, are expressed as products of fixed prime-power-th roots of unity (with exponent 1) given at the top of the table. These decompositions of the values of characters are practical for the gradual calculation of norms $N_\psi(\Theta(\psi))$ according to the structure of P_ψ. For the ψ_{p^ρ} the value $\psi_{p^\rho}(1+p)$ is set to be equal to a given fixed (primitive) $p^{\rho-1}$-th root of unity.

[169] (Translator's remark) See the contributions $N_{\chi_p}(\Theta(\chi_p))$ for $f = 37,\ 59$ in Table 3.1, 1. Prime-power conductors.

the quantity of f). The second part contains the composite conductors f; these are arranged by the quantity of f for lack of a better systematically practical principle.

Table of Relative Class Numbers

Column 1 gives again the conductors $f \leqq 100$ together with their decomposition into prime-powers $f = \prod_p p^\rho$. The classification by prime-power conductors and composite conductors as well as the ordering of conductors in these two parts of Table 3.2 corresponds to the ones in Table 3.1.

Column 2 gives the serial numbers for the individual imaginary abelian number fields K that belong to a fixed conductor f. When the field K' arises from K by the transformation to an algebraic-conjugate component in the character system, as appeared by the previously emphasized conductor $f = 63, 80, 85, 91$, the K' are characterized by adding a prime to the number of K.

Column 3 gives a system of generating characters of K by which K is uniquely determined. This system is in general selected by corresponding to the prime-power invariants of the character group X of K; so, for example, for $f = 35$ in No. 6 instead of a generating character $\chi_5^2 \cdot \chi_7$ of order 6, a pair of generating characters $\chi_5^2 \cdot \chi_7^3$, χ_7^2 of order 2, 3 is adopted. Only for the basic character χ_7 the character itself and its power are excluded from such decompositions. Characters with composite conductors are emphasized with characters with prime-power conductors by the attachment of a multiplication dot between the powers of basic characters; so, for example, $\psi_{2^3} \cdot \chi_3$ but $\chi_{2^2}\psi_{2^3}$. The order by which the individual fields K with a (fixed) conductor f are arranged lays no claim to strict systematic presentation, and not also the order by which the generating characters determining the fields K are arranged. As far as it is practicable, the procedure of lexicographical order with respect to the natural order of prime numbers in the subscript of basic characters is taken as a basis, but such a procedure fails indeed in the case of appearance of characters with composite conductor. Throughout the table every field K possibly appears before subfields $\tilde{K}$, which are important for the application of Theorem 3.29 (instead of Theorem 3.22) to the assurance of $Q = 1$ (see below).[170]

Column 4 gives the type, i.e., the system of prime-power invariants, of the character group X and then also of the Galois group $\mathfrak{G}$ of K. The ordering of prime-power invariants corresponds to that of previously given generating characters. One similar type (of fields) appears here often and also (appears) within the system of the fields K belonging to a fixed conductor f in different ordering, for example, $(2^2, 2)$ and $(2, 2^2)$, (which is) a situation to be noticed when one searches imaginary abelian number fields with a fixed type. To emphasize cyclic fields K, along with types of several powers of different primes, the corresponding one-term types are also given, for example, $(2^2, 3) = 12$, $(2, 3, 5) = 15$.

[170](Translator's remark) As mentioned in Theorem 3.29, one requires some conditions to use Theorem 3.29.

Column 5 gives the number w of roots of unity in K in the form of a prime-power decomposition.

Column 6 gives the unit index $Q = 1$ or 2 of K/K_0.

Column 7 gives the relative class number h^* of K/K_0 in the form of a prime-power decomposition, which was seen, by the calculation of the product of prime-powers, as arithmetic without deeper meaning.

To determine the numbers w, Q and h^*, one has to set up a complete system of representatives ψ of classes of algebraic-conjugate characters of K/K_0. As for the *cyclotomic fields* $K = P_f$—they always appear as No. 1—one obtains such a system by combining the characters of (conductor) f and of all conductors dividing f listed in Table 3.1; as for the *subfields K of P_f* belonging to the conductor f, one always obtains such a system by selecting the characters corresponding to K that can be assembled by the generating characters given in Column 3. In order to check the completeness of the system of representatives obtained in such a way, it can be useful that the sum of the degrees[171] $\varphi(n_\psi)$ of their characters ψ (see Table 3.1, Column 4) must always be equal to the half $n_0 = \frac{1}{2}n$ of the degree n of K, and then it is equal to the half of the product of prime-power invariants of K given in Column 4. In the following, for an example with $f = 77$, we give the scheme for the assembly of the system of representatives ψ and subsequently the determination of w, Q, and h^*, as we have to think of the system inserted between Columns 1–4 and Columns 5–7. The degrees $\varphi(n_\psi)$ are then given by the characters ψ, and the selection of the respective sub-system corresponding to K is therefore expressed in such a way that in the row of K the places corresponding to characters ψ are displayed by the enrollment of the contribution of ψ to h^* from Table 3.1.[172]

To determine the prime-power contribution p^ω to the number w of roots of unity in K, one only needs to examine for what prime-power p^ω the complete system of representatives χ_p, $\chi_p\psi_{p^\omega}$ of the characters of $P_{p^\omega}/P_{p^\omega,0}$ appears in the system of representatives ψ of K/K_0.

To determine the unit indices Q of K/K_0, one has Theorems 3.22–3.27 and 3.29 at one's disposal, among which Theorems 3.22 and 3.26 need the information of the system of representatives ψ. In the present fixed frame $f \leqq 100$, one can manage with the information of these systems; in this frame, there exists no unique case that requires the determination, by a special consideration, of the alternative whether K_0'/K_0 is of unit type or of class type. For the composite conductor it is compiled in the attached *auxiliary table* by which of the stated theorems the alternative $Q = 1$ or

---

[171](Translator's remark) In the original book, the degrees n_ψ on p. 136, lines 1 and 7 from the bottom, as well as in the table on p. 137 should be $\varphi(n_\psi)$.

[172](Translator's remark) In the original book the representatives $\chi_7, \chi_{11}^2; \chi_7^2, \chi_{11}; \chi_7^2, \chi_{11}^5; \chi_7^3, \chi_{11}^2$ in the second row in the table should be replaced by the products $\chi_7 \cdot \chi_{11}^2$, $\chi_7^2 \cdot \chi_{11}$, $\chi_7^2 \cdot \chi_{11}^5$, $\chi_7^3 \cdot \chi_{11}^2$, respectively.

In the following table "system of ψ and degrees $\varphi(n_\psi)$" should be translated as "system of representatives ψ together with degrees $\varphi(n_\psi)$." This abbreviation is due to the limited space.

1	2	3	4	System of ψ and degrees $\varphi(n_\psi)$								5	6	7
				2	1	4	1	8	8	2	4			
Conductor	No.	Gen.characters	Type	χ_7	χ_7^3	χ_{11}	χ_{11}^5	$\chi_7 \cdot \chi_{11}^2$	$\chi_7^2 \cdot \chi_{11}$	$\chi_7^2 \cdot \chi_{11}^5$	$\chi_7^3 \cdot \chi_{11}^2$	w	Q	h^*
$77 = 7 \cdot 11$	1	χ_7, χ_{11}	(2, 3, 2, 5)	$\frac{1}{7}$	$\frac{1}{2}$	$\frac{1}{11}$	$\frac{1}{2}$	2^4	2^4	1	5	$2 \cdot 7 \cdot 11$	2	$2^8 \cdot 5$
	2	χ_7, χ_{11}^2	(2, 3, 5) = 30	"	"			"			"	$2 \cdot 7$	1	$2^4 \cdot 5$
	3	χ_7, χ_{11}^5	(2, 3, 2)	"	"		"			"		$2 \cdot 7$	2	1
	4	χ_7^2, χ_{11}	(3, 2, 5) = 30			"	"		"	"		$2 \cdot 11$	1	2^4
	5	χ_7^2, χ_{11}^5	(3, 2) = 6				"			"		2	1	1
	6	χ_7^3, χ_{11}	(2, 2, 5)		"	"	"				"	$2 \cdot 11$	2	5
	7	χ_7^3, χ_{11}^2	(2, 5) = 10		"						"	2	1	5
	8	χ_7^3, χ_{11}^5	(2, 2)		"		"					2	2	1

2 can be determined (in a simplest way)[173]; for the prime-power conductor, $Q = 1$ always holds by Theorem 3.23.

To determine the relative class number h^* of K/K_0, one simply has to take by (3.9.1) the product of cited contributions of representatives ψ in the row corresponding to K in the arrangement above with the following given numbers w and Q.[174]

To show the mutual relations among the individual imaginary abelian number fields K with a fixed conductor f more clearly than this (showing), which is possible by the lists of the systems of generating characters determining the fields, we have illustrated the system corresponding to K by a *graph*[175] for every conductor f. In the graph the fields are expressed by points and inclusions by connecting lines where the upward direction (vertical or inclined) corresponds to the ascension from subfield to extension; the numbers by the lines denote the relative degree. In the graphs of our table the following three types of fields are classified:

1. Real abelian number fields are denoted by bold points ●.
2. Imaginary abelian number fields K with respective conductor f are denoted by small circles **O** with the numbers from Column 2 where the fields K with conductor f appear.
3. Imaginary abelian number fields $\tilde{K}$ whose conductor $\tilde{f}$ is a proper divisor of the relevant conductor f are denoted by small squares □ with the number from Column 2 under which the fields $\tilde{K}$ previously appeared; the sub-conductor $\tilde{f}$ is attached there as an index of the number.

The distribution of these latter fields $\tilde{K}$ with indicated numbers together with registered relative degrees can make us easily recognize which generating characters correspond (to the fields $\tilde{K}$) by the line coming from the lowest point—the rational number field P, so that we would not need to be anymore annoyed with the graph by introducing generating characters. Also we could limit the indication of the relative degrees that is always equal to 2 to a considered part,[176] just by giving the line

[173]Here we avoid the application of the complicated Theorem 3.22 by citing another theorem as often as we can.

[174](Translator's remark) For example, in the table just above for $f = 77$, the field K of No. 1 whose character group is generated by χ_7, χ_{11} has relative class number

$$h^* = Qw \prod_{\psi} N_\psi(\Theta(\psi)) = 2 \cdot (2 \cdot 7 \cdot 11) \cdot \frac{1}{7} \cdot \frac{1}{2} \cdot \frac{1}{11} \cdot \frac{1}{2} \cdot 2^4 \cdot 2^4 \cdot 1 \cdot 5 = 2^8 \cdot 5.$$

[175]This special form of graphic illustration of field-theoretic or group-theoretic relations I have been using for 20 years in my publications. Recently this form has been used in lattice theory.

[176](Translator's remark) For example, in the case of $f = 21$ the field of No. 3 and its subfields construct a parallelogram, which is "a considered part," and in the case of $f = 56$ the lower cube is "a considered part."

corresponding to the generating characters,[177] because one can read all the other relative degrees in a simple way by the famous rule.

In the individual cases, note further that one may imagine the type (2, 2, 2) as a cube (see $f = 24$) where together with the 8 vertices there exist the fields corresponding to the 6 "surface-middle-points" and to the 2 inner points on the "space-diagonal" between the lowest and the highest points. This structure also appears as a sub-structure in a series of broader graphs (see $f = 40, 48, 56, 60, 72, 80, 84, 88, 96$). The graph for $f = 80$ corresponding to type $(2, 2^2, 2^2)$ is particularly complicated among them; for this graph, an appearing subgraph of type $(2^2, 2^2)$ furthermore overlaps with the system of four such cubes between the two surfaces composed of four quadrates[178] (see, for example, the basic surface composed of four quadrates for $f = 65$), which in addition leads to the appearance of four more points inside the four cubes. For all the other graphs appearing hereafter, the structures are probably clear without explanation in detail.

The graphs also can really serve to read in any case the maximal real subfield K_0 of K quickly among others, whose enrollment was not taken in the actual table. For this purpose, among the lines flowing downward from the circle ○ corresponding to K, one has to search for the field corresponding to a point ● (neither ○ nor □) on the line flowing downward; the field K_0 corresponds to the point ● so appointed.

[177](Translator's remark) For the relative degree not equal to 2, we also refrain from the indication.

[178](Translator's remark) A *quadrate* is a diagram of type (2, 2).

Table 3.1 Contributions of characters to the relative class numbers

1. Prime-power conductors

1	2	3	4	5
Conductor	Character	Order	Degree	Contribution
$2^2 = 4$	χ_{2^2}	2	1	$\frac{1}{2^2}$
$2^3 = 8$	$\chi_{2^2}\psi_{2^3}$	2	1	$\frac{1}{2}$
$2^4 = 16$	$\chi_{2^2}\psi_{2^4}$	2^2	2	$\frac{1}{2}$
$2^5 = 32$	$\chi_{2^2}\psi_{2^5}$	2^3	4	$\frac{1}{2}$
$2^6 = 64$	$\chi_{2^2}\psi_{2^6}$	2^4	8	$\frac{1}{2} \cdot 17$
3	χ_3	2	1	$\frac{1}{2 \cdot 3}$
$3^2 = 9$	$\chi_3\psi_{3^2}$	$2 \cdot 3$	2	$\frac{1}{3}$
$3^3 = 27$	$\chi_3\psi_{3^3}$	$2 \cdot 3^2$	6	$\frac{1}{3}$
$3^4 = 81$	$\chi_3\psi_{3^4}$	$2 \cdot 3^3$	18	$\frac{1}{3} \cdot 2593$
5	χ_5	2^2	2	$\frac{1}{2 \cdot 5}$
$5^2 = 25$	$\chi_5\psi_{5^2}$	$2^2 \cdot 5$	8	$\frac{1}{5}$
7	χ_7	$2 \cdot 3$	2	$\frac{1}{7}$
	χ_7^3	2	1	$\frac{1}{2}$
$7^2 = 49$	$\chi_7\psi_{7^2}$	$2 \cdot 3 \cdot 7$	12	$\frac{1}{7} \cdot 43$
	$\chi_7^3\psi_{7^2}$	$2 \cdot 7$	6	1
11	χ_{11}	$2 \cdot 5$	4	$\frac{1}{11}$
	χ_{11}^5	2	1	$\frac{1}{2}$
13	χ_{13}	$2^2 \cdot 3$	4	$\frac{1}{13}$
	χ_{13}^3	2^2	2	$\frac{1}{2}$
17	χ_{17}	2^4	8	$\frac{1}{2 \cdot 17}$
19	χ_{19}	$2 \cdot 3^2$	6	$\frac{1}{19}$
	χ_{19}^3	$2 \cdot 3$	2	1
	χ_{19}^9	2	1	$\frac{1}{2}$
23	χ_{23}	$2 \cdot 11$	10	$\frac{1}{23}$
	χ_{23}^{11}	2	1	$\frac{1}{2} \cdot 3$
29	χ_{29}	$2^2 \cdot 7$	12	$\frac{1}{29} \cdot 2^3$
	χ_{29}^7	2^2	2	$\frac{1}{2}$
31	χ_{31}	$2 \cdot 3 \cdot 5$	8	$\frac{1}{31}$
	χ_{31}^3	$2 \cdot 5$	4	1
	χ_{31}^5	$2 \cdot 3$	2	3
	χ_{31}^{15}	2	1	$\frac{1}{2} \cdot 3$
37	χ_{37}	$2^2 \cdot 3^2$	12	$\frac{1}{37} \cdot 37$
	χ_{37}^3	$2^2 \cdot 3$	4	1
	χ_{37}^9	2^2	2	$\frac{1}{2}$
41	χ_{41}	$2^3 \cdot 5$	16	$\frac{1}{41} \cdot 11^2$
	χ_{41}^5	2^3	4	$\frac{1}{2}$

1	2	3	4	5
Conductor	Character	Order	Degree	Contribution
43	χ_{43}	$2 \cdot 3 \cdot 7$	12	$\frac{1}{43} \cdot 211$
	χ_{43}^3	$2 \cdot 7$	6	1
	χ_{43}^7	$2 \cdot 3$	2	1
	χ_{43}^{21}	2	1	$\frac{1}{2}$
47	χ_{47}	$2 \cdot 23$	22	$\frac{1}{47} \cdot 139$
	χ_{47}^{23}	2	1	$\frac{1}{2} \cdot 5$
53	χ_{53}	$2^2 \cdot 13$	24	$\frac{1}{53} \cdot 4889$
	χ_{53}^{13}	2^2	2	$\frac{1}{2}$
59	χ_{59}	$2 \cdot 29$	28	$\frac{1}{59} \cdot 59 \cdot 233$
	χ_{59}^{29}	2	1	$\frac{1}{2} \cdot 3$
61	χ_{61}	$2^2 \cdot 3 \cdot 5$	16	$\frac{1}{61} \cdot 1861$
	χ_{61}^3	$2^2 \cdot 5$	8	41
	χ_{61}^5	$2^2 \cdot 3$	4	1
	χ_{61}^{15}	2^2	2	$\frac{1}{2}$
67	χ_{67}	$2 \cdot 3 \cdot 11$	20	$\frac{1}{67} \cdot 12{,}739$
	χ_{67}^3	$2 \cdot 11$	10	67
	χ_{67}^{11}	$2 \cdot 3$	2	1
	χ_{67}^{33}	2	1	$\frac{1}{2}$
71	χ_{71}	$2 \cdot 5 \cdot 7$	24	$\frac{1}{71} \cdot 79{,}241$
	χ_{71}^5	$2 \cdot 7$	6	7
	χ_{71}^7	$2 \cdot 5$	4	1
	χ_{71}^{35}	2	1	$\frac{1}{2} \cdot 7$
73	χ_{73}	$2^3 \cdot 3^2$	24	$\frac{1}{73} \cdot 134{,}353$
	χ_{73}^3	$2^3 \cdot 3$	8	1
	χ_{73}^9	2^3	4	$\frac{1}{2} \cdot 89$
79	χ_{79}	$2 \cdot 3 \cdot 13$	24	$\frac{1}{79} \cdot 377{,}911$
	χ_{79}^3	$2 \cdot 13$	12	53
	χ_{79}^{13}	$2 \cdot 3$	2	1
	χ_{79}^{39}	2	1	$\frac{1}{2} \cdot 5$
83	χ_{83}	$2 \cdot 41$	40	$\frac{1}{83} \cdot 279{,}405{,}653$
	χ_{83}^{41}	2	1	$\frac{1}{2} \cdot 3$
89	χ_{89}	$2^3 \cdot 11$	40	$\frac{1}{89} \cdot 118{,}401{,}449$
	χ_{89}^{11}	2^3	4	$\frac{1}{2} \cdot 113$
97	χ_{97}	$2^5 \cdot 3$	32	$\frac{1}{97} \cdot 577 \cdot 206{,}209$
	χ_{97}^3	2^5	16	$\frac{1}{2} \cdot 3457$

(continued)

Table 3.1 (continued)

2. Composite conductors

1	2	3	4	5
Conductor	Character	Order	Degree	Contribution
$12 = 2^2 \cdot 3$	–	–	–	–
$15 = 3 \cdot 5$	$\chi_3 \cdot \chi_5^2$	2	1	1
$20 = 2^2 \cdot 5$	$\chi_{2^2} \cdot \chi_5^2$	2	1	1
$21 = 3 \cdot 7$	$\chi_3 \cdot \chi_7^2$	$2 \cdot 3$	2	1
$24 = 2^3 \cdot 3$	$\psi_{2^3} \cdot \chi_3$	2	1	1
$28 = 2^2 \cdot 7$	$\chi_{2^2} \cdot \chi_7^2$	$2 \cdot 3$	2	1
$33 = 3 \cdot 11$	$\chi_3 \cdot \chi_{11}^2$	$2 \cdot 5$	4	1
$35 = 5 \cdot 7$	$\chi_5 \cdot \chi_7^2$	$2^2 \cdot 3$	4	1
	$\chi_5^2 \cdot \chi_7$	$2 \cdot 3$	2	1
	$\chi_5^2 \cdot \chi_7^3$	2	1	1
$36 = 2^2 \cdot 3^2$	$\chi_{2^2} \cdot \psi_{3^2}$	$2 \cdot 3$	2	1
$39 = 3 \cdot 13$	$\chi_3 \cdot \chi_{13}^2$	$2 \cdot 3$	2	1
	$\chi_3 \cdot \chi_{13}^4$	$2 \cdot 3$	2	1
	$\chi_3 \cdot \chi_{13}^6$	2	1	2
$40 = 2^3 \cdot 5$	$\psi_{2^3} \cdot \chi_5$	2^2	2	1
	$\chi_{2^2}\psi_{2^3} \cdot \chi_5^2$	2	1	1
$44 = 2^2 \cdot 11$	$\chi_{2^2} \cdot \chi_{11}^2$	$2 \cdot 5$	4	1
$45 - 3^2 \cdot 5$	$\psi_{3^2} \cdot \chi_5$	$2^2 \cdot 3$	4	1
	$\chi_3\psi_{3^2} \cdot \chi_5^2$	$2 \cdot 3$	2	1
$48 = 2^4 \cdot 3$	$\psi_{2^4} \cdot \chi_3$	2^2	2	1
$51 = 3 \cdot 17$	$\chi_3 \cdot \chi_{17}^2$	2^3	4	1
	$\chi_3 \cdot \chi_{17}^4$	2^2	2	5
	$\chi_3 \cdot \chi_{17}^8$	2	1	1
$52 = 2^2 \cdot 13$	$\chi_{2^2} \cdot \chi_{13}^2$	$2 \cdot 3$	2	1
	$\chi_{2^2} \cdot \chi_{13}^4$	$2 \cdot 3$	2	3
	$\chi_{2^2} \cdot \chi_{13}^6$	2	1	1
$55 = 5 \cdot 11$	$\chi_5 \cdot \chi_{11}^2$	$2^2 \cdot 5$	8	5
	$\chi_5^2 \cdot \chi_{11}$	$2 \cdot 5$	4	1
	$\chi_5^2 \cdot \chi_{11}^5$	2	1	2
$56 = 2^3 \cdot 7$	$\psi_{2^3} \cdot \chi_7$	$2 \cdot 3$	2	1
	$\chi_{2^2}\psi_{2^3} \cdot \chi_7^2$	$2 \cdot 3$	2	1
	$\psi_{2^3} \cdot \chi_7^3$	2	1	2
$57 = 3 \cdot 19$	$\chi_3 \cdot \chi_{19}^2$	$2 \cdot 3^2$	6	3
	$\chi_3 \cdot \chi_{19}^6$	$2 \cdot 3$	2	3
$60 = 2^2 \cdot 3 \cdot 5$	$\chi_{2^2} \cdot \chi_3 \cdot \chi_5$	2^2	2	2
$63 = 3^2 \cdot 7$	$\psi_{3^2} \cdot \chi_7$	$2 \cdot 3$	2	1
	$\psi_{3^2} \cdot \overline{\chi}_7$	$2 \cdot 3$	2	7
	$\chi_3\psi_{3^2} \cdot \chi_7^2$	$2 \cdot 3$	2	1
	$\chi_3\psi_{3^2} \cdot \overline{\chi}_7^2$	$2 \cdot 3$	2	1
	$\psi_{3^2} \cdot \chi_7^3$	$2 \cdot 3$	2	1
$65 = 5 \cdot 13$	$\chi_5 \cdot \chi_{13}^2$	$2^2 \cdot 3$	4	2^2
	$\chi_5 \cdot \chi_{13}^4$	$2^2 \cdot 3$	4	2^2
	$\chi_5 \cdot \chi_{13}^6$	2^2	2	1
	$\chi_5^2 \cdot \chi_{13}$	$2^2 \cdot 3$	4	2^2
	$\chi_5^2 \cdot \chi_{13}^3$	2^2	2	1
$68 = 2^2 \cdot 17$	$\chi_{2^2} \cdot \chi_{17}^2$	2^3	4	2
	$\chi_{2^2} \cdot \chi_{17}^4$	2^2	2	2
	$\chi_{2^2} \cdot \chi_{17}^8$	2	1	2
$69 = 3 \cdot 23$	$\chi_3 \cdot \chi_{23}^2$	$2 \cdot 11$	10	23
$72 = 2^3 \cdot 3^2$	$\chi_{2^2}\psi_{2^3} \cdot \psi_{3^2}$	$2 \cdot 3$	2	3
	$\psi_{2^3} \cdot \chi_3\psi_{3^2}$	$2 \cdot 3$	2	1
$75 = 3 \cdot 5^2$	$\chi_3 \cdot \psi_{5^2}$	$2 \cdot 5$	4	11
	$\chi_3 \cdot \chi_5^2\psi_{5^2}$	$2 \cdot 5$	4	1
$76 = 2^2 \cdot 19$	$\chi_{2^2} \cdot \chi_{19}^2$	$2 \cdot 3^2$	6	19
	$\chi_{2^2} \cdot \chi_{19}^6$	$2 \cdot 3$	2	1
$77 = 7 \cdot 11$	$\chi_7 \cdot \chi_{11}^2$	$2 \cdot 3 \cdot 5$	8	2^4
	$\chi_7^2 \cdot \chi_{11}$	$2 \cdot 3 \cdot 5$	8	2^4
	$\chi_7^2 \cdot \chi_{11}^5$	$2 \cdot 3$	2	1
	$\chi_7^3 \cdot \chi_{11}^2$	$2 \cdot 5$	4	5
$80 = 2^4 \cdot 5$	$\psi_{2^4} \cdot \chi_5$	2^2	2	1
	$\psi_{2^4} \cdot \overline{\chi}_5$	2^2	2	5
	$\chi_{2^2}\psi_{2^4} \cdot \chi_5^2$	2^2	2	1
$84 = 2^2 \cdot 3 \cdot 7$	$\chi_{2^2} \cdot \chi_3 \cdot \chi_7$	$2 \cdot 3$	2	1
	$\chi_{2^2} \cdot \chi_3 \cdot \chi_7^3$	2	1	2
$85 = 5 \cdot 17$	$\chi_5 \cdot \chi_{17}^2$	2^3	4	1
	$\chi_5 \cdot \overline{\chi}_{17}^2$	2^3	4	73
	$\chi_5 \cdot \chi_{17}^4$	2^2	2	1
	$\chi_5 \cdot \overline{\chi}_{17}^4$	2^2	2	5
	$\chi_5 \cdot \chi_{17}^8$	2^2	2	1
	$\chi_5^2 \cdot \chi_{17}$	2^4	8	17

(continued)

Table 3.1 (continued)

2. Composite conductors

1	2	3	4	5
Conductor	Character	Order	Degree	Contribution
$87 = 3 \cdot 29$	$\chi_3 \cdot \chi_{29}^2$	$2 \cdot 7$	6	2^3
	$\chi_3 \cdot \chi_{29}^4$	$2 \cdot 7$	6	2^3
	$\chi_3 \cdot \chi_{29}^{14}$	2	1	3
$88 = 2^3 \cdot 11$	$\psi_{2^3} \cdot \chi_{11}$	$2 \cdot 5$	4	11
	$\chi_{2^2}\psi_{2^3} \cdot \chi_{11}^2$	$2 \cdot 5$	4	5
	$\psi_{2^3} \cdot \chi_{11}^5$	2	1	1
$91 = 7 \cdot 13$	$\chi_7 \cdot \chi_{13}^2$	$2 \cdot 3$	2	1
	$\chi_7 \cdot \overline{\chi}_{13}^2$	$2 \cdot 3$	2	1
	$\chi_7 \cdot \chi_{13}^4$	$2 \cdot 3$	2	1
	$\chi_7 \cdot \overline{\chi}_{13}^4$	$2 \cdot 3$	2	13
	$\chi_7 \cdot \chi_{13}^6$	$2 \cdot 3$	2	2^2
	$\chi_7^2 \cdot \chi_{13}$	$2^2 \cdot 3$	4	1
	$\chi_7^2 \cdot \overline{\chi}_{13}$	$2^2 \cdot 3$	4	37
	$\chi_7^2 \cdot \chi_{13}^3$	$2^2 \cdot 3$	4	2^2
	$\chi_7^3 \cdot \chi_{13}^2$	$2 \cdot 3$	2	7
	$\chi_7^3 \cdot \chi_{13}^4$	$2 \cdot 3$	2	1
	$\chi_7^3 \cdot \chi_{13}^6$	2	1	1
$92 = 2^2 \cdot 23$	$\chi_{2^2} \cdot \chi_{23}^2$	$2 \cdot 11$	10	67
$93 = 3 \cdot 31$	$\chi_3 \cdot \chi_{31}^2$	$2 \cdot 3 \cdot 5$	8	151
	$\chi_3 \cdot \chi_{31}^6$	$2 \cdot 5$	4	5
	$\chi_3 \cdot \chi_{31}^{10}$	$2 \cdot 3$	2	1
$95 = 5 \cdot 19$	$\chi_5 \cdot \chi_{19}^2$	$2^2 \cdot 3^2$	12	109
	$\chi_5 \cdot \chi_{19}^6$	$2^2 \cdot 3$	4	13
	$\chi_5^2 \cdot \chi_{19}$	$2 \cdot 3^2$	6	19
	$\chi_5^2 \cdot \chi_{19}^3$	$2 \cdot 3$	2	1
	$\chi_5^2 \cdot \chi_{19}^9$	2	1	2^2
$96 = 2^5 \cdot 3$	$\psi_{2^5} \cdot \chi_3$	2^3	4	3^2
$99 = 3^2 \cdot 11$	$\psi_{3^2} \cdot \chi_{11}$	$2 \cdot 3 \cdot 5$	8	31
	$\chi_3\psi_{3^2} \cdot \chi_{11}^2$	$2 \cdot 3 \cdot 5$	8	31
	$\psi_{3^2} \cdot \chi_{11}^5$	$2 \cdot 3$	2	3
$100 = 2^2 \cdot 5^2$	$\chi_{2^2} \cdot \psi_{5^2}$	$2 \cdot 5$	4	5
	$\chi_{2^2} \cdot \chi_5^2\psi_{5^2}$	$2 \cdot 5$	4	11

Auxiliary table: values of the basic characters

Fixed roots of unity: $i^2 = -1,\ j^2 = i,\ \rho^3 = 1,\ \varepsilon^3 = \rho,\ \eta^5 = 1,\ \theta^7 = 1$

In addition ζ is always especially fixed on the top of the relevant column

f	2^2	2^3	2^4	2^5	2^6	3	3^2	3^3	3^4	5	5^2	7	7^2	11	13	17
ζ					$\zeta^2 = j$				$\zeta^3 = \varepsilon$							$\zeta^2 = j$
1	1	1	1	1	1	$\underline{1}$	1	1	1	1	1	1	1	1	1	1
2	—					-1	ρ^2	$\rho\varepsilon^2$	$\rho\varepsilon\zeta^2$	$\underline{i}$	η^2	ρ^2	θ^5	$-\eta$	$i\rho^2$	$-j$
3	-1	-1	$-i$	ij	$j\zeta$	===				$-i$	η^4	$\underline{-\rho}$	θ	η^3	ρ^2	$-j\zeta$
4	===	—				$\underline{1}$	$\underline{\rho}$	ε	ζ	-1	η^4	ρ	θ^3	η^2	$-\rho$	i
5	1	-1	i	j	ζ	-1	ρ	$\rho^2\varepsilon$	$\rho^2\varepsilon^2\zeta$	===		$-\rho^2$	θ	$\underline{\eta^4}$	i	$ij\zeta$
6	—					===				1	η	-1	θ^6	$-\eta^4$	$\underline{i\rho}$	$i\zeta$
7	-1	1	-1	i	$-j$	$\underline{1}$	ρ^2	$\rho^2\varepsilon^2$	$\varepsilon^2\zeta^2$	$\underline{i}$	1	===		$-\eta^2$	$-i\rho$	$-\zeta$
8	===	===	—			-1	1	ρ^2	$\rho\varepsilon^2$	$-i$	η	1	θ	$-\eta^3$	$-i$	$\underline{-ij}$
9	1	1	-1	$-i$	ij	===	===			-1	η^3	ρ^2	θ^2	η	ρ	ij
10	—					$\underline{1}$	1	ρ	$\rho\varepsilon$	===		$\underline{-\rho}$	θ^6	-1	$-\rho^2$	ζ
11	-1	-1	i	$-j$	$i\zeta$	-1	ρ^2	ε^2	$\rho^2\zeta^2$	1	η^2	ρ	θ^5	===	$-i\rho^2$	$-i\zeta$
12	===	—				===				$\underline{i}$	$\underline{\eta^3}$	$-\rho^2$	θ^4	1	-1	$-ij\zeta$
13	1	-1	$-i$	$-ij$	$-ij\zeta$	$\underline{1}$	$\underline{\rho}$	$\underline{\rho\varepsilon}$	$\varepsilon\zeta$	$-i$	η^3	-1	θ^5	$-\eta$	===	$-i$
14	—					-1	ρ	$\rho\varepsilon$	$\rho^2\varepsilon\zeta$	-1	η^2	===		η^3	1	$j\zeta$
15	-1	1	1	-1	i	===				===		1	θ^2	η^2	$i\rho^2$	j
16	===	===	====	—		$\underline{1}$	ρ^2	ε^2	ζ^2	1	η^3	ρ^2	θ^6	$\underline{\eta^4}$	ρ^2	-1
17	1	1	1	-1	$-i$	-1	1	ρ	ε	$\underline{i}$	η	$\underline{-\rho}$	θ^4	$-\eta^4$	$-\rho$	===
18	—					===	===			$-i$	1	ρ	1	$-\eta^2$	i	1

(continued)

f	2^2	2^3	2^4	2^5	2^6	3	3^2	3^3	3^4	5	5^2	7	7^2	11	13	17
ζ					$\zeta^2=j$				$\zeta^3=\varepsilon$							$\zeta^2=j$
19	-1	-1	$-i$	$-ij$	$ij\zeta$	$\underline{1}$	1	ρ^2	$\rho^2\varepsilon^2$	-1	η	$-\rho^2$	1	$-\eta^3$	$\underline{i\rho}$	$-j$
20	===	—				-1	ρ^2	$\rho^2\varepsilon^2$	$\rho^2\varepsilon^2\zeta^2$	===		-1	θ^4	η	$-i\rho$	$-j\zeta$
21	1	-1	i	$-j$	$-i\zeta$	===				1	η^4	===		-1	$-i$	i
22	—					$\underline{1}$	$\underline{\rho}$	$\rho^2\varepsilon$	$\varepsilon^2\zeta$	$\underline{i}$	η^4	1	θ^3	===	ρ	$ij\zeta$
23	-1	1	-1	$-i$	$-ij$	-1	ρ	ε	$\rho^2\zeta$	$-i$	η^2	ρ^2	θ^3	1	$-\rho^2$	$i\zeta$
24	===	===	—			===				-1	1	$\underline{-\rho}$	$\underline{\theta^2}$	$-\eta$	$-i\rho^2$	$-\zeta$
25	1	1	-1	i	j	$\underline{1}$	ρ^2	$\rho\varepsilon^2$	$\rho^2\varepsilon\zeta^2$	===	===	ρ	θ^2	η^3	-1	$\underline{-ij}$
26	—					-1	1	1	ρ^2	1	1	$-\rho^2$	θ^3	η^2	===	ij
27	-1	-1	i	j	$-\zeta$	===	===	===		$\underline{i}$	η^2	-1	θ^3	$\underline{\eta^4}$	1	ζ
28	===	—				$\underline{1}$	1	1	ρ	$-i$	η^4	===		$-\eta^4$	$i\rho^2$	$-i\zeta$
29	1	-1	$-i$	ij	$-j\zeta$	-1	ρ^2	$\rho\varepsilon^2$	$\varepsilon\zeta^2$	-1	η^4	1	θ^4	$-\eta^2$	ρ^2	$-ij\zeta$
30	—					===				===		ρ^2	1	$-\eta^3$	$-\rho$	$-i$
31	-1	1	1	1	-1	$\underline{1}$	$\underline{\rho}$	ε	$\rho\zeta$	1	η	$\underline{-\rho}$	1	η	i	$j\zeta$
32	===	===	===	===	—	-1	ρ	$\rho^2\varepsilon$	$\rho\varepsilon^2\zeta$	$\underline{i}$	1	ρ	θ^4	-1	$\underline{i\rho}$	j

f	19	23	29	31	37	41	43	47	53	59	61	67	71	73	79	83	89	97
ζ		$\zeta^{11}=1$						$\zeta^{23}=1$	$\zeta^{13}=1$	$\zeta^{29}=1$		$\zeta^{11}=1$			$\zeta^{13}=1$	$\zeta^{41}=1$	$\zeta^{11}=1$	$\zeta^{4}=j$
1	1	1	1	1	1	1	1	1	1	1	1	1	1	1	1	1	1	1
2	$-\rho^2\varepsilon^2$	ζ^8	$-i\theta^4$	η^2	$-i\varepsilon^2$	$i\eta$	$-\theta^4$	ζ^7	$i\zeta^{12}$	$-\zeta^{25}$	$-i\rho^2\eta^2$	$-\rho^2\zeta^7$	$\eta^3\theta^2$	$\rho^2\varepsilon^2$	$\rho^2\zeta^{11}$	$-\zeta^3$	ζ^6	$-\rho^2 j\zeta^2$
3	$-\rho\varepsilon^2$	ζ^9	$-i\theta^6$	$-\rho\eta^3$	$-\rho^2\varepsilon$	$-ij$	$-\rho^2\theta^3$	ζ^{18}	$i\zeta^9$	ζ^3	$-\eta^2$	$-\zeta^9$	$\eta^3\theta^4$	$-i\rho^2$	$-\rho^2\zeta^6$	ζ^{11}	$-ij\zeta^{10}$	$\rho^2\zeta^2$
4	$\rho^2\varepsilon$	ζ^5	$-\theta$	η^4	$-\rho\varepsilon$	$-\eta^2$	θ	ζ^{14}	$-\zeta^{11}$	ζ^{21}	$-\rho\eta^4$	$\rho\zeta^3$	$\eta\theta^4$	$\rho^2\varepsilon$	$\rho\zeta^9$	ζ^6	ζ	ρij
5	ε^2	$-\zeta^4$	$-\theta^4$	ρ^2	$i\varepsilon$	$-i\eta^2$	$-\rho^2\theta^5$	$-\zeta^{17}$	$-i\zeta^5$	ζ^5	$-\rho^2\eta^4$	$-\zeta^6$	η^4	$j\varepsilon$	$\rho\zeta^8$	$-\zeta^{40}$	$i\zeta^7$	$\rho^2 i\zeta^3$
6	$\rho\varepsilon$	ζ^6	$-\theta^3$	$-\rho$	i	$j\eta$	ρ^2	ζ^2	$-\zeta^8$	$-\zeta^{28}$	$i\rho^2\eta^4$	$\rho^2\zeta^5$	$\eta\theta^6$	$-i\rho\varepsilon^2$	$-\rho\zeta^4$	$-\zeta^{14}$	$-ij\zeta^5$	$-\rho i$
7	ρ	$-\zeta^{10}$	θ^6	$\rho\eta^4$	ε	$-ij\eta^4$	$-\rho$	ζ^{15}	$-\zeta^{12}$	ζ^{15}	$-i\rho^2\eta^3$	$-\rho\zeta^7$	$-\eta^3\theta^5$	$j\rho^2$	$-\rho\zeta^6$	ζ^{24}	$-ij\zeta^7$	$-\rho^2 j\zeta$
8	$-\rho^2$	ζ^2	$i\theta^5$	η	$i\rho^2$	$-i\eta^3$	$-\theta^5$	ζ^{21}	$-i\zeta^{10}$	$-\zeta^{17}$	$i\eta$	$-\zeta^{10}$	$\eta^4\theta^6$	ρ^2	ζ^7	$-\zeta^9$	ζ^7	ζ^2
9	$\underline{\varepsilon}$	ζ^7	$-\theta^5$	$\rho^2\eta$	$\rho\varepsilon^2$	$-i$	$\rho\theta^6$	ζ^{13}	$-\zeta^5$	ζ^6	η^4	ζ^7	$\eta\theta$	$-\rho$	$\rho\zeta^{12}$	ζ^{22}	$-i\zeta^9$	ρj
10	$-\varepsilon$	$-\zeta$	$i\theta$	$\rho^2\eta^2$	ρ	η^3	$\rho^2\theta^2$	$-\zeta$	ζ^4	$-\zeta$	$i\rho\eta$	$\rho^2\zeta^2$	$\eta^2\theta^2$	j	ζ^6	ζ^2	$i\zeta^2$	$\rho\zeta$
11	ρ^2	$\underline{-\zeta^3}$	$-i\theta^2$	$-\rho^2\eta^4$	$-\rho^2$	$ij\eta^3$	θ^6	$-\zeta^4$	$-\zeta^7$	$-\zeta^{16}$	i	$-\rho\zeta^6$	$-\eta^3\theta$	$-ij\varepsilon$	$\rho\zeta^5$	ζ^{31}	$-\zeta^4$	$-\rho\zeta^2$
12	$-\rho$	ζ^3	i	$-\rho\eta^2$	ε^2	$ij\eta^2$	$-\rho^2\theta^4$	ζ^9	$-i\zeta^7$	ζ^{24}	$\rho\eta$	$-\rho\zeta$	$\eta^4\theta$	$-i\rho\varepsilon$	$-\zeta^2$	ζ^{17}	$-ij$	$ij\zeta^2$
13	$-\rho\varepsilon$	ζ	$-\theta^2$	$-\rho^2\eta^3$	$i\rho\varepsilon$	$-ij\eta$	$\rho\theta^5$	$-\zeta^3$	ζ^2	$-\zeta^{23}$	ρ^2	$-\rho^2\zeta$	$-\eta^2\theta^6$	$-ij\rho\varepsilon^2$	$\rho^2\zeta^9$	$-\zeta^{26}$	$j\zeta^{10}$	$-\rho^2\zeta^3$
14	$-\varepsilon^2$	$-\zeta^7$	$\underline{-i\theta^3}$	$\rho\eta$	$-i\rho$	j	$\rho\theta^4$	ζ^{22}	$-i\zeta^{1}$[illegible]	$-\zeta^{11}$	$-\rho$	ζ^3	$-\eta$	$j\rho\varepsilon^2$	$-\zeta^4$	$-\zeta^{27}$	$-ij\zeta^2$	$\rho i\zeta^3$
15	$-\rho^2\varepsilon$	$-\zeta^2$	$i\theta^3$	$\underline{-\eta^3}$	$-i\rho^2\varepsilon^2$	$-j\eta^2$	$\rho\theta$	$-\zeta^{12}$	ζ	ζ^8	$\rho^2\eta$	ζ^4	$\eta^2\theta^4$	$-ij\rho^2\varepsilon$	$-\zeta$	$-\zeta^{10}$	$j\zeta^6$	$\rho ij\zeta$
16	$\rho\varepsilon^2$	ζ^{10}	θ^2	η^3	$\rho^2\varepsilon^2$	η^4	θ^2	ζ^5	ζ^9	ζ^{13}	$\rho^2\eta^3$	$\rho^2\zeta^6$	$\eta^2\theta$	$\rho\varepsilon^2$	$\rho^2\zeta^5$	ζ^{12}	ζ^2	$-\rho^2 i$
17	$\rho^2\varepsilon^2$	$-\zeta^6$	$-i$	$-\rho\eta$	$i\rho\varepsilon^2$	$j\eta^3$	$\rho\theta^2$	ζ^{19}	$-\zeta^3$	ζ^{14}	$i\rho\eta^4$	$\rho^2\zeta^8$	$-\eta^2$	$-j\rho$	$-\zeta^9$	ζ^4	$i\zeta^5$	$-\rho\zeta^3$
18	-1	ζ^4	$i\theta^2$	$\rho^2\eta^3$	$\underline{-i\rho^2\varepsilon}$	η	$-\rho\theta^3$	ζ^{20}	$-i\zeta^4$	$-\zeta^2$	$-i\rho^2\eta$	$-\rho^2\zeta^3$	$\eta^4\theta^3$	$-\varepsilon^2$	ζ^{10}	$-\zeta^{25}$	$-i\zeta^4$	$-i\zeta^2$

(continued)

f	19	23	29	31	37	41	43	47	53	59	61	67	71	73	79	83	89	97
ζ		$\zeta^{11}=1$						$\zeta^{23}=1$	$\zeta^{13}=1$	$\zeta^{29}=1$		$\zeta^{11}=1$			$\zeta^{13}=1$	$\zeta^{41}=1$	$\zeta^{11}=1$	$\zeta^4=j$
19	===	$-\zeta^5$	$-i\theta$	$\rho\eta^2$	$i\rho^2\varepsilon$	$j\eta^4$	$-\rho^2\theta$	$-\zeta^6$	$i\zeta^2$	ζ^{22}	$-\rho\eta^2$	$\rho^2\zeta^4$	$\eta^3\theta^3$	$-i\rho^2\varepsilon^2$	$\rho\zeta^{10}$	$-\zeta^{18}$	$-j\zeta^9$	$-i\zeta^3$
20	1	$-\zeta^9$	θ^5	$\rho^2\eta^4$	$-i\rho\varepsilon^2$	$\underline{i\eta^4}$	$-\rho^2\theta^6$	$-\zeta^8$	$i\zeta^3$	ζ^{26}	η^3	$-\rho\zeta^9$	θ^4	$j\rho^2\varepsilon^2$	$\rho^2\zeta^4$	$-\zeta^5$	$i\zeta^8$	$-j\zeta^3$
21	$-\rho^2\varepsilon^2$	$-\zeta^8$	$-i\theta^5$	$-\rho^2\eta^2$	$-\rho^2\varepsilon^2$	$-i\eta^4$	$\underline{\theta^3}$	ζ^{10}	$-i\zeta^8$	ζ^{18}	$i\rho^2$	$\rho\zeta^5$	$-\eta\theta^2$	$-ij\rho$	ζ^{12}	ζ^{35}	$-i\zeta^6$	$-\rho j\zeta^3$
22	$-\rho\varepsilon^2$	-1	$-\theta^6$	$-\rho^2\eta$	$i\rho^2\varepsilon^2$	$-j\eta^4$	$-\theta^3$	$-\zeta^{11}$	$-i\zeta^6$	ζ^{12}	$\rho^2\eta^2$	ζ^2	$-\eta\theta^3$	$-ij$	ζ^3	$-\zeta^{34}$	$-\zeta^{10}$	i
23	$\rho^2\varepsilon$	===	θ^3	$-\eta$	$i\rho$	$-\eta$	$\rho^2\theta^6$	$\underline{-\zeta^{16}}$	$-i$	$-\zeta^{27}$	$-i\eta^4$	$\rho^2\zeta^9$	$-\theta^5$	$-i\varepsilon$	ρ	ζ^{16}	$-ij\zeta^9$	$\rho ij\zeta^3$
24	ε^2	1	θ^4	$-\rho\eta^4$	$-i\rho\varepsilon$	$-j\eta^3$	$\rho^2\theta$	ζ^{16}	ζ^6	$-\zeta^{20}$	$-i\eta^3$	ζ^8	$\eta^2\theta^3$	$-i\rho$	$-\rho^2$	$-\zeta^{20}$	$-ij\zeta^6$	$\rho^2 j$
25	$\rho\varepsilon$	ζ^8	θ	ρ	$-\varepsilon^2$	$-\eta^4$	$\rho\theta^3$	ζ^{11}	$-\zeta^{10}$	ζ^{10}	$\rho\eta^3$	ζ	η^3	$i\varepsilon^2$	$\rho^2\zeta^3$	ζ^{39}	$-\zeta^3$	$-\rho j\zeta^2$
26	ρ	ζ^9	$i\theta^6$	$-\rho^2$	ρ^2	$j\eta^2$	$-\rho\theta^2$	$-\zeta^{10}$	$\underline{i\zeta}$	ζ^{19}	$-i\rho\eta^2$	$\rho\zeta^8$	$-\theta$	$ij\rho\varepsilon$	$\rho\zeta^7$	ζ^{29}	$j\zeta^5$	$\rho i\zeta$
27	$-\rho^2$	ζ^5	$i\theta^4$	$-\eta^4$	$-\rho$	$-j$	$-\theta^2$	ζ^8	$-i\zeta$	ζ^9	$-\eta$	$-\zeta^5$	$\eta^4\theta^5$	i	$-\zeta^5$	ζ^{33}	$-j\zeta^8$	$j\zeta^2$
28	$\underline{\varepsilon}$	$-\zeta^4$	-1	$\rho\eta^3$	$-\rho\varepsilon^2$	$ij\eta$	$-\rho\theta$	ζ^6	ζ^{10}	ζ^7	$i\eta^2$	$-\rho^2\zeta^{10}$	$-\eta^4\theta^2$	$j\rho\varepsilon$	$-\rho^2\zeta^2$	ζ^{30}	$-ij\zeta^8$	ζ
29	$-\varepsilon$	ζ^6	===	$-\eta^2$	$-i\rho^2$	$-ij\eta^2$	$-\rho\theta^4$	$-\zeta^{20}$	$-\zeta^6$	$\underline{\zeta^4}$	$i\rho$	ρ	$\eta^4\theta^4$	$ij\rho^2\varepsilon^2$	$-\rho\zeta$	ζ^{36}	$-j\zeta^7$	$\rho^2 ij\zeta^3$
30	ρ^2	$-\zeta^{10}$	1	-1	$-\varepsilon$	$-ij\eta^3$	$-\rho\theta^5$	$-\zeta^{19}$	i	$-\zeta^4$	$\underline{-i\rho\eta^3}$	$-\rho^2$	θ^6	$-ij\rho^2$	$-\rho^2\zeta^{12}$	ζ^{13}	$j\zeta$	ζ^3
31	$-\rho$	ζ^2	$-i\theta^4$	===	$-i$	$-\eta^3$	$\rho^2\theta^4$	$-\zeta^5$	$i\zeta^6$	$-\zeta^7$	$i\rho\eta^3$	$-\rho\zeta^{10}$	$-\eta^3\theta^6$	$ij\varepsilon^2$	$\rho\zeta^{11}$	ζ^{32}	$j\zeta^2$	$-\rho^2 i\zeta^2$
32	$-\rho\varepsilon$	ζ^7	$-i\theta^6$	1	$-i\varepsilon$	i	$-\theta^6$	ζ^{12}	$i\zeta^8$	$-\zeta^9$	$-i\rho$	$-\rho\zeta^2$	θ^3	$\rho\varepsilon$	$\rho\zeta^3$	$-\zeta^{15}$	ζ^8	$\rho ij\zeta^2$

f	2^2	2^3	2^4	2^5	2^6	3	3^2	3^3	3^4	5	5^2	7	7^2	11	13	17
ζ					$\zeta^2 = j$				$\zeta^3 = \varepsilon$							$\zeta^2 = j$
33	1	1	1	1	−1	===				$-i$	η	$-\rho^2$	θ^6	===	$-i\rho$	−1
34	—					$\underline{1}$	ρ^2	$\rho^2\varepsilon^2$	$\rho\varepsilon^2\zeta^2$	−1	η^3	−1	θ^2	1	$-i$	===
35	−1	−1	$-i$	ij	$-j\zeta$	−1	1	ρ^2	ε^2	===		===		$-\eta$	ρ	1
36	===	—				===	===			1	η^2	1	θ^5	η^3	$-\rho^2$	$-j$
37	1	−1	i	j	$-\zeta$	$\underline{1}$	1	ρ	$\rho^2\varepsilon$	$\underline{i}$	$\underline{\eta^3}$	ρ^2	θ^4	η^2	$-i\rho^2$	$-j\zeta$
38	—					−1	ρ^2	ε^2	$\rho\zeta^2$	$-i$	η^3	$\underline{-\rho}$	θ^5	$\underline{\eta^4}$	−1	i
39	−1	1	−1	i	j	===				−1	η^2	ρ	θ^6	$-\eta^4$	===	$ij\zeta$
40	===	===	—			$\underline{1}$	$\underline{\rho}$	$\underline{\rho\varepsilon}$	$\underline{\rho\varepsilon\zeta}$	===		$-\rho^2$	θ^2	$-\eta^2$	1	$i\zeta$
41	1	1	−1	$-i$	$-ij$	−1	ρ	$\rho\varepsilon$	$\rho\varepsilon\zeta$	1	η^3	−1	θ	$-\eta^3$	$i\rho^2$	$-\zeta$
42	—					===				$\underline{i}$	η	===		η	ρ^2	$\underline{-ij}$
43	−1	−1	i	$-j$	$-i\zeta$	$\underline{1}$	ρ^2	ε^2	$\rho\zeta^2$	$-i$	1	1	θ^6	−1	$-\rho$	ij
44	===	—				−1	1	ρ	$\rho^2\varepsilon$	−1	η	ρ^2	θ	===	i	ζ
45	1	−1	$-i$	$-ij$	$ij\zeta$	===	===			===		$\underline{-\rho}$	θ^3	1	$\underline{i\rho}$	$-i\zeta$
46	—					$\underline{1}$	1	ρ^2	ε^2	1	η^4	ρ	θ	$-\eta$	$-i\rho$	$-ij\zeta$
47	−1	1	1	−1	$-i$	−1	ρ^2	$\rho^2\varepsilon^2$	$\rho\varepsilon^2\zeta^2$	$\underline{i}$	η^4	$-\rho^2$	θ^5	η^3	$-i$	$-i$
48	===	===	===	—		===				$-i$	η^2	−1	1	η^2	ρ	$j\zeta$
49	1	1	1	−1	i	$\underline{1}$	$\underline{\rho}$	$\rho^2\varepsilon$	$\rho\varepsilon^2\zeta$	−1	1	===	===	$\underline{\eta^4}$	$-\rho^2$	j
50	—					−1	ρ	ε	$\rho\zeta$	===	===	1	1	$-\eta^4$	$-i\rho^2$	−1

(continued)

f	2^2	2^3	2^4	2^5	2^6	3	3^2	3^3	3^4	5	5^2	7	7^2	11	13	17
ζ					$\zeta^2 = j$				$\zeta^3 = \varepsilon$							$\zeta^2 = j$
51	-1	-1	$-i$	$-ij$	$-ij\zeta$	===				1	1	ρ^2	θ^5	$-\eta^2$	-1	===
52	===	—				$\underline{1}$	ρ^2	$\rho\varepsilon^2$	$\varepsilon\zeta^2$	$\underline{i}$	η^2	$\underline{-\rho}$	θ	$-\eta^3$	===	1
53	1	-1	i	$-j$	$i\zeta$	-1	1	1	ρ	$-i$	η^4	ρ	θ^3	η	1	$-j$
54	—					===	===	===		-1	η^4	$-\rho^2$	θ	-1	$i\rho^2$	$-j\zeta$
55	-1	1	-1	$-i$	ij	$\underline{1}$	1	1	ρ^2	===		-1	θ^6	===	ρ^2	i
56	===	===	—			-1	ρ^2	$\rho\varepsilon^2$	$\rho^2\varepsilon\zeta^2$	1	η	===		1	$-\rho$	$ij\zeta$
57	1	1	-1	i	$-j$	===				$\underline{i}$	1	1	θ	$-\eta$	i	$i\zeta$
58	—					$\underline{1}$	$\underline{\rho}$	ε	$\rho^2\zeta$	$-i$	η	ρ^2	θ^2	η^3	$\underline{i\rho}$	$-\zeta$
59	-1	-1	i	j	ζ	-1	ρ	$\rho^2\varepsilon$	$\varepsilon^2\zeta$	-1	η^3	$\underline{-\rho}$	θ^6	η^2	$-i\rho$	$\underline{-ij}$
60	===	—				===				===		ρ	θ^5	$\underline{\eta^4}$	$-i$	ij
61	1	-1	$-i$	ij	$j\zeta$	$\underline{1}$	ρ^2	$\rho^2\varepsilon^2$	$\rho^2\varepsilon^2\zeta^2$	1	η^2	$-\rho^2$	θ^4	$-\eta^4$	ρ	ζ
62	—					-1	1	ρ^2	$\rho^2\varepsilon^2$	$\underline{i}$	$\underline{\eta^3}$	-1	θ^5	$-\eta^2$	$-\rho^2$	$-i\zeta$
63	-1	1	1	1	1	===	===			$-i$	η^3	===		$-\eta^3$	$-i\rho^2$	$-ij\zeta$
64	===	===	===	===	===	$\underline{1}$	1	ρ	ε	-1	η^2	1	θ^2	η	-1	$-i$
65	1	1	1	1	1	-1	ρ^2	ε^2	ζ^2	===		ρ^2	θ^6	-1	===	$j\zeta$
66	—					===				1	η^3	$\underline{-\rho}$	θ^4	===	1	j

f	19	23	29	31	37	41	43	47	53	59	61	67	71	73	79	83	89	97
ζ		$\zeta^{11}=1$						$\zeta^{23}=1$	$\zeta^{13}=1$	$\zeta^{29}=1$		$\zeta^{11}=1$			$\zeta^{13}=1$	$\zeta^{41}=1$	$\zeta^{11}=1$	$\zeta^4=j$
33	$-\varepsilon^2$	$-\zeta$	$-\theta$	η^2	$\rho\varepsilon$	$i\eta^3$	$-\rho^2\theta^2$	$-\zeta^{22}$	$-i\zeta^3$	$-\zeta^{19}$	$-i\eta^2$	$\underline{\rho\zeta^4}$	$-\eta\theta^5$	$-j\rho^2\varepsilon$	$-\zeta^{11}$	ζ	$ij\zeta^3$	$-j$
34	$-\rho^2\varepsilon$	$\underline{-\zeta^3}$	$-\theta^4$	$-\rho\eta^3$	$\rho^2\varepsilon$	$ij\eta^4$	$-\rho\theta^6$	ζ^3	$-i\zeta^2$	$-\zeta^{10}$	η	$-\rho\zeta^4$	$-\theta^2$	$-j\varepsilon^2$	$-\rho^2\zeta^7$	$-\zeta^7$	i	$i\zeta$
35	$\rho\varepsilon^2$	ζ^3	$-\theta^3$	η^4	$i\varepsilon^2$	$-j\eta$	θ^5	$-\zeta^9$	$i\zeta^4$	ζ^{20}	$i\rho\eta^2$	$\rho\zeta^2$	$\underline{-\eta^2\theta^5}$	$i\rho^2\varepsilon$	$-\rho^2\zeta$	$-\zeta^{23}$	$j\zeta^3$	ρ
36	$\rho^2\varepsilon^2$	ζ	θ^6	ρ^2	-1	$i\eta^2$	ρ	ζ^4	ζ^3	ζ^{27}	$-\rho\eta^3$	$\rho\zeta^{10}$	$\eta^2\theta^5$	$\underline{-\varepsilon}$	$\rho^2\zeta^8$	ζ^{28}	$-i\zeta^{10}$	$-\rho^2$
37	-1	$-\zeta^7$	$i\theta^5$	$-\rho$	===	η^2	$-\rho^2$	ζ	$-\zeta^9$	$-\zeta^{12}$	$i\eta^3$	ρ^2	θ^2	ε	$-\rho^2\zeta^{10}$	ζ^{19}	$-j$	$\rho^2 i\zeta$
38	===	$-\zeta^2$	$-\theta^5$	$\rho\eta^4$	1	ij	$\rho^2\theta^5$	$-\zeta^{13}$	$-\zeta$	$-\zeta^{18}$	$i\eta^4$	$-\rho$	$\eta\theta^5$	$-i\rho^2\varepsilon$	ζ^8	ζ^{21}	$-j\zeta^4$	$-\rho^2\zeta$
39	1	ζ^{10}	$i\theta$	η	$-i\varepsilon^2$	$-i\eta$	$-\theta$	$-\zeta^{21}$	$i\zeta^{11}$	$-\zeta^{26}$	$-\rho^2\eta^2$	$\rho^2\zeta^{10}$	$-\theta^3$	$j\varepsilon^2$	$\underline{-\rho\zeta^2}$	$-\zeta^{37}$	ζ^9	$-\rho j\zeta$
40	$-\rho^2\varepsilon^2$	$-\zeta^6$	$-i\theta^2$	$\rho^2\eta$	$-\rho^2\varepsilon$	-1	$\rho^2\theta^3$	$-\zeta^{15}$	$-\zeta^2$	$-\zeta^{22}$	$-i\rho^2$	ζ^5	$\eta^3\theta^6$	$i\rho^2\varepsilon$	$\rho\zeta^2$	ζ^8	$i\zeta^3$	$\rho^2 ij\zeta$
41	$-\rho\varepsilon^2$	ζ^4	i	$\rho^2\eta^2$	$-\rho\varepsilon$	===	θ^4	$-\zeta^2$	$i\zeta^7$	ζ^2	$-\eta^3$	$-\rho\zeta^8$	$-\theta^6$	$-\rho\varepsilon$	$-\zeta^8$	$\underline{\zeta^{38}}$	$ij\zeta$	$\rho^2 j\zeta^3$
42	$\rho^2\varepsilon$	$-\zeta^5$	$-\theta^2$	$-\rho^2\eta^4$	$i\varepsilon$	1	-1	ζ^{17}	ζ^7	$-\zeta^{14}$	$\rho\eta^2$	$-\zeta$	$-\eta^4\theta^4$	$-ij\varepsilon^2$	$\rho^2\zeta^{10}$	$-\zeta^{38}$	$-i\zeta$	$ij\zeta$
43	ε^2	$-\zeta^9$	$\underline{-i\theta^3}$	$-\rho\eta^2$	i	$i\eta$	===	$-\zeta^{14}$	$-\zeta^4$	$-\zeta^{13}$	$i\rho^2\eta$	$-\zeta^8$	$\eta^4\theta^2$	$ij\rho^2$	$-\rho^2\zeta^8$	$-\zeta^8$	$ij\zeta^4$	$\rho^2 ij$
44	$\rho\varepsilon$	$-\zeta^8$	$i\theta^3$	$-\rho^2\eta^3$	ε	$-ij$	1	$-\zeta^{18}$	ζ^5	$-\zeta^8$	$-i\rho\eta^4$	$-\rho^2\zeta^9$	$-\eta^4\theta^5$	$-ij\rho^2\varepsilon^2$	$\rho^2\zeta$	ζ^{37}	$\underline{-\zeta^5}$	$-\rho^2 ij\zeta^2$
45	ρ	-1	θ^2	$\rho\eta$	$i\rho^2$	$-\eta^2$	$-\theta^4$	$-\zeta^7$	$i\zeta^{10}$	ζ^{11}	$-\rho^2\eta^3$	$-\zeta^2$	θ	$-j\rho\varepsilon$	$\rho^2\zeta^7$	$-\zeta^{21}$	ζ^5	$ij\zeta^3$
46	$-\rho^2$	===	$-i$	$\underline{-\eta^3}$	$\rho\varepsilon^2$	$-i\eta^2$	$-\rho^2\theta^3$	-1	ζ^{12}	ζ^{23}	$-\rho^2\eta$	$-\rho\zeta^5$	$-\eta^3$	$-i$	ζ^{11}	$-\zeta^{19}$	$-ij\zeta^4$	$j\zeta$
47	$\underline{\varepsilon}$	1	$i\theta^2$	η^3	ρ	$j\eta$	θ	===	ζ^8	$-\zeta^{24}$	ρ	$\rho\zeta^9$	$-\eta^2\theta^3$	$-ij\rho\varepsilon$	$-\rho\zeta^3$	$-\zeta^{28}$	$i\zeta$	$-ij$

(continued)

f	19	23	29	31	37	41	43	47	53	59	61	67	71	73	79	83	89	97
ζ		$\zeta^{11}=1$						$\zeta^{23}=1$	$\zeta^{13}=1$	$\zeta^{29}=1$		$\zeta^{11}=1$			$\zeta^{13}=1$	$\zeta^{41}=1$	$\zeta^{11}=1$	$\zeta^4=j$
48	$-\varepsilon$	ζ^8	$-i\theta$	$-\rho\eta$	$-\rho^2$	$-ij\eta^4$	$-\rho^2\theta^5$	1	$i\zeta^5$	ζ^{16}	$-\rho^2$	$-\rho^2\zeta^4$	θ^5	$-i\varepsilon^2$	$-\rho\zeta^{11}$	ζ^{23}	$-ij\zeta$	$\underline{-\rho i\zeta^2}$
49	ρ^2	ζ^9	θ^5	$\rho^2\eta^3$	ε^2	$-i\eta^3$	ρ^2	ζ^7	ζ^{11}	ζ	$-\rho\eta$	$\rho^2\zeta^3$	$\eta\theta^3$	$i\rho$	$\rho^2\zeta^{12}$	ζ^7	$-i\zeta^3$	$\rho i\zeta^2$
50	$-\rho$	ζ^5	$-i\theta^5$	$\rho\eta^2$	$i\rho\varepsilon$	$-i$	$-\rho$	ζ^{18}	$-i\zeta^9$	$-\zeta^6$	$-i$	$-\rho^2\zeta^8$	$\eta\theta^2$	$i\varepsilon$	$\rho\zeta$	$-\zeta$	$-\zeta^9$	ij
51	$-\rho\varepsilon$	$-\zeta^4$	$-\theta^6$	$\rho^2\eta^4$	$-i\rho$	η^3	$-\theta^5$	ζ^{14}	$-i\zeta^{12}$	ζ^{17}	$-i\rho\eta$	$-\rho^2\zeta^6$	$-\theta^4$	ij	$\rho^2\zeta^2$	ζ^{15}	$j\zeta^4$	$-j\zeta$
52	$-\varepsilon^2$	ζ^6	θ^3	$-\rho^2\eta^2$	$-i\rho^2\varepsilon^2$	$ij\eta^3$	$\rho\theta^6$	$-\zeta^{17}$	-1	$-\zeta^{15}$	$-\eta^4$	$-\zeta^4$	$-\eta^3\theta^3$	$ij\rho$	ζ^5	$-\zeta^{32}$	j	$-ij\zeta^3$
53	$-\rho^2\varepsilon$	$-\zeta^{10}$	θ^4	$-\rho^2\eta$	$\rho^2\varepsilon^2$	$ij\eta^2$	$\rho^2\theta^2$	ζ^2	===	ζ^{28}	$-i\eta$	$-\zeta^3$	$-\eta^4\theta^3$	$-j\rho^2\varepsilon^2$	$-\rho\zeta^7$	$-\zeta^{13}$	$i\zeta^{10}$	$\rho^2 ij\zeta^2$
54	$\rho\varepsilon^2$	ζ^2	θ	$-\eta$	$i\rho\varepsilon^2$	$-ij\eta$	θ^6	ζ^{15}	1	$-\zeta^5$	$i\rho^2\eta^3$	$\rho^2\zeta$	η^2	$i\rho^2\varepsilon^2$	$-\rho^2\zeta^3$	$-\zeta^{36}$	$-j\zeta^3$	$-\rho^2 ij$
55	$\rho^2\varepsilon^2$	ζ^7	$i\theta^6$	$-\rho\eta^4$	$\underline{-i\rho^2\varepsilon}$	j	$-\rho^2\theta^4$	ζ^{21}	$i\zeta^{12}$	$-\zeta^{21}$	$-i\rho^2\eta^4$	$\rho\zeta$	$-\eta^2\theta$	ε^2	ρ^2	$-\zeta^{30}$	$-i$	$-ij\zeta$
56	-1	$-\zeta$	$i\theta^4$	ρ	$i\rho^2\varepsilon$	$-j\eta^2$	$\rho\theta^5$	ζ^{13}	$i\zeta^9$	$-\zeta^3$	$\rho^2\eta^4$	$\rho\zeta^6$	$-\eta^2\theta^4$	$j\rho$	$-\rho$	$-\zeta^{33}$	$-ij\zeta^3$	$-\rho^2 j\zeta^3$
57	===	$\underline{-\zeta^3}$	-1	$-\rho^2$	$-i\rho\varepsilon^2$	η^4	$\rho\theta^4$	$-\zeta$	$-\zeta^{11}$	ζ^{25}	$\rho\eta^4$	$-\rho^2\zeta^2$	η	$-\rho\varepsilon^2$	$-\zeta^3$	$-\zeta^{29}$	$-\zeta^8$	$-\rho^2 ij\zeta$
58	1	ζ^3	===	$-\eta^4$	$-\rho^2\varepsilon^2$	$j\eta^3$	$\rho\theta$	$-\zeta^4$	$-i\zeta^5$	-1	η^2	$-\zeta^7$	$\eta^2\theta^6$	$ij\rho^2\varepsilon$	$-\zeta^{12}$	$-\zeta^{39}$	$-j\zeta^2$	$\rho j\zeta$
59	$-\rho^2\varepsilon^2$	ζ	1	$\rho\eta^3$	$i\rho^2\varepsilon^2$	η	θ^2	ζ^9	$-\zeta^8$	===	$i\rho^2\eta^2$	ζ^{10}	$-\eta^4\theta$	$-j\rho\varepsilon^2$	$-\rho^2\zeta^4$	ζ^{20}	$-j\zeta$	$\rho^2\zeta$
60	$-\rho\varepsilon^2$	$-\zeta^7$	$-i\theta^4$	$-\eta^2$	$i\rho$	$j\eta^4$	$\rho\theta^2$	$-\zeta^3$	$-\zeta^{12}$	1	-1	$\rho\zeta^7$	$\eta^3\theta$	$-ij\rho\varepsilon^2$	$-\rho\zeta^{10}$	$-\zeta^{16}$	$j\zeta^7$	$-\rho^2 i\zeta$
61	$\rho^2\varepsilon$	$-\zeta^2$	$-i\theta^6$	-1	$-i\rho\varepsilon$	$\underline{i\eta^4}$	$-\rho\theta^3$	ζ^{22}	$-i\zeta^{10}$	$-\zeta^{25}$	===	$-\rho^2\zeta^5$	$-\eta^2\theta^2$	$i\rho\varepsilon$	$-\zeta^{10}$	ζ^{34}	$ij\zeta^8$	ρ^2
62	ε^2	ζ^{10}	$-\theta$	===	$-\varepsilon^2$	$-i\eta^4$	$-\rho^2\theta$	$-\zeta^{12}$	$-\zeta^5$	ζ^3	1	ζ^6	$-\eta\theta$	$ij\varepsilon$	ζ^9	$-\zeta^{35}$	$j\zeta^8$	$-\rho$
63	$\rho\varepsilon$	$-\zeta^6$	$-\theta^4$	1	ρ^2	$-j\eta^4$	$-\rho^2\theta^6$	ζ^5	ζ^4	ζ^{21}	$-i\rho^2\eta^2$	$-\rho\zeta^3$	$-\eta^4\theta^6$	$-j$	$-\rho^2\zeta^5$	ζ^5	$-j\zeta^5$	$-i\zeta$
64	ρ	ζ^4	$-\theta^3$	η^2	$-\rho$	$-\eta$	$\underline{\theta^3}$	ζ^{19}	$-\zeta^7$	ζ^5	$-\eta^2$	ζ^9	$\eta^3\theta^5$	ρ	ζ	ζ^{18}	ζ^3	j
65	$-\rho^2$	$-\zeta^5$	θ^6	$-\rho\eta^3$	$-\rho\varepsilon^2$	$-j\eta^3$	$-\theta^3$	ζ^{20}	$-i\zeta^7$	$-\zeta^{28}$	$-\rho\eta^4$	$\rho^2\zeta^7$	$-\eta\theta^6$	$-\rho^2$	ζ^4	ζ^{25}	$ij\zeta^6$	$-\rho ij\zeta^2$
66	$\underline{\varepsilon}$	$-\zeta^9$	$i\theta^5$	η^4	$-i\rho^2$	$-\eta^4$	$\rho^2\theta^6$	$-\zeta^6$	ζ^2	ζ^{15}	$-\rho^2\eta^4$	-1	$-\eta^4$	$-j\rho^2$	$-\rho^2\zeta^9$	$-\zeta^4$	$ij\zeta^9$	$\rho^2 i\zeta^2$

f	2^2	2^3	2^4	2^5	2^6	3	3^2	3^3	3^4	5	5^2	7	7^2	11	13	17
ζ					$\zeta^2=j$				$\zeta^3=\varepsilon$							$\zeta^2=j$
67	-1	-1	$-i$	ij	$j\zeta$	$\underline{1}$	$\underline{\rho}$	$\underline{\rho\varepsilon}$	$\rho^2\varepsilon\zeta$	$\underline{i}$	η	ρ	1	1	$i\rho^2$	-1
68	===	—				-1	ρ	$\rho\varepsilon$	$\varepsilon\zeta$	$-i$	1	$-\rho^2$	1	$-\eta$	ρ^2	===
69	1	-1	i	j	ζ	===				-1	η	-1	θ^4	η^3	$-\rho$	1
70	—					$\underline{1}$	ρ^2	ε^2	$\rho^2\zeta^2$	===		===		η^2	i	$-j$
71	-1	1	-1	i	$-j$	-1	1	ρ	$\rho\varepsilon$	1	η^4	1	θ^3	$\underline{\eta^4}$	$\underline{i\rho}$	$-j\zeta$
72	===	===	—			===	===			$\underline{i}$	η^4	ρ^2	θ^3	$-\eta^4$	$-i\rho$	i
73	1	1	-1	$-i$	ij	$\underline{1}$	1	ρ^2	$\rho\varepsilon^2$	$-i$	η^2	$\underline{-\rho}$	$\underline{\theta^2}$	$-\eta^2$	$-i$	$ij\zeta$
74	—					-1	ρ^2	$\rho^2\varepsilon^2$	$\varepsilon^2\zeta^2$	-1	1	ρ	θ^2	$-\eta^3$	ρ	$i\zeta$
75	-1	-1	i	$-j$	$i\zeta$	===				===	===	$-\rho^2$	θ^3	η	$-\rho^2$	$-\zeta$
76	===	—				$\underline{1}$	$\underline{\rho}$	$\rho^2\varepsilon$	$\rho^2\varepsilon^2\zeta$	1	1	-1	θ^3	-1	$-i\rho^2$	$\underline{-ij}$
77	1	-1	$-i$	$-ij$	$-ij\zeta$	-1	ρ	ε	ζ	$\underline{i}$	η^2	===		===	-1	ij
78	—					===				$-i$	η^4	1	θ^4	1	===	ζ
79	-1	1	1	-1	i	$\underline{1}$	ρ^2	$\rho\varepsilon^2$	$\rho\varepsilon\zeta^2$	-1	η^4	ρ^2	1	$-\eta$	1	$-i\zeta$
80	===	===	===	—		-1	1	1	1	===		$\underline{-\rho}$	1	η^3	$i\rho^2$	$-ij\zeta$
81	1	1	1	-1	$-i$	===	===	===	===	1	η	ρ	θ^4	η^2	ρ^2	$-i$
82	—					$\underline{1}$	1	1	1	$\underline{i}$	1	$-\rho^2$	θ^6	$\underline{\eta^4}$	$-\rho$	$j\zeta$
83	-1	-1	$-i$	$-ij$	$ij\zeta$	-1	ρ^2	$\rho\varepsilon^2$	$\rho\varepsilon\zeta^2$	$-i$	η	-1	θ^2	$-\eta^4$	i	j
84	===	—				===				-1	η^3	===		$-\eta^2$	$\underline{i\rho}$	-1
85	1	-1	i	$-j$	$-i\zeta$	$\underline{1}$	$\underline{\rho}$	ε	ζ	===		1	θ^5	$-\eta^3$	$-i\rho$	===

(continued)

f	2^2	2^3	2^4	2^5	2^6	3	3^2	3^3	3^4	5	5^2	7	7^2	11	13	17
ζ					$\zeta^2 = j$				$\zeta^3 = \varepsilon$							$\zeta^2 = j$
86	—					-1	ρ	$\rho^2\varepsilon$	$\rho^2\varepsilon^2\zeta$	1	η^2	ρ^2	θ^4	η	$-i$	1
87	-1	1	-1	$-i$	$-ij$	===				$\underline{i}$	$\underline{\eta^3}$	$\underline{-\rho}$	θ^5	-1	ρ	$-j$
88	===	===	—			$\underline{1}$	ρ^2	$\rho^2\varepsilon^2$	$\varepsilon^2\zeta^2$	$-i$	η^3	ρ	θ^6	===	$-\rho^2$	$-j\zeta$
89	1	1	-1	i	j	-1	1	ρ^2	$\rho\varepsilon^2$	-1	η^2	$-\rho^2$	θ^2	1	$-i\rho^2$	i
90	—					===	===			===		-1	θ	$-\eta$	-1	$ij\zeta$
91	-1	-1	i	j	$-\zeta$	$\underline{1}$	1	ρ	$\rho\varepsilon$	1	η^3	===		η^3	===	$i\zeta$
92	===	—				-1	ρ^2	ε^2	$\rho^2\zeta^2$	$\underline{i}$	η	1	θ^6	η^2	1	$-\zeta$
93	1	-1	$-i$	ij	$-j\zeta$	====				$-i$	1	ρ^2	θ	$\underline{\eta^4}$	$i\rho^2$	$\underline{-ij}$
94	—					$\underline{1}$	$\underline{\rho}$	$\underline{\rho\varepsilon}$	$\varepsilon\zeta$	-1	η	$\underline{-\rho}$	θ^3	$-\eta^4$	ρ^2	ij
95	-1	1	1	1	-1	-1	ρ	$\rho\varepsilon$	$\rho^2\varepsilon\zeta$	===		ρ	θ	$-\eta^2$	$-\rho$	ζ
96	===	===	===	===	—	===				1	η^4	$-\rho^2$	θ^5	$-\eta^3$	i	$-i\zeta$
97	1	1	1	1	-1	$\underline{1}$	ρ^2	ε^2	ζ^2	$\underline{i}$	η^4	-1	1	η	$\underline{i\rho}$	$-ij\zeta$
98	—					-1	1	ρ	ε	$-i$	η^2	===	===	-1	$-i\rho$	$-i$
99	-1	-1	$-i$	ij	$-j\zeta$	===	===			-1	1	1	1	===	$-i$	$j\zeta$
100	===	—				$\underline{1}$	1	ρ^2	$\rho^2\varepsilon^2$	===	===	ρ^2	θ^5	1	ρ	j

f	19	23	29	31	37	41	43	47	53	59	61	67	71	73	79	83	89	97
ζ		$\zeta^{11}=1$						$\zeta^{23}=1$	$\zeta^{13}=1$	$\zeta^{29}=1$		$\zeta^{11}=1$			$\zeta^{13}=1$	$\zeta^{41}=1$	$\zeta^{11}=1$	$\zeta^4=j$
67	$-\varepsilon$	$-\zeta^8$	$-\theta^5$	ρ^2	$-\varepsilon$	$j\eta^2$	$\rho^2\theta$	$-\zeta^8$	$-i\zeta^{11}$	$-\zeta^{17}$	$i\rho^2\eta^4$	===	$-\eta\theta^4$	$i\rho\varepsilon^2$	ζ^2	$-\zeta^{12}$	ζ^{10}	$-\zeta^3$
68	ρ^2	-1	$i\theta$	$-\rho$	$-i$	$-j$	$\rho\theta^3$	ζ^{10}	ζ	ζ^6	$-i\rho^2\eta^3$	1	$-\eta^3\theta^4$	$-j\varepsilon$	$-\rho\zeta^5$	ζ^{10}	$i\zeta^6$	$-\rho^2ij\zeta^3$
69	$-\rho$	===	$-i\theta^2$	$\rho\eta^4$	$-i\varepsilon$	$ij\eta$	$-\rho\theta^2$	$-\zeta^{11}$	ζ^9	$-\zeta$	$i\eta$	$-\rho^2\zeta^7$	$-\eta^3\theta^2$	$-\rho^2\varepsilon$	$-\zeta^6$	ζ^{27}	$-i\zeta^8$	$-\zeta$
70	$-\rho\varepsilon$	1	i	η	$\rho\varepsilon$	$-ij\eta^2$	$-\theta^2$	$\underline{-\zeta^{16}}$	$-\zeta^3$	$-\zeta^{16}$	η^4	$-\zeta^9$	-1	$i\rho^2$	$-\rho\zeta^{12}$	ζ^{26}	$j\zeta^9$	$-j\zeta^2$
71	$-\varepsilon^2$	ζ^8	$-\theta^2$	$\rho^2\eta$	$\rho^2\varepsilon$	$-ij\eta^3$	$-\rho\theta$	ζ^{16}	$-i\zeta^4$	ζ^{24}	$i\rho\eta$	$\rho\zeta^3$	===	$-\rho^2\varepsilon^2$	$-\zeta^7$	$-\zeta^{17}$	$i\zeta^4$	$-\rho i\zeta$
72	$-\rho^2\varepsilon$	ζ^9	$\underline{-i\theta^3}$	$\rho^2\eta^2$	$i\varepsilon^2$	$-\eta^3$	$-\rho\theta^4$	ζ^{11}	$i\zeta^2$	$-\zeta^{23}$	i	$-\zeta^6$	1	-1	$\rho\zeta^6$	$-\zeta^{31}$	$-i\zeta^5$	$\rho j\zeta^2$
73	$\rho\varepsilon^2$	ζ^5	$i\theta^3$	$-\rho^2\eta^4$	-1	i	$-\rho\theta^5$	$-\zeta^{10}$	$i\zeta^3$	$-\zeta^{11}$	$\rho\eta$	$\rho^2\zeta^5$	$\eta^3\theta^2$	===	$\rho\zeta^4$	$-\zeta^2$	$-\zeta^2$	$-\rho^2j$
74	$\rho^2\varepsilon^2$	$-\zeta^4$	θ^2	$-\rho\eta^2$	===	$i\eta^3$	$\rho^2\theta^4$	ζ^8	$-i\zeta^8$	ζ^8	ρ^2	$-\rho\zeta^7$	$\eta^3\theta^4$	1	$-\rho\zeta^8$	$-\zeta^{22}$	$-j\zeta^6$	$-\rho ij\zeta^3$
75	-1	ζ^6	$-i$	$-\rho^2\eta^3$	1	$ij\eta^4$	$-\theta^6$	ζ^6	$-i\zeta^6$	ζ^{13}	$-\rho$	$-\zeta^{10}$	$\eta\theta^4$	$\rho^2\varepsilon^2$	$-\rho\zeta^9$	ζ^9	$ij\zeta^2$	$-i$
76	===	$-\zeta^{10}$	$i\theta^2$	$\rho\eta$	$-i\varepsilon^2$	$-j\eta$	$-\rho^2\theta^2$	$-\zeta^{20}$	$-i$	ζ^{14}	$\rho^2\eta$	ζ^7	η^4	$-i\rho^2$	$\rho^2\zeta^6$	$-\zeta^{24}$	$-j\zeta^{10}$	$\rho j\zeta^3$
77	1	ζ^2	$-i\theta$	$\underline{-\eta^3}$	$-\rho^2\varepsilon$	$i\eta^2$	$-\rho\theta^6$	$-\zeta^{19}$	ζ^6	$-\zeta^2$	$\rho^2\eta^3$	$\rho^2\zeta^2$	$\eta\theta^6$	$\rho^2\varepsilon$	$-\rho^2\zeta^{11}$	ζ^{14}	ij	$j\zeta^3$
78	$-\rho^2\varepsilon^2$	ζ^7	θ^5	η^3	$-\rho\varepsilon$	η^2	θ^5	$-\zeta^5$	$-\zeta^{10}$	ζ^{22}	$i\rho\eta^4$	$-\rho\zeta^6$	$-\eta^3\theta^5$	$j\varepsilon$	-1	ζ^{40}	ζ^4	$i\zeta^3$
79	$-\rho\varepsilon^2$	$-\zeta$	$-i\theta^5$	$-\rho\eta$	$i\varepsilon$	ij	ρ	ζ^{12}	$\underline{i\zeta}$	ζ^{26}	$-i\rho^2\eta$	$-\rho\zeta$	$\eta^4\theta^6$	$-i\rho\varepsilon^2$	===	$-\zeta^6$	$-i\zeta^2$	$i\zeta^2$
80	$\rho^2\varepsilon$	$\underline{-\zeta^3}$	$-\theta^6$	$\rho^2\eta^3$	i	$-i\eta$	$-\rho^2$	$-\zeta^{22}$	$-i\zeta$	ζ^{18}	$-\rho\eta^2$	$-\rho^2\zeta$	$\eta\theta$	$j\rho^2$	1	$-\zeta^{11}$	$i\zeta^9$	$\rho\zeta^3$
81	ε^2	ζ^3	θ^3	$\rho\eta^2$	ε	-1	$\rho^2\theta^5$	ζ^3	ζ^{10}	ζ^{12}	η^3	ζ^3	$\eta^2\theta^2$	ρ^2	$\rho^2\zeta^{11}$	ζ^3	$-\zeta^7$	ρ^2i
82	$\rho\varepsilon$	ζ	θ^4	$\rho^2\eta^4$	$i\rho^2$	===	$-\theta$	$-\zeta^9$	$-\zeta^6$	$-\zeta^{27}$	$i\rho^2$	ζ^4	$-\eta^3\theta$	$-\rho$	$-\rho^2\zeta^6$	-1	$ij\zeta^7$	$-\rho ij\zeta$
83	ρ	$-\zeta^7$	θ	$-\rho^2\eta^2$	$\rho\varepsilon^2$	1	$\rho^2\theta^3$	ζ^4	i	$-\zeta^{20}$	$\rho^2\eta^2$	$\rho^2\zeta^6$	$\eta^4\theta$	j	$\rho\zeta^9$	===	$ij\zeta^5$	$-\rho i\zeta^3$
84	$-\rho^2$	$-\zeta^2$	$i\theta^6$	$-\rho^2\eta$	ρ	$i\eta$	θ^4	ζ	$i\zeta^6$	ζ^{10}	$-i\eta^4$	$\rho^2\zeta^8$	$-\eta^2\theta^6$	$-ij\varepsilon$	$\rho\zeta^8$	1	$-i\zeta^7$	$\rho^2\zeta^3$
85	$\underline{\varepsilon}$	ζ^{10}	$i\theta^4$	$-\eta$	$-\rho^2$	$-ij$	-1	$-\zeta^{13}$	$i\zeta^8$	ζ^{19}	$-i\eta^3$	$-\rho^2\zeta^3$	$-\eta$	$-i\rho\varepsilon$	$-\rho\zeta^4$	$-\zeta^3$	$-\zeta$	$-ij\zeta^2$

(continued)

f	19	23	29	31	37	41	43	47	53	59	61	67	71	73	79	83	89	97
ζ		$\zeta^{11}=1$						$\zeta^{23}=1$	$\zeta^{13}=1$	$\zeta^{29}=1$		$\zeta^{11}=1$			$\zeta^{13}=1$	$\zeta^{41}=1$	$\zeta^{11}=1$	$\zeta^4=j$
86	$-\varepsilon$	$-\zeta^6$	-1	$-\rho\eta^4$	ε^2	$-\eta^2$	===	$-\zeta^{21}$	$-i\zeta^3$	ζ^9	$\rho\eta^3$	$\rho^2\zeta^4$	$\eta^2\theta^4$	$ij\rho\varepsilon^2$	$-\rho\zeta^6$	ζ^{11}	$ij\zeta^{10}$	$\rho\zeta^2$
87	ρ^2	ζ^4	===	ρ	$i\rho\varepsilon$	$-i\eta^2$	1	$-\zeta^{15}$	$-i\zeta^2$	ζ^7	$-i\rho\eta^2$	$-\rho\zeta^9$	$\eta^2\theta$	$j\rho\varepsilon^2$	ζ^7	ζ^6	$-\zeta^6$	$-\rho\zeta$
88	$-\rho$	$-\zeta^5$	1	$-\rho^2$	$-i\rho$	$j\eta$	$-\theta^4$	$-\zeta^2$	$i\zeta^4$	$\underline{\zeta^4}$	$-\eta$	$\rho\zeta^5$	$-\eta^2$	$-ij\rho^2\varepsilon$	$\rho\zeta^{12}$	$-\zeta^{40}$	-1	$-\rho j$
89	$-\rho\varepsilon$	$-\zeta^9$	$-i\theta^4$	$-r^4$	$-i\rho^2\varepsilon^2$	$-ij\eta^4$	$-\rho^2\theta^3$	ζ^{17}	ζ^3	$-\zeta^4$	$i\eta^2$	ζ^2	$\eta^4\theta^3$	$\rho\varepsilon^2$	ζ^6	$-\zeta^{14}$	===	$-\zeta^2$
90	$-\varepsilon^2$	$-\zeta^8$	$-i\theta^6$	$\rho\eta^3$	$\rho^2\varepsilon^2$	$-i\eta^3$	θ	$-\zeta^{14}$	$-\zeta^9$	$-\zeta^7$	$i\rho$	$\rho^2\zeta^9$	$\eta^3\theta^3$	$-j\rho$	$\rho\zeta^5$	ζ^{24}	1	$\rho^2 j\zeta$
91	$-\rho^2\varepsilon$	-1	$-\theta$	$-r^2$	$i\rho\varepsilon^2$	$-i$	$-\rho^2\theta^5$	$-\zeta^{18}$	$-\zeta$	$-\zeta^9$	$\underline{-i\rho\eta^3}$	ζ^8	θ^4	$-\varepsilon^2$	$-\zeta^2$	$-\zeta^9$	ζ^6	ρi
92	$\rho\varepsilon^2$	===	$-\theta^4$	-1	$\underline{-i\rho^2\varepsilon}$	η^3	ρ^2	$-\zeta^7$	$i\zeta^{11}$	$-\zeta^{19}$	$i\rho\eta^3$	ζ	$-\eta\theta^2$	$-i\rho^2\varepsilon^2$	$\rho^2\zeta^9$	ζ^{22}	$-ij\zeta^{10}$	$-\rho^2 i\zeta^3$
93	$\rho^2\varepsilon^2$	1	$-\theta^3$	===	$i\rho^2\varepsilon$	$ij\eta^3$	$-\rho$	-1	$-\zeta^2$	$-\zeta^{10}$	$-i\rho$	$\rho\zeta^8$	$-\eta\theta^3$	$j\rho^2\varepsilon^2$	$-\zeta^4$	ζ^2	ζ	$-\rho ij$
94	-1	ζ^8	θ^6	1	$-i\rho\varepsilon^2$	$ij\eta^2$	$-\theta^5$	===	$i\zeta^7$	ζ^{20}	$-i\eta^2$	$-\zeta^5$	$-\theta^5$	$-ij\rho$	$-\zeta$	ζ^{31}	$i\zeta^7$	$-\rho^2\zeta^2$
95	===	ζ^9	$i\theta^5$	η^2	$-\rho^2\varepsilon^2$	$-ij\eta$	$\rho\theta^6$	1	ζ^7	ζ^{27}	η	$-\rho^2\zeta^{10}$	$\eta^2\theta^3$	$-ij$	$\rho^2\zeta^5$	ζ^{17}	$-ij\zeta^5$	$\rho^2 j\zeta^2$
96	1	ζ^5	$-\theta^5$	$-\rho\eta^3$	$i\rho^2\varepsilon^2$	j	$\rho^2\theta^2$	ζ^7	$-\zeta^4$	$-\zeta^{12}$	$i\rho\eta^2$	ρ	η^3	$-i\varepsilon$	$-\zeta^9$	$-\zeta^{26}$	$-ij\zeta^7$	-1
97	$-\rho^2\varepsilon^2$	$-\zeta^4$	$i\theta$	η^4	$i\rho$	$-j\eta^2$	θ^6	ζ^{18}	ζ^5	$-\zeta^{18}$	$-\rho\eta^3$	$-\rho^2$	$-\theta$	$-i\rho$	ζ^{10}	$-\zeta^{27}$	ζ^7	===
98	$-\rho\varepsilon^2$	ζ^6	$-i\theta^2$	ρ^2	$-i\rho\varepsilon$	η^4	$-\rho^2\theta^4$	ζ^{14}	$i\zeta^{10}$	$-\zeta^{26}$	$i\eta^3$	$-\rho\zeta^{10}$	$\eta^4\theta^5$	$i\varepsilon^2$	$\rho\zeta^{10}$	$-\zeta^{10}$	$-i\zeta^9$	1
99	$\rho^2\varepsilon$	$-\zeta^{10}$	i	$-\rho$	$-\varepsilon^2$	$j\eta^3$	$\rho\theta^5$	$-\zeta^{17}$	ζ^{12}	$-\zeta^{22}$	$i\eta^4$	$-\rho\zeta^2$	$-\eta^4\theta^2$	$ij\rho\varepsilon$	$\rho^2\zeta^4$	ζ^{12}	$i\zeta^2$	$-\rho^2 j\zeta^2$
100	ε^2	ζ^2	$-\theta^2$	ρr^4	ρ^2	η	$\rho\theta^4$	ζ^2	ζ^8	ζ^2	$-\rho^2\eta^2$	$\underline{\rho\zeta^4}$	$\eta^4\theta^4$	i	ζ^{12}	ζ^4	$-\zeta^4$	$\rho^2\zeta^2$

Table 3.2 Relative class numbers

1. Prime-power conductor

1	2	3	4	5	6	7
Conductor	No.	Gen. characters	Type	w	Q	h^*
$2^2 = 4$	1	χ_{2^2}	2	2^2	1	1
$2^3 = 8$	1	χ_{2^2}, ψ_{2^3}	$(2, 2)$	2^3	1	1
	2	$\chi_{2^2}\psi_{2^3}$	2	2	1	1
$2^4 = 16$	1	χ_{2^2}, ψ_{2^4}	$(2, 2^2)$	2^4	1	1
	2	$\chi_{2^2}\psi_{2^4}$	2^2	2	1	1
$2^5 = 32$	1	χ_{2^2}, ψ_{2^5}	$(2, 2^3)$	2^5	1	1
	2	$\chi_{2^2}\psi_{2^5}$	2^3	2	1	1
$2^6 = 64$	1	χ_{2^2}, ψ_{2^6}	$(2, 2^4)$	2^6	1	17
	2	$\chi_{2^2}\psi_{2^6}$	2^4	2	1	17
3	1	χ_3	2	$2 \cdot 3$	1	1
$3^2 = 9$	1	χ_3, ψ_{3^2}	$(2, 3) = 6$	$2 \cdot 3^2$	1	1
$3^3 = 27$	1	χ_3, ψ_{3^3}	$(2, 3^2) = 18$	$2 \cdot 3^3$	1	1
$3^4 = 81$	1	χ_3, ψ_{3^4}	$(2, 3^3) = 54$	$2 \cdot 3^4$	1	2593
5	1	χ_5	2^2	$2 \cdot 5$	1	1
$5^2 = 25$	1	χ_5, ψ_{5^2}	$(2^2, 5) = 20$	$2 \cdot 5^2$	1	1
7	1	χ_7	$(2, 3) = 6$	$2 \cdot 7$	1	1
	2	χ_7^3	2	2	1	1
$7^2 = 49$	1	χ_7, ψ_{7^2}	$(2, 3, 7) = 42$	$2 \cdot 7^2$	1	43
	2	χ_7^3, ψ_{7^2}	$(2, 7) = 14$	2	1	1

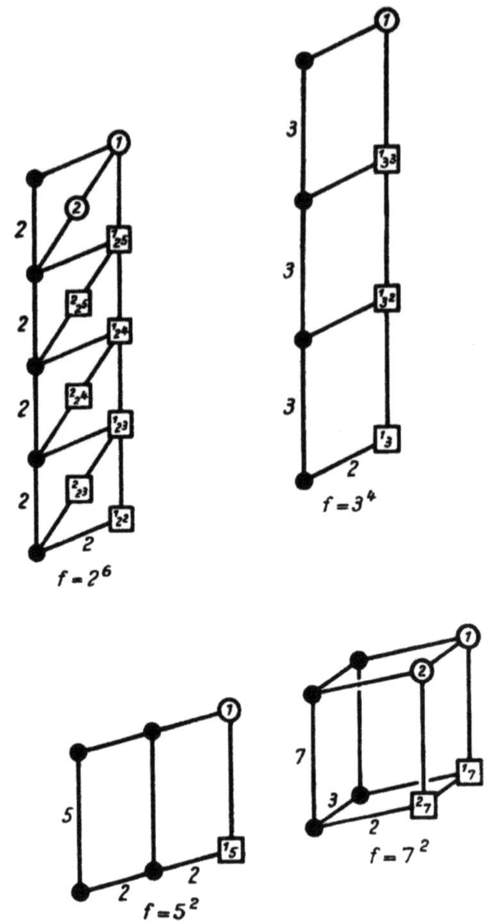
f = 2^6
f = 3^4
f = 5^2
f = 7^2

1	2	3	4	5	6	7
Conductor	No.	Gen. characters	Type	w	Q	h^*
11	1	χ_{11}	$(2, 5) = 10$	$2 \cdot 11$	1	1
	2	χ_{11}^5	2	2	1	1
13	1	χ_{13}	$(2^2, 3) = 12$	$2 \cdot 13$	1	1
	2	χ_{13}^3	2^2	2	1	1
17	1	χ_{17}	2^4	$2 \cdot 17$	1	1
19	1	χ_{19}	$(2, 3^2) = 18$	$2 \cdot 19$	1	1
	2	χ_{19}^3	$(2, 3) = 6$	2	1	1
	3	χ_{19}^9	2	2	1	1
23	1	χ_{23}	$(2, 11) = 22$	$2 \cdot 23$	1	3
	2	χ_{23}^{11}	2	2	1	3
29	1	χ_{29}	$(2^2, 7) = 28$	$2 \cdot 29$	1	2^3
	2	χ_{29}^7	2^2	2	1	1
31	1	χ_{31}	$(2, 3, 5) = 30$	$2 \cdot 31$	1	3^2
	2	χ_{31}^3	$(2, 5) = 10$	2	1	3
	3	χ_{31}^5	$(2, 3) = 6$	2	1	3^2
	4	χ_{31}^{15}	2	2	1	3
37	1	χ_{37}	$(2^2, 3^2) = 36$	$2 \cdot 37$	1	37
	2	χ_{37}^3	$(2^2, 3) = 12$	2	1	1
	3	χ_{37}^9	2^2	2	1	1

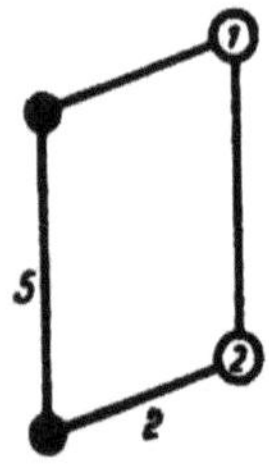

f = 11

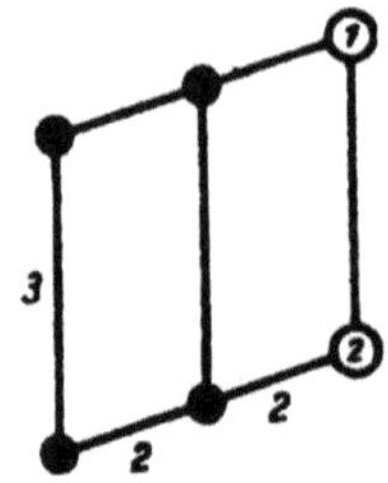

f = 13

f = 17

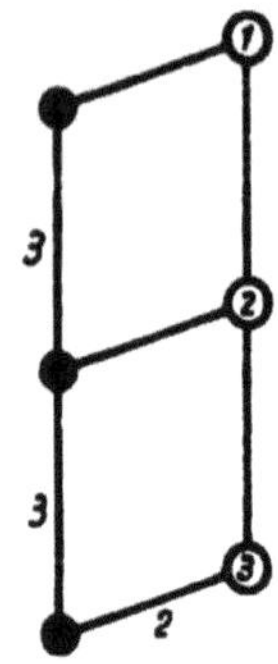

f = 19

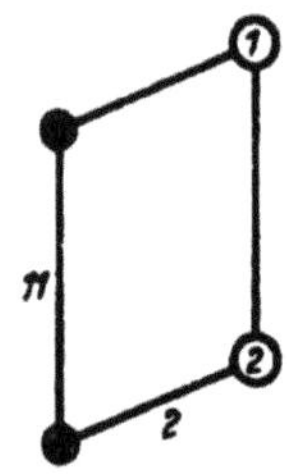

f = 23

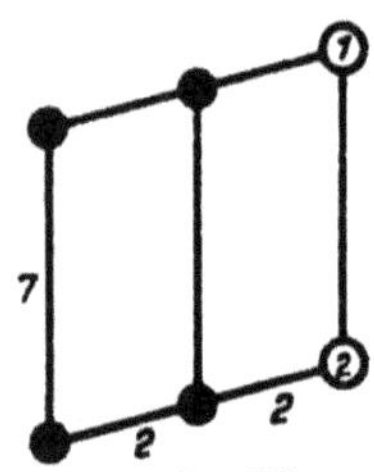

f = 29

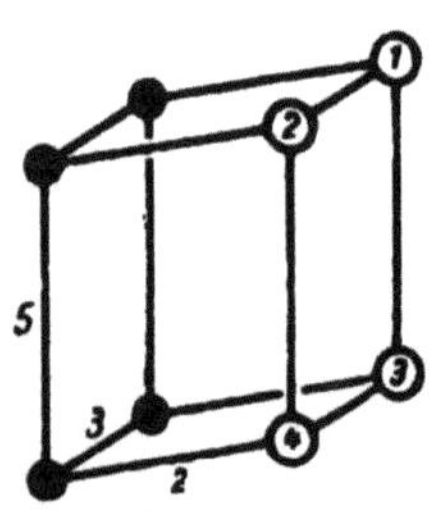

f = 31

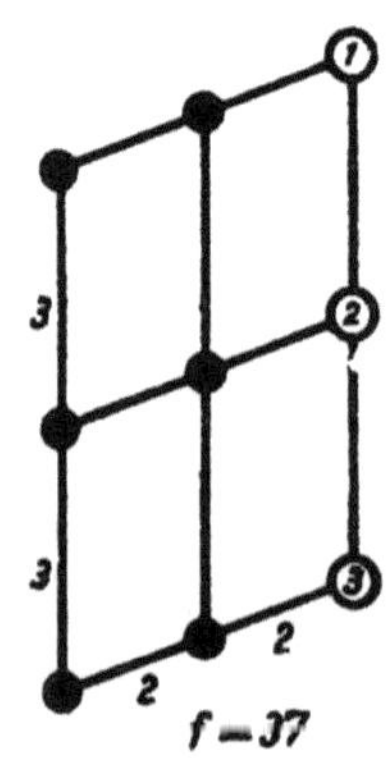

f = 37

1	2	3	4	5	6	7
Conductor	No.	Gen. characters	Type	w	Q	h^*
41	1	χ_{41}	$(2^3, 5) = 40$	$2 \cdot 41$	1	11^2
	2	χ_{41}^5	2^3	2	1	1
43	1	χ_{43}	$(2, 3, 7) = 42$	$2 \cdot 43$	1	211
	2	χ_{43}^3	$(2, 7) = 14$	2	1	1
	3	χ_{43}^7	$(2, 3) = 6$	2	1	1
	4	χ_{43}^{21}	2	2	1	1
47	1	χ_{47}	$(2, 23) = 46$	$2 \cdot 47$	1	$5 \cdot 139$
	2	χ_{47}^{23}	2	2	1	5
53	1	χ_{53}	$(2^2, 13) = 52$	$2 \cdot 53$	1	4889
	2	χ_{53}^{13}	2^2	2	1	1
59	1	χ_{59}	$(2, 29) = 58$	$2 \cdot 59$	1	$3 \cdot 59 \cdot 233$
	2	χ_{59}^{29}	2	2	1	3
61	1	χ_{61}	$(2^2, 3, 5) = 60$	$2 \cdot 61$	1	$41 \cdot 1861$
	2	χ_{61}^3	$(2^2, 5) = 20$	2	1	41
	3	χ_{61}^5	$(2^2, 3) = 12$	2	1	1
	4	χ_{61}^{15}	2^2	2	1	1
67	1	χ_{67}	$(2, 3, 11) = 66$	$2 \cdot 67$	1	$67 \cdot 12{,}739$
	2	χ_{67}^3	$(2, 11) = 22$	2	1	67
	3	χ_{67}^{11}	$(2, 3) = 6$	2	1	1
	4	χ_{67}^{33}	2	2	1	1

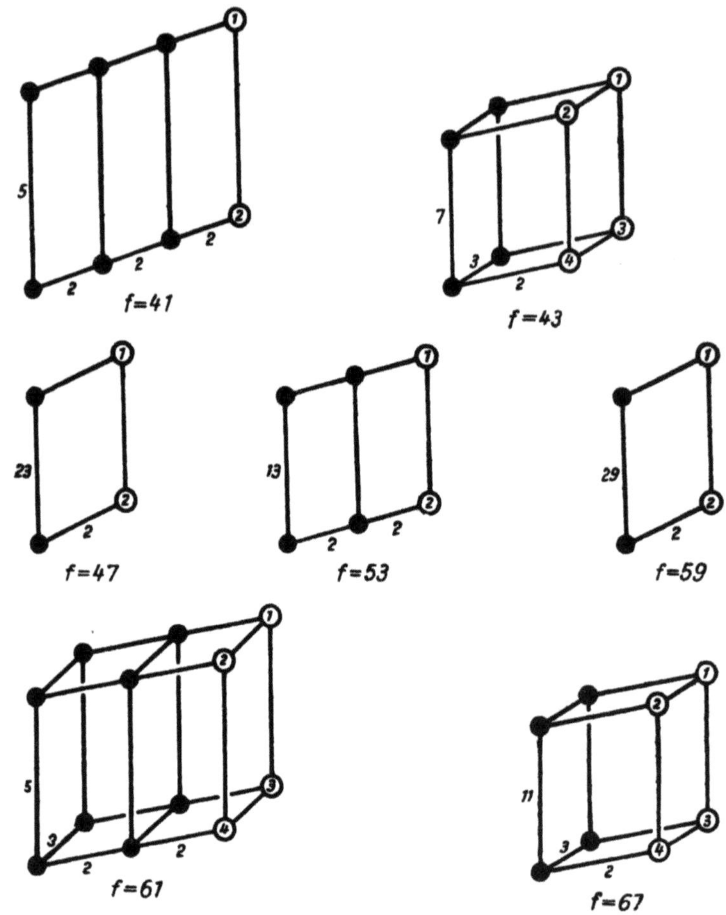

f=41
5
2
2
2
f=43
7
3
2
f=47
23
2
f=53
13
2
2
f=59
29
2
f=61
5
3
2
2
f=67
11
3
2

1	2	3	4	5	6	7
Conductor	No.	Gen. characters	Type	w	Q	h^*
71	1	χ_{71}	$(2, 5, 7) = 70$	$2 \cdot 71$	1	$7^2 \cdot 79{,}241$
	2	χ_{71}^5	$(2, 7) = 14$	2	1	7^2
	3	χ_{71}^7	$(2, 5) = 10$	2	1	7
	4	χ_{71}^{35}	2	2	1	7
73	1	χ_{73}	$(2^3, 3^2) = 72$	$2 \cdot 73$	1	$89 \cdot 134{,}353$
	2	χ_{73}^3	$(2^3, 3) = 24$	2	1	89
	3	χ_{73}^9	2^3	2	1	89
79	1	χ_{79}	$(2, 3, 13) = 78$	$2 \cdot 79$	1	$5 \cdot 53 \cdot 377{,}911$
	2	χ_{79}^3	$(2, 13) = 26$	2	1	$5 \cdot 53$
	3	χ_{79}^{13}	$(2, 3) = 6$	2	1	5
	4	χ_{79}^{39}	2	2	1	5
83	1	χ_{83}	$(2, 41) = 82$	$2 \cdot 83$	1	$3 \cdot 279{,}405{,}653$
	2	χ_{83}^{41}	2	2	1	3
89	1	χ_{89}	$(2^3, 11) = 88$	$2 \cdot 89$	1	$113 \cdot 118{,}401{,}449$
	2	χ_{89}^{11}	2^3	2	1	113
97	1	χ_{97}	$(2^5, 3) = 96$	$2 \cdot 97$	1	$577 \cdot 3457 \cdot 206{,}209$
	2	χ_{97}^3	2^5	2	1	3457

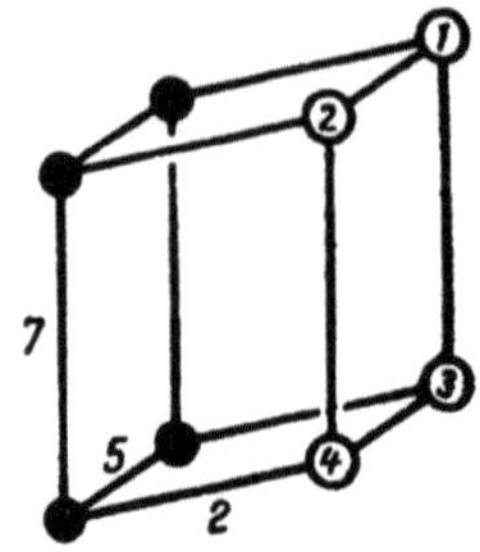

$f = 71$

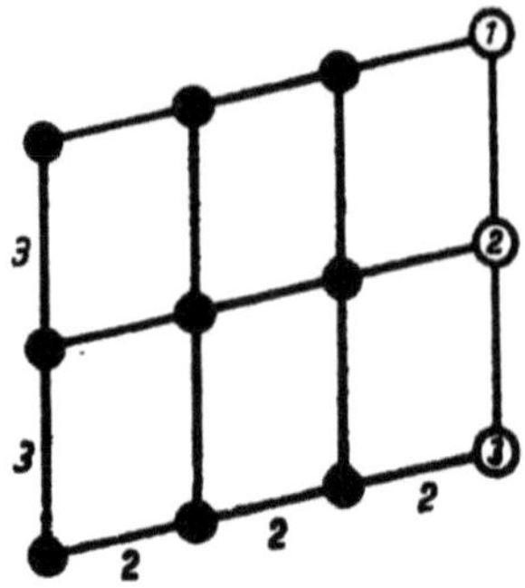

$f = 73$

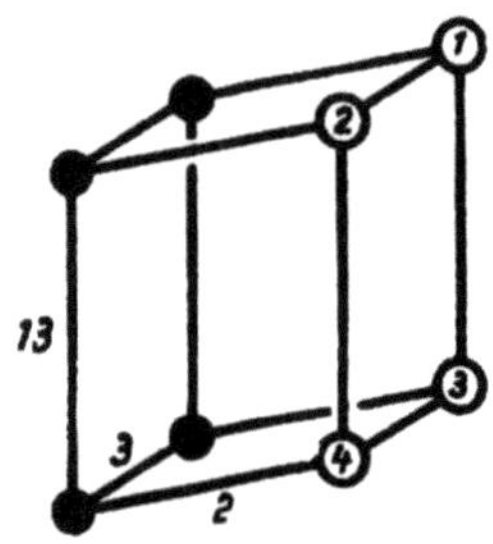

$f = 79$

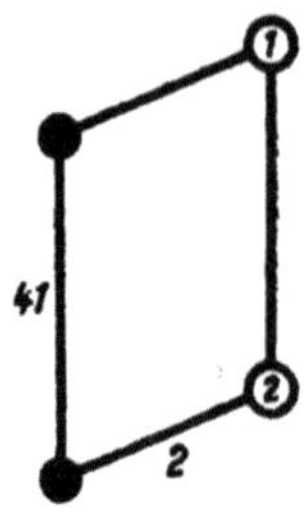

$f = 83$

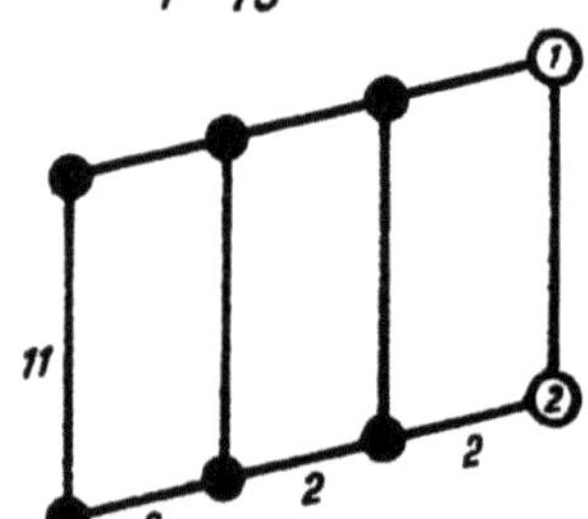

$f = 89$

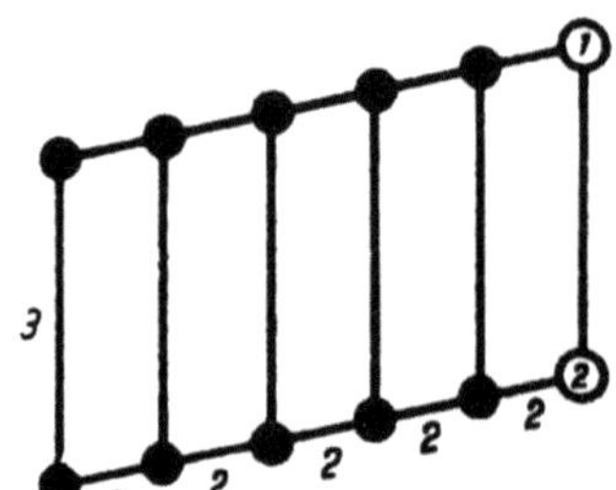

$f = 97$

2. Composite conductor

1	2	3	4	5	6	7
Conductor	No.	Gen. characters	Type	w	Q	h^*
$12 = 2^2 \cdot 3$	1	χ_{2^2}, χ_3	(2, 2)	$2^2 \cdot 3$	2	1
$15 = 3 \cdot 5$	1	χ_3, χ_5	$(2, 2^2)$	$2 \cdot 3 \cdot 5$	2	1
	2	χ_3, χ_5^2	(2, 2)	$2 \cdot 3$	1	1
	3	$\chi_3 \cdot \chi_5^2$	2	2	1	2
$20 = 2^2 \cdot 5$	1	χ_{2^2}, χ_5	$(2, 2^2)$	$2^2 \cdot 5$	2	1
	2	χ_{2^2}, χ_5^2	(2, 2)	2^2	1	1
	3	$\chi_{2^2} \cdot \chi_5^2$	2	2	1	2
$21 = 3 \cdot 7$	1	χ_3, χ_7	(2, 2, 3)	$2 \cdot 3 \cdot 5$	2	1
	2	χ_3, χ_7^2	$(2, 3) = 6$	$2 \cdot 3$	1	1
	3	χ_3, χ_7^3	(2, 2)	$2 \cdot 3$	2	1
$24 = 2^3 \cdot 3$	1	$\chi_{2^2}, \psi_{2^3}; \chi_3$	(2, 2, 2)	$2^3 \cdot 3$	2	1
	2	ψ_{2^3}, χ_3	(2, 2)	$2 \cdot 3$	1	1
	3	$\chi_{2^2}\psi_{2^3}, \chi_3$	(2, 2)	$2 \cdot 3$	2	1
	4	$\chi_{2^2}, \psi_{2^3} \cdot \chi_3$	(2, 2)	$2 \cdot 3$	2	2
	5	$\chi_{2^2}\psi_{2^3}, \psi_{2^3} \cdot \chi_3$	(2, 2)	2	2	2
	6	$\psi_{2^3} \cdot \chi_3$	2	2	1	2
$28 = 2^2 \cdot 7$	1	χ_{2^2}, χ_7	(2, 2, 3)	$2^2 \cdot 7$	2	1
	2	χ_{2^2}, χ_7^2	$(2, 3) = 6$	2^2	1	1
	3	χ_{2^2}, χ_7^3	(2, 2)	2^2	2	1

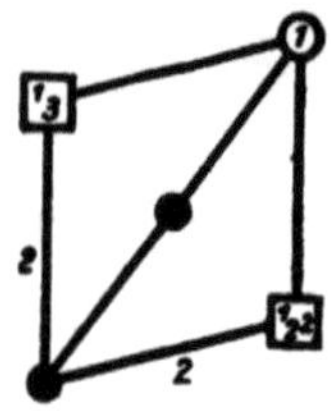

f=12

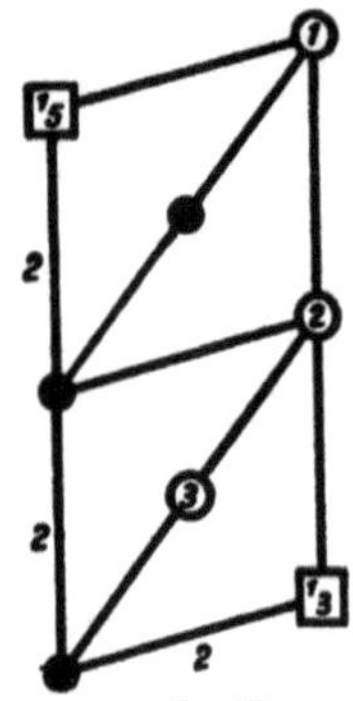

f=15

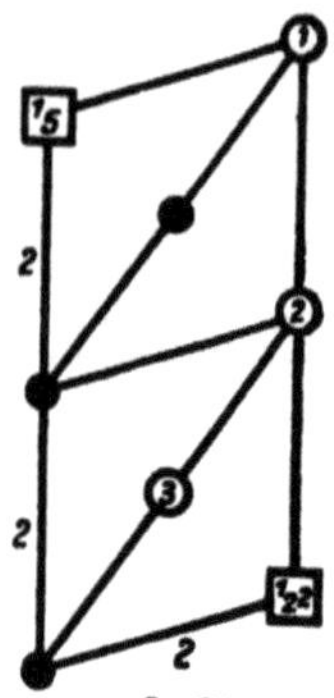

f=20

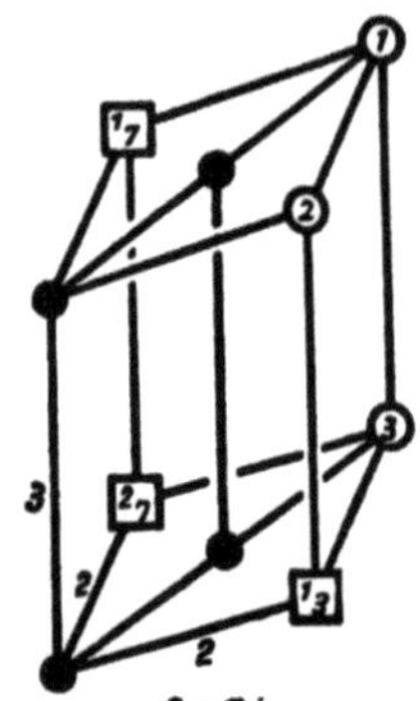

f=21

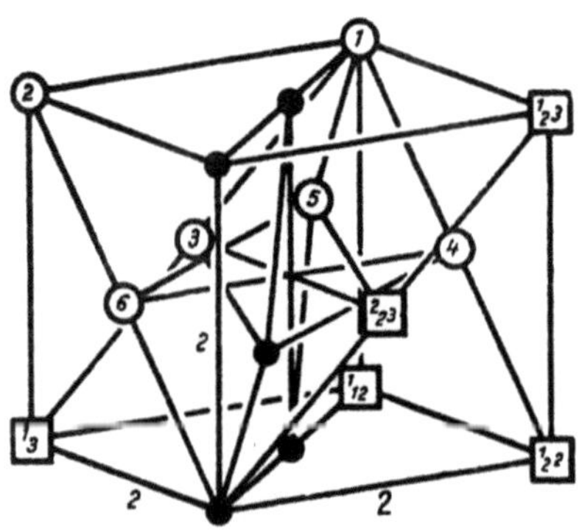

f=24

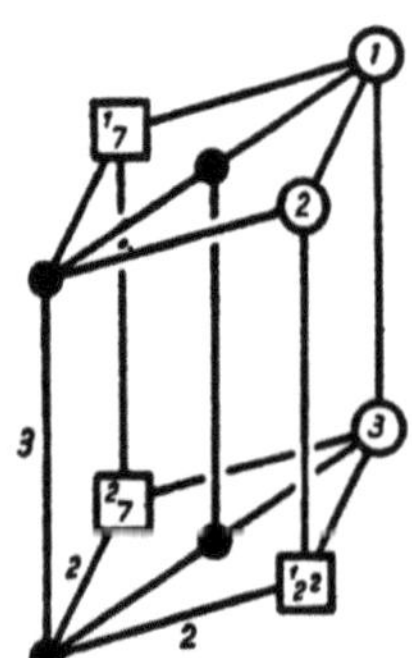

f=28

1	2	3	4	5	6	7
Conductor	No.	Gen. characters	Type	w	Q	h^*
$33 = 3 \cdot 11$	1	χ_3, χ_{11}	(2, 2, 5)	$2 \cdot 3 \cdot 11$	2	1
	2	χ_3, χ_{11}^2	(2, 5) = 10	$2 \cdot 3$	1	1
	3	χ_3, χ_{11}^5	(2, 2)	$2 \cdot 3$	2	1
$35 = 5 \cdot 7$	1	χ_5, χ_7	$(2^2, 2, 3)$	$2 \cdot 5 \cdot 7$	2	1
	2	χ_5, χ_7^2	$(2^2, 3) = 12$	$2 \cdot 5$	1	1
	3	χ_5, χ_7^3	$(2^2, 2)$	$2 \cdot 5$	2	1
	4	χ_5^2, χ_7	(2, 2, 3)	$2 \cdot 7$	1	1
	5	χ_5^2, χ_7^3	(2, 2)	2	1	1
	6	$\chi_5^2 \cdot \chi_7^3, \chi_7^2$	(2, 3) = 6	2	1	2
	7	$\chi_5^2 \cdot \chi_7^3$	2	2	1	2
$36 = 2^2 \cdot 3^2$	1	$\chi_{2^2}; \chi_3, \psi_{3^2}$	(2, 2, 3)	$2^2 \cdot 3^2$	2	1
	2	χ_{2^2}, ψ_{3^2}	(2, 3) = 6	2^2	1	1
$39 = 3 \cdot 13$	1	χ_3, χ_{13}	$(2, 2^2, 3)$	$2 \cdot 3 \cdot 13$	2	2
	2	χ_3, χ_{13}^2	(2, 2, 3)	$2 \cdot 3$	1	2
	3	χ_3, χ_{13}^3	$(2, 2^2)$	$2 \cdot 3$	2	2
	4	χ_3, χ_{13}^4	(2, 3) = 6	$2 \cdot 3$	1	1
	5	χ_3, χ_{13}^6	(2, 2)	$2 \cdot 3$	1	2
	6	$\chi_3 \cdot \chi_{13}^6, \chi_{13}^4$	(2, 3) = 6	2	1	2^2
	7	$\chi_3 \cdot \chi_{13}^6$	2	2	1	2^2

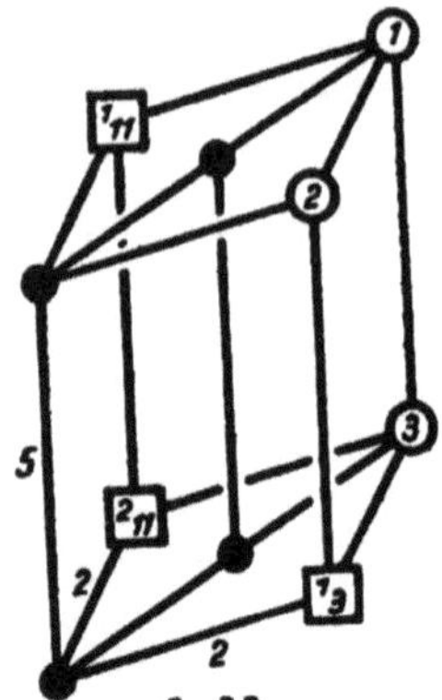

$f = 33$

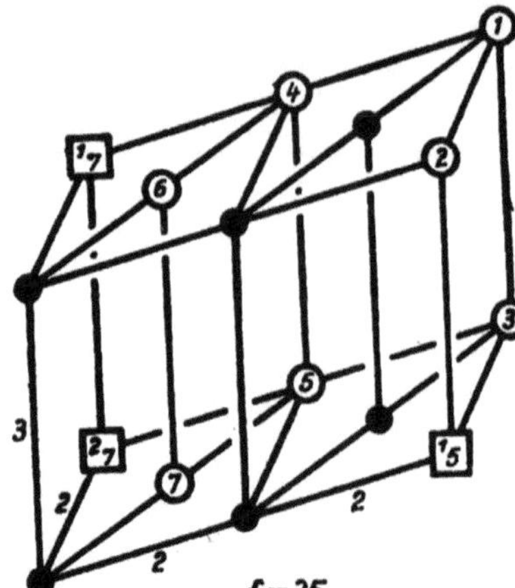

$f = 35$

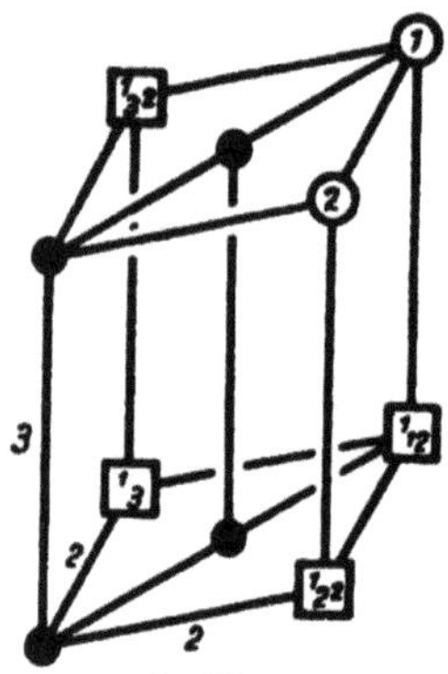

$f = 36$

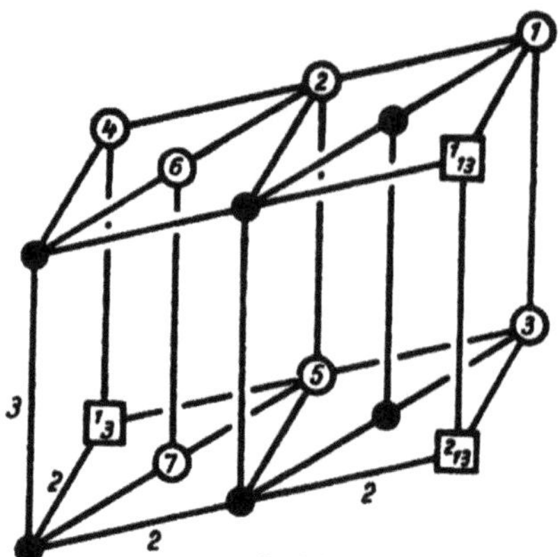

$f = 39$

1	2	3	4	5	6	7
Conductor	No.	Gen. characters	Type	w	Q	h^*
$40 = 2^3 \cdot 5$	1	$\chi_{2^2}, \psi_{2^3}; \chi_5$	$(2, 2, 2^2)$	$2^3 \cdot 5$	2	1
	2	$\chi_{2^2}, \psi_{2^3}; \chi_5^2$	$(2, 2, 2)$	2^3	1	1
	3	ψ_{2^3}, χ_5	$(2, 2^2)$	$2 \cdot 5$	1	1
	4	$\chi_{2^2}\psi_{2^3}, \chi_5$	$(2, 2^2)$	$2 \cdot 5$	2	1
	5	$\chi_{2^2}\psi_{2^3}, \chi_5^2$	$(2, 2)$	2	1	1
	6	$\chi_{2^2}, \psi_{2^3} \cdot \chi_5$	$(2, 2^2)$	2^2	2	2
	7	$\chi_{2^2}, \psi_{2^3} \cdot \chi_5^2$	$(2, 2)$	2^2	1	1
	8	$\chi_{2^2} \cdot \chi_5^2, \psi_{2^3}$	$(2, 2)$	2	1	2
	9	$\chi_{2^2}\psi_{2^3}, \chi_{2^2} \cdot \chi_5$	$(2, 2^2)$	2	2	2
	10	$\chi_{2^2}\psi_{2^3}, \chi_{2^2} \cdot \chi_5^2$	$(2, 2)$	2	1	1
	11	$\psi_{2^3} \cdot \chi_5$	2^2	2	1	2
	12	$\chi_{2^2}\psi_{2^3} \cdot \chi_5^2$	2	2	1	2
$44 = 2^2 \cdot 11$	1	χ_{2^2}, χ_{11}	$(2, 2, 5)$	$2^2 \cdot 11$	2	1
	2	χ_{2^2}, χ_{11}^2	$(2, 5) = 10$	2^2	1	1
	3	χ_{2^2}, χ_{11}^5	$(2, 2)$	2^2	2	1
$45 = 3^2 \cdot 5$	1	$\chi_3, \psi_{3^2}; \chi_5$	$(2, 3, 2^2)$	$2 \cdot 3^2 \cdot 5$	2	1
	2	$\chi_3, \psi_{3^2}; \chi_5^2$	$(2, 3, 2)$	$2 \cdot 3^2$	1	1
	3	ψ_{3^2}, χ_5	$(3, 2^2) = 12$	$2 \cdot 5$	1	1
	4	$\chi_3 \cdot \chi_5^2, \psi_{3^2}$	$(2, 3) = 6$	2	1	2

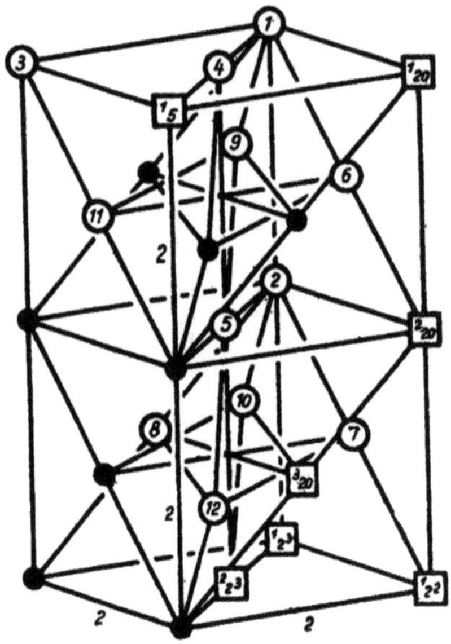

f = 40

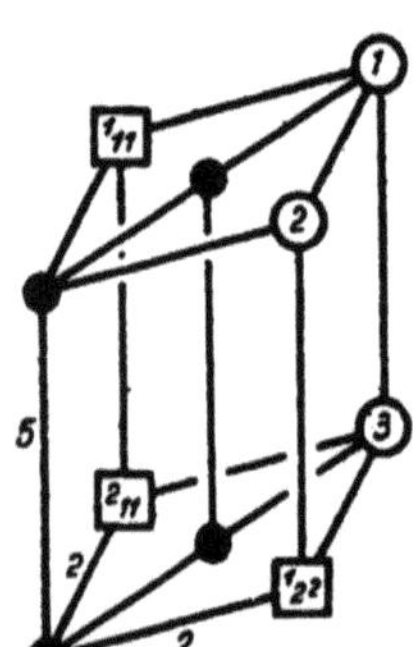

f = 44

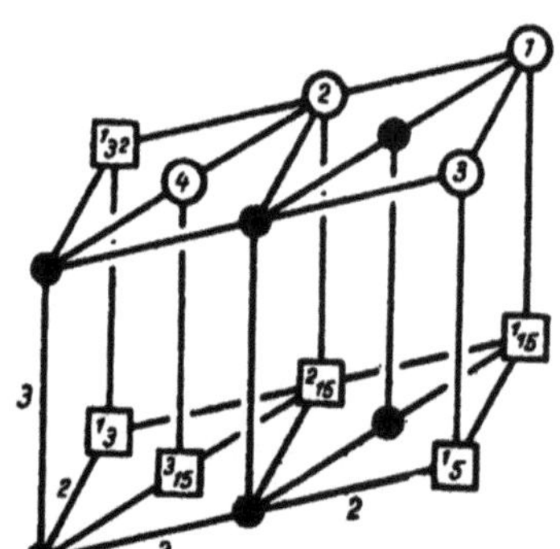

f = 45

1	2	3	4	5	6	7
Conductor	No.	Gen. characters	Type	w	Q	h^*
$48 = 2^4 \cdot 3$	1	$\chi_{2^2}, \psi_{2^4}; \chi_3$	$(2, 2^2, 2)$	$2^4 \cdot 3$	2	1
	2	ψ_{2^4}, χ_3	$(2^2, 2)$	$2 \cdot 3$	1	1
	3	$\chi_{2^2}\psi_{2^4}, \chi_3$	$(2^2, 2)$	$2 \cdot 3$	2	1
	4	$\chi_{2^2}, \psi_{2^4} \cdot \chi_3$	$(2, 2^2)$	2^3	2	2
	5	$\chi_{2^2}\psi_{2^4}, \chi_{2^2} \cdot \chi_3$	$(2^2, 2)$	2	2	2
	6	$\psi_{2^4} \cdot \chi_3$	2^2	2	1	2
$51 = 3 \cdot 17$	1	χ_3, χ_{17}	$(2, 2^4)$	$2 \cdot 3 \cdot 17$	2	5
	2	χ_3, χ_{17}^2	$(2, 2^3)$	$2 \cdot 3$	1	5
	3	χ_3, χ_{17}^4	$(2, 2^2)$	$2 \cdot 3$	1	5
	4	χ_3, χ_{17}^8	$(2, 2)$	$2 \cdot 3$	1	1
	5	$\chi_3 \cdot \chi_{17}^2$	2^3	2	1	2
	6	$\chi_3 \cdot \chi_{17}^4$	2^2	2	1	$2 \cdot 5$
	7	$\chi_3 \cdot \chi_{17}^8$	2	2	1	2
$52 = 2^2 \cdot 13$	1	χ_{2^2}, χ_{13}	$(2, 2^2, 3)$	$2^2 \cdot 13$	2	3
	2	χ_{2^2}, χ_{13}^2	$(2, 2, 3)$	2^2	1	3
	3	χ_{2^2}, χ_{13}^3	$(2, 2^2)$	2^2	2	1
	4	χ_{2^2}, χ_{13}^4	$(2, 3) = 6$	2^2	1	3
	5	χ_{2^2}, χ_{13}^6	$(2, 2)$	2^2	1	1
	6	$\chi_{2^2} \cdot \chi_{13}^6, \chi_{13}^4$	$(2, 3) = 6$	2	1	2
	7	$\chi_{2^2} \cdot \chi_{13}^6$	2	2	1	2

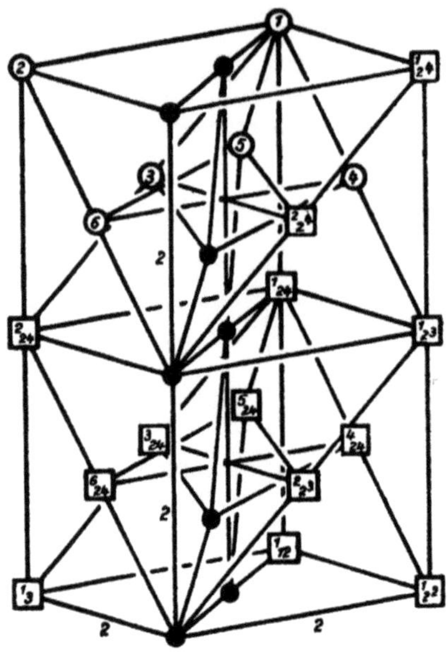

f=48

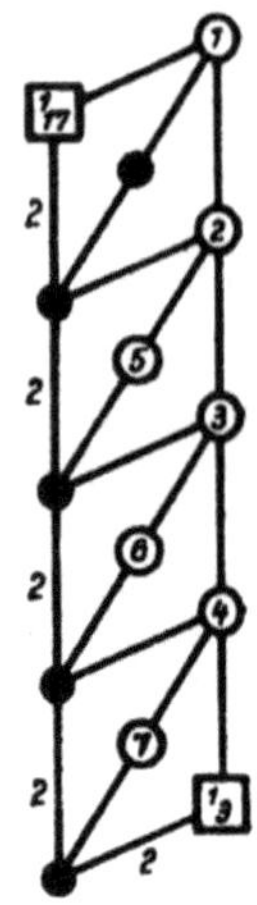

f=51

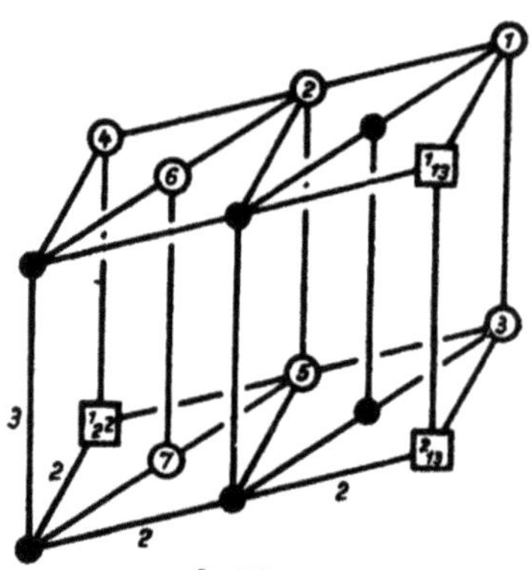

f=52

1	2	3	4	5	6	7
Conductor	No.	Gen. characters	Type	w	Q	h^*
$55 = 5 \cdot 11$	1	χ_5, χ_{11}	$(2^2, 2, 5)$	$2 \cdot 5 \cdot 11$	2	$2 \cdot 5$
	2	χ_5, χ_{11}^2	$(2^2, 5) = 20$	$2 \cdot 5$	1	5
	3	χ_5, χ_{11}^5	$(2^2, 2)$	$2 \cdot 5$	2	2
	4	χ_5^2, χ_{11}	$(2, 2, 5)$	$2 \cdot 11$	1	2
	5	χ_5^2, χ_{11}^5	$(2, 2)$	2	1	2
	6	$\chi_5^2 \cdot \chi_{11}^5, \chi_{11}^2$	$(2, 5) = 10$	2	1	2^2
	7	$\chi_5^2 \cdot \chi_{11}^5$	2	2	1	2^2
$56 = 2^3 \cdot 7$	1	$\chi_{2^2}, \psi_{2^3}; \chi_7$	$(2, 2, 2, 3)$	$2^3 \cdot 17$	2	2
	2	$\chi_{2^2}, \psi_{2^3}; \chi_7^2$	$(2, 2, 3)$	2^3	1	1
	3	$\chi_{2^2}, \psi_{2^3}; \chi_7^3$	$(2, 2, 2)$	2^3	2	2
	4	ψ_{2^3}, χ_7	$(2, 2, 3)$	$2 \cdot 7$	1	2
	5	ψ_{2^3}, χ_7^3	$(2, 2)$	2	1	2
	6	$\chi_{2^2}\psi_{2^3}, \chi_7$	$(2, 2, 3)$	$2 \cdot 7$	2	1
	7	$\chi_{2^2}\psi_{2^3}, \chi_7^2$	$(2, 3) = 6$	2	1	1
	8	$\chi_{2^2}\psi_{2^3}, \chi_7^3$	$(2, 2)$	2	2	1
	9	$\chi_{2^2}, \psi_{2^3} \cdot \chi_7^3, \chi_7^2$	$(2, 2, 3)$	2^2	2	2^2
	10	$\chi_{2^2}, \psi_{2^3} \cdot \chi_7^3$	$(2, 2)$	2^2	2	2^2
	11	$\chi_{2^2}\psi_{2^3}, \psi_{2^3} \cdot \chi_7^3, \chi_7^2$	$(2, 2, 3)$	2	2	2^2
	12	$\chi_{2^2}\psi_{2^3}, \psi_{2^3} \cdot \chi_7^3$	$(2, 2)$	2	2	2^2
	13	$\psi_{2^3} \cdot \chi_7^3, \chi_7^2$	$(2, 3) = 6$	2	1	2^2
	14	$\psi_{2^3} \cdot \chi_7^3$	2	2	1	2^2

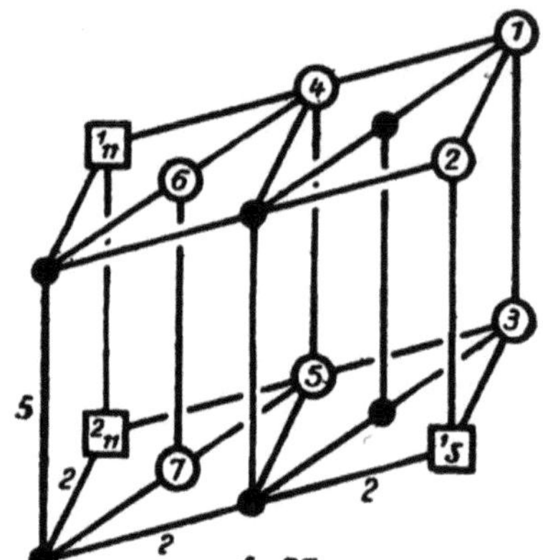

f=55

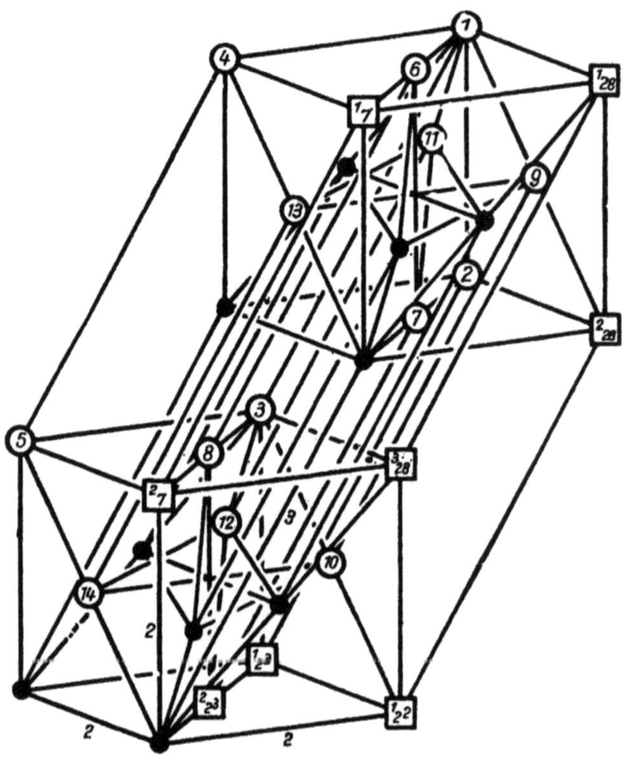

f=56

1	2	3	4	5	6	7
Conductor	No.	Gen. characters	Type	w	Q	h^*
$57 = 3 \cdot 19$	1	χ_3, χ_{19}	$(2, 2, 3^2)$	$2 \cdot 3 \cdot 19$	2	3^2
	2	χ_3, χ_{19}^2	$(2, 3^2) = 18$	$2 \cdot 3$	1	3^2
	3	χ_3, χ_{19}^3	$(2, 2, 3)$	$2 \cdot 3$	2	3
	4	χ_3, χ_{19}^6	$(2, 3) = 6$	$2 \cdot 3$	1	3
	5	χ_3, χ_{19}^9	$(2, 2)$	$2 \cdot 3$	2	1
$60 = 2^2 \cdot 3 \cdot 5$	1	$\chi_{2^2}, \chi_3, \chi_5$	$(2, 2, 2^2)$	$2^2 \cdot 3 \cdot 5$	2	1
	2	$\chi_{2^2}, \chi_3, \chi_5^2$	$(2, 2, 2)$	$2^2 \cdot 3$	2	1
	3	$\chi_{2^2} \cdot \chi_3, \chi_5$	$(2, 2^2)$	$2 \cdot 5$	1	2
	4	$\chi_{2^2}, \chi_3 \cdot \chi_5$	$(2, 2^2)$	2^2	1	2
	5	$\chi_{2^2}, \chi_3 \cdot \chi_5^2$	$(2, 2)$	2^2	1	1
	6	$\chi_3, \chi_{2^2} \cdot \chi_5$	$(2, 2^2)$	$2 \cdot 3$	1	2
	7	$\chi_3, \chi_{2^2} \cdot \chi_5^2$	$(2, 2)$	$2 \cdot 3$	1	1
	8	$\chi_{2^2} \cdot \chi_3 \cdot \chi_5$	2^2	2	1	2^2
	9	$\chi_{2^2} \cdot \chi_3, \chi_{2^2} \cdot \chi_5^2$	$(2, 2)$	2	1	2

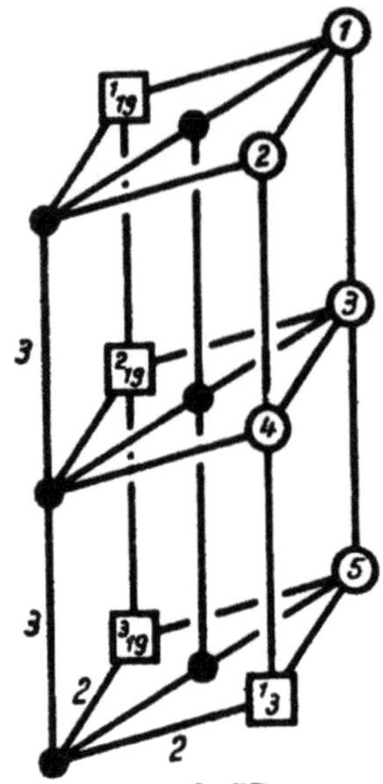

f=57

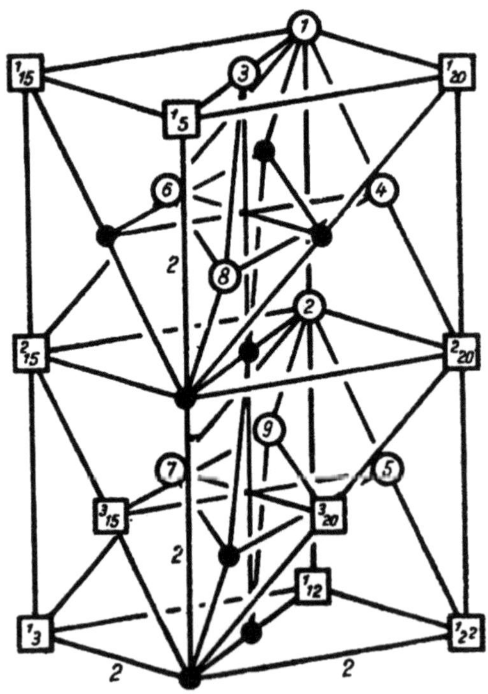

f=60

1	2	3	4	5	6	7
Conductor	No.	Gen. characters	Type	w	Q	h^*
$63 = 3^2 \cdot 7$	1	$\chi_3, \psi_{3^2}; \chi_7$	(2, 3, 2, 3)	$2 \cdot 3^2 \cdot 7$	2	7
	2	$\chi_3, \psi_{3^2}; \chi_7^2$	(2, 3, 3)	$2 \cdot 3^2$	1	1
	3	$\chi_3, \psi_{3^2}; \chi_7^3$	(2, 3, 2)	$2 \cdot 3^2$	2	1
	4	$\chi_3, \psi_{3^2} \cdot \chi_7^2, \chi_7^3$	(2, 3, 2)	$2 \cdot 3$	2	7
	4′	$\chi_3, \psi_{3^2} \cdot \overline{\chi}_7^2, \chi_7^3$	(2, 3, 2)	$2 \cdot 3$	2	1
	5	$\chi_3, \psi_{3^2} \cdot \chi_7^2$	$(2, 3) = 6$	$2 \cdot 3$	1	1
	5′	$\chi_3, \psi_{3^2} \cdot \overline{\chi}_7^2$	$(2, 3) = 6$	$2 \cdot 3$	1	1
	6	ψ_{3^2}, χ_7	(3, 2, 3)	$2 \cdot 7$	1	7
	7	ψ_{3^2}, χ_7^3	$(3, 2) = 6$	2	1	1
	8	$\psi_{3^2} \cdot \chi_7^2, \chi_7^3$	$(3, 2) = 6$	2	1	7
	8′	$\psi_{3^2} \cdot \overline{\chi}_7^2, \chi_7^3$	$(3, 2) = 6$	2	1	1
$65 = 5 \cdot 13$	1	χ_5, χ_{13}	$(2^2, 2^2, 3)$	$2 \cdot 5 \cdot 13$	2	2^6
	2	χ_5, χ_{13}^2	$(2^2, 2, 3)$	$2 \cdot 5$	1	2^4
	3	χ_5, χ_{13}^3	$(2^2, 2^2)$	$2 \cdot 5$	2	1
	4	χ_5, χ_{13}^4	$(2^2, 3) = 12$	$2 \cdot 5$	1	2^2
	5	χ_5, χ_{13}^6	$(2^2, 2)$	$2 \cdot 5$	1	1
	6	χ_5^2, χ_{13}	$(2, 2^2, 3)$	$2 \cdot 13$	1	2^2
	7	χ_5^2, χ_{13}^3	$(2, 2^2)$	2	1	1
	8	$\chi_5 \cdot \chi_{13}^2$	$(2^2, 3) = 12$	2	1	2^3
	9	$\chi_5 \cdot \chi_{13}^6$	2^2	2	1	2
	10	$\chi_5^2 \cdot \chi_{13}$	$(2^2, 3) = 12$	2	1	2^3
	11	$\chi_5^2 \cdot \chi_{13}^3$	2^2	2	1	2

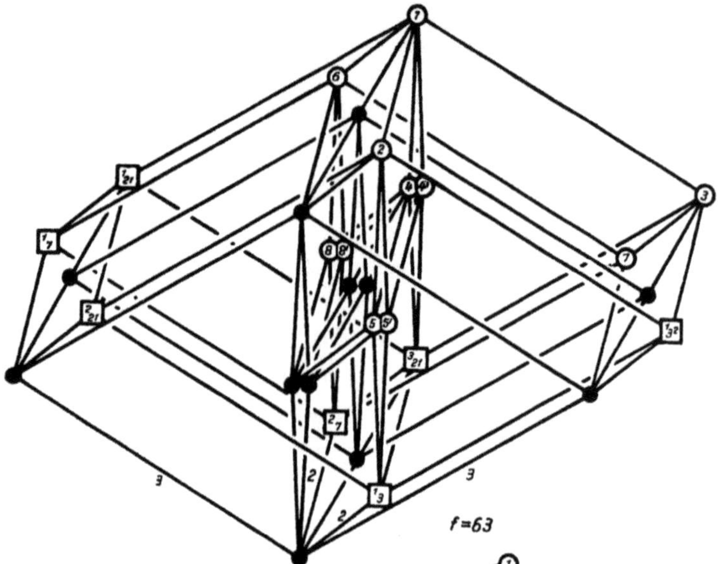

f=63

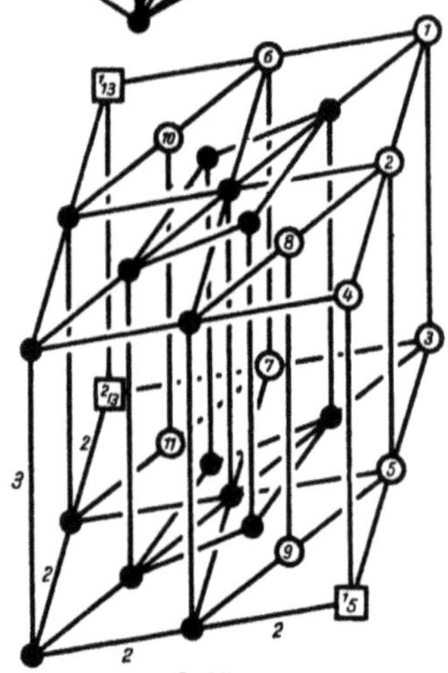

f=65

1	2	3	4	5	6	7
Conductor	No.	Gen. characters	Type	w	Q	h^*
$68 = 2^2 \cdot 17$	1	χ_{2^2}, χ_{17}	$(2, 2^4)$	$2^2 \cdot 17$	2	2^3
	2	χ_{2^2}, χ_{17}^2	$(2, 2^3)$	2^2	1	2^3
	3	χ_{2^2}, χ_{17}^4	$(2, 2^2)$	2^2	1	2^2
	4	χ_{2^2}, χ_{17}^8	$(2, 2)$	2^2	1	2
	5	$\chi_{2^2} \cdot \chi_{17}^2$	2^3	2	1	2^2
	6	$\chi_{2^2} \cdot \chi_{17}^4$	2^2	2	1	2^2
	7	$\chi_{2^2} \cdot \chi_{17}^8$	2	2	1	2^2
$69 = 3 \cdot 23$	1	χ_3, χ_{23}	$(2, 2, 11)$	$2 \cdot 3 \cdot 23$	2	$3 \cdot 23$
	2	χ_3, χ_{23}^2	$(2, 11) = 22$	$2 \cdot 3$	1	23
	3	χ_3, χ_{23}^{11}	$(2, 2)$	$2 \cdot 3$	2	3
$72 = 2^3 \cdot 3^2$	1	$\chi_{2^2}, \psi_{2^3}; \chi_3, \psi_{3^2}$	$(2, 2, 2, 3)$	$2^3 \cdot 3^2$	2	3
	2	$\chi_{2^2}, \psi_{2^3}; \psi_{3^2}$	$(2, 2, 3)$	2^3	1	3
	3	$\psi_{2^3}; \chi_3, \psi_{3^2}$	$(2, 2, 3)$	$2 \cdot 3^2$	1	1
	4	$\chi_{2^2}\psi_{2^3}, \chi_3, \psi_{3^2}$	$(2, 2, 3)$	$2 \cdot 3^2$	2	3
	5	$\chi_{2^2}, \psi_{2^3} \cdot \chi_3, \psi_{3^2}$	$(2, 2, 3)$	2^2	2	2
	6	$\chi_{2^2}\psi_{2^3}, \psi_{2^3} \cdot \chi_3, \psi_{3^2}$	$(2, 2, 3)$	2	2	$2 \cdot 3$
	7	$\psi_{2^3} \cdot \chi_3, \psi_{3^2}$	$(2, 3) = 6$	2	1	2
	8	$\chi_{2^2}\psi_{2^3}, \psi_{3^2}$	$(2, 3) = 6$	2	1	3

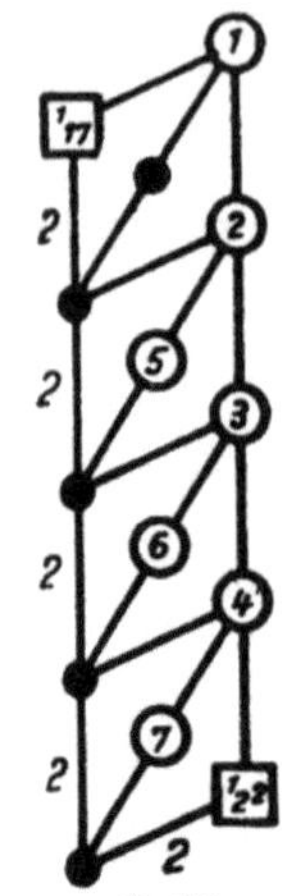

$f=68$

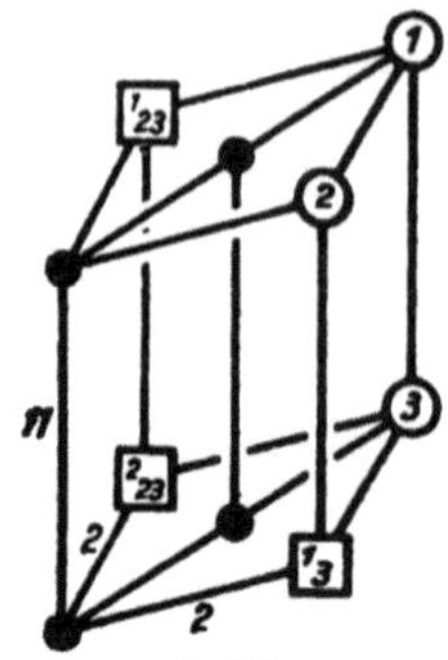

$f=69$

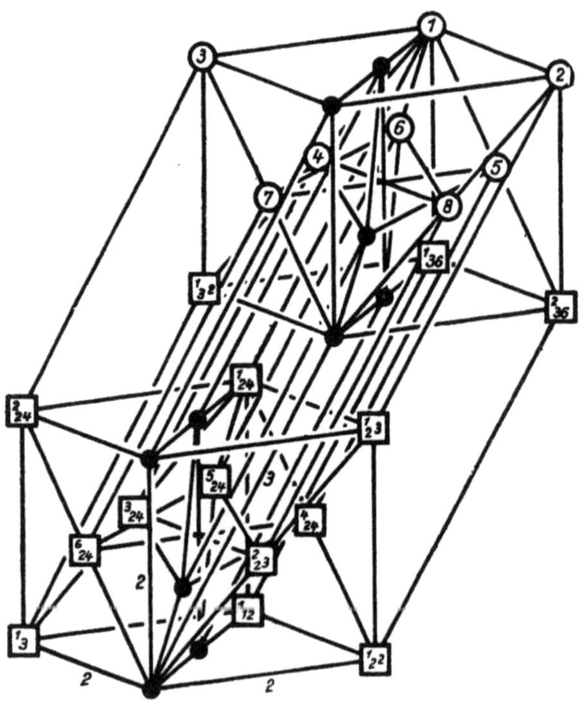

$f=72$

1	2	3	4	5	6	7
Conductor	No.	Gen. characters	Type	w	Q	h^*
$75 = 3 \cdot 5^2$	1	$\chi_3;\ \chi_5, \psi_{5^2}$	$(2, 2^2, 5)$	$2 \cdot 3 \cdot 5^2$	2	11
	2	$\chi_3;\ \chi_5^2, \psi_{5^2}$	$(2, 2, 5)$	$2 \cdot 3$	1	11
	3	χ_3, ψ_{5^2}	$(2, 5) = 10$	$2 \cdot 3$	1	11
	4	$\chi_3 \cdot \chi_5^2, \psi_{5^2}$	$(2, 5) = 10$	2	1	2
$76 = 2^2 \cdot 19$	1	χ_{2^2}, χ_{19}	$(2, 2, 3^2)$	$2^2 \cdot 19$	2	19
	2	χ_{2^2}, χ_{19}^2	$(2, 3^2) = 18$	2^2	1	19
	3	χ_{2^2}, χ_{19}^3	$(2, 2, 3)$	2^2	2	1
	4	χ_{2^2}, χ_{19}^6	$(2, 3) = 6$	2^2	1	1
	5	χ_{2^2}, χ_{19}^9	$(2, 2)$	2^2	2	1
$77 = 7 \cdot 11$	1	χ_7, χ_{11}	$(2, 3, 2, 5)$	$2 \cdot 7 \cdot 11$	2	$2^8 \cdot 5$
	2	χ_7, χ_{11}^2	$(2, 3, 5) = 30$	$2 \cdot 7$	1	$2^4 \cdot 5$
	3	χ_7, χ_{11}^5	$(2, 3, 2)$	$2 \cdot 7$	2	1
	4	χ_7^2, χ_{11}	$(3, 2, 5) = 30$	$2 \cdot 11$	1	2^4
	5	χ_7^2, χ_{11}^5	$(3, 2) = 6$	2	1	1
	6	χ_7^3, χ_{11}	$(2, 2, 5)$	$2 \cdot 11$	2	5
	7	χ_7^3, χ_{11}^2	$(2, 5) = 10$	2	1	5
	8	χ_7^3, χ_{11}^5	$(2, 2)$	2	2	1

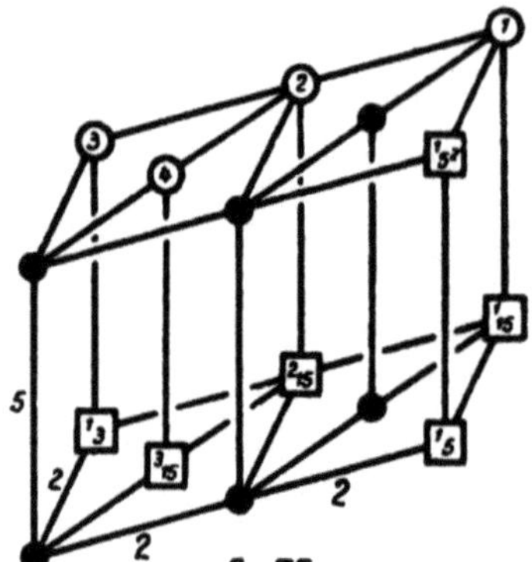

f = 75

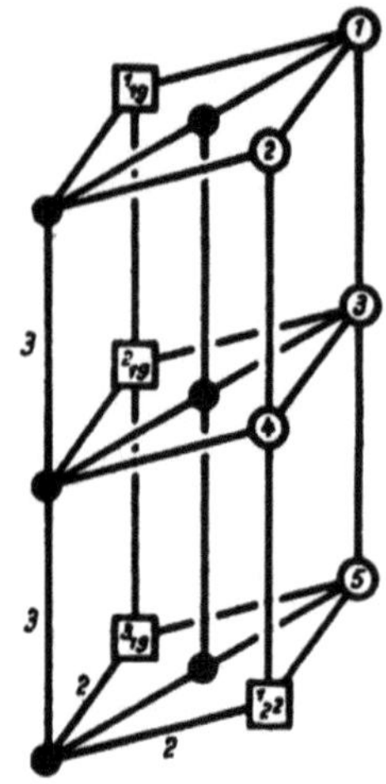

f = 76

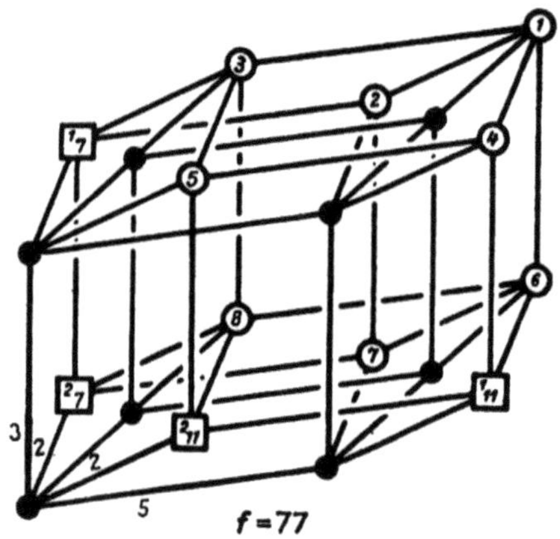

f = 77

1	2	3	4	5	6	7
Conductor	No.	Gen. characters	Type	w	Q	h^*
$80 = 2^4 \cdot 5$	1	$\chi_{2^2}, \psi_{2^4}; \chi_5$	$(2, 2^2, 2^2)$	$2^4 \cdot 5$	2	5
	2	$\chi_{2^2}, \psi_{2^4}; \chi_5^2$	$(2, 2^2, 2)$	2^4	1	1
	3	ψ_{2^4}, χ_5	$(2^2, 2^2)$	$2 \cdot 5$	1	5
	4	$\chi_{2^2}\psi_{2^4}, \chi_5$	$(2^2, 2^2)$	$2 \cdot 5$	2	1
	5	$\chi_{2^2}\psi_{2^4}, \chi_5^2$	$(2^2, 2)$	2	1	1
	6	$\chi_{2^2}, \psi_{2^4} \cdot \chi_5, \chi_5^2$	$(2, 2^2, 2)$	2^3	2	$2 \cdot 5$
	7	$\chi_{2^2}, \psi_{2^4} \cdot \chi_5$	$(2, 2^2)$	2^2	2	2
	7′	$\chi_{2^2}, \psi_{2^4} \cdot \overline{\chi}_5$	$(2, 2^2)$	2^2	2	$2 \cdot 5$
	8	$\chi_{2^2}, \psi_{2^4} \cdot \chi_5^2$	$(2, 2^2)$	2^3	1	1
	9	$\psi_{2^4}, \chi_{2^2} \cdot \chi_5^2$	$(2^2, 2)$	2	1	2
	10	$\chi_{2^2}\psi_{2^4}, \chi_{2^2} \cdot \chi_5$	$(2^2, 2^2)$	2	2	$2 \cdot 5$
	11	$\chi_{2^2}\psi_{2^4}, \chi_{2^2} \cdot \chi_5^2$	$(2^2, 2)$	2	1	1
	12	$\chi_5^2, \psi_{2^4} \cdot \chi_5$	$(2, 2^2)$	2	1	$2 \cdot 5$
	13	$\chi_{2^2} \cdot \chi_5^2, \psi_{2^4} \cdot \chi_5$	$(2, 2^2)$	2	2	2
	13′	$\chi_{2^2} \cdot \chi_5^2, \psi_{2^4} \cdot \overline{\chi}_5$	$(2, 2^2)$	2	2	$2 \cdot 5$
	14	$\psi_{2^4} \cdot \chi_5$	2^2	2	1	2
	14′	$\psi_{2^4} \cdot \overline{\chi}_5$	2^2	2	1	$2 \cdot 5$
	15	$\chi_{2^2}\psi_{2^4} \cdot \chi_5^2$	2^2	2	1	2

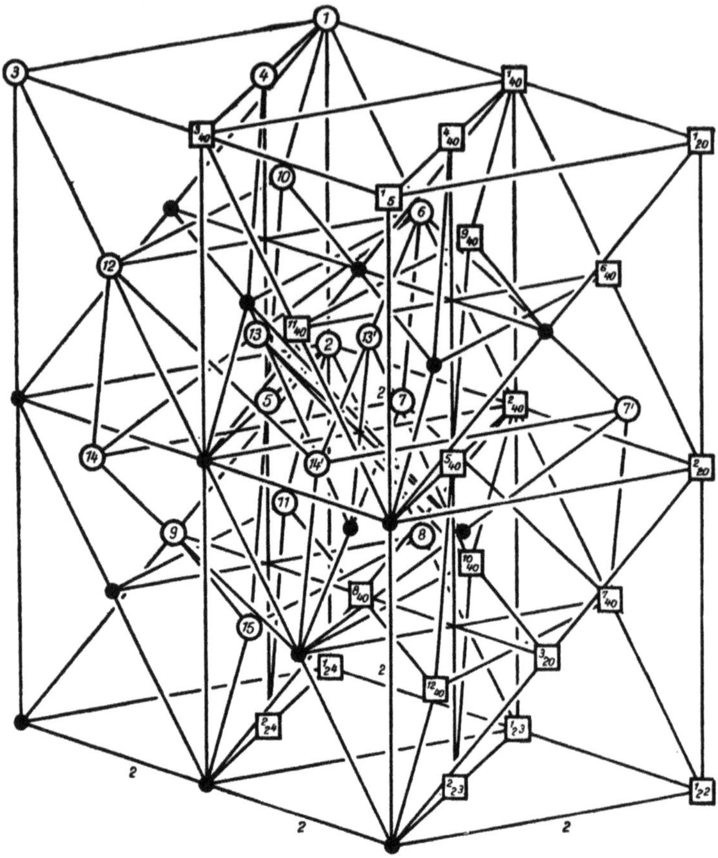

$f=80$

1	2	3	4	5	6	7
Conductor	No.	Gen. characters	Type	w	Q	h^*
$84 = 2^2 \cdot 3 \cdot 7$	1	$\chi_{2^2}, \chi_3, \chi_7$	(2, 2, 2, 3)	$2^2 \cdot 3 \cdot 7$	2	1
	2	$\chi_{2^2}, \chi_3, \chi_7^2$	(2, 2, 3)	$2^2 \cdot 3$	2	1
	3	$\chi_{2^2}, \chi_3, \chi_7^3$	(2, 2, 2)	$2^2 \cdot 3$	2	1
	4	$\chi_{2^2} \cdot \chi_3, \chi_7$	(2, 2, 3)	$2 \cdot 7$	1	2
	5	$\chi_{2^2} \cdot \chi_3, \chi_7^3$	(2, 2)	2	1	2
	6	$\chi_{2^2}, \chi_3 \cdot \chi_7^3, \chi_7^2$	(2, 2, 3)	2^2	1	2
	7	$\chi_{2^2}, \chi_3 \cdot \chi_7^3$	(2, 2)	2^2	1	2
	8	$\chi_{2^2} \cdot \chi_7^3, \chi_3, \chi_7^2$	(2, 2, 3)	$2 \cdot 3$	1	2
	9	$\chi_{2^2} \cdot \chi_7^3, \chi_3$	(2, 2)	$2 \cdot 3$	1	2
	10	$\chi_{2^2} \cdot \chi_3 \cdot \chi_7^3, \chi_7^2$	$(2, 3) = 6$	2	1	2^2
	11	$\chi_{2^2} \cdot \chi_3 \cdot \chi_7^3$	2	2	1	2^2
$85 = 5 \cdot 17$	1	χ_5, χ_{17}	$(2^2, 2^4)$	$2 \cdot 5 \cdot 17$	2	$5 \cdot 17 \cdot 73$
	2	χ_5, χ_{17}^2	$(2^2, 2^3)$	$2 \cdot 5$	1	$5 \cdot 73$
	3	χ_5, χ_{17}^4	$(2^2, 2^2)$	$2 \cdot 5$	1	5
	4	χ_5, χ_{17}^8	$(2^2, 2)$	$2 \cdot 5$	1	1
	5	χ_5^2, χ_{17}	$(2, 2^4)$	$2 \cdot 17$	1	17
	6	$\chi_5^2, \chi_5 \cdot \chi_{17}^2$	$(2, 2^3)$	2	1	$2 \cdot 73$
	7	$\chi_5^2, \chi_5 \cdot \chi_{17}^4$	$(2, 2^2)$	2	1	$2 \cdot 5$
	8	$\chi_5 \cdot \chi_{17}^2$	2^3	2	1	2
	8′	$\chi_5 \cdot \overline{\chi}_{17}^2$	2^3	2	1	$2 \cdot 73$
	9	$\chi_5 \cdot \chi_{17}^4$	2^2	2	1	2
	9′	$\chi_5 \cdot \overline{\chi}_{17}^4$	2^2	2	1	$2 \cdot 5$
	10	$\chi_5 \cdot \chi_{17}^8$	2^2	2	1	2
	11	$\chi_5^2 \cdot \chi_{17}$	2^4	2	1	$2 \cdot 17$

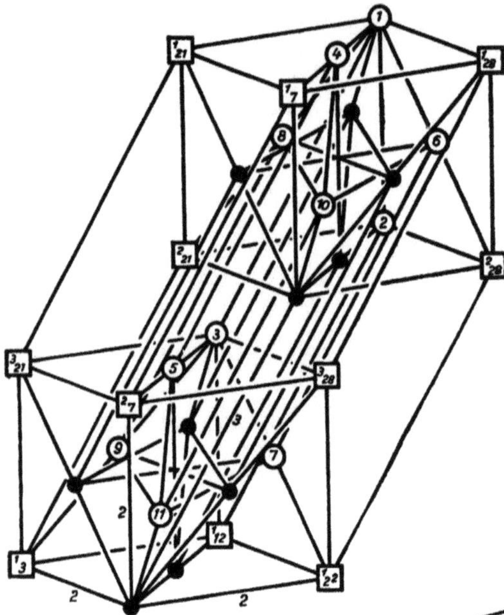

f=84

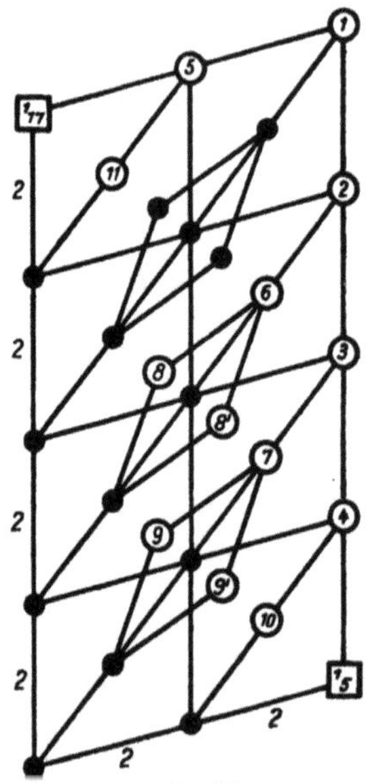

f=85

1	2	3	4	5	6	7
Conductor	No.	Gen. characters	Type	w	Q	h^*
$87 = 3 \cdot 29$	1	χ_3, χ_{29}	$(2, 2^2, 7)$	$2 \cdot 3 \cdot 29$	2	$2^9 \cdot 3$
	2	χ_3, χ_{29}^2	$(2, 2, 7)$	$2 \cdot 3$	1	$2^6 \cdot 3$
	3	χ_3, χ_{29}^4	$(2, 7) = 14$	$2 \cdot 3$	1	2^3
	4	χ_3, χ_{29}^7	$(2, 2^2)$	$2 \cdot 3$	2	3
	5	χ_3, χ_{29}^{14}	$(2, 2)$	$2 \cdot 3$	1	3
	6	$\chi_3 \cdot \chi_{29}^{14}, \chi_{29}^4$	$(2, 7) = 14$	2	1	$2^4 \cdot 3$
	7	$\chi_3 \cdot \chi_{29}^{14}$	2	2	1	$2 \cdot 3$
$88 = 2^3 \cdot 11$	1	$\chi_{2^2}, \psi_{2^3}; \chi_{11}$	$(2, 2, 2, 5)$	$2^3 \cdot 11$	2	$5 \cdot 11$
	2	$\chi_{2^2}, \psi_{2^3}; \chi_{11}^2$	$(2, 2, 5)$	2^3	1	5
	3	$\chi_{2^2}, \psi_{2^3}; \chi_{11}^5$	$(2, 2, 2)$	2^3	2	1
	4	ψ_{2^3}, χ_{11}	$(2, 2, 5)$	$2 \cdot 11$	1	11
	5	ψ_{2^3}, χ_{11}^5	$(2, 2)$	2	1	1
	6	$\chi_{2^2}\psi_{2^3}, \chi_{11}$	$(2, 2, 5)$	$2 \cdot 11$	2	5
	7	$\chi_{2^2}\psi_{2^3}, \chi_{11}^2$	$(2, 5) = 10$	2	1	5
	8	$\chi_{2^2}\psi_{2^3}, \chi_{11}^5$	$(2, 2)$	2	2	1
	9	$\chi_{2^2}, \psi_{2^3} \cdot \chi_{11}^5, \chi_{11}^2$	$(2, 2, 5)$	2^2	2	$2 \cdot 11$
	10	$\chi_{2^2}, \psi_{2^3} \cdot \chi_{11}^5$	$(2, 2)$	2^2	2	2
	11	$\chi_{2^2}\psi_{2^3}, \psi_{2^3} \cdot \chi_{11}^5, \chi_{11}^2$	$(2, 2, 5)$	2	2	$2 \cdot 5 \cdot 11$
	12	$\chi_{2^2}\psi_{2^3}, \psi_{2^3} \cdot \chi_{11}^5$	$(2, 2)$	2	2	2
	13	$\psi_{2^3} \cdot \chi_{11}^5, \chi_{11}^2$	$(2, 5) = 10$	2	1	$2 \cdot 11$
	14	$\psi_{2^3} \cdot \chi_{11}^5$	2	2	1	2

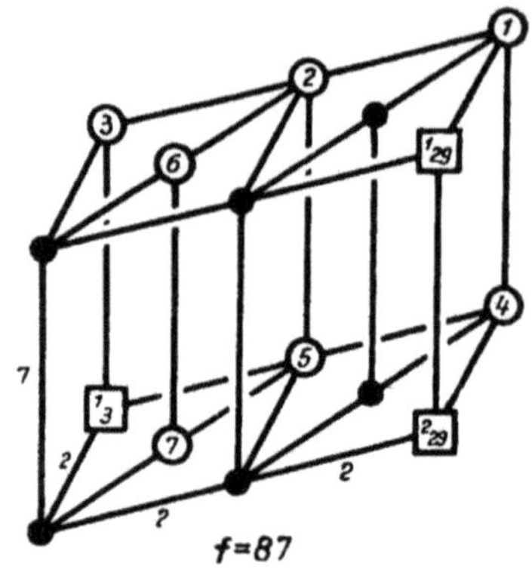

f=87

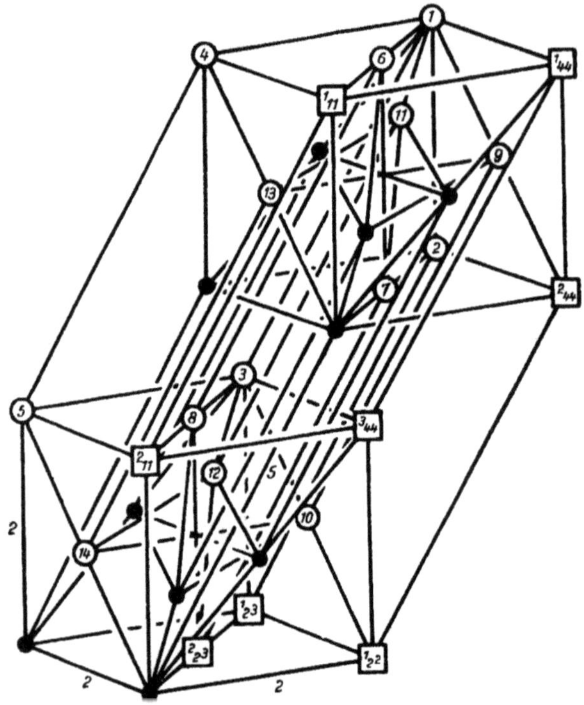

f=88

1	2	3	4	5	6	7
Conductor	No.	Gen. characters	Type	w	Q	h^*
$91 = 7 \cdot 13$	1	χ_7, χ_{13}	$(2, 3, 2^2, 3)$	$2 \cdot 7 \cdot 13$	2	$2^4 \cdot 5 \cdot 13 \cdot 37$
	2	χ_7, χ_{13}^2	$(2, 3, 2, 3)$	$2 \cdot 7$	1	$2^2 \cdot 7 \cdot 13$
	3	χ_7, χ_{13}^3	$(2, 3, 2^2)$	$2 \cdot 7$	2	2^4
	4	χ_7, χ_{13}^4	$(2, 3, 3)$	$2 \cdot 7$	1	13
	5	χ_7, χ_{13}^6	$(2, 3, 2)$	$2 \cdot 7$	1	2^2
	6	χ_7^2, χ_{13}	$(3, 2^2, 3)$	$2 \cdot 13$	1	$2^2 \cdot 37$
	7	χ_7^2, χ_{13}^3	$(3, 2^2) = 12$	2	1	2^2
	8	χ_7^3, χ_{13}	$(2, 2^2, 3)$	$2 \cdot 13$	2	7
	9	χ_7^3, χ_{13}^2	$(2, 2, 3)$	2	1	7
	10	χ_7^3, χ_{13}^3	$(2, 2^2)$	2	2	1
	11	χ_7^3, χ_{13}^4	$(2, 3) = 6$	2	1	1
	12	χ_7^3, χ_{13}^6	$(2, 2)$	2	1	1
	13	$\chi_7^2, \chi_7^3 \cdot \chi_{13}^6, \chi_{13}^4$	$(3, 2, 3)$	2	1	$2^3 \cdot 7$
	14	$\chi_7^2, \chi_7^3 \cdot \chi_{13}^6$	$(3, 2) = 6$	2	1	2^3
	15	$\chi_7^3, \chi_7^2 \cdot \chi_{13}^4, \chi_{13}^3$	$(2, 3, 2^2)$	2	2	13
	15′	$\chi_7^3, \chi_7^2 \cdot \overline{\chi}_{13}^4, \chi_{13}^3$	$(2, 3, 2^2)$	2	2	37
	16	$\chi_7^3, \chi_7^2 \cdot \chi_{13}^4, \chi_{13}^6$	$(2, 3, 2)$	2	1	13
	16′	$\chi_7^3, \chi_7^2 \cdot \overline{\chi}_{13}^4, \chi_{13}^6$	$(2, 3, 2)$	2	1	1
	17	$\chi_7^3, \chi_7^2 \cdot \chi_{13}^4$	$(2, 3) = 6$	2	1	13
	17′	$\chi_7^3, \chi_7^2 \cdot \overline{\chi}_{13}^4$	$(2, 3) = 6$	2	1	1
	18	$\chi_7^2 \cdot \chi_{13}^4, \chi_{13}^3$	$(3, 2^2) = 12$	2	1	1
	18′	$\chi_7^2 \cdot \overline{\chi}_{13}^4, \chi_{13}^3$	$(3, 2^2) = 12$	2	1	37
	19	$\chi_7^3 \cdot \chi_{13}^6, \chi_{13}^4$	$(2, 3) = 6$	2	1	$2 \cdot 7$
	20	$\chi_7^3 \cdot \chi_{13}^6, \chi_7^2 \cdot \chi_{13}^4$	$(2, 3) = 6$	2	1	2
	20′	$\chi_7^3 \cdot \chi_{13}^6, \chi_7^2 \cdot \overline{\chi}_{13}^4$	$(2, 3) = 6$	2	1	2
	21	$\chi_7^3 \cdot \chi_{13}^6$	2	2	1	2

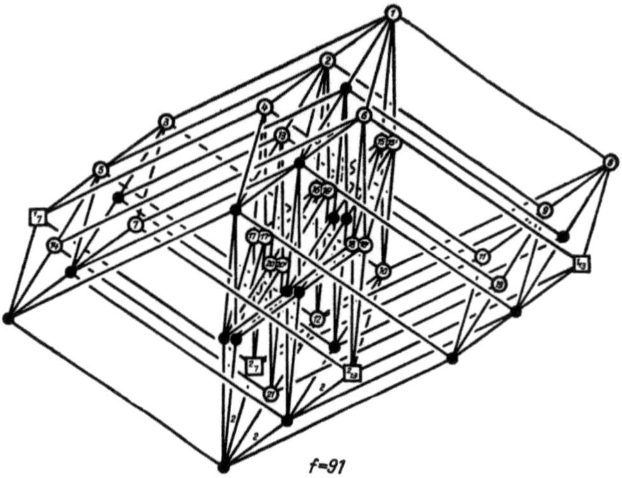

f=91

1	2	3	4	5	6	7
Conductor	No.	Gen. characters	Type	w	Q	h^*
$92 = 2^2 \cdot 23$	1	χ_{2^2}, χ_{23}	(2, 2, 11)	$2^2 \cdot 23$	2	$3 \cdot 67$
	2	χ_{2^2}, χ_{23}^2	$(2, 11) = 22$	2^2	1	67
	3	$\chi_{2^2}, \chi_{23}^{11}$	(2, 2)	2^2	2	3
$93 = 3 \cdot 31$	1	χ_3, χ_{31}	(2, 2, 3, 5)	$2 \cdot 3 \cdot 31$	2	$3^2 \cdot 5 \cdot 151$
	2	χ_3, χ_{31}^2	$(2, 3, 5) = 30$	$2 \cdot 3$	1	$5 \cdot 151$
	3	χ_3, χ_{31}^3	(2, 2, 5)	$2 \cdot 3$	2	$3 \cdot 5$
	4	χ_3, χ_{31}^5	(2, 2, 3)	$2 \cdot 3$	2	3^2
	5	χ_3, χ_{11}^6	$(3, 5) = 10$	$2 \cdot 3$	1	5
	6	χ_3, χ_{11}^{10}	$(2, 3) = 6$	$2 \cdot 3$	1	1
	7	χ_3, χ_{11}^{15}	(2, 2)	$2 \cdot 3$	2	3
$95 = 5 \cdot 19$	1	χ_5, χ_{19}	$(2^2, 2, 3^2)$	$2 \cdot 5 \cdot 19$	2	$2^2 \cdot 13 \cdot 19 \cdot 109$
	2	χ_5, χ_{19}^2	$(2^2, 3^2) = 36$	$2 \cdot 5$	1	$13 \cdot 109$
	3	χ_5, χ_{19}^3	$(2^2, 2, 3)$	$2 \cdot 5$	2	$2^2 \cdot 13$
	4	χ_5, χ_{19}^6	$(2^2, 3) = 12$	$2 \cdot 5$	1	13
	5	χ_5, χ_{19}^9	$(2^2, 2)$	$2 \cdot 5$	2	2^2
	6	χ_5^2, χ_{19}	$(2, 2, 3^2)$	$2 \cdot 19$	1	$2^2 \cdot 19$
	7	χ_5^2, χ_{19}^3	(2, 2, 3)	2	1	2^2
	8	χ_5^2, χ_{19}^9	(2, 2)	2	1	2^2
	9	$\chi_5^2 \cdot \chi_{19}^9, \chi_{19}^2$	$(2, 3^2) = 18$	2	1	$2^3 \cdot 19$
	10	$\chi_5^2 \cdot \chi_{19}^9, \chi_{19}^6$	$(2, 3) = 6$	2	1	2^3
	11	$\chi_5^2 \cdot \chi_{19}^9$	2	2	1	2^3

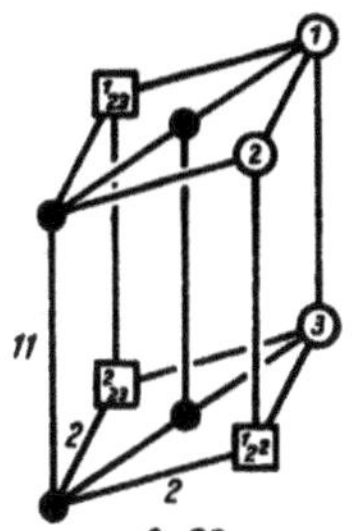

f=92

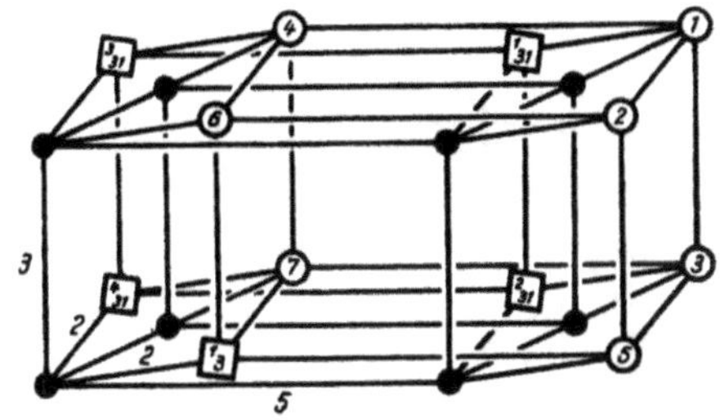

f=93

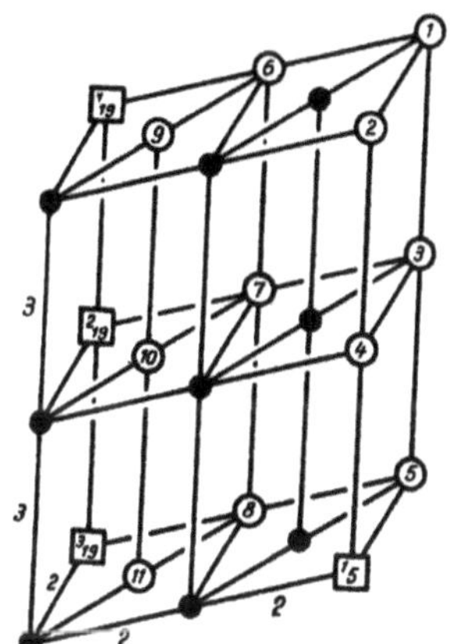

f=95

1	2	3	4	5	6	7
Conductor	No.	Gen. characters	Type	w	Q	h^*
$96 = 2^5 \cdot 3$	1	$\chi_{2^2}, \psi_{2^5}; \chi_3$	$(2, 2^3, 2)$	$2^5 \cdot 3$	2	3^2
	2	ψ_{2^5}, χ_3	$(2^3, 2)$	$2 \cdot 3$	1	3^2
	3	$\chi_{2^2}\psi_{2^5}, \chi_3$	$(2^3, 2)$	$2 \cdot 3$	2	1
	4	$\chi_{2^2}, \psi_{2^5} \cdot \chi_3$	$(2, 2^3)$	2^4	2	$2 \cdot 3^2$
	5	$\chi_{2^2}\psi_{2^5}, \chi_{2^2} \cdot \chi_3$	$(2^3, 2)$	2	2	$2 \cdot 3^2$
	6	$\psi_{2^5} \cdot \chi_3$	2^3	2	1	$2 \cdot 3^2$
$99 = 3^2 \cdot 11$	1	$\chi_3, \psi_{3^2}; \chi_{11}$	$(2, 3, 2, 5)$	$2 \cdot 3^2 \cdot 11$	2	$3 \cdot 31^2$
	2	$\chi_3, \psi_{3^2}; \chi_{11}^2$	$(2, 3, 5) = 30$	$2 \cdot 3^2$	1	31
	3	$\chi_3, \psi_{3^2}; \chi_{11}^5$	$(2, 3, 2)$	$2 \cdot 3^2$	2	3
	4	ψ_{3^2}, χ_{11}	$(3, 2, 5) = 30$	$2 \cdot 11$	1	$3 \cdot 31$
	5	ψ_{3^2}, χ_{11}^5	$(3, 2) = 6$	2	1	3
$100 = 2^2 \cdot 5^2$	1	$\chi_{2^2}; \chi_5, \psi_{5^2}$	$(2, 2^2, 5)$	$2^2 \cdot 5^2$	2	$5 \cdot 11$
	2	$\chi_{2^2}; \chi_5^2, \psi_{5^2}$	$(2, 2, 5)$	2^2	1	$5 \cdot 11$
	3	χ_{2^2}, ψ_{5^2}	$(2, 5) = 10$	2^2	1	5
	4	$\chi_{2^2} \cdot \chi_5^2, \psi_{5^2}$	$(2, 5) = 10$	2	1	$2 \cdot 11$

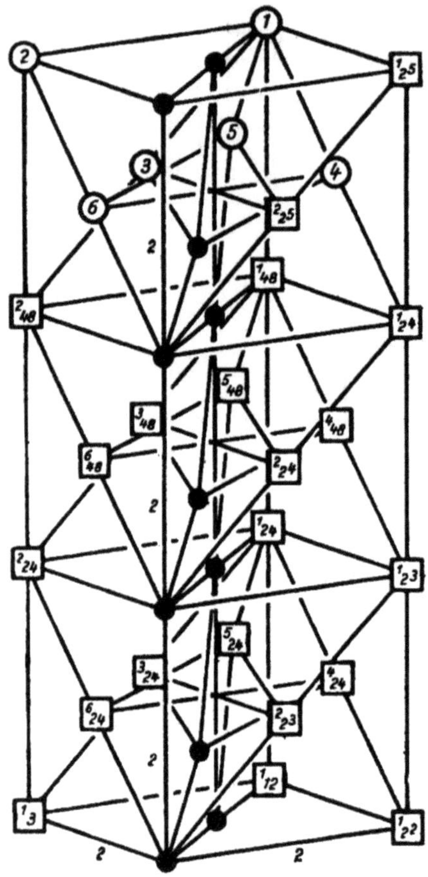

f=96

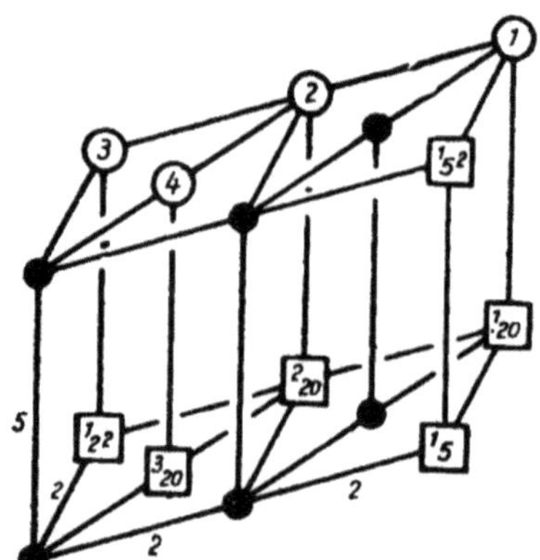

f=100

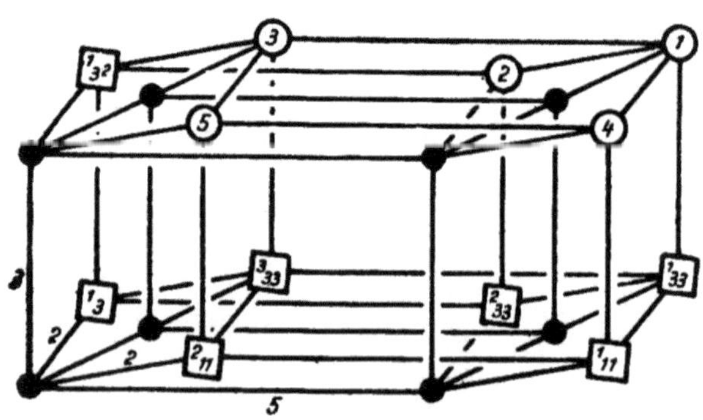

f=99

Auxiliary table: values of the unit indices[179]

f	No.	Q	Theorem
12	1	2	3.27
15	1	2	3.27
	2	1	3.25
	3	1	3.24
20	1	2	3.27
	2	1	3.25
	3	1	3.24
21	1	2	3.27
	2	1	3.24
	3	2	3.26
24	1	2	3.27
	2	1	3.25
	3	2	3.26
	4	2	3.26
	5	2	3.26
	6	1	3.24
28	1	2	3.27
	2	1	3.24
	3	2	3.26
33	1	2	3.27
	2	1	3.24
	3	2	3.26
35	1	2	3.27
	2	1	3.24
	3	2	3.26
	4	1	3.22
	5	1	3.25 [3.29]
	6	1	3.24
	7	1	3.24
36	1	2	3.27
	2	1	3.24
	1	2	3.27
	2	1	3.22
	3	2	3.26

f	No.	Q	Theorem
39	4	1	3.24
	5	1	3.25 [3.29]
	6	1	3.24
	7	1	3.24
40	1	2	3.27
	2	1	3.22
	3	1	3.22
	4	2	3.26
	5	1	3.25 [3.29]
	6	2	3.26
	7	1	3.25 [3.29]
	8	1	3.25, 3.29
	9	2	3.26
	10	1	3.25 [3.29]
	11	1	3.24
	12	1	3.24
44	1	2	3.27
	2	1	3.24
	3	2	3.26
45	1	2	3.27
	2	1	3.22
	3	1	3.24
	4	1	3.24
48	1	2	3.27
	2	1	3.25, 2.6
	3	2	3.26
	4	2	3.26
	5	2	3.26
	6	1	3.24
51	1	2	3.27
	2	1	3.25, 2.7
	3	1	3.29 [*3.22]
	4	1	3.29 [*3.22]
	5	1	3.24
	6	1	3.24
	7	1	3.24

f	No.	Q	Theorem
52	1	2	3.27
	2	1	3.22
	3	2	3.26
	4	1	3.24
	5	1	3.25 [3.29]
	6	1	3.24
	7	1	3.24
55	1	2	3.27
	2	1	3.22
	3	2	3.26
	4	1	3.22
	5	1	3.25 [3.29]
	6	1	3.24
	7	1	3.24
56	1	2	3.27
	2	1	3.22
	3	2	3.26 [3.29]
	4	1	3.22
	5	1	3.25, 3.29
	6	2	3.26 [3.29]
	7	1	3.24
	8	2	3.26 [Sect. 3.8,3.15]
	9	2	3.24
	10	2	3.26 [Sect. 3.8,3.15]
	11	2	3.26 [3.29]
	12	2	3.26 [Sect. 3.8,3.15]
	13	1	3.24
	14	1	3.24
57	1	2	3.27
	2	1	3.24
	3	2	3.26
	4	1	3.24
	5	2	3.26

[179](Translator's remark) The number in bold found in brackets [ ] indicates the number of Theorem by which we also determine the unit index. The inclusion of $*$ in the brackets [ ] means that it has been added by the translator, because Theorem 29 is not correct in general. For the fields No. 8 and 11 with conductor 80, see Theorem 2 in M. Hirabayashi and K. Yoshino, Remarks on unit indices imaginary abelian number fields, *Manuscripta Math.* **60**, 423–436.

f	No.	Q	Theorem
60	1	2	3.27
	2	2	3.27, 3.29 [Sect. 3.15]
	3	1	3.22
	4	1	3.22
	5	1	3.22
	6	1	3.22
	7	1	3.22
	8	1	3.24
	9	1	3.22
63	1	2	3.27
	2	1	3.22
	3	2	3.26 [3.29]
	4, 4′	2	3.26 [3.29]
	5, 5′	1	3.24
	6	1	3.24
	7	1	3.24
	8, 8′	1	3.24
65	1	2	3.27
	2	1	3.22
	3	2	3.26
	4	1	3.24
	5	1	3.29 [*3.22]
	6	1	3.22
	7	1	3.29 [*3.22]
	8	1	3.24
	9	1	3.24
	10	1	3.24
	11	1	3.24
68	1	2	3.27
	2	1	3.25, 2.7
	3	1	3.29 [*3.22]
	4	1	3.29 [*3.22]
	5	1	3.24
	6	1	3.24
	7	1	3.24
69	1	2	3.27
	2	1	3.24
	3	2	3.26
72	1	2	3.27
	2	1	3.22
	3	1	3.22
	4	2	3.26 [3.29]
	5	2	3.26 [3.29]
	6	2	3.26 [3.29]
	7	1	3.24
	8	1	3.24
75	1	2	3.27
	2	1	3.22
	3	1	3.24
	4	1	3.24
76	1	2	3.27
	2	1	3.24
	3	2	3.26
	4	1	3.24
	5	2	3.26
77	1	2	3.27
	2	1	3.24
	3	2	3.26
	4	1	3.24
	5	1	3.24
	6	2	3.26
	7	1	3.24
	8	2	3.26
80	1	2	3.27
	2	1	3.22
	3	1	3.22
	4	2	3.26
	5	1	3.29 [*3.22]
	6	2	3.26
	7, 7′	2	3.26
	8	1	3.29 [*3.25]
	9	1	3.29 [*3.22]
	10	2	3.26
	11	1	3.29 [*3.25]
	12	1	3.29 [*3.22]
	13, 13′	2	3.26
	14, 14′	1	3.24
	15	1	3.24
84	1	2	3.27
	2	2	3.27, 3.29 [Sect. 3.15]
	3	2	3.26
	4	1	3.22
	5	1	3.29 [*3.22]
	6	1	3.22
	7	1	3.29 [*3.22]
	8	1	3.22
	9	1	3.29 [*3.22]
	10	1	3.24
	11	1	3.24

f	No.	Q	Theorem
85	1	2	3.27
	2	1	3.22
	3	1	3.29 [*3.22]
	4	1	3.29 [*3.22]
	5	1	3.22
	6	1	3.29 [*3.22]
	7	1	3.29 [*3.22]
	8, 8′	1	3.24
	9, 9′	1	3.24
	10	1	3.24
	11	1	3.24
87	1	2	3.27
	2	1	3.22
	3	1	3.24
	4	2	3.26
	5	1	3.25 [3.29]
	6	1	3.24
	7	1	3.24
88	1	2	3.27
	2	1	3.22
	3	2	3.26
	4	1	3.22
	5	1	3.25 [3.29]
	6	2	3.26
	7	1	3.24
	8	2	3.26
	9	2	3.26
	10	2	3.26
	11	2	3.26
	12	2	3.26
	13	1	3.24
	14	1	3.24

f	No.	Q	Theorem
91	1	2	3.27
	2	1	3.22
	3	2	3.26
	4	1	3.29 [*3.22]
	5	1	3.29 [*3.22]
	6	1	3.22
	7	1	3.24
	8	2	3.26
	9	1	3.29 [*3.22]
	10	2	3.26
	11	1	3.24
	12	1	3.25 [3.29]
	13	1	3.29 [*3.22]
	14	1	3.24
	15, 15′	2	3.26
	16, 16′	1	3.29 [*3.22]
	17, 17′	1	3.24
	18, 18′	1	3.24
	19	1	3.24
	20, 20′	1	3.24
	21	1	3.24
92	1	2	3.27
	2	1	3.24
	3	2	3.26
93	1	2	3.27
	2	1	3.22
	3	2	3.26
	4	2	3.26
	5	1	3.24
	6	1	3.24
	7	2	3.26
95	1	2	3.27
	2	1	3.24
	3	2	3.26
	4	1	3.24
	5	2	3.26
	6	1	3.22
	7	1	3.29 [*3.22]
	8	1	3.25 [3.29]
	9	1	3.24
	10	1	3.24
	11	1	3.24

f	No.	Q	Theorem
96	1	2	3.27
	2	1	3.25, 2.6
	3	2	3.26
	4	2	3.26
	5	2	3.26
	6	1	3.24
99	1	2	3.27
	2	1	3.24
	3	2	3.26 [3.29]
	4	1	3.24
	5	1	3.24

f	No.	Q	Theorem
100	1	2	3.27
	2	1	3.22
	3	1	3.24
	4	1	3.24

References[180]

1. E.J. Amberg, Über den Körper, dessen Zahlen sich rational aus zwei Quadratwurzeln zusammensetzen. Diss. Zürich, 1897 [Introduction, **3.8**]
2. E. Artin, Idealklassen in Oberköpern und allgemeines Reziprozitätsgesetz. Abhandl. Math. Sem. Hamburg **7** (1930) [**3.2**]
3. P. Bachmann, Zur Theorie der komplexen Zahlen. J. f. d. reine und angew. Math. **67** (1867) [Introduction, **3.8**]
4. G.L. Dirichlet, Recherches sur les formes quadratilques à coefficients et à indéterminées complexes. J. f. d. reine und angew. Math. **24** (1842) = Collected Works **1**, 533–618 [Introduction, **3.8**]
5. H. Hasse, Bericht über neuere Untersuchungen und Probleme aus der Theroie der algbraichen Zahlkörper. Part I: Klassenörpertheorie. Jahresbericht D.M.-V. **35** (1926); H. Hasse, Bericht über neuere Untersuchungen und Probleme aus der Theroie der algbraichen Zahlkörper. Part Ia: Beweise zu Teil I. Jahresbericht D.M.-V. **36** (1927); H. Hasse, Bericht über neuere Untersuchungen und Probleme aus der Theroie der algbraichen Zahlkörper. Part II: Reziprozitätsgesetz. Jahresbericht D.M.-V. Supplemental Ed. **6** (1930). Cited as "Klassenkörperbericht." [ntroduction, **1.2**, **1.3**, **2.3**, **2.12**, **3.1**, **3.2**, **3.4**, **3.5**, **3.9**, **3.19**]
6. H. Hasse, Aufgabe 327. Jahresbericht D.M.-V. **53** (1943) [**3.12**]
7. H. Hasse, *Zahlentheorie* (Akademie-Verlag, Berlin, 1949) [Introduction, **2.9**, **3.1**, **3.2**, **3.5**, **3.9**]
8. G. Herglotz, Über einen Dirichletschen Satz. Math. Z. **12** (1922) [Introduction, **3.8**]
9. D. Hilbert, Über den Dirichletschen biquadratischen Zahlkörper. Math. Ann. **45** (1894) [Introduction, **3.8**]
10. D. Hilbert, Die Theorie der algebraischen Zahlkörper. Jahresbericht D.M.-V. **4** (1897). Cited as "Zahlbericht." [Preface, Introduction, **1.2**, **1.6**, **2.9**, **3.2**, **3.5**, **3.8**, **3.19**]
11. C.G. Jacobi, *Canon Arithmeticus* (Akademie-Verlag, Berlin, 1839) [**3.16**, Appendix]
12. L. Kronecker, Bemerkung über dek Klassenzahl der aus Einheitswurzeln gebildeten komplexen Zahlen. Monatsber. Akad. d. Wissensch. Berlin (1863) = Collected Works **1**, 123–131 [Introduction, **1.6**, **3.7**]

[180]The bold-typed numbers attached in the square brackets denote the sections of this book in which the individual works are cited.

13. L. Kronecker, Auseinandersetzung einer Einschaften der Klassenzahl idealer komplexer Zahlen. Monatsber. Akad. d. Wissensch. Berlin (1870) = Collected Works **1**, 271–282 [Introduction, **1.6**, **3.1**]
14. E. Kummer, Bestimmung der Anzahl nicht-äquivalenter Klassen für die aus λ-ten Wurzeln der Einheit gebildeten komplexen Zahlen und die idealen Faktoren derselben. J. f. d. reine und angew. Math. **40** (1850) [Introduction, **1.6**, **3.20**]
15. E. Kummer, Zwei besondere Untersuchungen über die Klassenzahl und über die Einheiten der aus λ-ten Wurzeln der Einheit gebildeten komplexen Zahlen. J. f. d. reine und angew. Math. **40** (1850) [Introduction, **3.19**]
16. E. Kummer, Allgemeiner Beweis des Fermatschen Satzes, daß die Gleichung $x^\lambda + y^\lambda = z^\lambda$ durch ganze Zahlen unlösbar ist, für alle diejenigen Potenzexponenten λ, welche ungerade Primzahlen sind und in den Zählern der ersten $\frac{1}{2}(\lambda - 3)$ Bernoullischen Zahlen als Faktoren nicht vorkommen. J. f. d. reine und angew. Math. **40** (1850) [**3.19**]
17. E. Kummer, Sur la théorie des nombres complexes composés de racines de l'unité et de nombres entiers. J. de Math. **16** (1851) (French summary laying the foundation of the theory of ideal numbers from. J. f. d. reine und angew. Math. **35** (1847), as well as three works cited above) [Introduction, **1.6**, **3.19**, **3.20**, Appendix]
18. E. Kummer, Über die Irregularität der Determinanten. Monatsber. Akad. d. Wissensch. Berlin (1853) (Extract from the letter to Dirichlet.) [Introduction, **3.1**, **3.15**]
19. E. Kummer, Über die Klassenanzahl der aus n-ten Einheitenwurzeln gebildeten komplexen Zahlen. Monatsber. Akad. d. Wissensch. Berlin (1861) [Introduction, **3.16**]
20. E. Kummer, Über die Klassenanzahl der aus zusammengesetzten Einheitenwurzeln gebildeten idealen komplexen Zahlen. Monatsber. Akad. d. Wissensch. Berlin (1863) [Introduction, **1.6**, **3.10**, **3.12**, **3.15**, **3.20**, Appendix]
21. E. Kummer, Über eine Eigenschaft der Einheiten der aus den Wurzeln der Gleichung $\alpha^\lambda = 1$ gebildeten komplexen Zahlen und über den zweiten Faktor der Klassenzahl. Monatsber. Akad. d. Wissensch. Berlin (1870) [Introduction, **3.19**, **3.20**]
22. C.G. Reuschle, *Tafeln komplexer Primzahlen, whelche aus Wurzeln der Einheit gebildet sind.* Berlin (1875) [Preface, Appendix]
23. J. Sommer, *Vorlesungen über Zahlentheorie. Einführung in die Theorie der algebraischen Zahlkörpern.* (Druck Und Verlag, Leipzig-Berlin, 1907) [Preface, **3.8**]
24. H. Weber, Theorie der Abelschen Zahlkörper. Acta Math. **8** (1886) [Introduction, **2.1**, **2.6**, **2.10**, **3.16**]
25. H. Weber, *Lehrbuch der Algebra*, vol. 2, 2nd edn. (Wentworth Press, Braunschweig, 1899), pp. 219–223 [Introduction, **2.1**, **2.6**, **2.10**, **3.16**]
26. J.M. Winogradow, *Grundzüge der Zahlentheorie*, 5th ed. (Moskow-Leningrad, 1949), p. 38, 75 [**3.11**]

Part II

Chapter 4
On the Relative Class Number of the Imaginary Abelian Number Field I

© Springer Nature Switzerland AG 2019

H. Hasse, *On the Class Number of Abelian Number Fields*,
https://doi.org/10.1007/978-3-030-01512-1_4

On the Relative Class Number of the Imaginary Abelian Number Field I

Ken-ichi Yoshino* and Mikihito Hirabayashi**

Abstract In Section 1 we show that the relative class number of any imaginary abelian field is expressed as a product of some unit indices and determinants and their inverses. In Section 2, we determine the relative class number of all imaginary abelian number fields with conductor f, $100 < f \leqq 150$, and tabulate them with Hasse diagrams of such fields.

Introduction

In his monograph [5] Hasse had acquired class number formulae of the abelian number fields over $\boldsymbol{Q}$, the field of rational numbers, and investigated the arithmetical meanings of those formulae. Moreover in its appendix he gave the complete list of the relative class number h^*, the unit index Q, the number w of roots of unity, the generating characters and the diagrams for the lattices of subfields of the abelian number fields whose conductor f does not exceed 100.

Historically, for the f-th cyclotomic field $\boldsymbol{Q}(\zeta_f)$, Kummer had obtained the following formulae (1), (2) and had calculated the relative class number $h^*=h^*(f)$ for all $f<100$ and all primes $f=p$, $100<f\leqq 163$. (Kummer [1], [2], [3]) After Kummer, Hasse made a grand contribution to this theory. Subsequently Schrutka v. Rechtenstamm [8] followed this method and computed the relative class number of the f-th cyclotomic field $\boldsymbol{Q}(\zeta_f)$, $\varphi(f)<256$, where φ is the Euler's φ-function. Recently the computation of the value $h^*(p)$ are carried out for all primes p, $p<521$. (Newman [6], Lehmer and Masley [7])

Now, we have intended to extend Hasse's table. In this paper we give some elementary formulae of the relative class number in § 1 and determine the relative class numbers of all abelian number fields with conductor f, $100 < f \leqq 150$, in § 2. We note that our formulae are more elementary than Hasse's method.

We express our deep gratitude to Professor M. Suzuki and Mr. J. Kose who gave us a lot of valuable suggestions and much aid on the occasion of calculating determinants

---

* Department of mathematics
** Kanazawa Institute of Technology

by the computer.

§ 1. Relative Class Number Formulae

The class number h of an abelian number field K of finite degree over $\boldsymbol{Q}$ is decomposed as follows;

(1) $$h = h_o h^*$$

where h_o is the class number of the maximal real subfield K_o of K. This number h^* is an integer (Hasse [55] p.49, Satz 10) and is given by the formula

(2) $$h^* = Qw \prod_{\chi_1} \frac{1}{2f(\chi_1)} \sum_{\substack{x \bmod f(\chi_1) \\ 0<x<f(\chi_1)}} (-\chi_1(x)x)$$

where Q is the unit index of K,w the number of roots of unity in K, $f(\chi)$ the conductor of χ and χ_1 runs through all odd characters of K. (Hasse [5] p. 12) h^* and h_0 are called the first and second factor of the class number h of an abelian number field K respectively. h^* is also called the relative class number of K.

From now on, let f be a positive integer such that $f \not\equiv 2 \bmod 4$.

Let K be an imaginary cyclic field of degree $2a$ with conductor f, χ the generating character of K and ε a primitive $2a$-th root of unity. Further let $A_i = \sum_{\substack{\chi(x)=\varepsilon^i \\ 0<x<f(\chi)}} x$ where the summation is over all representatives x mod f of the irreducible residue class group mod f such that $\chi(x) = \varepsilon^i$. And let

$$F(X) = \sum_{i=0}^{2a-1} A_i X^i$$

Defining the symmetric matrix

$$M = (A_{i+j} - A_{a+i+j})_{0 \leq i, j < a},$$

the same method as Borevich and Shafarevich's book ([4] p.393) shows the following proposition.

Proposition 1. *Let K be an imaginary cyclic field over* $\boldsymbol{Q}$ *of degree $2a$ with conductor f, χ the generating character of K. Suppose that $f(\chi^{2j+1}) = f$ for every j, $j = 0, 1, \cdots, a-1$. Then*

(3) $$h^* = Qw \frac{1}{(2f)^a} |\det M|$$

Secondly assume that there exists an integer c such that $f(\chi^{2c+1}) \neq f$ and $f(\chi^{2j+1}) = f$ for any j, $j = 0, 1, \cdots, c-1$. We let now k denote the imaginary cyclic subfield of K with generating character χ^{2c+1} of degree $2b$. We split the product on the right side of (2) into two parts; that is, χ_2 belongs to k and χ_3 does not. Then

$$h_k^* = Q_k w_k \prod_{\chi_2} \frac{1}{2f(\chi_2)} \sum_{\substack{x \bmod f(\chi_2) \\ 0<x<f(\chi_2)}} (-\chi_2(x)x)$$

and so we have

$$h_K^* = \frac{Q_K w_K}{Q_k w_k} h_k^* \prod_{\chi_3} \frac{1}{2f(\chi_3)} \sum_{\substack{x \bmod f(\chi_3) \\ 0<x<f(\chi_3)}} (-\chi_3(x)x).$$

Therefore

$$(4) \qquad h_K^* = \frac{Q_K w_K}{Q_k w_k} \frac{h_k^*}{(2f)^{a-b}} (-1)^{a-b} \prod_{2c+1 \nmid 2j+1} F(\varepsilon^{2j+1})$$

Let $B_j = \sum_{\substack{i \equiv j \bmod 2b \\ 0 \leq i < 2a}} A_i$ and we define the symmetric matrix $N = (B_{i+j} - B_{b+i+j})_{0 \leq i,j < b}$.
Thus, from (4), we have directly

Proposition 2. *Let K be an imaginary cyclic field of degree $2a$ with conductor f and χ the generating character of K. Assume that there exists a positive integer c such that $f(\chi^{2c+1}) \neq f$ and $f(\chi^{2j+1}) = f$ for any integer j, $j = 0, 1, \cdots, c-1$. Further let k be the imaginary cyclic subfield of K with generating character χ^{2c+1} and $[k : \mathbf{Q}] = 2b$. If $\det N \neq 0$, then we have*

$$(5) \qquad h_K^* = \frac{Q_K w_K}{Q_k w_k} \frac{h_k^*}{(2f)^{a-b}} \frac{|\det M|}{|\det N|}.$$

Remark 1. In the case that $\det N = 0$, we may use the above Proposition 2 by a little modification as follows. In fact, setting $\eta = \varepsilon^{2c+1}$, $\det N = 0$ implies that there exists a non-negative integer j, $0 \leq j < b$, such that $F(\eta^{2j+1}) = 0$.

First suppose $F(\eta) = 0$; i.e. $j = 0$. Putting new polynomial

$$R(X) = \sum_{j=0}^{2c} (-1)^j X^{bj} = \sum_{j=0}^{2a-1} C_j X^j,$$

it follows that $R(\varepsilon^{2j+1}) = 0$ for any non-negative integer j such that $2c+1 \nmid 2j+1$ and $R(\eta^{2j+1}) \neq 0$ for any integer j, $0 \leq j < b$. Therefore we replace $F(\varepsilon)$ by $F(\varepsilon) + \gamma R(\varepsilon)$ where γ is $\underset{0 \leq j \leq a-1}{g.c.d}(A_j - A_{a+j})$; that is, M, N, by

$$M' = (A_{i+j} + \gamma C_{i+j} - A_{a+i+j} - \gamma C_{a+i+j})_{0 \leq i,j < a},$$
$$N' = (B'_{i+j} - B'_{b+i+j})_{0 \leq i,\, j < b}$$

respectively, where $B'_j = \sum_{\substack{i \equiv j \bmod 2b \\ 0 \leq i < 2a}} (A_i + \gamma C_i)$. Then we may show easily that $\det N' \neq 0$ and

$$h_K^* = \frac{Q_K w_K}{Q_k w_k} \frac{h_k^*}{(2f)^{a-b}} \frac{|\det M'|}{|\det N'|}.$$

Secondly we consider the case that there exists a positive integer d such that $F(\eta^{2j+1}) \neq 0$ for any non-negative integer $j < d$. and $F(\eta^{2d+1}) = 0$. Put $e = a/(2c+1)(2d+1)$. In this case, we set

$$R(X)=\sum_{j=0}^{(2c+1)(2d+1)-1}(-1)^j X^{ej}=\sum_{i=0}^{2a-1} C_i X^i.$$

Then by same modification as in the former case, we have also det $N' \neq 0$.

In the sequel we shall study the imaginary abelian field which is not cyclic over $\boldsymbol{Q}$. In the first place, we prove the following proposition.

Proposition 3. *Let L be an imaginary abelian field over $\boldsymbol{Q}$ and K its real subfield such that the Galois group $G(L/K)$ is of type (2,2). Further let M_1, M_2 and M_3 be the proper intermediate fields such that M_2 is real and the others are imaginary. Then*

$$(6)\qquad h_L^* = \frac{Q_L}{Q_1 Q_3}\cdot\frac{w_L}{w_1 w_3}\cdot h_{M_1}^* h_{M_3}^*$$

where w_i is the number of roots of unity in M_i and Q_i the unit index of M_i ($i = 1, 3$).

Proof. We denote by X_M the character of $G(L/K)$ induced by the principal character of $G(L/M)$, where M is any intermediate field of L/K. Then, as usual, induced characters of the Galois group $G(L/K)$ satisfy the relation

$$X_L+2X_K=\sum_{i=1}^{3} X_{M_i},$$

so that through the theory of Artin's L-function we obtain

$$\zeta_L(s)\zeta_K^2(s)=\prod_{i=1}^{3}\zeta_{M_i}(s)$$

where $\zeta_L(s)$, $\zeta_K(s)$ and $\zeta_{M_i}(s)$ are the Dedekind zeta function of L, K and M_i respectively. Comparing the residue at $s=1$ on both sides, we have

$$\frac{h_L R_L}{w_L}\left(\frac{h_K R_K}{w_K}\right)^2=\prod_{i=1}^{3}\frac{h_{M_i}R_i}{w_i}$$

where R_L, R_K and R_i are the regulator of L, K and M_i respectively. Now it is well known that $Q_1R_1 = 2^{n-1}R_K = Q_3R_3$ and $Q_LR_L = 2^{2n-1}R_2$ where $n = [K:\boldsymbol{Q}]$. Therefore we have

$$h_L h_K^2=\frac{Q_L}{2Q_1Q_3}\cdot\frac{w_L w_K^2}{w_1 w_2 w_3}\prod_{i=1}^{3}h_{M_i}.$$

Thus in view of the definition of the relative class number, we have obtained the desired formula. q.e.d.

Remark 2. Proposition 3 remains true under the weaker conditions that (i) L, K and M_i are Galois fields over $\boldsymbol{Q}$ and (ii) K and M_2 are totally real over $\boldsymbol{Q}$ and L, M_1 and M_3 are totally imaginary over $\boldsymbol{Q}$.

Finally we state the following proposition.

Proposition 4. *Let L be an imaginary abelian field over* $\boldsymbol{Q}$ *and K its imaginary subfield such that the Galois group G(L/K) is of type (l, l), l a prime number. Then*

$$(7) \qquad h_L^* h_K^{*l} = \frac{w_L w_K^l}{\prod_M w_M} \frac{Q_L Q_K^l}{\prod_M Q_M} \prod_M h_M^*$$

where M runs through all proper intermediate fields of the extension L/K.

Proof. Among the induced characters of the Galois group $G(L/K)$ there exists a relation

$$(8) \qquad X_L + lX_K = \sum_M X_M.$$

Indeed let M_i $(i=1,2,\cdots,l+1)$ be all proper intermediate fields of the extension L/K and denote by G_i the subgroup of $G(L/K)$ which is associated to the field M_i. Then

$$G(L/K) = \{1\} \cup (\bigcup_{i=1}^{l+1} (G_i - \{1\})) \qquad \text{(disjoint)}$$

From this equation the above relation (8) is deduced. On the other hand, putting L_o, K_o and M_o the maximal real subfield of L, K and M respectively, the Galois group $G(L_o/K_o)$ is also of type (l, l) and

$$X_{L_o} + lX_{K_o} = \sum_{M_o} X_{M_0}$$

so that by means of the theory of Artin's L-function it follows that

$$\zeta_L(s)\,\zeta_K^l(s) = \prod_M \zeta_M(s),$$

$$\zeta_{L_o}(s)\,\zeta_{K_o}^l(s) = \prod_{M_o} \zeta_{M_o}(s).$$

As before we compare the residue at $s = 1$ on both sides, and then

$$\frac{h_L R_L}{w_L}\left(\frac{h_K R_K}{w_K}\right)^l = \prod_M \frac{h_M R_M}{w_M},$$

$$h_{L_o} R_{L_o} (h_{K_o} R_{K_o})^l = \prod_{M_o} h_{M_o} R_{M_o}.$$

Taking the quotient of these two equations and applying the well known fact about the unit index, we can prove this Proposition 4 completely. q.e.d.

Corollary. *If* $l \neq 2$, *then*

$$h_L^* h_K^{*l} = \frac{Q_L Q_K^l}{\prod_M Q_M} \prod_M h_M^*.$$

Proof. It is known that $w_L w_K^l / \prod_M w_M$ is an integer of some power of *2*. (Brauer

[10]) For any extension K/k with $[K:k]=l \neq 2$, $2^{\omega} \| w_K$ implies $2^{\omega} \| w_k$, so we have $w_L w_K^l / \prod_M w_M = 1$.

Remark 3. Proposition 4 also remains true under the weaker conditions

(i) L, L_0, K and any proper intermediate field M are Galois fields over $\boldsymbol{Q}$ and

(ii) L, K and M are totally imaginary over $\boldsymbol{Q}$.

Combining Proposition 3 with Proposition 4, we may deal with any imaginary abelian field which is not cyclic over $\boldsymbol{Q}$. Thus in view of Proposition 1 and 2 we obtain the

Theorem. *The relative class number of any imaginary abelian field over* $\boldsymbol{Q}$ *can be expressed as a product of some unit indices and determinants attached to its imaginary cyclic subfields and their inverses.*

§ 2. Table of Unit Indices and Relative Class Numbers

Hasse had used the following formula (9) in order to calculate the value h^*

$$(9) \qquad h^* = Qw \prod_{\psi} N_{\psi}(\Theta(\psi))$$

where the character ψ runs through all representatives of the Frobenius partition of all odd characters of K,

$$\Theta(\psi) = \frac{1}{2f(\psi)} \sum_{\substack{x \bmod f(\psi) \\ 0<x<f(\psi)}} (-\psi(x)x)$$

and N_{ψ} denotes the absolute norm for the field of n_{ψ}-th root of unity, n_{ψ} being the order of the character ψ. In fact he had computed the value $N_{\psi}(\Theta(\psi))$ and tabulated them in the appendix in [5]

Now, we compute the value h^* by formulae (3), (5), (6) and (7). In fact we construct the matrices as above and calculate its determinants. For the unit index Q, we use properties of Q which appear in Hasse's book [5] and our notes [9] .

We adopted the Hasse's symbol, arrangements and diagrams in constructing our table and diagram, that is, ○ denotes an imaginary field, ● a real field and etc. For complicated diagrams we often omit some segments which denote the relation of inclusion. In the case $f = 120$,the Galois group $G(\boldsymbol{Q}(\zeta_f)/\boldsymbol{Q})$ is of type $(2, 2, 2, 2^2)$ and its Hasse's diagram is so complicated that we divide them easily to see the relation of inclusion. In the round bracket of the column of Q we write the number of theorem in [5] by which we determine the value Q.

f=101					
No.	gen. Character	Type	w	Q	h^*
1	χ_{101}	$(2^2, 5^2)$	$2 \cdot 101$	1 (27)	$5^5 \cdot 101 \cdot 601 \cdot 18701$
2	χ_{101}^5	$(2^2, 5)$	2	1 (23)	5^3
3	χ_{101}^{25}	2^2	2	1 (23)	5

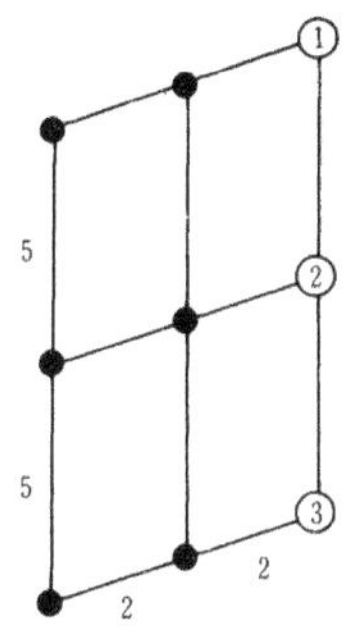

f=103					
No.	gen. Character	Type	w	Q	h^*
1	χ_{103}	(2, 3, 17)	$2 \cdot 103$	1 (27)	$5 \cdot 103 \cdot 1021 \cdot$ 172 47691
2	χ_{103}^3	(2, 17)	2	1 (23)	$5 \cdot 1021$
3	χ_{103}^{17}	(2, 3)	2	1 (23)	5
4	χ_{103}^{51}	2	2	1 (23)	5

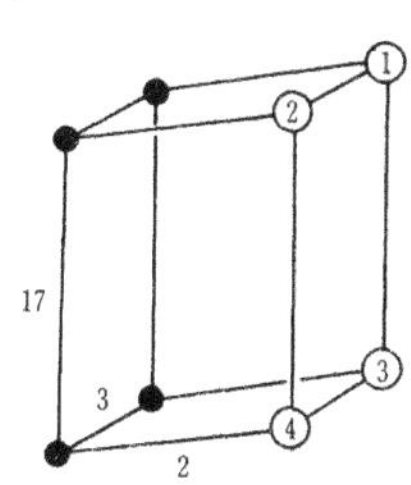

f=104					
No.	gen. Character	Type	w	Q	h^*
1	χ_{2^2}, ψ_{2^3}, χ_{13}	(2, 2, 2^2. 3)	$2^3 \cdot 13$	2 (27)	$3^3 \cdot 13$
2	χ_{2^2}, ψ_{2^3}, χ_{13}^2	(2, 2, 2, 3)	2^3	1 (22)	3^3
3	χ_{2^2}, ψ_{2^3}, χ_{13}^3	(2, 2, 2^2)	2^3	2 (29)	3
4	χ_{2^2}, ψ_{2^3}, χ_{13}^4	(2, 2, 3)	2^3	1 (22)	3
5	χ_{2^2}, ψ_{2^3}, χ_{13}^6	(2, 2, 2)	2^3	1 (22)	3
6	ψ_{2^3}, χ_{13}	(2, 2^2, 3)	$2 \cdot 13$	1 (22)	13
7	$\chi_{2^2}\psi_{2^3}$, χ_{13}	(2, 2^2, 3)	$2 \cdot 13$	2 (26)	3^2
8	$\chi_{2^2}\psi_{2^3}$, $\psi_{2^3}\chi_{13}$	(2, 2^2, 3)	2	2 (26)	$2 \cdot 3^2 \cdot 13$
9	χ_{2^2}, $\psi_{2^3}\chi_{13}$	(2, 2^2, 3)	2^2	2 (26)	$2 \cdot 3 \cdot 13$
10	χ_{2^2}, $\psi_{2^3}\chi_{13}^2$	(2, 2, 3)	2^2	1 (29)	3^3
11	ψ_{2^3}, $\chi_{2^2}\chi_{13}^2$	(2, 2, 3)	2	1 (29)	$2 \cdot 3^2$
12	$\chi_{2^2}\psi_{2^3}$, χ_{13}^2	(2, 2, 3)	2	1 (22)	3^2
13	$\chi_{2^2}\psi_{2^3}$, $\chi_{2^2}\chi_{13}^2$	(2, 2, 3)	2	1 (29)	1
14	$\psi_{2^3}\chi_{13}$	(2^2, 3)	2	1 (24)	$2 \cdot 13$
15	$\chi_{2^2}\psi_{2^3}\chi_{13}^2$	(2, 3)	2	1 (24)	$2 \cdot 3^2$
16	$\chi_{2^2}\psi_{2^3}\chi_{13}^4$	(2, 3)	2	1 (24)	1
17	ψ_{2^3}, χ_{13}^3	(2, 2^2)	2	1 (29)	1
18	$\chi_{2^2}\psi_{2^3}$, χ_{13}^3	(2, 2^2)	2	2 (26)	3
19	$\chi_{2^2}\psi_{2^3}$, $\chi_{2^2}\chi_{13}^3$	(2, 2^2)	2	2 (26)	$2 \cdot 3$
20	χ_{2^2}, $\psi_{2^3}\chi_{13}^3$	(2, 2^2)	2^2	2 (26)	2
21	χ_{2^2}, $\psi_{2^3}\chi_{13}^6$	(2, 2)	2^2	1 (29)	3
22	ψ_{2^3}, $\chi_{2^2}\chi_{13}^6$	(2, 2)	2	1 (25)	$2 \cdot 3$
23	$\chi_{2^2}\psi_{2^3}$, χ_{13}^6	(2, 2)	2	1 (29)	3
24	$\chi_{2^2}\psi_{2^3}$, $\chi_{2^2}\chi_{13}^6$	(2, 2)	2	1 (29)	1
25	$\psi_{2^3}\chi_{13}^3$	2^2	2	1 (24)	2
26	$\chi_{2^2}\psi_{2^3}\chi_{13}^6$	2	2	1 (24)	$2 \cdot 3$

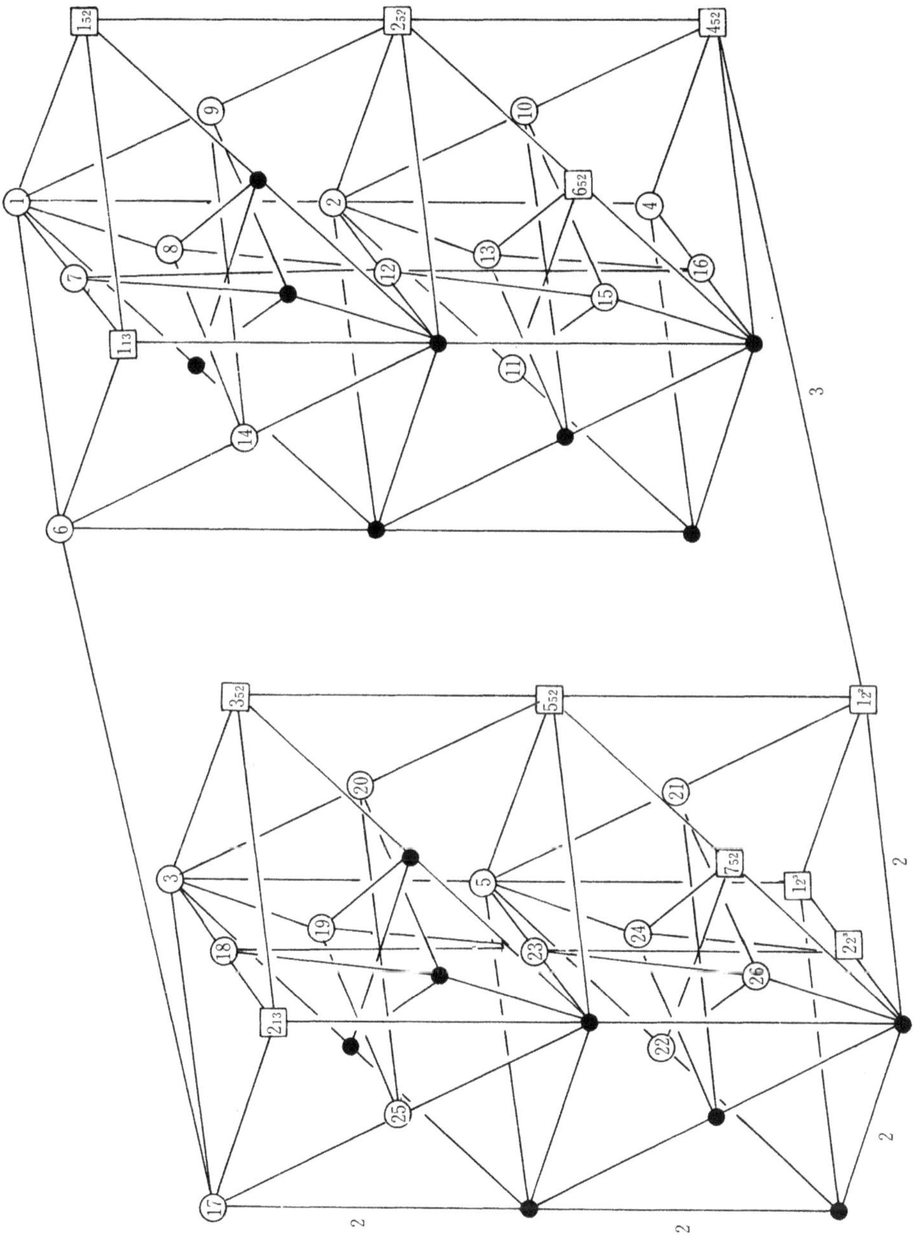
1_{52}
2_{52}
4_{52}
1_{13}
6_{52}
3_{52}
5_{52}
7_{52}
2_{13}
1_{2^2}
1_{2^3}
2_{2^3}
1
2
3
4
5
6
7
8
9
10
11
12
13
14
15
16
17
18
19
20
21
22
23
24
25
26
3
2
2
2
2

f=105					
No.	gen. Character	Type	w	Q	h^*
1	$\chi_3,\ \chi_5,\ \chi_7$	$(2, 2^2, 2, 3)$	$2\cdot 3\cdot 5\cdot 7$	2 (27)	13
2	$\chi_3,\ \chi_5,\ \chi_7^2$	$(2, 2^2, 3)$	$2\cdot 3\cdot 5$	2 (29)	1
3	$\chi_3,\ \chi_5,\ \chi_7^3$	$(2, 2^2, 2)$	$2\cdot 3\cdot 5$	2 (29)	1
4	$\chi_3,\ \chi_5^2,\ \chi_7$	$(2, 2, 2, 3)$	$2\cdot 3\cdot 7$	2 (29)	1
5	$\chi_5,\ \chi_3\chi_7$	$(2^2, 2, 3)$	$2\cdot 5$	1 (22)	$2\cdot 13$
6	$\chi_3,\ \chi_5\chi_7$	$(2, 2^2, 3)$	$2\cdot 3$	1 (22)	$2\cdot 13$
7	$\chi_3\chi_5,\ \chi_7$	$(2^2, 2, 3)$	$2\cdot 7$	1 (22)	$2\cdot 13$
8	$\chi_3,\ \chi_5^2,\ \chi_7^2$	$(2, 2, 3)$	$2\cdot 3$	1 (22)	1
9	$\chi_3\chi_5\chi_7$	$(2^2, 3)$	2	1 (24)	$2^2\cdot 13$
10	$\chi_3,\ \chi_5^2\chi_7^3,\ \chi_7^2$	$(2, 2, 3)$	$2\cdot 3$	1 (**)	1
11	$\chi_3\chi_5^2,\ \chi_7$	$(2, 2, 3)$	$2\cdot 7$	1 (**)	1
12	$\chi_3\chi_5^2,\ \chi_5^2\chi_7^3,\ \chi_7^2$	$(2, 2, 3)$	2	1 (22)	2
13	$\chi_3\chi_5^2\chi_7^2$	$(2, 3)$	2	1 (24)	2
14	$\chi_5,\ \chi_3\chi_7^3$	$(2^2, 2)$	$2\cdot 5$	1 (22)	2
15	$\chi_3,\ \chi_5\chi_7^3$	$(2, 2^2)$	$2\cdot 3$	1 (22)	2
16	$\chi_3\chi_5,\ \chi_7^3$	$(2^2, 2)$	2	1 (22)	2
17	$\chi_3\chi_5\chi_7^3$	2^2	2	1 (24)	2^2
18	$\chi_3,\ \chi_5^2,\ \chi_7^3$	$(2, 2, 2)$	$2\cdot 3$	2 (29)	1
19	$\chi_3\chi_5^2,\ \chi_5^2\chi_7^3$	$(2, 2)$	2	1 (22)	2
20	$\chi_3,\ \chi_5^2\chi_7^3$	$(2, 2)$	$2\cdot 3$	1 (*)	1
21	$\chi_3\chi_5^2,\ \chi_7^3$	$(2, 2)$	2	1 (*)	1

*) see [5] § 26　　　**) see [9]

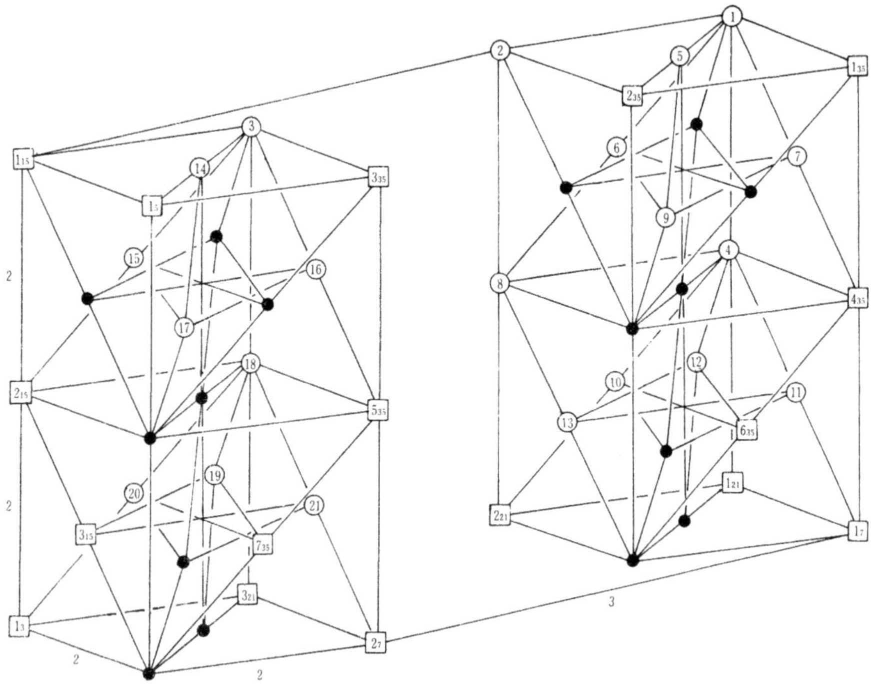

f=107					
No.	gen. Character	Type	w	Q	h^*
1	χ_{107}	(2, 53)	2 · 107	1 (27)	3 · 743 · 9859 · 28 86593
2	χ_{107}^{53}	2	2	1 (23)	3

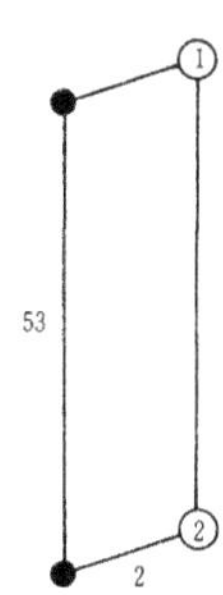

f=108					
No.	gen. Character	Type	w	Q	h^*
1	χ_{2^2}, χ_3, ψ_{3^3}	(2, 2, 3^2)	$2^2 \cdot 3^3$	2 (27)	19
2	χ_{2^2}, ψ_{3^3}	(2, 3^2)	2^2	1 (24)	19

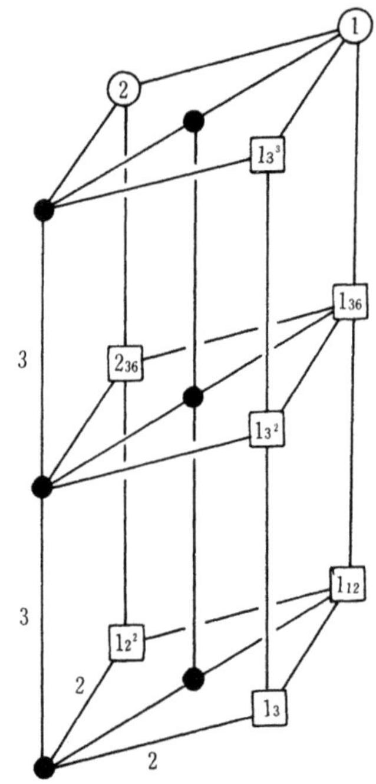

f=109					
No.	gen. Character	Type	w	Q	h^*
1	χ_{109}	$(2^2, 3^3)$	$2 \cdot 109$	1 (27)	$17 \cdot 1009 \cdot$ 94318 66153
2	χ_{109}^3	$(2^2, 3^2)$	2	1 (23)	$17 \cdot 1009$
3	χ_{109}^9	$(2^2, 3)$	2	1 (23)	17
4	χ_{109}^{27}	2^2	2	1 (23)	17

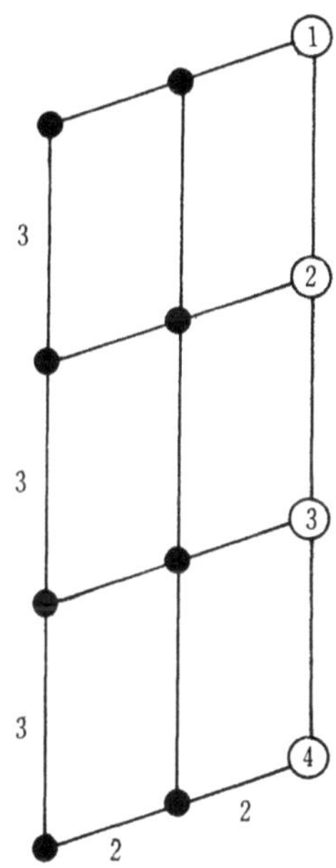

f=111					
No.	gen. Character	Type	w	Q	h^*
1	χ_3, χ_{37}	$(2, 2^2, 3^2)$	$2 \cdot 3 \cdot 37$	2 (27)	$2^2 \cdot 3^2 \cdot 19^2 \cdot 37$
2	χ_3, χ_{37}^2	$(2, 2, 3^2)$	$2 \cdot 3$	1 (22)	$2^2 \cdot 3^2 \cdot 19^2$
3	χ_3, χ_{37}^3	$(2, 2^2, 3)$	$2 \cdot 3$	2 (26)	$2^2 \cdot 3$
4	$\chi_3\chi_{37}^4$	$(2, 3^2)$	$2 \cdot 3$	1 (24)	$3^2 \cdot 19$
5	χ_3, χ_{37}^6	$(2, 2, 3)$	$2 \cdot 3$	1 (22)	$2^2 \cdot 3$
6	χ_3, χ_{37}^9	$(2, 2^2)$	$2 \cdot 3$	2 (26)	2^2
7	$\chi_3\chi_{37}^{12}$	$(2, 3)$	$2 \cdot 3$	1 (24)	3
8	χ_3, χ_{37}^{18}	$(2, 2)$	$2 \cdot 3$	1 (29)	2^2
9	$\chi_3\chi_{37}^2$	$(2, 3^2)$	2	1 (24)	$2^3 \cdot 19$
10	$\chi_3\chi_{37}^6$	$(2, 3)$	2	1 (24)	2^3
11	$\chi_3\chi_{37}^{18}$	2	2	1 (24)	2^3

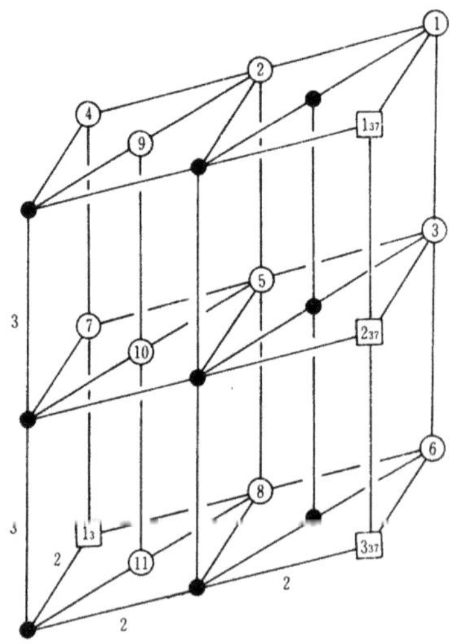

f=112					
No.	gen. Character	Type	w	Q	h^*
1	$\chi_{2^2},\ \psi_{2^4},\ \chi_7$	(2, 2^2, 2, 3)	$2^4\cdot 7$	2 (27)	$2^2\cdot 3^2\cdot 13$
2	$\chi_{2^2},\ \psi_{2^4},\ \chi_7^2$	(2, 2^2, 3)	2^4	1 (22)	3^2
3	$\chi_{2^2},\ \psi_{2^4},\ \chi_7^3$	(2, 2^2, 2)	2^4	2 (29)	2^2
4	$\chi_{2^2}\psi_{2^4},\ \chi_7$	(2^2, 2, 3)	$2\cdot 7$	2 (26)	$2\cdot 3^2$
5	$\chi_{2^2}\psi_{2^4}\chi_7^2$	(2^2, 3)	2	1 (24)	3^2
6	$\chi_{2^2}\psi_{2^4},\ \chi_7^3$	(2^2, 2)	2	2 (26)	2
7	$\psi_{2^4},\ \chi_7$	(2^2, 2, 3)	$2\cdot 7$	1 (22)	$2^2\cdot 13$
8	$\psi_{2^4},\ \chi_7^3$	(2^2, 2)	2	1 (29)	2^2
9	$\chi_{2^2},\ \psi_{2^4}\chi_7^3,\ \chi_7^2$	(2, 2^2, 3)	2^3	2 (26)	$2^2\cdot 13$
10	$\chi_{2^2}\psi_{2^4},\ \psi_{2^4}\chi_7^3,\ \chi_7^2$	(2^2, 2, 3)	2	2 (26)	$2^2\cdot 3^2\cdot 13$
11	$\psi_{2^4}\chi_7$	(2^2, 3)	2	1 (24)	$2^2\cdot 13$
12	$\chi_{2^2},\ \psi_{2^4}\chi_7^3$	(2, 2^2)	2^3	2 (26)	2^2
13	$\chi_{2^2}\psi_{2^4},\ \psi_{2^4}\chi_7^3$	(2^2, 2)	2	2 (26)	2^2
14	$\psi_{2^4}\chi_7^3$	2^2	2	1 (24)	2^2

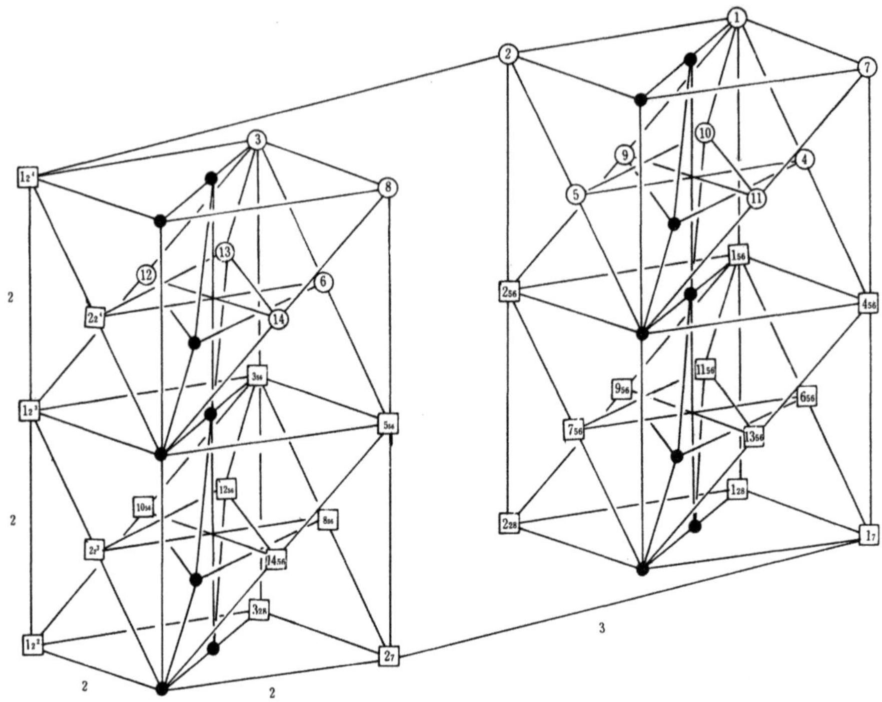

f=113					
No.	gen. Character	Type	w	Q	h^*
1	χ_{113}	$(2^4,\ 7)$	$2\cdot 113$	1 (27)	$2^3\cdot 17\cdot$ 1185 34705 98257
2	χ_{113}^7	2^4	2	1 (23)	17

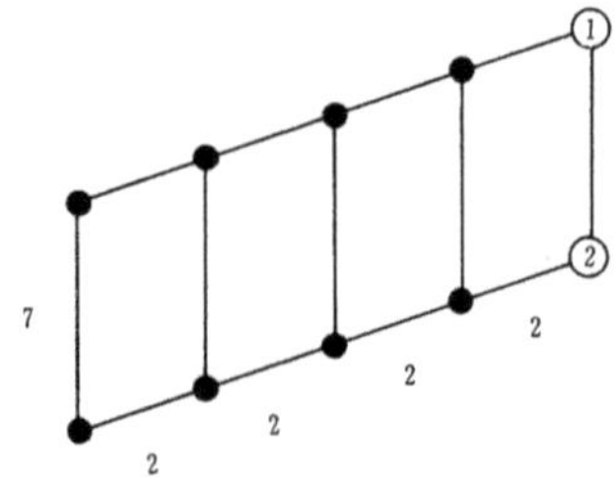

f=115					
No.	gen. Character	Type	w	Q	h^*
1	$\chi_5, \ \chi_{23}$	$(2^2, 2, 11)$	$2\cdot 5\cdot 23$	2 (27)	$3\cdot 331\cdot 45013$
2	$\chi_5\chi_{23}^2$	$(2^2, 11)$	$2\cdot 5$	1 (24)	45013
3	$\chi_5, \ \chi_{23}^{11}$	$(2^2, 2)$	$2\cdot 5$	2 (26)	3
4	$\chi_5^2, \ \chi_{23}$	$(2, 2, 11)$	$2\cdot 23$	1 (22)	$3\cdot 331$
5	$\chi_5^2, \ \chi_{23}^{11}$	$(2, 2)$	2	1 (29)	3
6	$\chi_5^2\chi_{23}$	$(2, 11)$	2	1 (24)	$2\cdot 331$
7	$\chi_5^2\chi_{23}^{11}$	2	2	1 (24)	2

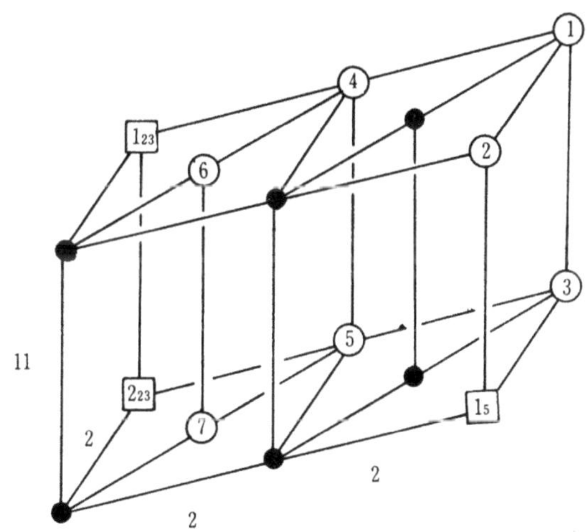

f=116					
No.	gen. Character	Type	w	Q	h^*
1	$\chi_{2^2},\ \chi_{29}$	(2, 2^2, 7)	$2^2\cdot 29$	2 (27)	$2^9\cdot 3\cdot 7$
2	$\chi_{2^2},\ \chi_{29}^2$	(2, 2, 7)	2^2	1 (22)	$2^6\cdot 3\cdot 7$
3	$\chi_{2^2}\chi_{29}^4$	(2, 7)	2^2	1 (24)	$2^3\cdot 7$
4	$\chi_{2^2},\ \chi_{29}^7$	(2, 2^2)	2^2	2 (26)	3
5	$\chi_{2^2},\ \chi_{29}^{14}$	(2, 2)	2^2	1 (22)	3
6	$\chi_{2^2}\chi_{29}^2$	(2, 7)	2	1 (24)	$2^4\cdot 3$
7	$\chi_{2^2}\chi_{29}^{14}$	2	2	1 (24)	$2\cdot 3$

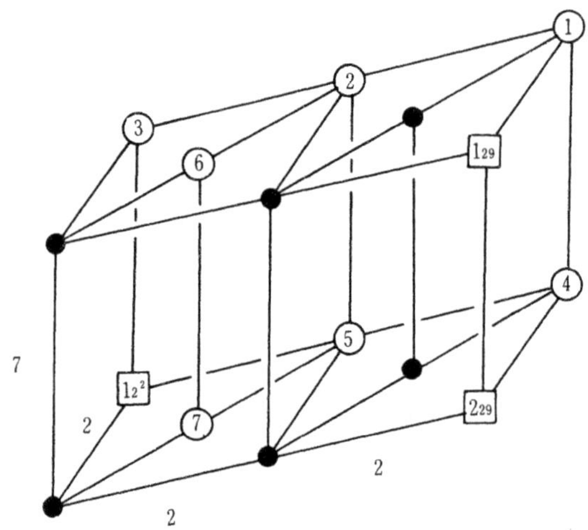

f=117					
No.	gen. Character	Type	w	Q	h^*
1	$\chi_3,\ \psi_{3^2},\ \chi_{13}$	$(2, 3, 2^2, 3)$	$2\cdot 3^2\cdot 13$	2 (27)	$2\cdot 3^6\cdot 7\cdot 13$
2	$\chi_3,\ \psi_{3^2},\ \chi_{13}^2$	$(2, 3, 2, 3)$	$2\cdot 3^2$	1 (22)	$2\cdot 7\cdot 13$
3	$\chi_3,\ \psi_{3^2},\ \chi_{13}^3$	$(2, 3, 2^2)$	$2\cdot 3^2$	2 (26)	$2\cdot 3^2$
4	$\chi_3,\ \psi_{3^2},\ \chi_{13}^4$	$(2, 3, 3)$	$2\cdot 3^2$	1 (22)	7
5	$\chi_3,\ \psi_{3^2},\ \chi_{13}^6$	$(2, 3, 2)$	$2\cdot 3^2$	1 (22)	2
6	$\psi_{3^2},\ \chi_{13}$	$(3, 2^2, 3)$	$2\cdot 13$	1 (22)	3^6
7	$\psi_{3^2}\chi_{13}^3$	$(3, 2^2)$	2	1 (24)	3^2
8	$\psi_{3^2},\ \chi_3\chi_{13}^6,\ \chi_{13}^4$	$(3, 2, 3)$	2	1 (22)	$2^2\cdot 13$
9	$\psi_{3^2},\ \chi_3\chi_{13}^6$	$(3, 2)$	2	1 (24)	2^2
10	$\chi_3,\ \psi_{3^2}\chi_{13}^4,\ \chi_{13}^3$	$(2, 3, 2^2)$	$2\cdot 3$	2 (29)	$2\cdot 3^2\cdot 13$
10′	$\chi_3,\ \psi_{3^2}\bar{\chi}_{13}^4,\ \chi_{13}^3$	$(2, 3, 2^2)$	$2\cdot 3$	2 (29)	$2\cdot 3^2\cdot 7$
11	$\chi_3,\ \psi_{3^2}\chi_{13}^4,\ \chi_{13}^6$	$(2, 3, 2)$	$2\cdot 3$	1 (22)	$2\cdot 13$
11′	$\chi_3,\ \psi_{3^2}\bar{\chi}_{13}^4,\ \chi_{13}^6$	$(2, 3, 2)$	$2\cdot 3$	1 (22)	$2\cdot 7$
12	$\chi_3,\ \psi_{3^2}\chi_{13}^4$	$(2, 3)$	$2\cdot 3$	1 (24)	1
12′	$\chi_3,\ \psi_{3^2}\bar{\chi}_{13}^4$	$(2, 3)$	$2\cdot 3$	1 (24)	7
13	$\psi_{3^2}\chi_{13}^4,\ \chi_{13}^3$	$(3, 2^2)$	2	1 (24)	3^2
13′	$\psi_{3^2}\bar{\chi}_{13}^4,\ \chi_{13}^3$	$(3, 2^2)$	2	1 (24)	3^2
14	$\chi_3\chi_{13}^6,\ \psi_{3^2}\chi_{13}^4$	$(2, 3)$	2	1 (24)	$2^2\cdot 13$
14′	$\chi_3\chi_{13}^6,\ \psi_{3^2}\bar{\chi}_{13}^4$	$(2, 3)$	2	1 (24)	2^2

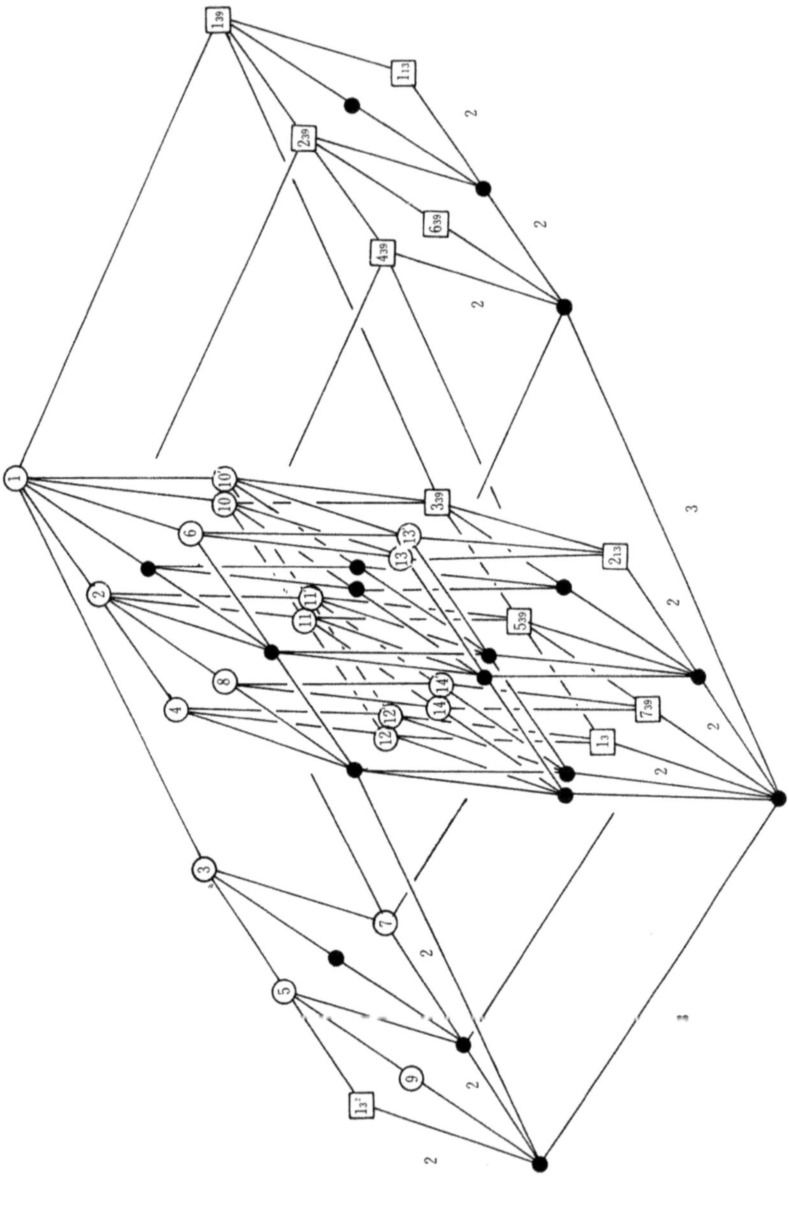

f=119					
No.	gen. Character	Type	w	Q	h^*
1	$\chi_7,\ \chi_{17}$	$(2,\ 3,\ 2^4)$	$2\cdot 7\cdot 17$	2 (27)	$3^4\cdot 5^3\cdot 13\cdot 97^2$
2	$\chi_7,\ \chi_{17}^2$	$(2,\ 3,\ 2^3)$	$2\cdot 7$	1 (22)	$3^4\cdot 5^3\cdot 13$
3	$\chi_7,\ \chi_{17}^4$	$(2,\ 3,\ 2^2)$	$2\cdot 7$	1 (22)	$5\cdot 13$
4	$\chi_7,\ \chi_{17}^8$	$(2,\ 3,\ 2)$	$2\cdot 7$	1 (22)	5
5	$\chi_7^3,\ \chi_{17}$	$(2,\ 2^4)$	$2\cdot 17$	2 (26)	$3^2\cdot 5$
6	$\chi_7^3,\ \chi_{17}^2$	$(2,\ 2^3)$	2	1 (29)	$3^2\cdot 5$
7	$\chi_7^3,\ \chi_{17}^4$	$(2,\ 2^2)$	2	1 (29)	5
8	$\chi_7^3,\ \chi_{17}^8$	$(2,\ 2)$	2	1 (29)	5
9	$\chi_7^2\chi_{17}$	$(3,\ 2^4)$	$2\cdot 17$	1 (24)	97^2
10	$\chi_7\chi_{17}^2$	$(3,\ 2^3)$	2	1 (24)	$2\cdot 3^4\cdot 5^2$
11	$\chi_7\chi_{17}^4$	$(3,\ 2^2)$	2	1 (24)	$2\cdot 13$
12	$\chi_7\chi_{17}^8$	$(3,\ 2)$	2	1 (24)	$2\cdot 5$
13	$\chi_7^3\chi_{17}^2$	2^3	2	1 (24)	$2\cdot 3^2$
14	$\chi_7^3\chi_{17}^4$	2^2	2	1 (24)	2
15	$\chi_7^3\chi_{17}^8$	2	2	1 (24)	$2\cdot 5$

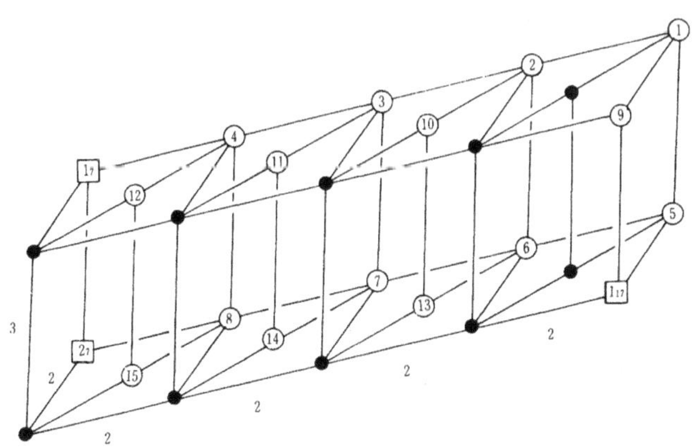

f=120					
No.	gen. Character	Type	w	Q	h^*
1	$\chi_{2^2}, \psi_{2^3}, \chi_3, \chi_5$	$(2, 2, 2, 2^2)$	$2^3 \cdot 3 \cdot 5$	2 (27)	2^2
2	$\psi_{2^3}, \chi_3, \chi_5$	$(2, 2, 2^2)$	$2 \cdot 3 \cdot 5$	2 (29)	2
3	$\chi_{2^2}\psi_{2^3}, \chi_3, \chi_5$	$(2, 2, 2^2)$	$2 \cdot 3 \cdot 5$	2 (29)	1
4	$\chi_{2^2}, \psi_{2^3}\chi_3, \chi_5$	$(2, 2, 2^2)$	$2^2 \cdot 5$	2 (29)	2^2
5	$\chi_{2^2}\psi_{2^3}, \psi_{2^3}\chi_3, \chi_5$	$(2, 2, 2^2)$	$2 \cdot 5$	2 (29)	2^2
6	$\psi_{2^3}\chi_3, \chi_5$	$(2, 2^2)$	$2 \cdot 5$	1 (25)**	2
7	$\psi_{2^3}, \chi_{2^2}\chi_3, \chi_5$	$(2, 2, 2^2)$	$2 \cdot 5$	1 (22)	2^2
8	$\chi_{2^2}\psi_{2^3}\chi_3, \chi_5$	$(2, 2^2)$	$2 \cdot 5$	1 (22)	2
9	$\psi_{2^3}, \chi_3, \chi_{2^2}\chi_5$	$(2, 2, 2^2)$	$2 \cdot 3$	1 (22)	2^3
10	$\chi_{2^2}, \psi_{2^3}, \chi_3\chi_5$	$(2, 2, 2^2)$	2^3	1 (22)	2^2
11	$\psi_{2^3}, \chi_{2^2}\chi_3\chi_5$	$(2, 2^2)$	2	1 (22)	2^3
12	$\chi_{2^2}, \chi_3, \psi_{2^3}\chi_5$	$(2, 2, 2^2)$	$2^2 \cdot 3$	2 (29)	2
13	$\chi_{2^2}, \psi_{2^3}\chi_3, \chi_3\chi_5$	$(2, 2, 2^2)$	2^2	2 (29)	2^3
14	$\chi_{2^2}, \psi_{2^3}\chi_3\chi_5$	$(2, 2^2)$	2^2	1 (29)	2
15	$\chi_{2^2}\psi_{2^3}, \chi_3, \chi_{2^2}\chi_5$	$(2, 2, 2^2)$	$2 \cdot 3$	2 (29)	2
16	$\chi_3, \psi_{2^3}\chi_5$	$(2, 2^2)$	$2 \cdot 3$	1 (25)**	1
17	$\chi_3, \chi_{2^2}\psi_{2^3}\chi_5$	$(2, 2^2)$	$2 \cdot 3$	1 (22)	2
18	$\psi_{2^3}\chi_3, \chi_{2^2}\psi_{2^3}, \chi_{2^2}\chi_5$	$(2, 2, 2^2)$	2	2 (29)	2^3
19	$\psi_{2^3}\chi_3, \chi_3\chi_5$	$(2, 2^2)$	2	1 (22)	2^2
20	$\psi_{2^3}\chi_3, \chi_{2^2}\chi_5$	$(2, 2^2)$	2	1 (22)	2^3
21	$\psi_{2^3}\chi_3, \chi_{2^2}\psi_{2^3}\chi_5$	$(2, 2^2)$	2	1 (22)	2^3
22	$\chi_{2^2}\psi_{2^3}\chi_3\chi_5$	2^2	2	1 (24)	2^2
23	$\chi_{2^2}\psi_{2^3}, \chi_3\chi_5$	$(2, 2^2)$	2	1 (22)	2
24	$\chi_{2^2}\psi_{2^3}, \psi_{2^3}\chi_3\chi_5$	$(2, 2^2)$	2	1 (29)	2
25	$\chi_{2^2}\chi_3, \psi_{2^3}\chi_5$	$(2, 2^2)$	2	1 (22)	2^2
26	$\chi_{2^2}\psi_{2^3}\chi_3, \psi_{2^3}\chi_5$	$(2, 2^2)$	2	1 (22)	2^2
27	$\chi_{2^2}, \psi_{2^3}, \chi_3, \chi_5^2$	$(2, 2, 2, 2)$	$2^3 \cdot 3$	2 (29)	2
28	$\psi_{2^3}, \chi_3, \chi_5^2$	$(2, 2, 2)$	$2 \cdot 3$	1 (22)	2
29	$\chi_{2^2}\psi_{2^3}, \chi_3, \chi_5^2$	$(2, 2, 2)$	$2 \cdot 3$	2 (29)	1

30	$\chi_{2^2},\ \psi_{2^3}\chi_3,\ \chi_5^2$	(2, 2, 2)	2^2	2 (29)	2^2
31	$\chi_{2^2}\psi_{2^3},\ \psi_{2^3}\chi_3,\ \chi_5^2$	(2, 2, 2)	2	2 (29)	2^2
32	$\psi_{2^3}\chi_3,\ \chi_5^2$	(2, 2)	2	1 (22)	2^2
33	$\psi_{2^3},\ \chi_3,\ \chi_{2^2}\chi_5^2$	(2, 2, 2)	$2\cdot 3$	2 (29)	2
34	$\chi_{2^2},\ \psi_{2^3},\ \chi_3\chi_5^2$	(2, 2, 2)	2^3	1 (22)	2
35	$\psi_{2^3},\ \chi_{2^2}\chi_3,\ \chi_3\chi_5^2$	(2, 2, 2)	2	1 (22)	2^2
36	$\psi_{2^3},\ \chi_3\chi_5^2$	(2, 2)	2	1 (22)	2^2
37	$\chi_{2^2},\ \chi_3,\ \psi_{2^3}\chi_5^2$	(2, 2, 2)	$2^2\cdot 3$	2 (29)	2
38	$\chi_{2^2},\ \psi_{2^3}\chi_3,\ \chi_3\chi_5^2$	(2, 2, 2)	2^2	2 (29)	2
39	$\chi_{2^2},\ \psi_{2^3}\chi_3\chi_5^2$	(2, 2)	2^2	1 (29)	2
40	$\chi_{2^2}\psi_{2^3},\ \chi_3,\ \chi_{2^2}\chi_5^2$	(2, 2, 2)	$2\cdot 3$	2 (29)	2
41	$\chi_3,\ \psi_{2^3}\chi_5^2$	(2, 2)	$2\cdot 3$	1 (22)	2
42	$\chi_3,\ \chi_{2^2}\psi_{2^3}\chi_5^2$	(2, 2)	$2\cdot 3$	1 (*)	1
43	$\psi_{2^3}\chi_3,\ \chi_3\chi_5^2$	(2, 2)	2	1 (22)	2
44	$\psi_{2^3}\chi_3\chi_5^2$	2	2	1 (24)	2^2
45	$\chi_{2^2}\psi_{2^3},\ \psi_{2^3}\chi_3,\ \chi_{2^2}\chi_5^2$	(2, 2, 2)	2	2 (29)	2
46	$\psi_{2^3}\chi_3,\ \chi_{2^2}\chi_5^2$	(2, 2)	2	2 (*)	2^2
47	$\psi_{2^3}\chi_3,\ \chi_{2^2}\psi_{2^3}\chi_5^2$	(2, 2)	2	2 (*)	2^2
48	$\chi_{2^2}\psi_{2^3}\chi_3,\ \chi_{2^2}\chi_5^2$	(2, 2)	2	1 (22)	2^2
49	$\chi_{2^2}\psi_{2^3},\ \chi_3\chi_5^2$	(2, 2)	2	1 (29)	1
50	$\chi_{2^2}\psi_{2^3},\ \psi_{2^3}\chi_3\chi_5^2$	(2, 2)	2	1 (29)	2
51	$\chi_{2^2}\psi_{2^3}\chi_3,\ \chi_3\chi_5^2$	(2, 2)	2	1 (29)	2
52	$\chi_{2^2}\chi_3,\ \chi_{2^2}\psi_{2^3}\chi_5^2$	(2, 2)	2	1 (29)	2^2

*See [5] § 26 **See [12]

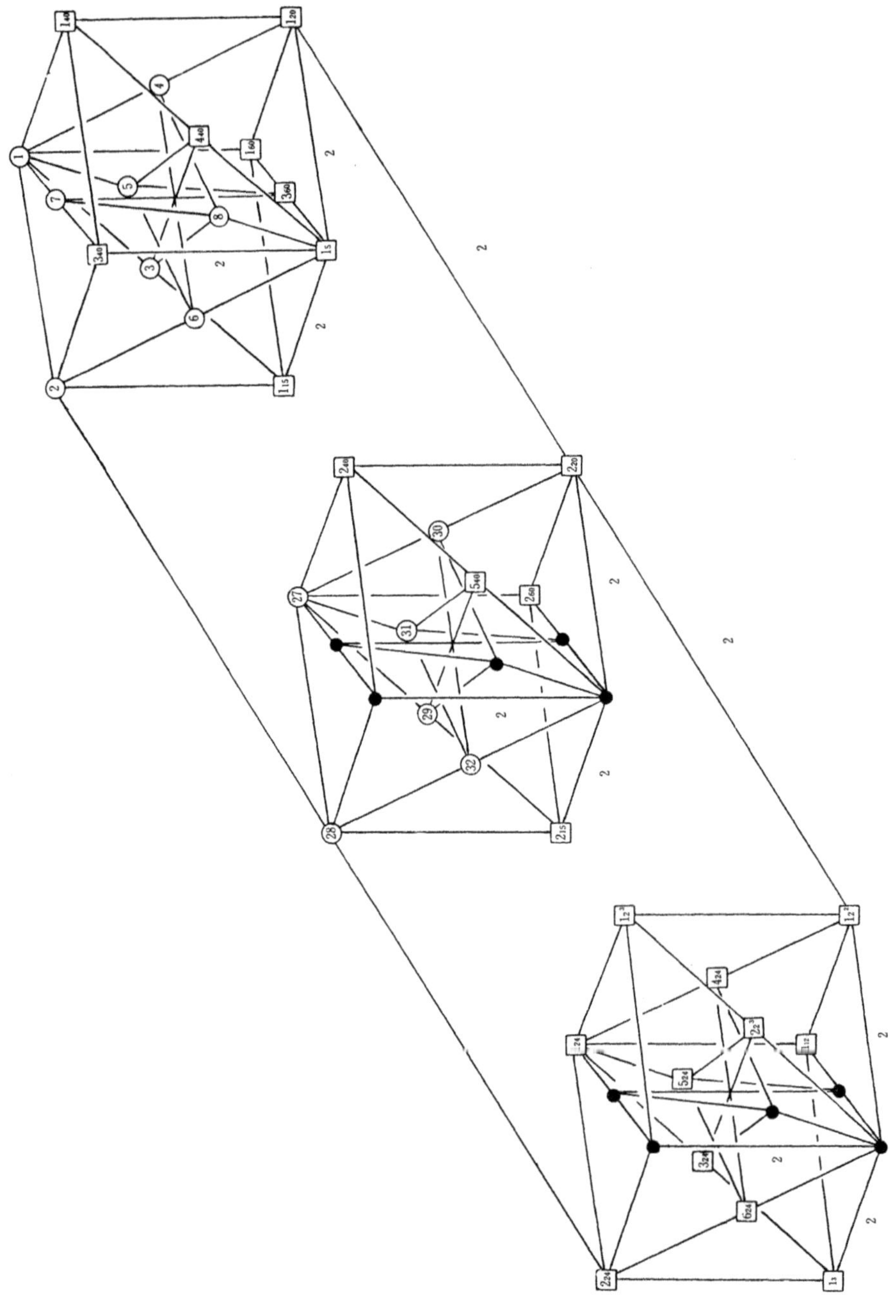
1
2
3
4
5
6
7
8
27
28
29
30
31
32

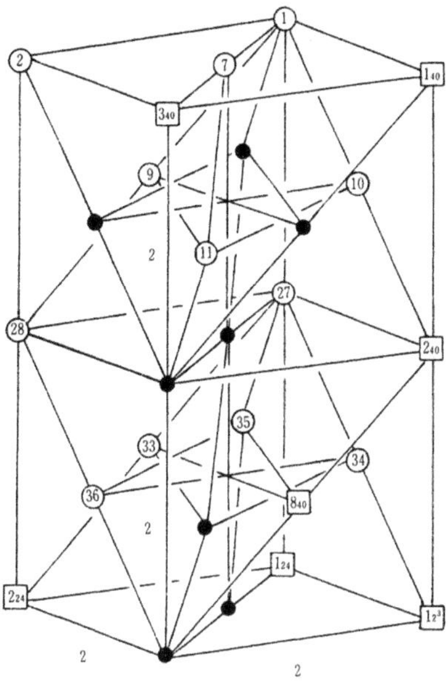
1
2
7
1_{40}
3_{40}
9
10
11
2
27
28
2_{40}
35
33
34
36
8_{40}
2
1_{24}
2_{24}
1_{2^3}
2
2

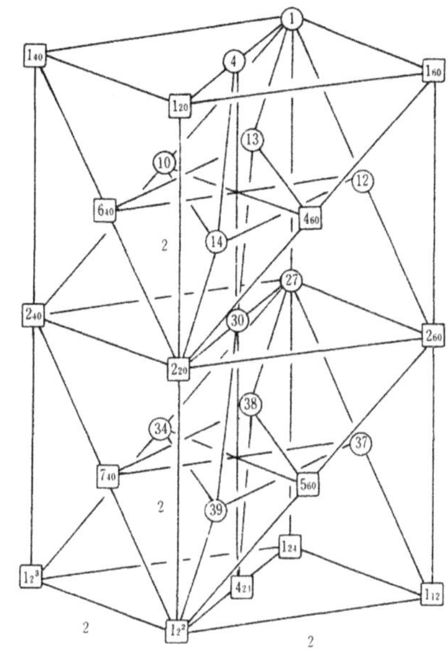
1
1_{40}
4
1_{60}
1_{20}
13
10
12
6_{40}
4_{60}
2
14
27
2_{40}
30
2_{60}
2_{20}
38
34
37
7_{40}
5_{60}
2
39
1_{24}
1_{2^3}
4_{21}
1_{12}
2
1_{2^2}
2

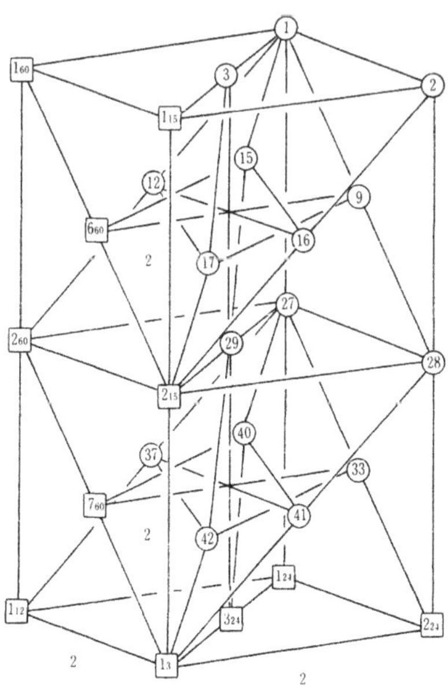
1
1_{60}
3
2
1_{15}
15
12
9
6_{60}
16
2
17
27
29
2_{60}
28
2_{15}
40
37
33
7_{60}
41
2
42
1_{24}
1_{12}
3_{21}
2_{21}
2
1_3
2

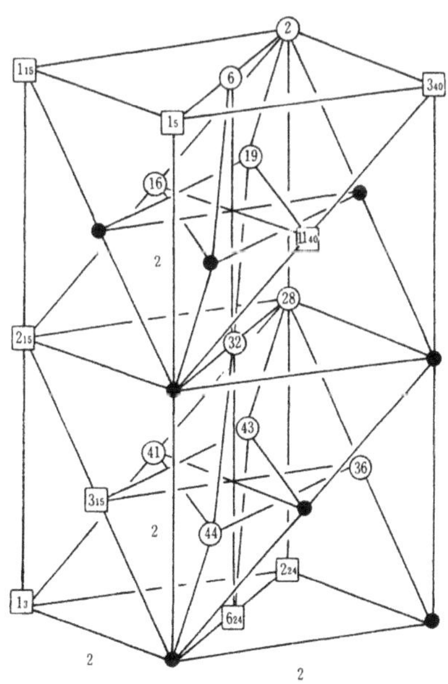
2
1_{15}
6
3_{40}
1_5
19
16
11_{40}
2
28
2_{15}
32
43
41
36
3_{15}
2
44
2_{24}
1_3
6_{24}
2
2

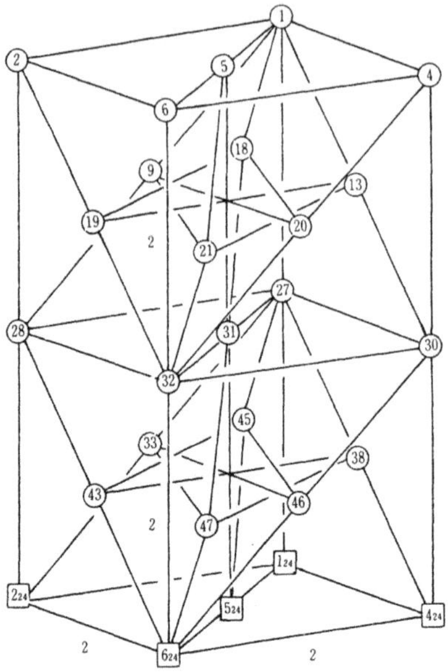

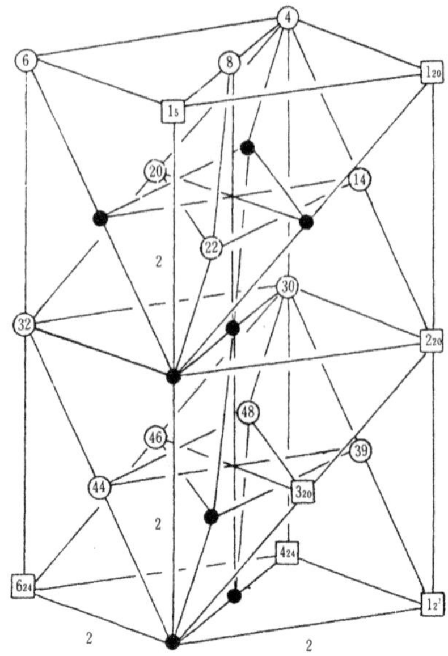

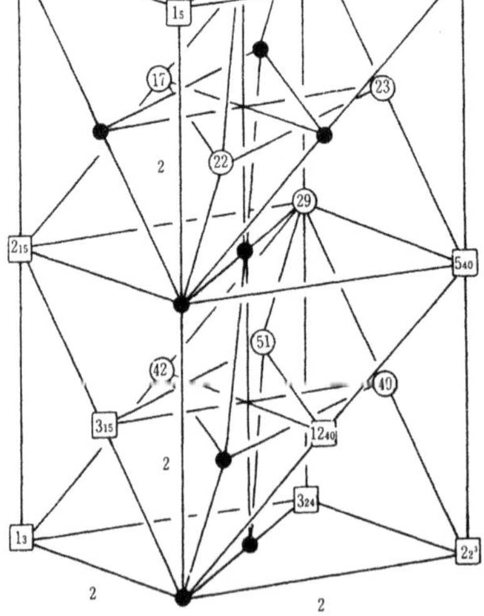

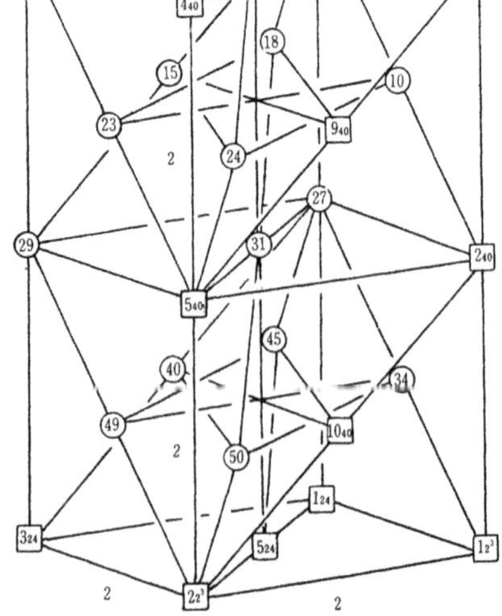

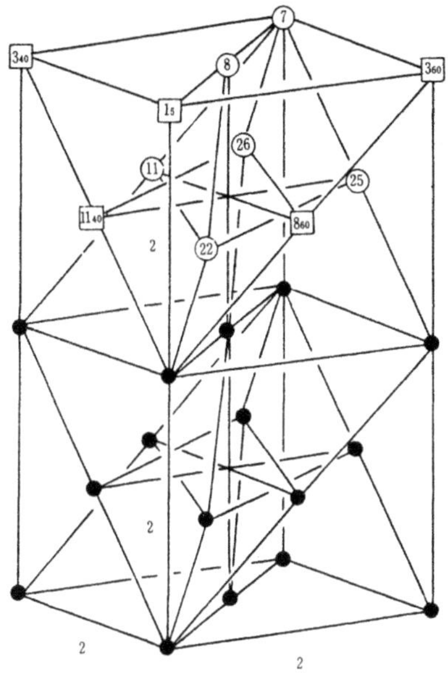
7
3_{40}
8
3_{60}
1_5
26
11
25
11_{40}
8_{60}
2
22
2
2
2

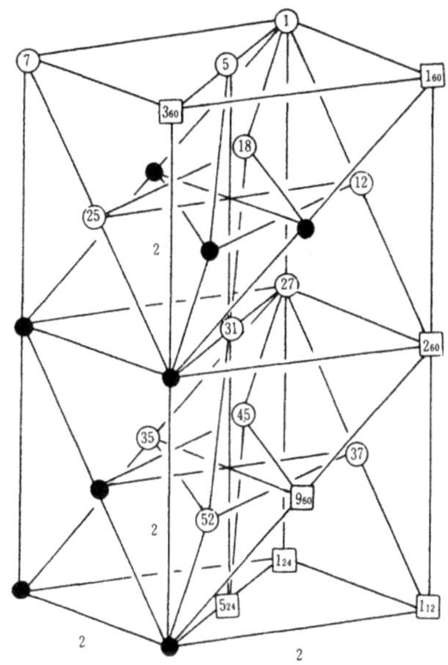
1
7
5
1_{60}
3_{60}
18
12
25
2
27
31
2_{60}
45
35
37
9_{60}
52
2
1_{24}
5_{24}
1_{12}
2
2

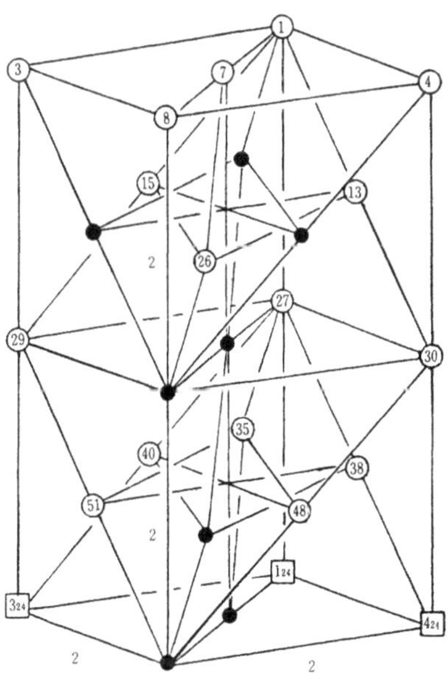
1
3
7
4
8
15
13
2
26
27
29
30
35
40
38
51
48
2
1_{24}
3_{24}
4_{24}
2
2

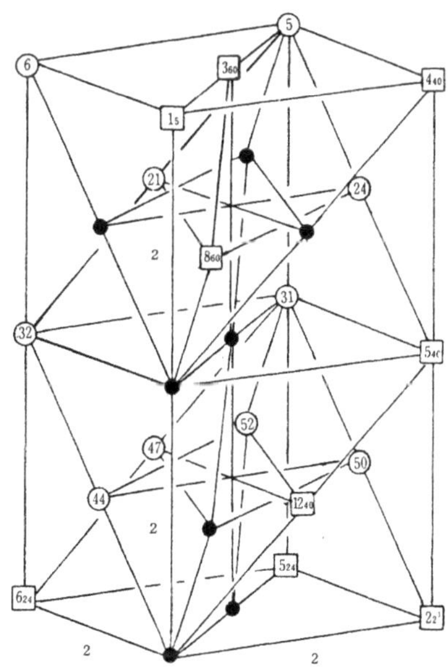
5
6
3_{60}
4_{40}
1_5
21
24
2
8_{60}
31
32
52
47
50
44
12_{40}
2
5_{24}
6_{24}
2
2

f=121					
No.	gen. Character	Type	w	Q	h^*
1	χ_{11}, ψ_{11^2}	(2, 5, 11)	$2 \cdot 11^2$	1 (27)	67·353·20021·25741
2	χ_{11}^5, ψ_{11^2}	(2, 11)	2	1 (23)	67 · 353

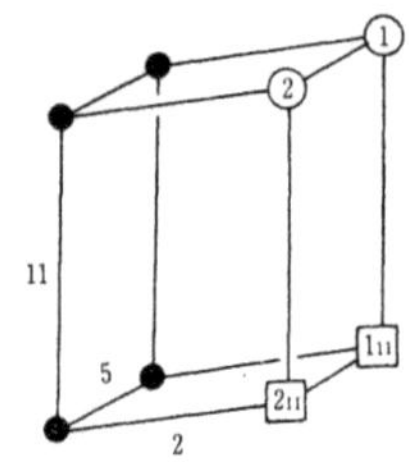

f=123					
No.	gen. Character	Type	w	Q	h^*
1	χ_3, χ_{41}	(2, 2^3, 5)	2 · 3 · 41	2 (27)	$2^{12} \cdot 11^2 \cdot 17$
2	χ_3, χ_{41}^2	(2, 2^2, 5)	2 · 3	1 (22)	$2^{12} \cdot 17$
3	χ_3, χ_{41}^4	(2, 2, 5)	2 · 3	1 (22)	2^8
4	χ_3, χ_{41}^5	(2, 2^3)	2 · 3	2 (26)	17
5	χ_3, χ_{41}^8	(2, 5)	2 · 3	1 (24)	2^4
6	χ_3, χ_{41}^{10}	(2, 2^2)	2 · 3	1 (22)	17
7	χ_3, χ_{41}^{20}	(2, 2)	2 · 3	1 (22)	1
8	$\chi_3\chi_{41}^2$	(2^2, 5)	2	1 (24)	$2^5 \cdot 17$
9	$\chi_3\chi_{41}^4$	(2, 5)	2	1 (24)	2^5
10	$\chi_3\chi_{41}^{10}$	2^2	2	1 (24)	2 · 17
11	$\chi_3\chi_{41}^{20}$	2	2	1 (24)	2

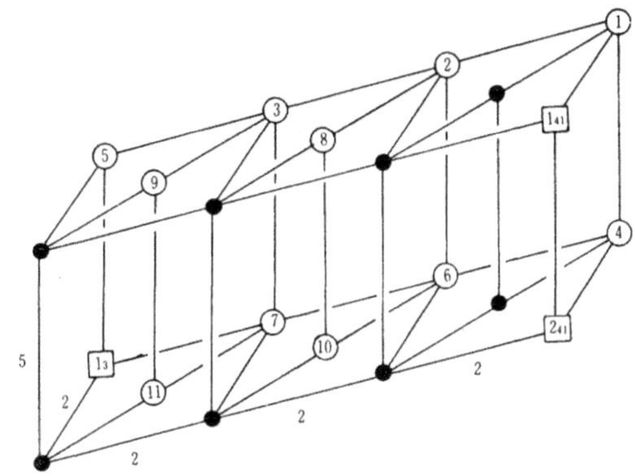

f=124					
No.	gen. Character	Type	w	Q	h^*
1	$\chi_{2^2},\ \chi_{31}$	(2, 2, 3, 5)	$2^2 \cdot 31$	2 (27)	$2^2 \cdot 3^2 \cdot 31 \cdot 41$
2	$\chi_{2^2},\ \chi_{31}^2$	(2, 3, 5)	2^2	1 (24)	$2^2 \cdot 31 \cdot 41$
3	$\chi_{2^2},\ \chi_{31}^3$	(2, 2, 5)	2^2	2 (29)	$3 \cdot 41$
4	$\chi_{2^2},\ \chi_{31}^5$	(2, 2, 3)	2^2	2 (29)	$2^2 \cdot 3^2$
5	$\chi_{2^2},\ \chi_{31}^6$	(2, 5)	2^2	1 (24)	41
6	$\chi_{2^2},\ \chi_{31}^{10}$	(2, 3)	2^2	1 (24)	2^7
7	$\chi_{2^2},\ \chi_{31}^{15}$	(2, 2)	2^2	2 (26)	3

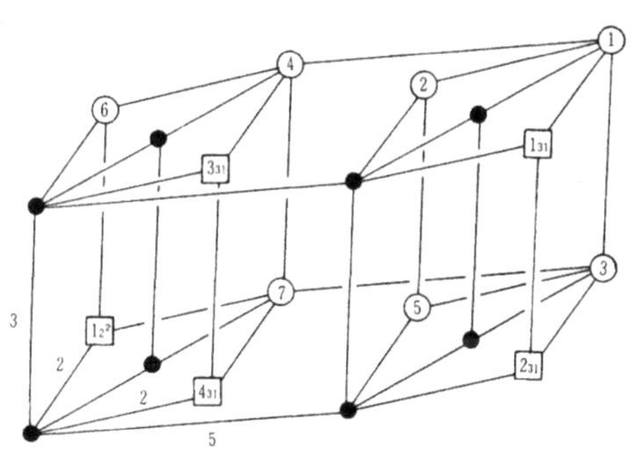

f=125					
No.	gen. Character	Type	w	Q	h^*
1	$\chi_5, \ \psi_{5^3}$	$(2^2, 5^2)$	$2 \cdot 5^3$	1 (27)	2801 · 206 02801

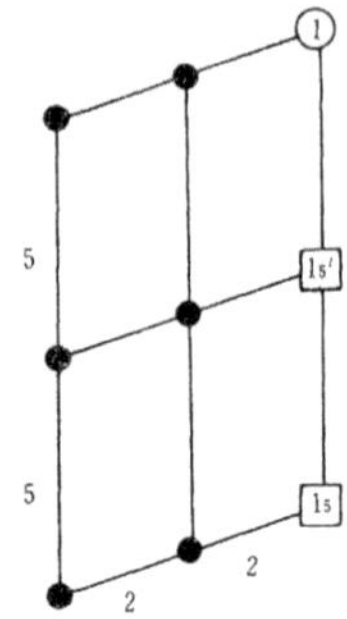

f=127					
No.	gen. Character	Type	w	Q	h^*
1	χ_{127}	$(2, 3^2, 7)$	$2 \cdot 127$	1 (27)	5·13·43·547·883·3079 · 6 26599
2	χ_{127}^{3}	(2, 3, 7)	2	1 (23)	5 · 13 · 43 · 547
3	χ_{127}^{7}	$(2, 3^2)$	2	1 (23)	5 · 13 · 3079
4	χ_{127}^{9}	(2, 7)	2	1 (23)	5 · 43
5	χ_{127}^{21}	(2, 3)	2	1 (23)	5 · 13
6	χ_{127}^{63}	2	2	1 (23)	5

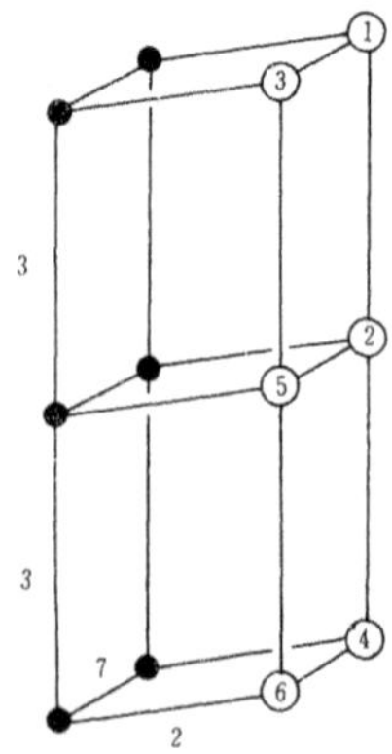

f=128					
No.	gen. Character	Type	w	Q	h^*
1	χ_{2^2}, ψ_{2^7}	$(2,\ 2^5)$	2^7	1 (27)	17 · 21121
2	$\chi_{2^2}\psi_{2^7}$	2^5	2	1 (23)	21121

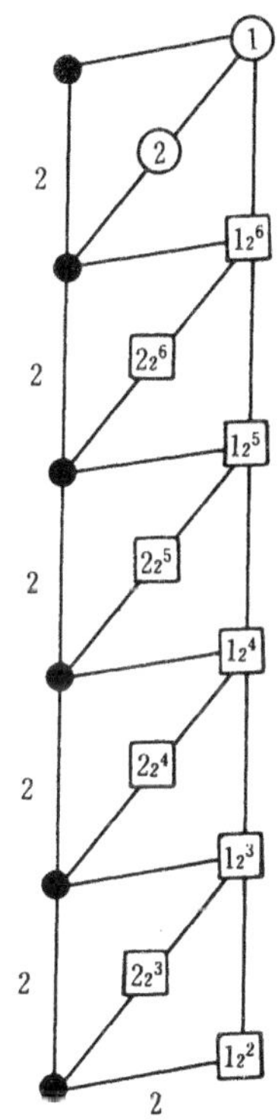

f=129					
No.	gen. Character	Type	w	Q	h^*
1	$\chi_3,\ \chi_{43}$	(2, 2, 3, 7)	2・3・43	2 (27)	7・29・211・883
2	$\chi_3,\ \chi_{43}^2$	(2, 3, 7)	2・3	1 (24)	7・29・883
3	$\chi_3,\ \chi_{43}^3$	(2, 2, 7)	2・3	2 (26)	7・29
4	$\chi_3,\ \chi_{43}^6$	(2, 7)	2・3	1 (24)	7・29
5	$\chi_3,\ \chi_{43}^7$	(2, 2, 3)	2・3	2 (26)	1
6	$\chi_3,\ \chi_{43}^{14}$	(2, 3)	2・3	1 (24)	1
7	$\chi_3,\ \chi_{43}^{21}$	(2, 2)	2・3	2 (26)	1

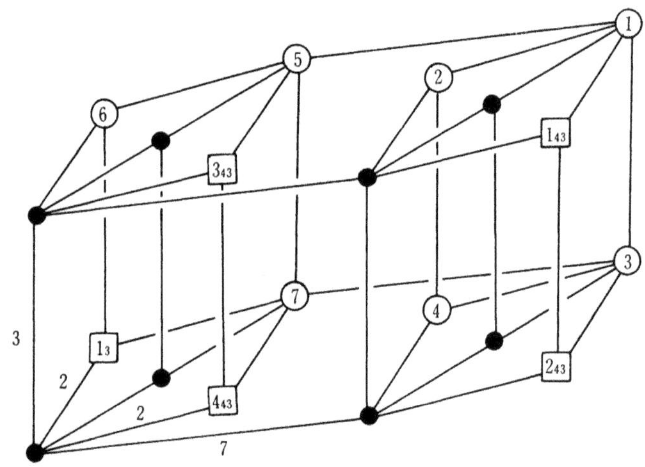

f=131					
No.	gen. Character	Type	w	Q	h^*
1	χ_{131}	(2, 5, 13)	$2 \cdot 131$	1 (27)	$3^3 \cdot 5^2 \cdot 53 \cdot 131 \cdot 1301 \cdot$ 46737 06701
2	χ_{131}^{5}	(2, 13)	2	1 (23)	$3^3 \cdot 5 \cdot 53$
3	χ_{131}^{13}	(2, 5)	2	1 (23)	5^2
4	χ_{131}^{65}	2	2	1 (23)	5

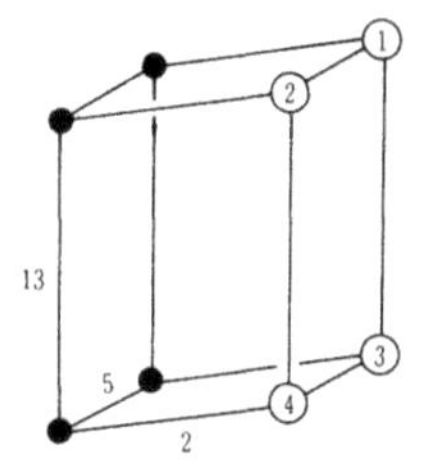

f=132					
No.	gen. Character	Type	w	Q	h^*
1	$\chi_{2^2}, \chi_3, \chi_{11}$	(2, 2, 2, 5)	$2^2 \cdot 3 \cdot 11$	2 (27)	11
2	$\chi_{2^2}, \chi_3, \chi_{11}^{2}$	(2, 2, 5)	$2^2 \cdot 3$	2 (29)	1
3	$\chi_{2^2}, \chi_3, \chi_{11}^{5}$	(2, 2, 2)	$2^2 \cdot 3$	2 (29)	1
4	$\chi_3, \chi_{2^2}\chi_{11}$	(2, 2, 5)	$2 \cdot 3$	1 (22)	$2 \cdot 11$
5	$\chi_{2^2}, \chi_3\chi_{11}$	(2, 2, 5)	2^2	1 (22)	$2 \cdot 11$
6	$\chi_{2^2}\chi_3, \chi_{11}$	(2, 2, 5)	$2 \cdot 11$	1 (22)	$2 \cdot 11$
7	$\chi_{2^2}\chi_3\chi_{11}$	(2, 5)	2	1 (24)	$2^2 \cdot 11$
8	$\chi_3, \chi_{2^2}\chi_{11}^{5}$	(2, 2)	$2 \cdot 3$	1 (29)	2
9	$\chi_{2^2}, \chi_3\chi_{11}^{5}$	(2, 2)	2^2	1 (29)	2
10	$\chi_{2^2}\chi_3, \chi_{11}^{5}$	(2, 2)	2	1 (29)	2
11	$\chi_{2^2}\chi_3\chi_{11}^{5}$	2	2	1 (24)	2^2

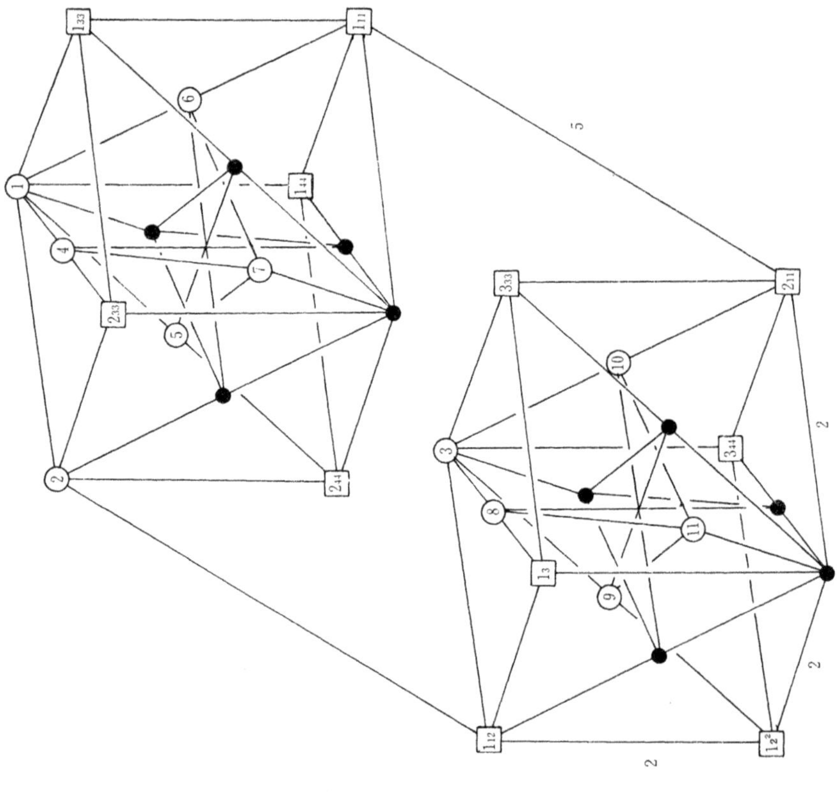

f=133					
No.	gen. Character	Type	w	Q	h^*
1	$\chi_7,\ \chi_{19}$	$(2, 3, 2, 3^2)$	$2\cdot7\cdot19$	2 (27)	$2^2\cdot3^{10}\cdot13\cdot19\cdot37\cdot73$
2	$\chi_7,\ \chi_{19}^2$	$(2, 3, 3^2)$	$2\cdot7$	1 (22)	$2^2\cdot13\cdot19\cdot37\cdot73$
3	$\chi_7,\ \chi_{19}^3$	$(2, 3, 2, 3)$	$2\cdot7$	2 (29)	$2^2\cdot3^3\cdot13$
4	$\chi_7,\ \chi_{19}^6$	$(2, 3, 3)$	$2\cdot7$	1 (22)	$2^2\cdot13$
5	$\chi_7,\ \chi_{19}^9$	$(2, 3, 2)$	$2\cdot7$	2 (26)	3
6	$\chi_7^2,\ \chi_{19}$	$(3, 2, 3^2)$	$2\cdot19$	1 (22)	3^{10}
7	$\chi_7^2,\ \chi_{19}^3$	$(3, 2, 3)$	2	1 (22)	3^3
8	$\chi_7^2,\ \chi_{19}^9$	$(3, 2)$	2	1 (24)	3
9	$\chi_7^3,\ \chi_{19}$	$(2, 2, 3^2)$	$2\cdot19$	2 (26)	$2^2\cdot37$
10	$\chi_7^3,\ \chi_{19}^2$	$(2, 3^2)$	2	1 (24)	$2^2\cdot37$
11	$\chi_7^3,\ \chi_{19}^3$	$(2, 2, 3)$	2	2 (26)	2^2
12	$\chi_7^3,\ \chi_{19}^6$	$(2, 3)$	2	1 (24)	2^2
13	$\chi_7^3,\ \chi_{19}^9$	$(2, 2)$	2	2 (26)	1
14	$\chi_7^3,\ \chi_7\chi_{19}^2,\ \chi_{19}^3$	$(2, 3^2, 2)$	2	2 (26)	$2^2\cdot3^4\cdot73$
14′	$\chi_7^3,\ \chi_7\bar{\chi}_{19}^2,\ \chi_{19}^3$	$(2, 3^2, 2)$	2	2 (26)	$2^2\cdot3^3\cdot19$
15	$\chi_7^2\chi_{19},\ \chi_{19}^3$	$(2, 3^2)$	2	1 (22)	3^4
15′	$\chi_7^2\bar{\chi}_{19},\ \chi_{19}^3$	$(2, 3^2)$	2	1 (22)	3^3
16	$\chi_7^3,\ \chi_7\chi_{19}^2,\ \chi_{19}^6$	$(2, 3^2)$	2	1 (29)	$2^2\cdot73$
16′	$\chi_7^3,\ \chi_7\bar{\chi}_{19}^2,\ \chi_{19}^6$	$(2, 3^2)$	2	1 (29)	$2^2\cdot19$
17	$\chi_7^3,\ \chi_7\chi_{19}^6,\ \chi_{19}^9$	$(2, 3, 2)$	2	2 (26)	3
17′	$\chi_7^3,\ \chi_7\bar{\chi}_{19}^6,\ \chi_{19}^9$	$(2, 3, 2)$	2	2 (26)	$3\cdot13$
18	$\chi_7^2\chi_{19}^3$	$(3, 2)$	2	1 (24)	3
18′	$\chi_7^2\bar{\chi}_{19}^3$	$(3, 2)$	2	1 (24)	3
19	$\chi_7\chi_{19}^6$	$(2, 3)$	2	1 (24)	1
19′	$\chi_7\bar{\chi}_{19}^6$	$(2, 3)$	2	1 (24)	13

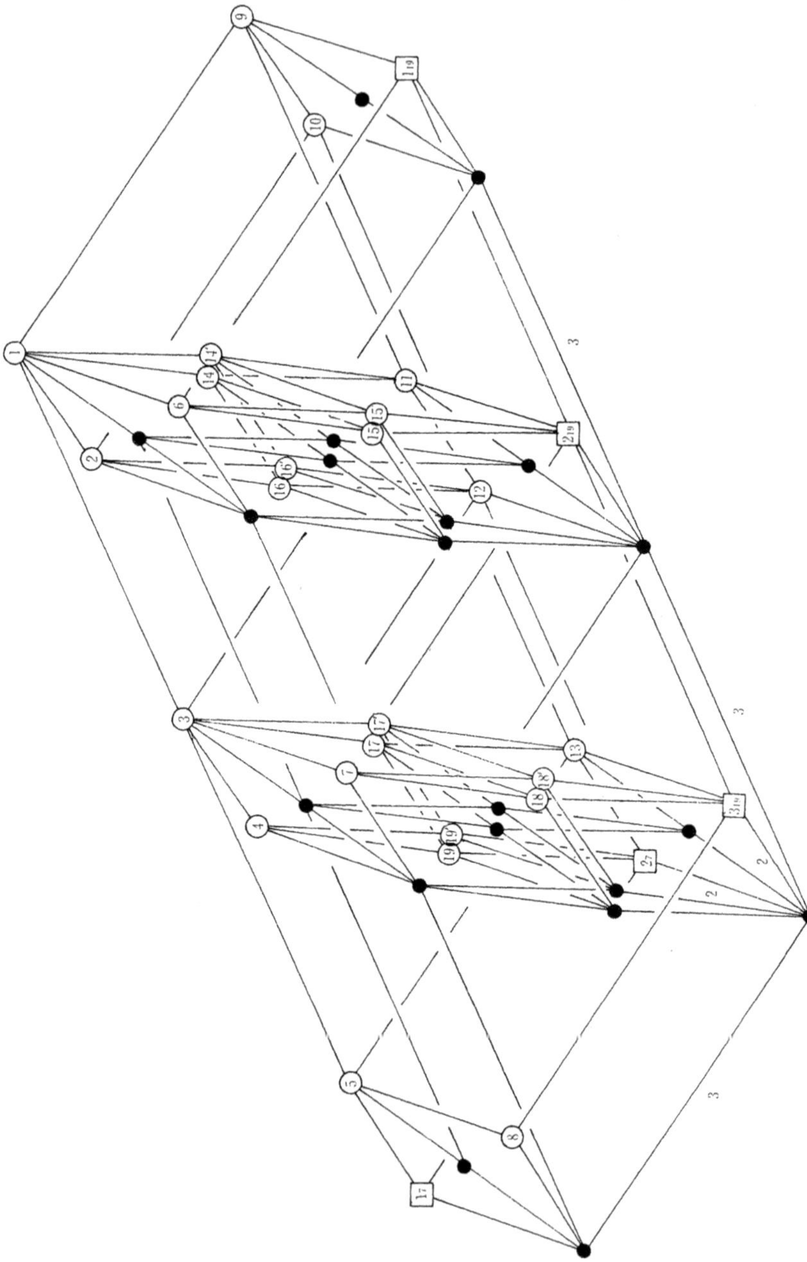
9
10
1
2
3
4
5
6
7
8
11
12
13
3
3
3
2
2

f=135					
No.	gen. Character	Type	w	Q	h^*
1	χ_3, ψ_{3^3}, χ_5	$(2, 3^2, 2^2)$	$2 \cdot 3^3 \cdot 5$	2 (27)	$37 \cdot 2053$
2	χ_3, ψ_{3^3}, χ_5^2	$(2, 3^2, 2)$	$2 \cdot 3^3$	1 (22)	37
3	ψ_{3^3}, χ_5	$(3^2, 2^2)$	$2 \cdot 5$	1 (24)	2053
4	ψ_{3^3}, $\chi_3\chi_5^2$	$(3^2, 2)$	2	1 (24)	$2 \cdot 37$

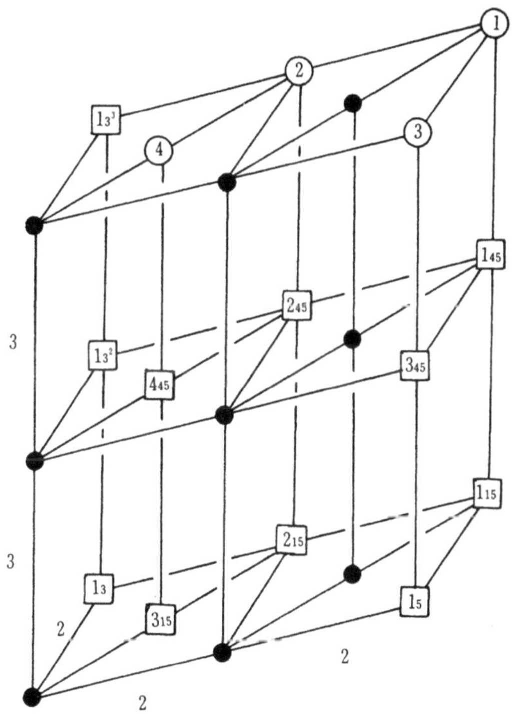

f=136					
No.	gen. Character	Type	w	Q	h^*
1	χ_{2^2}, ψ_{2^3}, χ_{17}	$(2, 2, 2^4)$	$2^3 \cdot 17$	2 (27)	$2^7 \cdot 3^2 \cdot 97$
2	χ_{2^2}, ψ_{2^3}, χ_{17}^2	$(2, 2, 2^3)$	2^3	1 (22)	$2^6 \cdot 3^2$
3	χ_{2^2}, ψ_{2^3}, χ_{17}^4	$(2, 2, 2^2)$	2^3	1 (22)	2^4
4	χ_{2^2}, ψ_{2^3}, χ_{17}^8	$(2, 2, 2)$	2^3	1 (22)	2^2
5	ψ_{2^3}, χ_{17}	$(2, 2^4)$	$2 \cdot 17$	1 (22)	$2 \cdot 97$
6	χ_{2^2}, $\psi_{2^3}\chi_{17}$	$(2, 2^4)$	2^2	2 (26)	$2^5 \cdot 97$
7	$\chi_{2^2}\psi_{2^3}$, $\chi_{2^2}\chi_{17}$	$(2, 2^4)$	2	2 (26)	$2^5 \cdot 3^2 \cdot 97$
8	$\chi_{2^2}\psi_{2^3}$, χ_{17}	$(2, 2^4)$	$2 \cdot 17$	2 (26)	$2^3 \cdot 3^2$
9	$\psi_{2^3}\chi_{17}$	2^4	2	1 (24)	$2^2 \cdot 97$
10	$\chi_{2^2}\psi_{2^3}$, χ_{17}^2	$(2, 2^3)$	2	1 (22)	$2^3 \cdot 3^2$
11	$\chi_{2^2}\psi_{2^3}$, $\chi_{2^2}\chi_{17}^2$	$(2, 2^3)$	2	1 (29)	2^3
12	$\chi_{2^2}\psi_{2^3}$, χ_{17}^4	$(2, 2^2)$	2	1 (22)	2^2
13	$\chi_{2^2}\psi_{2^3}$, $\chi_{2^2}\chi_{17}^4$	$(2, 2^2)$	2	1 (29)	2^2
14	$\chi_{2^2}\psi_{2^3}$, χ_{17}^8	$(2, 2)$	2	1 (22)	2
15	$\chi_{2^2}\psi_{2^3}$, $\chi_{2^2}\chi_{17}^8$	$(2, 2)$	2	1 (29)	2
16	ψ_{2^3}, $\chi_{2^2}\chi_{17}^2$	$(2, 2^3)$	2	1 (22)	$2^3 \cdot 3^2$
17	χ_{2^2}, $\psi_{2^3}\chi_{17}^2$	$(2, 2^3)$	2^2	1 (29)	$2^3 \cdot 3^2$
18	$\chi_{2^2}\psi_{2^3}\chi_{17}^2$	2^3	2	1 (24)	$2^2 \cdot 3^2$
19	ψ_{2^3}, $\chi_{2^2}\chi_{17}^4$	$(2, 2^2)$	2	1 (22)	2^3
20	χ_{2^2}, $\psi_{2^3}\chi_{17}^4$	$(2, 2^2)$	2^2	1 (29)	2^2
21	$\chi_{2^2}\psi_{2^3}\chi_{17}^4$	2^2	2	1 (24)	2^2
22	ψ_{2^3}, $\chi_{2^2}\chi_{17}^8$	$(2, 2)$	2	1 (25)	2^3
23	χ_{2^2}, $\psi_{2^3}\chi_{17}^8$	$(2, 2)$	2^2	1 (29)	2
24	$\chi_{2^2}\psi_{2^3}\chi_{17}^8$	2	2	1 (24)	2^2

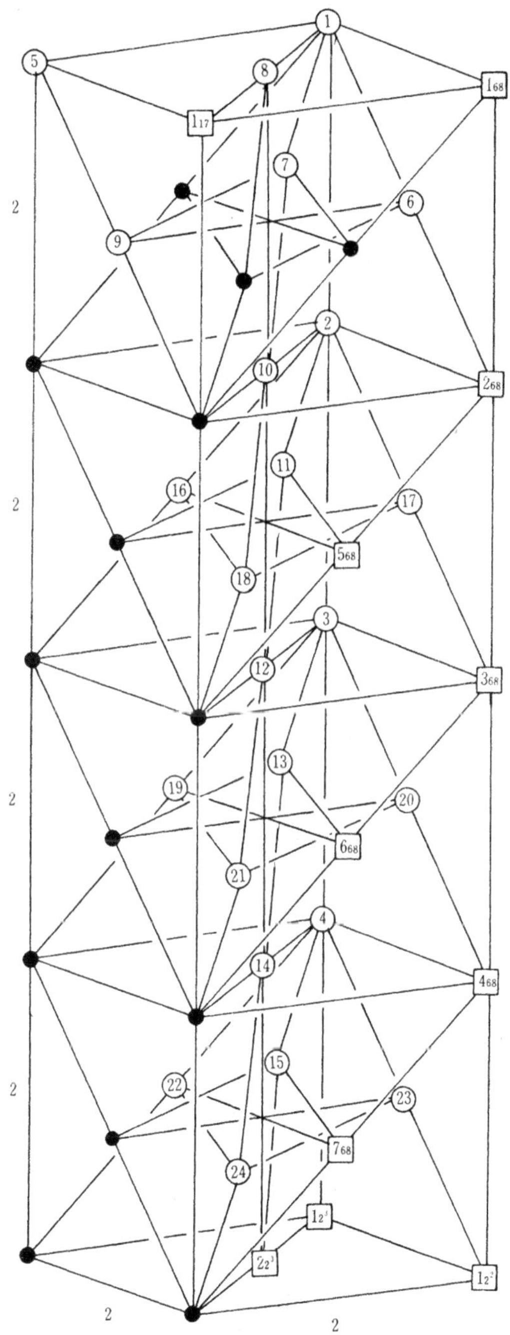
1
5
8
1_{68}
1_{17}
7
6
9
2
2
10
2_{68}
11
16
17
2
5_{68}
18
3
12
3_{68}
13
19
20
2
6_{68}
21
4
14
4_{68}
15
22
23
2
7_{68}
24
1_{2^3}
2_{2^3}
1_{2^2}
2
2

f=137					
No.	gen. Character	Type	w	Q	h^*
1	χ_{137}	$(2^3,\ 17)$	$2\cdot 137$	1 (27)	$17^2\cdot 47737\cdot$ 4689 05406 21121
2	χ_{137}^{17}	2^3	2	1 (23)	17

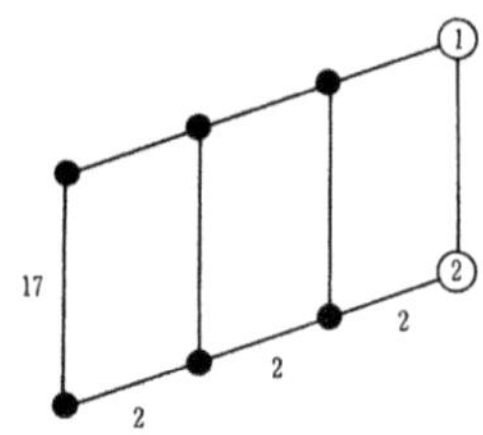

f=139					
No.	gen. Character	Type	w	Q	h^*
1	χ_{139}	(2, 3, 23)	$2\cdot 139$	1 (27)	$3^2\cdot 47^2\cdot 277^2\cdot 967\cdot$ 11889 61909
2	χ_{139}^{3}	(2, 23)	2	1 (23)	$3\cdot 47^2\cdot 277$
3	χ_{139}^{23}	(2, 3)	2	1 (23)	3^2
4	χ_{139}^{69}	2	2	1 (23)	3

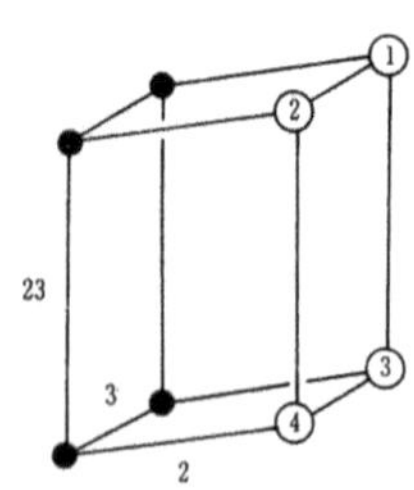

f=140					
No.	gen. Character	Type	w	Q	h^*
1	χ_{2^2}, χ_5, χ_7	$(2, 2^2, 2, 3)$	$2^2 \cdot 5 \cdot 7$	2 (27)	$3 \cdot 13$
2	χ_{2^2}, χ_5, χ_7^2	$(2, 2^2, 3)$	$2^2 \cdot 5$	2 (29)	3
3	χ_{2^2}, χ_5, χ_7^3	$(2, 2^2, 2)$	$2^2 \cdot 5$	2 (29)	1
4	χ_{2^2}, χ_5^2, χ_7	$(2, 2, 2, 3)$	$2^2 \cdot 7$	2 (29)	3
5	χ_5, $\chi_{2^2}\chi_7$	$(2^2, 2, 3)$	$2 \cdot 5$	1 (22)	$2 \cdot 13$
6	χ_{2^2}, χ_5^2, χ_7^3	$(2, 2, 2)$	2^2	2 (29)	1
7	χ_{2^2}, $\chi_5\chi_7$	$(2, 2^2, 3)$	2^2	1 (22)	$2 \cdot 3 \cdot 13$
8	$\chi_{2^2}\chi_5$, χ_7	$(2^2, 2, 3)$	$2 \cdot 7$	1 (22)	$2 \cdot 13$
9	χ_{2^2}, χ_5^2, χ_7^2	$(2, 2, 3)$	2^2	1 (29)	3
10	$\chi_{2^2}\chi_5\chi_7^3$, χ_7^2	$(2^2, 3)$	2	1 (24)	$2^2 \cdot 13$
11	$\chi_{2^2}\chi_5^2$, χ_7	$(2, 2, 3)$	$2 \cdot 7$	1 (**)	3
12	χ_{2^2}, $\chi_5^2\chi_7$	$(2, 2, 3)$	2^2	1 (**)	1
13	$\chi_{2^2}\chi_5^2$, $\chi_5^2\chi_7$	$(2, 2, 3)$	2	1 (22)	$2 \cdot 3$
14	$\chi_{2^2}\chi_5^2$, χ_7^2	$(2, 3)$	2	1 (24)	$2 \cdot 3$
15	χ_5, $\chi_{2^2}\chi_7^3$	$(2^2, 2)$	$2 \cdot 5$	1 (29)	2
16	$\chi_{2^2}\chi_5$, χ_7^3	$(2^2, 2)$	2	1 (29)	2
17	χ_{2^2}, $\chi_5\chi_7^3$	$(2, 2^2)$	2^2	1 (29)	2
18	$\chi_{2^2}\chi_5\chi_7^3$	2^2	2	1 (24)	2^2
19	$\chi_{2^2}\chi_5^2$, χ_7^3	$(2, 2)$	2	1 (*)	1
20	χ_{2^2}, $\chi_5^2\chi_7^3$	$(2, 2)$	2^2	1 (*)	1
21	$\chi_{2^2}\chi_5^2$, $\chi_5^2\chi_7^3$	$(2, 2)$	2	1 (29)	2

*See [5] § 26 **See [9]

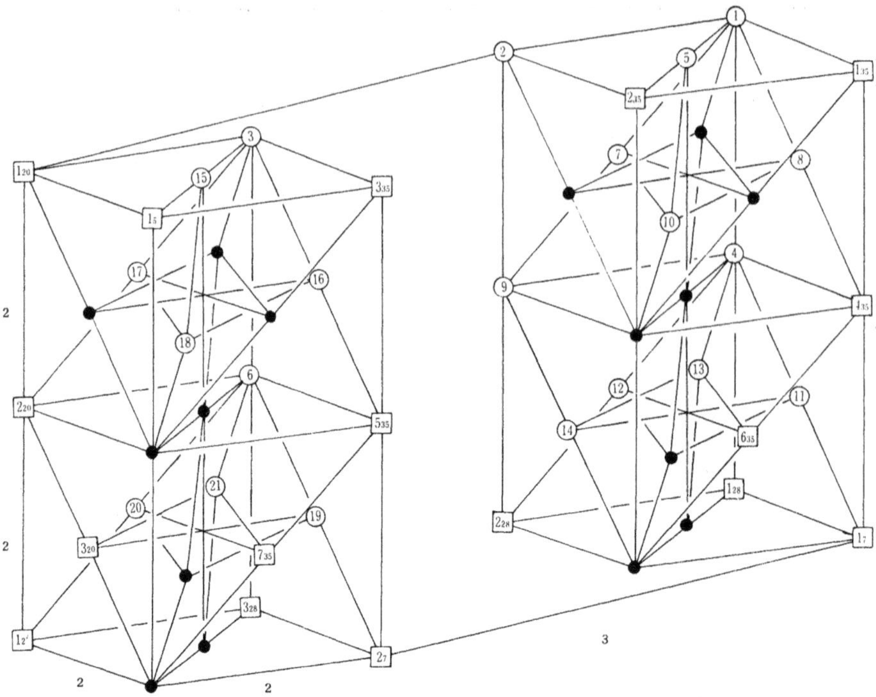

f=141					
No.	gen. Character	Type	w	Q	h^*
1	χ_3, χ_{47}	(2, 2, 23)	$2\cdot3\cdot47$	2 (27)	$5\cdot47\cdot139^2\cdot277$
2	χ_3, χ_{47}^2	(2, 23)	$2\cdot3$	1 (22)	$47\cdot139\cdot277$
3	χ_3, χ_{47}^{23}	(2, 2)	$2\cdot3$	2 (26)	5

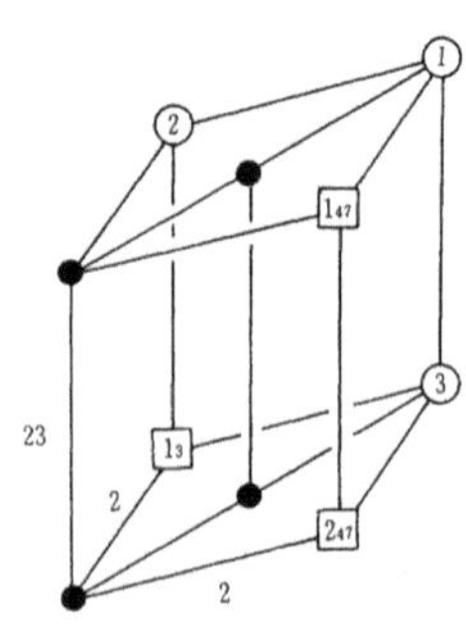

f=143					
No.	gen. Character	Type	w	Q	h^*
1	χ_{11}, χ_{13}	$(2, 2^2, 3, 5)$	$2\cdot 11\cdot 13$	2 (27)	$5^2\cdot 7\cdot 61^2\cdot 661\cdot 83701$
2	χ_{11}, χ_{13}^2	$(2, 2, 3, 5)$	$2\cdot 11$	1 (22)	$5^2\cdot 7\cdot 61\cdot 661$
3	χ_{11}, χ_{13}^3	$(2^2, 2, 5)$	$2\cdot 11$	2 (26)	$5^2\cdot 61$
4	χ_{11}, χ_{13}^4	$(2, 3, 5)$	$2\cdot 11$	1 (24)	$7\cdot 61$
5	χ_{11}, χ_{13}^6	$(2, 2, 5)$	$2\cdot 11$	1 (22)	5^2
6	χ_{11}^2, χ_{13}	$(2^2, 3, 5)$	$2\cdot 13$	1 (22)	$61\cdot 83701$
7	χ_{11}^2, χ_{13}^3	$(2^2, 5)$	2	1 (24)	61
8	χ_{11}^5, χ_{13}	$(2, 2^2, 3)$	$2\cdot 13$	2 (26)	$5\cdot 7$
9	χ_{11}^5, χ_{13}^2	$(2, 2, 3)$	2	1 (22)	$5\cdot 7$
10	χ_{11}^5, χ_{13}^3	$(2, 2^2)$	2	2 (26)	5
11	χ_{11}^5, χ_{13}^4	$(2, 3)$	2	1 (22)	7
12	χ_{11}^5, χ_{13}^6	$(2, 2)$	2	1 (22)	5
13	$\chi_{11}\chi_{13}^2$	$(2, 3, 5)$	2	1 (24)	$2\cdot 5^2\cdot 661$
14	$\chi_{11}\chi_{13}^6$	$(2, 5)$	2	1 (24)	$2\cdot 5^2$
15	$\chi_{11}^5\chi_{13}^2$	$(2, 3)$	2	1 (24)	$2\cdot 5$
16	$\chi_{11}^5\chi_{13}^6$	2	2	1 (24)	$2\cdot 5$

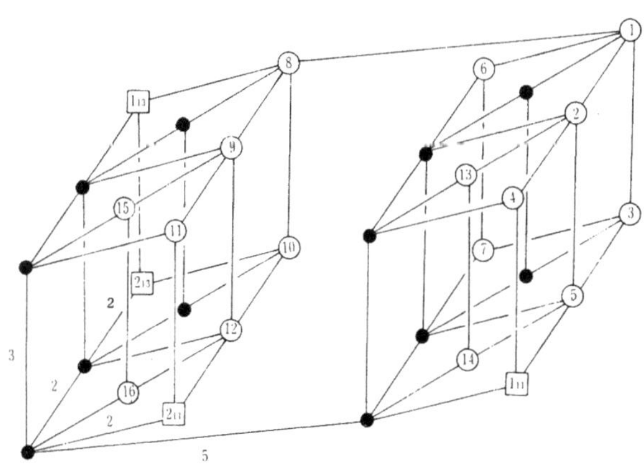

f=144					
No.	gen. Character	Type	w	Q	h^*
1	χ_{2^2}, ψ_{2^4}, χ_3, ψ_{3^2}	$(2, 2^2, 2, 3)$	$2^4 \cdot 3^2$	2 (27)	$3 \cdot 13^2$
2	χ_{2^2}, ψ_{2^4}, ψ_{3^2}	$(2, 2^2. 3)$	2^4	1 (22)	$3 \cdot 13$
3	ψ_{2^4}, χ_3, ψ_{3^2}	$(2^2, 2, 3)$	$2 \cdot 3^2$	1 (22)	13
4	$\chi_{2^2}\psi_{2^4}$, χ_3, ψ_{3^2}	$(2^2, 2, 3)$	$2 \cdot 3^2$	2 (29)	13
5	χ_{2^2}, $\psi_{2^4}\chi_3$, ψ_{3^2}	$(2, 2^2, 3)$	2^3	2 (29)	$2 \cdot 3 \cdot 13$
6	$\chi_{2^2}\psi_{2^4}$, $\chi_{2^2}\chi_3$, ψ_{3^2}	$(2^2, 2, 3)$	2	2 (29)	$2 \cdot 13^2$
7	$\psi_{2^4}\chi_3$, ψ_{3^2}	$(2^2, 3)$	2	1 (24)	$2 \cdot 13$
8	$\chi_{2^2}\psi_{2^4}$, ψ_{3^2}	$(2^2, 3)$	2	1 (24)	13

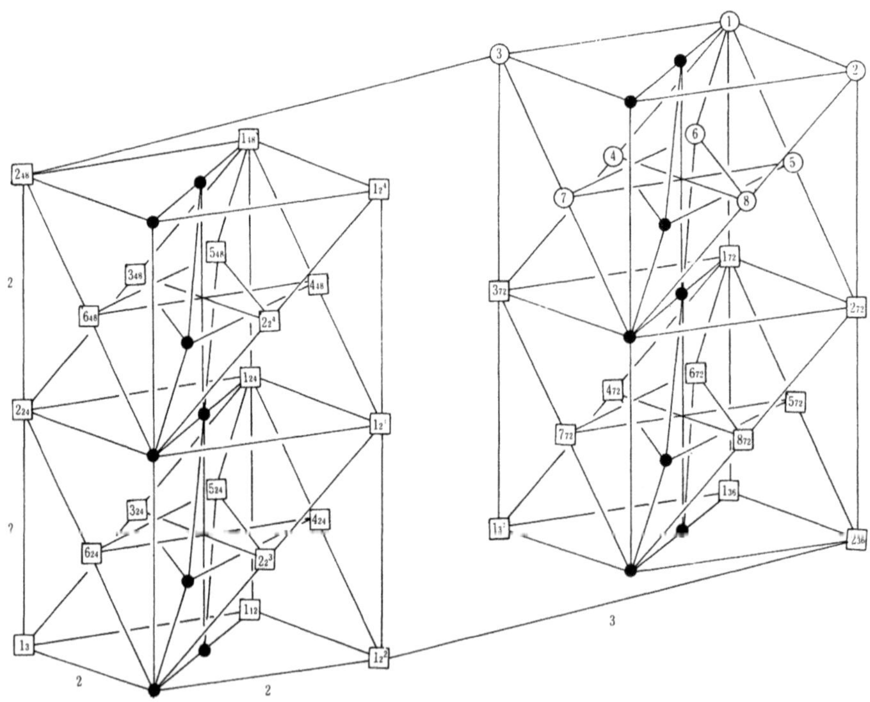

f=145					
No.	gen. Character	Type	w	Q	h^*
1	χ_5, χ_{29}	$(2^2,\ 2^2,\ 7)$	$2\cdot5\cdot29$	2 (27)	$2^{14}\cdot281\cdot421\cdot757$
2	χ_5, χ_{29}^2	$(2^2,\ 2,\ 7)$	$2\cdot5$	1 (22)	$2^7\cdot281\cdot421$
3	χ_5, χ_{29}^4	$(2^2,\ 7)$	$2\cdot5$	1 (24)	$2^3\cdot421$
4	χ_5, χ_{29}^7	$(2^2,\ 2^2)$	$2\cdot5$	2 (26)	2^2
5	χ_5, χ_{29}^{14}	$(2,\ 2^2)$	$2\cdot5$	1 (29)	2
6	χ_5^2, χ_{29}	$(2,\ 2^2,\ 7)$	$2\cdot29$	1 (22)	$2^7\cdot757$
7	χ_5^2, χ_{29}^7	$(2,\ 2^2)$	2	1 (29)	2
8	$\chi_5^2\chi_{29}$	$(2^2,\ 7)$	2	1 (24)	$2^5\cdot757$
9	$\chi_5^2\chi_{29}^7$	2^2	2	1 (24)	2^2
10	$\chi_5\chi_{29}^2$	$(2^2,\ 7)$	2	1 (24)	$2^5\cdot281$
11	$\chi_5\chi_{29}^{14}$	2^2	2	1 (24)	2^2

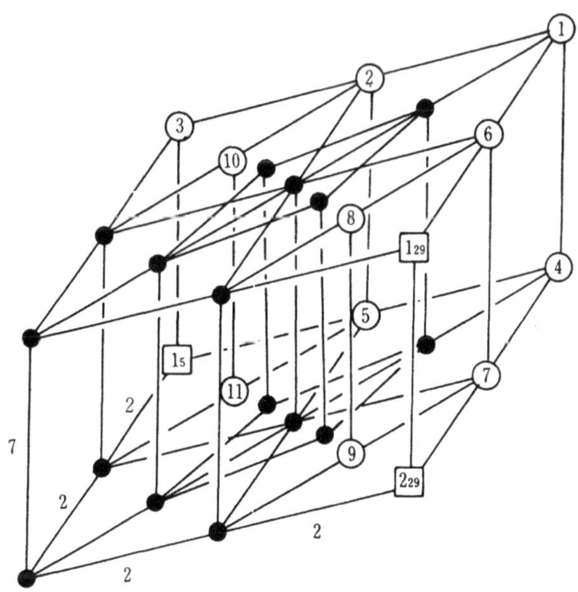

f=147					
No.	gen. Character	Type	w	Q	h^*
1	χ_3, χ_7, ψ_{7^2}	(2, 2, 3, 7)	$2\cdot3\cdot7^2$	2 (27)	$7\cdot29\cdot43\cdot673$
2	χ_3, χ_7^2, ψ_{7^2}	(2, 3, 7)	$2\cdot3$	1 (24)	$7\cdot29\cdot673$
3	χ_3, χ_7^3, ψ_{7^2}	(2, 2, 7)	$2\cdot3$	2 (26)	$7\cdot29$
4	χ_3, ψ_{7^2}	(2, 7)	$2\cdot3$	1 (24)	$7\cdot29$

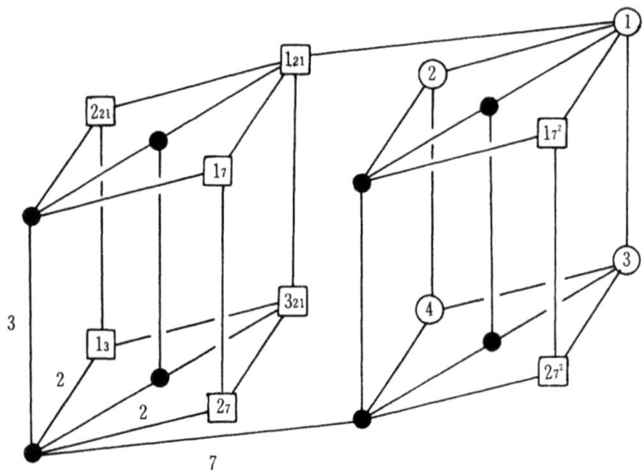

f=148					
No.	gen. Character	Type	w	Q	h^*
1	$\chi_{2^2},\ \chi_{37}$	$(2,\ 2^2,\ 3^2)$	$2^2\cdot 37$	2 (27)	$3^2\cdot 7\cdot 19\cdot 37\cdot 109$
2	$\chi_{2^2},\ \chi_{37}^2$	$(2,\ 2,\ 3^2)$	2^2	1 (22)	$3^2\cdot 7\cdot 19\cdot 109$
3	$\chi_{2^2},\ \chi_{37}^3$	$(2,\ 2^2,\ 3)$	2^2	2 (26)	$3\cdot 7$
4	$\chi_{2^2},\ \chi_{37}^4$	$(2,\ 3^2)$	2^2	1 (24)	$3^2\cdot 19$
5	$\chi_{2^2},\ \chi_{37}^6$	$(2,\ 2,\ 3)$	2^2	1 (29)	$3\cdot 7$
6	$\chi_{2^2},\ \chi_{37}^9$	$(2,\ 2^2)$	2^2	2 (26)	1
7	$\chi_{2^2},\ \chi_{37}^{12}$	$(2,\ 3)$	2^2	1 (24)	3
8	$\chi_{2^2},\ \chi_{37}^{18}$	$(2,\ 2)$	2^2	1 (29)	1
9	$\chi_{2^2}\chi_{37}^2$	$(2,\ 3^2)$	2	1 (24)	$2\cdot 7\cdot 109$
10	$\chi_{2^2}\chi_{37}^6$	$(2,\ 3)$	2	1 (24)	$2\cdot 7$
11	$\chi_{2^2}\chi_{37}^{18}$	2	2	1 (24)	2

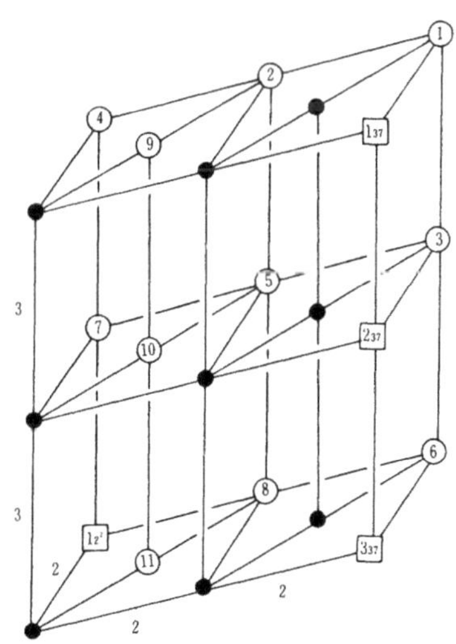

f=149					
No.	gen. Character	Type	w	Q	h^*
1	χ_{149}	$(2^2,\ 37)$	$2\cdot 149$	1 (27)	$3^2\cdot 149\cdot$ 5 12966 33832 00408 05461
2	χ_{149}^{37}	2^2	2	1 (23)	3^2

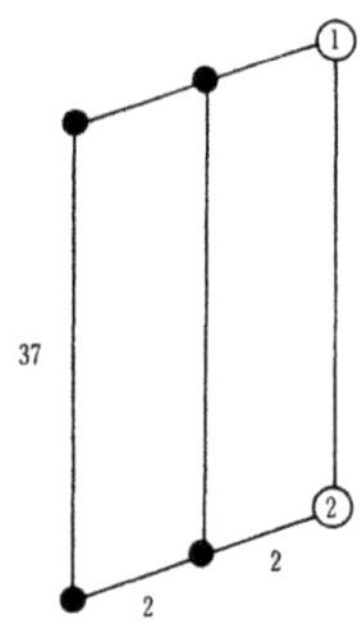

Bibliography

[1] E. E. Kummer, Mémoire sur la théorie des nombres complexes composés de racines de l'unité et nombres entires, Werke I, p. 363

[2] E. E. Kummer, Über die Klassenanzahl der aus n-ten Einheitenwurzeln gebildeten complexen Zahlen, Werke I, p. 883

[3] E. E. Kummer, Über diejenigen Primzahlen λ, für welche die Klassenzahl der aus λ -ten Einheitswurzeln gebildeten complexen Zahlen durch λ teilbar ist, Werke I, p. 945

[4] Z. I. Borevič and I. R. Šafarevič, Zahlentheorie, Birkhäuser Verlag, 1966

[5] H. Hasse, Über die Klassenzahl abelscher Zahlkörper, Akademie Verlag, 1952

[6] M. Newman, A Table of the First Factor for Prime Cyclotomic Fields, Math. Comp. vol. 24, p. 215-219, 1970

[7] D. H. Lehmer and J. M. Masley, Table of the Cyclotomic Class Number h* (p) and their Factors for 200<p <521, Math. Comp. vol. 32, p. 577-582, 1978

[8] G. Schrutka v. Rechtenstamm, Tabelle der (Relative-) Klassenzahlen der Kreiskörper, Abh. Deutsche Akad. Wiss. Math. Nat. Kl. Nr. 2, 1964

[9] M. Hirabayashi, On the unit indices of abelian number fields I, II (to appear)

[10] R. Brauer, Beziehungen zwischen Klassenzahlen von Teilkörpern eines galoischen Körpers, Math. Nachr. 4, 1951.

[11] C. G. Jacobi, Canon Arithmeticus, Akademie Verlag.1839

[12] M. N. Gras, Table numérique du nombre de classes et des unités des extensions cycliques réelles de degré 4 de **Q**. Publ. math. Fac. Sci. Besancon 1977/1978, fbsc. 2

(Received September 1, 1981)

Chapter 5
On the Relative Class Number of the Imaginary Abelian Number Field II

© Springer Nature Switzerland AG 2019

H. Hasse, *On the Class Number of Abelian Number Fields*,
https://doi.org/10.1007/978-3-030-01512-1_5

On the Relative Class Number of the Imaginary Abelian Number Field II

Ken-ichi YOSHINO* and Mikihito HIRABAYASHI**

This is the continuation of our previous paper [3] in which we treat all the imaginary abelian number fields over $\boldsymbol{Q}$, the field of rational numbers, with conductor f, $100<f\leqq 150$. In this paper we shall determine the unit indices and relative class numbers of the fields with conductor f, $150<f\leqq 200$.

We wish to express our hearty thanks to Mr. J. Kose for his much assistance in using the computer at Kanazawa Institute of Technology.

Now, the problems in this determination are

(1) how to draw the so-called Hasse's diagrams which denote the relations of inclusions of the abelian number fields with conductor f,

(2) how to determine the unit index Q,

(3) how to calculate the relative class number h^*.

For the problem (1), we also adopt the Hasse's method and symbols. We often omit some segments in Hasse's diagrams as before. In the cases of $f=160$, 168 and 195 we divide the Hasse's diagrams because they are too complicated to see the relations of inclusions.

For the problem (2), we got some results which determine the unit index after the previous paper was published. We summarize up them here.

Proposition 1 ([4]). Suppose that K is an imaginary abelian number field over $\boldsymbol{Q}$ and k its imaginary subfield. If the relative degree $[K:k]$ is odd, then $Q_K = Q_k$ where Q_K and Q_k are the unit indices of K and k respectively.

Proposition 2 ([4]). Suppose that K is an imaginary abelian number field with conductor f such that $f=4p^a$ or $f=p^aq^b$ where p, q are distinct odd primes and a, b are positive integers. Then $Q_K = 2$ if and only if the relative degree $[\boldsymbol{Q}(\zeta_f):K]$ is odd, where $\boldsymbol{Q}(\zeta_f)$ is the f-th cyclotomic field.

We note that Satz 29 in [1] is incorrect. Hasse's proof of Satz 29 only shows the next proposition.

*) Department of Mathematics **) Kanazawa Institute of Technology

Proposition 3. Let K and k be the same as in Proposition 1. If $\omega_k = \omega_K$, then Q_k divides Q_K where ω_k and ω_K are the 2-exponents of the numbers w_k and w_K of the roots of unity in k and K respectively.

In the case $\omega_k < \omega_K$ we found some examples that Q_k does not divide Q_K, i. e., $Q_k = 2$ and $Q_K = 1$ (See [5]). The technique used there is based on the following

Proposition 4 ([1] Satz 15). Let K be an imaginary abelian number field and K_o the maximal real subfield of K. Then $Q_K = 2$ if and only if there exists a unit ε in K_o such that $K_o' = K_o(\sqrt{\varepsilon})$ where $K' = K\mathbf{Q}(exp 2\pi i/2^{\omega+1})$, $\omega = \omega_K$ and K_o' is defined just as K_o.

As examples for the Proposition 4, let K be an imaginary abelian number field with conductor $f = 3 \cdot 5 \cdot 13 = 195$.

(i) Assume that the character group of K is generated by χ_3 and $\chi_5^2 \chi_{13}^3$. Then $K_o' = K_o(\sqrt{3})$ and $(3) = (\mathfrak{p}\mathfrak{p}')^2$ in K_o where $\mathfrak{p}$ and $\mathfrak{p}'$ are distinct prime ideals of K_o. Since the class number of K_o is 2 (Gras[2]), $\mathfrak{p}\mathfrak{p}'$ is principal in K_o. Setting $\mathfrak{p}\mathfrak{p}' = (\alpha)$, $\alpha \in K_o$, there exists a unit ε in K_o such that $3 = \alpha^2 \varepsilon$. Therefore by the above Proposition we obtain $Q_K = 2$ because $K_o' = K_o(\sqrt{\varepsilon})$.

(ii) If we replace χ_3 and $\chi_5^2 \chi_{13}^3$ by $\chi_3 \chi_5^2$ and χ_{13}^3, then $K_o' = K_o(\sqrt{15})$, $(3) = (\mathfrak{p}\mathfrak{p}')^2$ and $(5) = \mathfrak{q}^2$ in K_o, where $\mathfrak{p}$, $\mathfrak{p}'$ and $\mathfrak{q}$ are distinct prime ideals of K_o. So $Q_K = 2$ if and only if $\mathfrak{q}$ is principal in K_o. A defining equation for the maximal real subfield K_o of K is $f(x) = x^4 - 195x^2 + 2925 = 0$. Then the ideal class group of K_o is generated by the class which contains $\mathfrak{q}$. As there exists an unramified extension over K_o of degree 2, we see that $\mathfrak{q}$ is not principal. Therefore $Q_K = 1$.

For the problem (3), we got some formulae. The formula (7) in [3] is simplified by Proposition 1 as follows.

Proposition 5. Let K and k be the same as in Proposition 1 and let the Galois group $G(K/k)$ be of type (l, l), l an odd prime number. Then

$$h_K^* h_k^{*l} = \prod_M h_M^*$$

where M runs through all proper intermediate fields of the extension K/k.

Proposition 6. Suppose that K is an imaginary abelian number field and k its subfield such that the Galois group $G(K/k)$ is of type $(2, 2 \cdots, 2)$ (a-fold). Then

$$h_K^* h_k^{*\,2^a-2} = \frac{w_K w_k^{2^a-2}}{\prod_M w_M} \frac{Q_K Q_k^{2^a-2}}{\prod_M Q_M} \prod_M h_M^* \qquad \text{if } k \text{ is imaginary,}$$

$$h_K^* = \frac{w_K}{\prod_M w_M} \frac{Q_K}{\prod_M Q_M} \prod_M h_M^* \qquad \text{if } k \text{ is real,}$$

where M runs through all imaginary subfields of K/k such that $[M : k] = 2$.

After the previous paper was published, some unit indices in the table in [3] turn out to be incorrect because it is found that Satz 29 in [1] is incorrect. So we have to list the table of the correction of unit indices and relative class numbers.

f	No.	Q in [3]	Q correct	h^* correct
104	3	2 (29)	2 (**)	3
	10	1 (29)	1 (**)	3^3
	11	1 (29)	1 (**)	$2 \cdot 3^2$
	13	1 (29)	1 (**)	1
	21	1 (29)	1 (*)	3
	24	1 (29)	1 (*)	1
112	3	2 (29)	2 (**)	2^2
120	6	1 (25)**	1 (15)	2
	14	1 (29)	1 (15)	2
	16	1 (25)**	2 (15)	2
	24	1 (29)	1 (15)	2
	39	1 (29)	1 (*)	2
	49	1 (29)	1 (*)	1
	50	1 (29)	1 (*)	2
136	11	1 (29)	2 (15)	2^4
	13	1 (29)	2 (15)	2^3
	15	1 (29)	2 (15)	2^2
	17	1 (29)	2 (15)	$2^4 \cdot 3^2$
	20	1 (29)	2 (15)	2^3
	23	1 (29)	2 (15)	2^2

In this list (*) and (**) denote the following methods respectively.

(*) the facts in §26, [1]

(**) Proposition 1

In the following table we use these symbols. And we note that (29) denotes the determination of Q by using Proposition 3 in our whole table.

f=151					
No.	gen. Character	Type	w	Q	h^*
1	χ_{151}	$(2,\ 3,\ 5^2)$	2·151	1 (27)	$7\cdot 11^2\cdot 281\cdot 25951\cdot$ $\cdot 12\ 07501\cdot 3128\ 85301$
2	χ_{151}^{3}	$(2,\ 5^2)$	2	1 (23)	$7\cdot 281\cdot 25951$
3	χ_{151}^{5}	$(2,\ 3,\ 5)$	2	1 (23)	$7\cdot 11^2\cdot 281$
4	χ_{151}^{15}	$(2,\ 5)$	2	1 (23)	$7\cdot 281$
5	χ_{151}^{25}	$(2,\ 3)$	2	1 (23)	7
6	χ_{151}^{75}	2	2	1 (23)	7

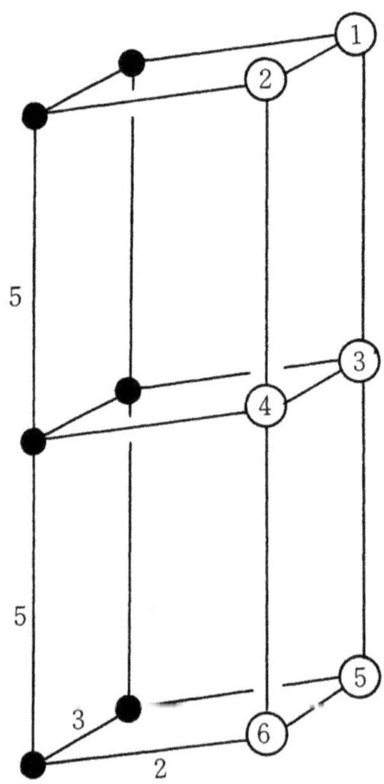

$f=152=2^3\cdot 19$					
No.	gen. Character	Type	w	Q	h^*
1	$\chi_{2^2},\ \psi_{2^3};\ \chi_{19}$	$(2,\ 2,\ 2,\ 3^2)$	$2^3\cdot 19$	2 (27)	$3^5\cdot 19^3$
2	$\chi_{2^2},\ \psi_{2^3};\ \chi_{19}^2$	$(2,\ 2,\ 3^2)$	2^3	1 (22)	$3^2\cdot 19^2$
3	$\chi_{2^2},\ \psi_{2^3};\ \chi_{19}^3$	$(2,\ 2,\ 2,\ 3)$	2^3	2 (29)	3^3
4	$\chi_{2^2},\ \psi_{2^3};\ \chi_{19}^6$	$(2,\ 2,\ 3)$	2^3	1 (22)	3
5	$\chi_{2^2},\ \psi_{2^3};\ \chi_{19}^9$	$(2,\ 2,\ 2)$	2^3	2 (26)	3
6	$\psi_{2^3},\ \chi_{19}$	$(2,\ 2,\ 3^2)$	$2\cdot 19$	1 (22)	$3^3\cdot 19$
7	$\psi_{2^3},\ \chi_{19}^3$	$(2,\ 2,\ 3)$	2	1 (22)	3^2
8	$\psi_{2^3},\ \chi_{19}^9$	$(2,\ 2)$	2	1 (22)	3
9	$\chi_{2^2}\psi_{2^3},\ \chi_{19}$	$(2,\ 2,\ 3^2)$	$2\cdot 19$	2 (26)	$3^2\cdot 19$
10	$\chi_{2^2}\psi_{2^3},\ \chi_{19}^2$	$(2,\ 3^2)$	2	1 (24)	$3^2\cdot 19$
11	$\chi_{2^2}\psi_{2^3},\ \chi_{19}^3$	$(2,\ 2,\ 3)$	2	2 (26)	3
12	$\chi_{2^2}\psi_{2^3},\ \chi_{19}^6$	$(2,\ 3)$	2	1 (24)	3
13	$\chi_{2^2}\psi_{2^3},\ \chi_{19}^9$	$(2,\ 2)$	2	2 (26)	1
14	$\chi_{2^2},\ \psi_{2^3}\chi_{19}$	$(2,\ 2,\ 3^2)$	2^2	2 (26)	$2\cdot 3^3\cdot 19^2$
15	$\chi_{2^2},\ \psi_{2^3}\chi_{19}^3$	$(2,\ 2,\ 3)$	2^2	2 (26)	$2\cdot 3^2$
16	$\chi_{2^2},\ \psi_{2^3}\chi_{19}^9$	$(2,\ 2)$	2^2	2 (26)	$2\cdot 3$
17	$\chi_{2^2}\psi_{2^3},\ \psi_{2^3}\chi_{19}$	$(2,\ 2,\ 3^2)$	2	2 (26)	$2\cdot 3^5\cdot 19^2$
18	$\chi_{2^2}\psi_{2^3},\ \psi_{2^3}\chi_{19}^3$	$(2,\ 2,\ 3)$	2	2 (26)	$2\cdot 3^3$
19	$\chi_{2^2}\psi_{2^3},\ \psi_{2^3}\chi_{19}^9$	$(2,\ 2)$	2	2 (26)	$2\cdot 3$
20	$\psi_{2^3}\chi_{19}$	$(2,\ 3^2)$	2	1 (24)	$2\cdot 3^3\cdot 19$
21	$\psi_{2^3}\chi_{19}^3$	$(2,\ 3)$	2	1 (24)	$2\cdot 3^2$
22	$\psi_{2^3}\chi_{19}^9$	2	2	1 (24)	$2\cdot 3$

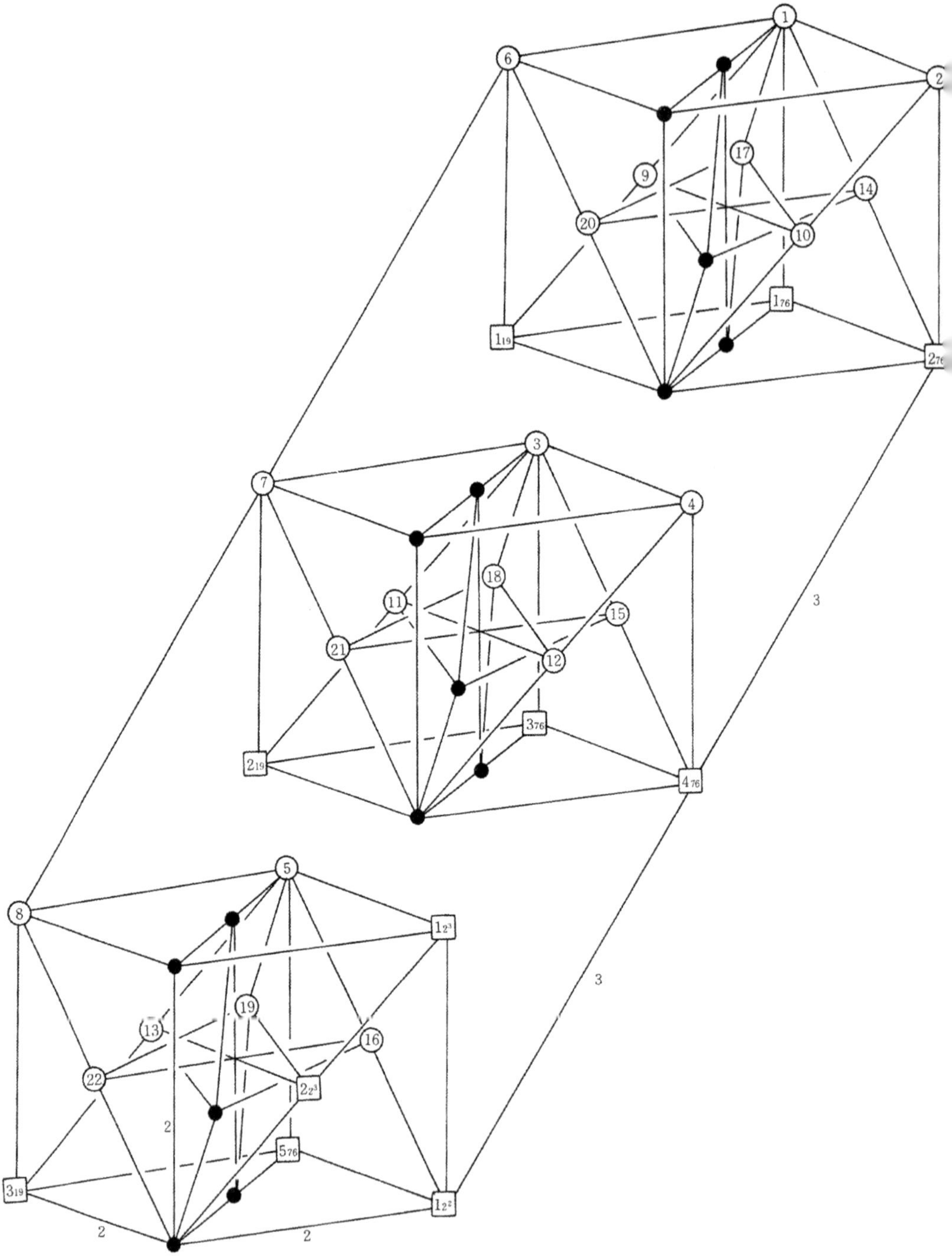
1
6
2
9
17
14
20
10
1_{19}
1_{76}
2_{76}
3
7
4
18
11
15
21
12
3_{76}
2_{19}
4_{76}
3
5
8
1_{2^3}
19
13
16
22
2_{2^3}
2
5_{76}
3_{19}
1_{2^2}
2
2
3

f=153=$3^2\cdot 17$					
No.	gen. Character	Type	w	Q	h^*
1	$\chi_3,\ \psi_{3^2}\ ;\ \chi_{17}$	$(2,\ 3,\ 2^4)$	$2\cdot 3^2\cdot 17$	2 (27)	$2^8\cdot 5\cdot 11^2\cdot 15601$
2	$\chi_3,\ \psi_{3^2}\ ;\ \chi_{17}^2$	$(2,\ 3,\ 2^3)$	$2\cdot 3^2$	1 (22)	$2^6\cdot 5\cdot 11^2$
3	$\chi_3,\ \psi_{3^2}\ ;\ \chi_{17}^4$	$(2,\ 3,\ 2^2)$	$2\cdot 3^2$	1 (22)	$2^4\cdot 5$
4	$\chi_3,\ \psi_{3^2}\ ;\ \chi_{17}^8$	$(2,\ 3,\ 2)$	$2\cdot 3^2$	1 (22)	2^2
5	$\chi_{17},\ \psi_{3^2}$	$(2^4,\ 3)$	$2\cdot 17$	1 (22)	$2^2\cdot 15601$
6	$\chi_3\chi_{17}^2,\ \psi_{3^2}$	$(2^3,\ 3)$	2	1 (24)	$2^3\cdot 11^2$
7	$\chi_3\chi_{17}^4,\ \psi_{3^2}$	$(2^2,\ 3)$	2	1 (24)	$2^3\cdot 5$
8	$\chi_3\chi_{17}^8,\ \psi_{3^2}$	$(2,\ 3)$	2	1 (24)	2^3

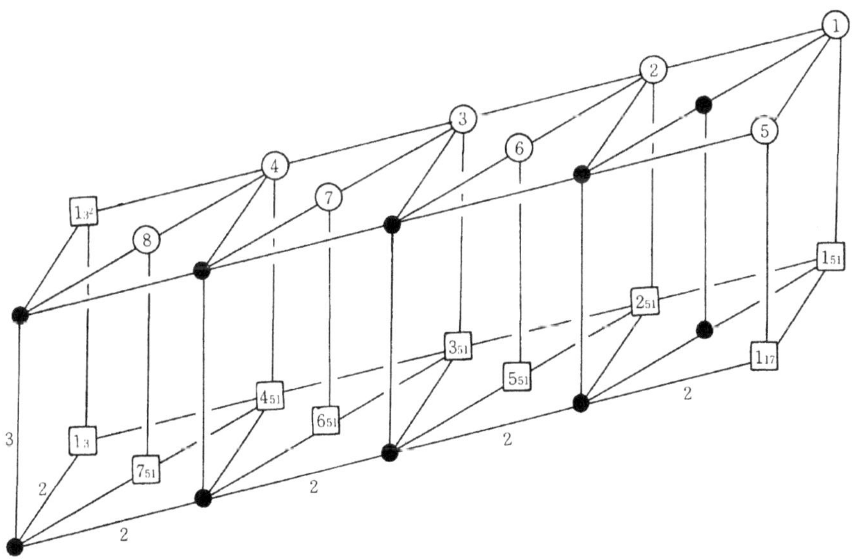

f=155=5・31					
No.	gen. Character	Type	w	Q	h^*
1	$\chi_5,\ \chi_{31}$	$(2^2,\ 2,\ 3,\ 5)$	$2\cdot5\cdot31$	2 (27)	$2^9\cdot3^4\cdot5^2\cdot631\cdot1\ 29121$
2	$\chi_5,\ \chi_{31}^2$	$(2^2,\ 3,\ 5)$	$2\cdot5$	1 (24)	$2^4\cdot3^2\cdot5^2\cdot1\ 29121$
3	$\chi_5,\ \chi_{31}^3$	$(2^2,\ 2,\ 5)$	$2\cdot5$	2 (26)	$2^9\cdot3\cdot5^2$
4	$\chi_5,\ \chi_{31}^5$	$(2^2,\ 2,\ 3)$	$2\cdot5$	2 (26)	$2\cdot3^4$
5	$\chi_5,\ \chi_{31}^6$	$(2^2,\ 5)$	$2\cdot5$	1 (24)	$2^4\cdot5^2$
6	$\chi_5,\ \chi_{31}^{10}$	$(2^2,\ 3)$	$2\cdot5$	1 (24)	3^2
7	$\chi_5,\ \chi_{31}^{15}$	$(2^2,\ 2)$	$2\cdot5$	2 (26)	$2\cdot3$
8	$\chi_5^2,\ \chi_{31}$	$(2,\ 2,\ 3,\ 5)$	$2\cdot31$	1 (22)	$2^5\cdot3^2\cdot631$
9	$\chi_5^2,\ \chi_{31}^3$	$(2,\ 2,\ 5)$	2	1 (22)	$2^5\cdot3$
10	$\chi_5^2,\ \chi_{31}^5$	$(2,\ 2,\ 3)$	2	1 (22)	$2\cdot3^2$
11	$\chi_5^2,\ \chi_{31}^{15}$	$(2,\ 2)$	2	1 (22)	$2\cdot3$
12	$\chi_5^2\chi_{31}$	$(2,\ 3,\ 5)$	2	1 (24)	$2^6\cdot631$
13	$\chi_5^2\chi_{31}^3$	$(2,\ 5)$	2	1 (24)	2^6
14	$\chi_5^2\chi_{31}^5$	$(2,\ 3)$	2	1 (24)	2^2
15	$\chi_5^2\chi_{31}^{15}$	2	2	1 (24)	2^2

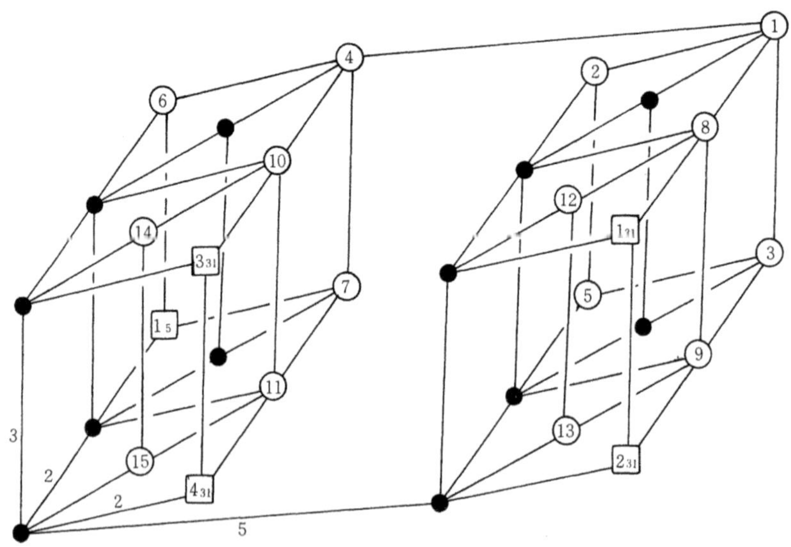

$f=156=2^2\cdot 3\cdot 13$					
No.	gen. Character	Type	w	Q	h^*
1	χ_{2^2}, χ_3, χ_{13}	$(2, 2, 2^2, 3)$	$2^2\cdot 3\cdot 13$	2 (27)	$2^2\cdot 3\cdot 13$
2	χ_{2^2}, χ_3, χ_{13}^2	$(2, 2, 2, 3)$	$2^2\cdot 3$	2 (29)	$2\cdot 3$
3	χ_{2^2}, χ_3, χ_{13}^3	$(2, 2, 2^2)$	$2^2\cdot 3$	2 (29)	2^2
4	χ_{2^2}, χ_3, χ_{13}^4	$(2, 2, 3)$	$2^2\cdot 3$	2 (29)	3
5	χ_{2^2}, χ_3, χ_{13}^6	$(2, 2, 2)$	$2^2\cdot 3$	2 (29)	2
6	$\chi_{2^2}\chi_3$, χ_{13}	$(2, 2^2, 3)$	$2\cdot 13$	1 (22)	$2^2\cdot 13$
7	$\chi_{2^2}\chi_3$, χ_{13}^3	$(2, 2^2)$	2	1 (22)	2^2
8	χ_{2^2}, $\chi_3\chi_{13}$	$(2, 2^2, 3)$	2^2	1 (22)	$2^2\cdot 3\cdot 13$
9	χ_{2^2}, $\chi_3\chi_{13}^2$	$(2, 2, 3)$	2^2	1 (**)	$2\cdot 3$
10	χ_{2^2}, $\chi_3\chi_{13}^3$	$(2, 2^2)$	2^2	1 (22)	2^2
11	χ_{2^2}, $\chi_3\chi_{13}^6$	$(2, 2)$	2^2	1 (*)	2
12	$\chi_{2^2}\chi_{13}$, χ_3	$(2^2, 3, 2)$	$2\cdot 3$	1 (22)	$2^3\cdot 13$
13	$\chi_{2^2}\chi_{13}^2$, χ_3	$(2, 3, 2)$	$2\cdot 3$	2 (**)	2
14	$\chi_{2^2}\chi_{13}^3$, χ_3	$(2^2, 2)$	$2\cdot 3$	1 (22)	2^3
15	$\chi_{2^2}\chi_{13}^6$, χ_3	$(2, 2)$	$2\cdot 3$	2 (*)	2
16	$\chi_{2^2}\chi_3\chi_{13}$	$(2^2, 3)$	2	1 (24)	$2^3\cdot 13$
17	$\chi_{2^2}\chi_3\chi_{13}^3$	2^2	2	1 (24)	2^3
18	$\chi_{2^2}\chi_3$, $\chi_3\chi_{13}^2$	$(2, 2, 3)$	2	1 (22)	2^2
19	$\chi_{2^2}\chi_3$, $\chi_3\chi_{13}^6$	$(2, 2)$	2	1 (22)	2^2

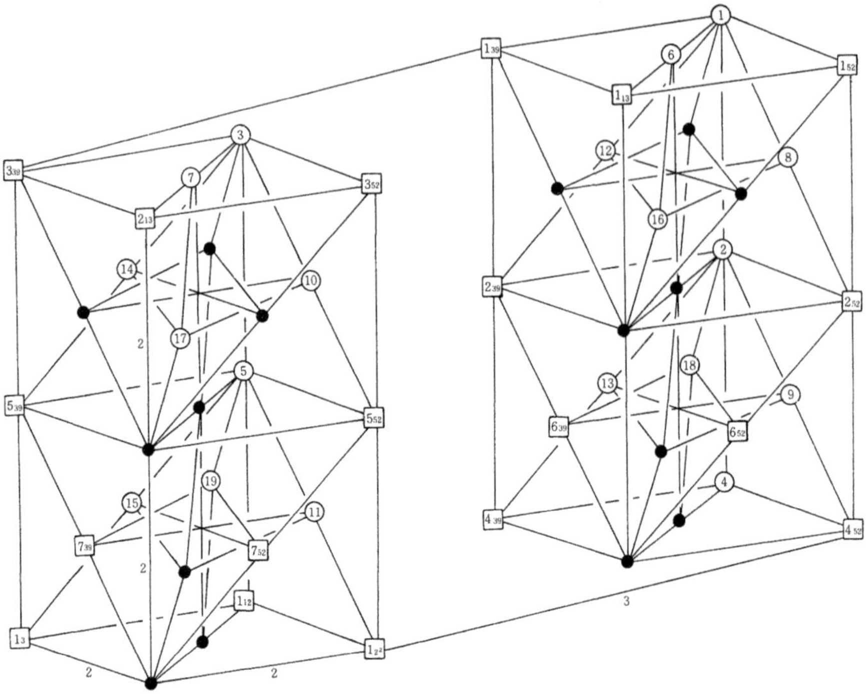

f=157					
No.	gen. Character	Type	w	Q	h^*
1	χ_{157}	$(2^2,\ 3,\ 13)$	$2\cdot 157$	1 (27)	$5\cdot 13^2\cdot 157^2\cdot 1093\cdot 1873$ $\cdot 4\ 18861\cdot 31\ 48601$
2	χ_{157}^{3}	$(2^2,\ 13)$	2	1 (23)	$5\cdot 31\ 48601$
3	χ_{157}^{13}	$(2^2,\ 3)$	2	1 (23)	$5\cdot 13$
4	χ_{157}^{39}	2^2	2	1 (23)	5

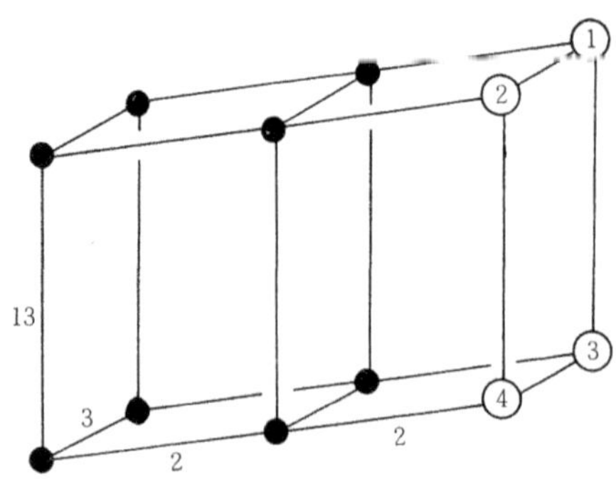

f=159=3・53					
No.	gen. Character	Type	w	Q	h^*
1	$\chi_3,\ \chi_{53}$	$(2,\ 2^2,\ 13)$	2・3・53	2 (27)	$5\cdot 53^2\cdot 3251\cdot 4889$
2	$\chi_3,\ \chi_{53}^2$	$(2,\ 2,\ 13)$	2・3	1 (22)	$5\cdot 53^2\cdot 3251$
3	$\chi_3,\ \chi_{53}^4$	$(2,\ 13)$	2・3	1 (24)	53^2
4	$\chi_3,\ \chi_{53}^{13}$	$(2,\ 2^2)$	2・3	2 (26)	5
5	$\chi_3,\ \chi_{53}^{26}$	$(2,\ 2)$	2・3	1 (22)	5
6	$\chi_3\chi_{53}^2$	$(2,\ 13)$	2	1 (24)	2・5・3251
7	$\chi_3\chi_{53}^{26}$	2	2	1 (24)	2・5

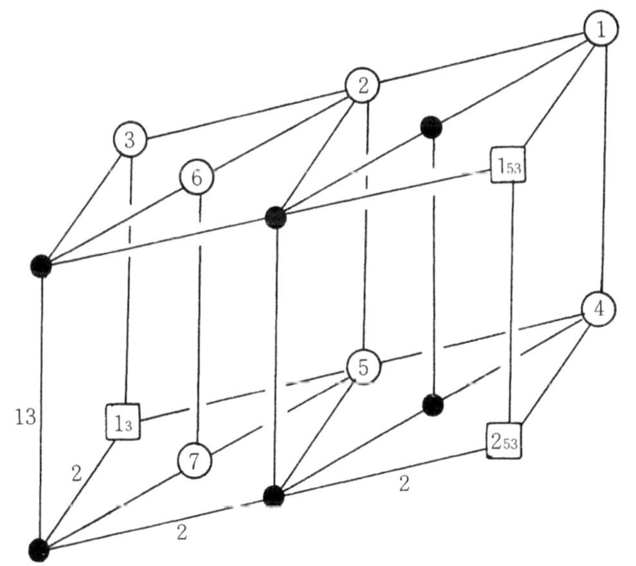

$f=160=2^5\cdot 5$					
No.	gen. Character	Type	w	Q	h^*
1	$\chi_{2^2},\ \psi_{2^5};\ \chi_5$	$(2,\ 2^3,\ 2^2)$	$2^5\cdot 5$	2 (27)	$3^2\cdot 5\cdot 17\cdot 41$
2	$\chi_{2^2},\ \psi_{2^5};\ \chi_5^2$	$(2,\ 2^3,\ 2)$	2^5	1 (22)	17
3	$\psi_{2^5},\ \chi_5$	$(2^3,\ 2^2)$	$2\cdot 5$	1 (22)	$3^2\cdot 5\cdot 41$
4	$\chi_{2^2}\psi_{2^5},\ \chi_5$	$(2^3,\ 2^2)$	$2\cdot 5$	2 (26)	$5\cdot 17$
5	$\chi_{2^2}\psi_{2^5},\ \chi_5^2$	$(2^3,\ 2)$	2	1 (22)	17
6	$\chi_{2^2},\ \psi_{2^5}\chi_5,\ \chi_5^2$	$(2,\ 2^3,\ 2)$	2^4	2 (26)	$2\cdot 3^2\cdot 41$
7	$\chi_{2^2},\ \psi_{2^5}\chi_5$	$(2,\ 2^3)$	2^3	2 (26)	$2\cdot 41$
7′	$\chi_{2^2},\ \psi_{2^5}\bar{\chi}_5$	$(2,\ 2^3)$	2^3	2 (26)	$2\cdot 3^2$
8	$\chi_{2^2},\ \psi_{2^5}\chi_5^2$	$(2,\ 2^3)$	2^4	1 (15)	17
9	$\psi_{2^5},\ \chi_{2^2}\chi_5^2$	$(2^3,\ 2)$	2	1 (22)	$2\cdot 17$
10	$\chi_{2^2}\psi_{2^5},\ \chi_{2^2}\chi_5$	$(2^3,\ 2^2)$	2	2 (26)	$2\cdot 3^2\cdot 17\cdot 41$
11	$\chi_{2^2}\psi_{2^5},\ \chi_{2^2}\chi_5^2$	$(2^3,\ 2)$	2	1 (15)	1
12	$\chi_5^2,\ \psi_{2^5}\chi_5$	$(2,\ 2^3)$	2	1 (22)	$2\cdot 3^2\cdot 41$
13	$\chi_{2^2}\chi_5^2,\ \psi_{2^5}\chi_5$	$(2,\ 2^3)$	2	2 (26)	$2\cdot 41$
13′	$\chi_{2^2}\chi_5^2,\ \psi_{2^5}\bar{\chi}_5$	$(2,\ 2^3)$	2	2 (26)	$2\cdot 3^2$
14	$\psi_{2^5}\chi_5$	2^3	2	1 (24)	$2\cdot 41$
14′	$\psi_{2^5}\bar{\chi}_5$	2^3	2	1 (24)	$2\cdot 3^2$
15	$\chi_{2^2}\psi_{2^5}\chi_5^2$	2^3	2	1 (24)	$2\cdot 17$

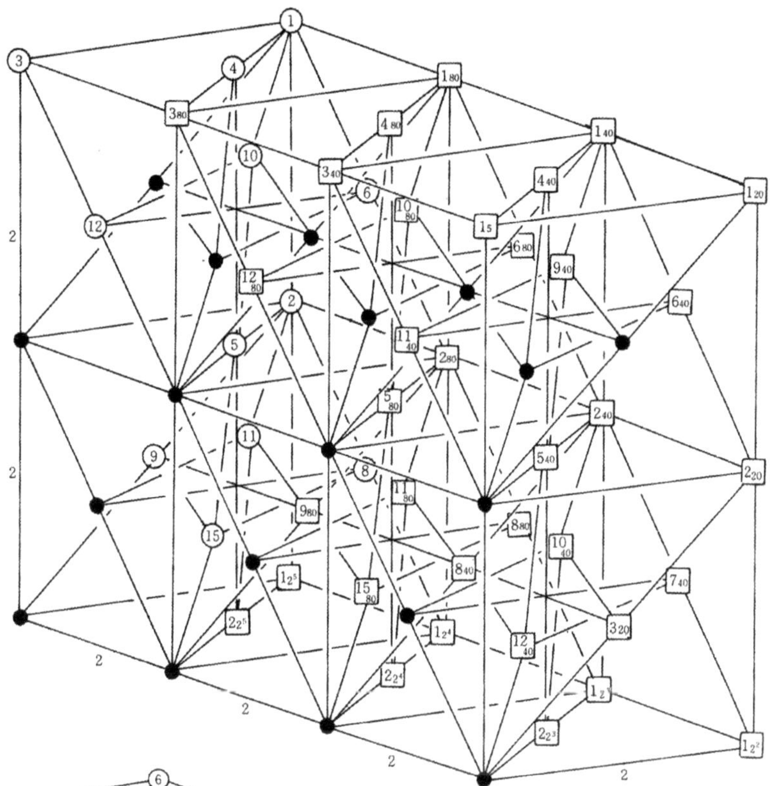

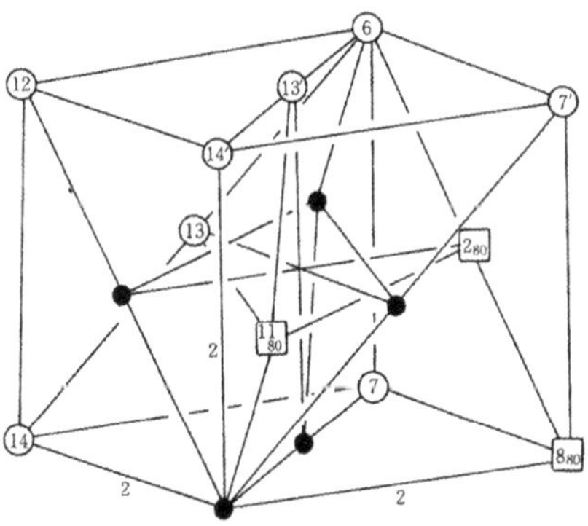

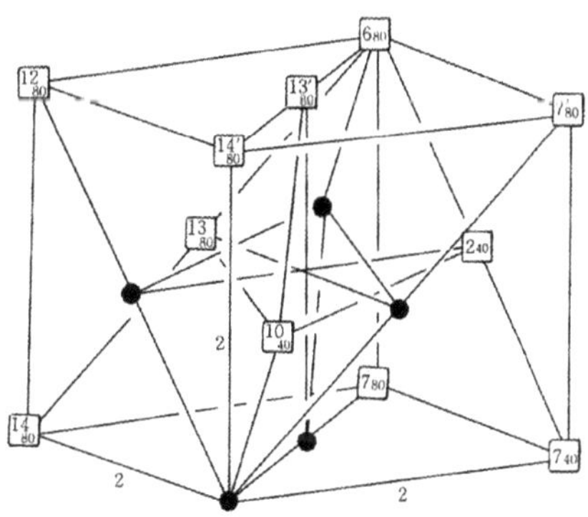

f=161=7・23					
No.	gen. Character	Type	w	Q	h^*
1	χ_7, χ_{23}	(2, 3, 2, 11)	2・7・23	2 (27)	3^2・11・67^3・22111・25873
2	χ_7, χ_{23}^2	(2, 3, 11)	2・7	1 (24)	11・67^2・22111
3	χ_7, χ_{23}^{11}	(2, 3, 2)	2・7	2 (26)	3^2
4	χ_7^2, χ_{23}	(3, 2, 11)	2・23	1 (24)	3^2・67・25873
5	χ_7^2, χ_{23}^{11}	(3, 2)	2	1 (24)	3^2
6	χ_7^3, χ_{23}	(2, 2, 11)	2・23	2 (26)	3・11・67
7	χ_7^3, χ_{23}^2	(2, 11)	2	1 (24)	11・67
8	χ_7^3, χ_{23}^{11}	(2, 2)	2	2 (26)	3

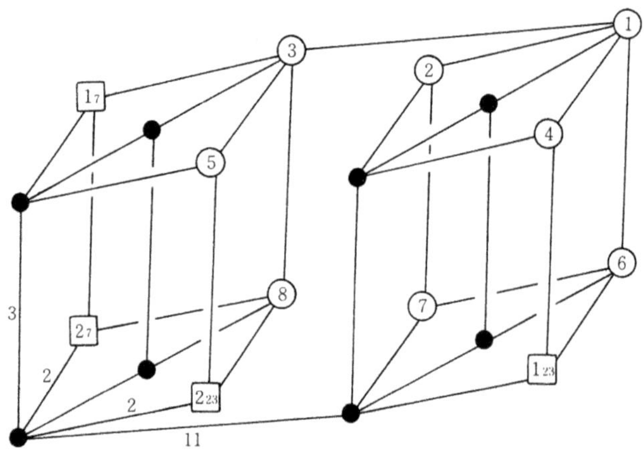

f=163					
No.	gen. Character	Type	w	Q	h^*
1	χ_{163}	(2, 3^4)	2・163	1 (27)	2^2・181・23167・3 65473 ・44184 58171 62679
2	χ_{163}^3	(2, 3^3)	2	1 (23)	2^2・181・3 65473
3	χ_{163}^9	(2, 3^2)	2	1 (23)	2^2・181
4	χ_{163}^{27}	(2, 3)	2	1 (23)	2^2
5	χ_{163}^{81}	2	2	1 (23)	1

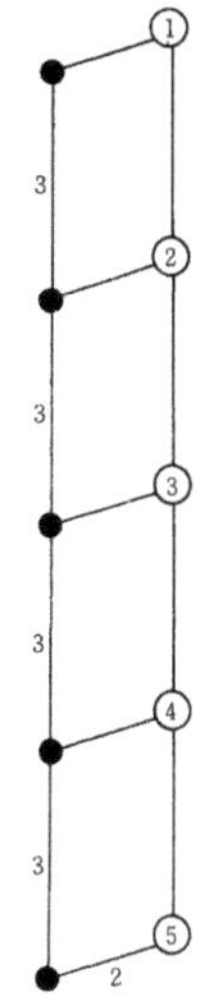

$f=164=2^2\cdot 41$					
No.	gen. Character	Type	w	Q	h^*
1	$\chi_{2^2},\ \chi_{41}$	$(2,\ 2^3,\ 5)$	$2^2\cdot 41$	2 (27)	$2^3\cdot 5\cdot 11^2\cdot 71\cdot 241$
2	$\chi_{2^2},\ \chi_{41}^2$	$(2,\ 2^2,\ 5)$	2^2	1 (22)	$2^3\cdot 5\cdot 71\cdot 241$
3	$\chi_{2^2},\ \chi_{41}^4$	$(2,\ 2,\ 5)$	2^2	1 (22)	$2^2\cdot 5\cdot 71$
4	$\chi_{2^2},\ \chi_{41}^5$	$(2,\ 2^3)$	2^2	2 (26)	2^3
5	$\chi_{2^2},\ \chi_{41}^8$	$(2,\ 5)$	2^2	1 (24)	5
6	$\chi_{2^2},\ \chi_{41}^{10}$	$(2,\ 2^2)$	2^2	1 (22)	2^3
7	$\chi_{2^2},\ \chi_{41}^{20}$	$(2,\ 2)$	2^2	1 (22)	2^2
8	$\chi_{2^2}\chi_{41}^2$	$(2^2,\ 5)$	2	1 (24)	$2^2\cdot 241$
9	$\chi_{2^2}\chi_{41}^4$	$(2,\ 5)$	2	1 (24)	$2^3\cdot 71$
10	$\chi_{2^2}\chi_{41}^{10}$	2^2	2	1 (24)	2^2
11	$\chi_{2^2}\chi_{41}^{20}$	2	2	1 (24)	2^3

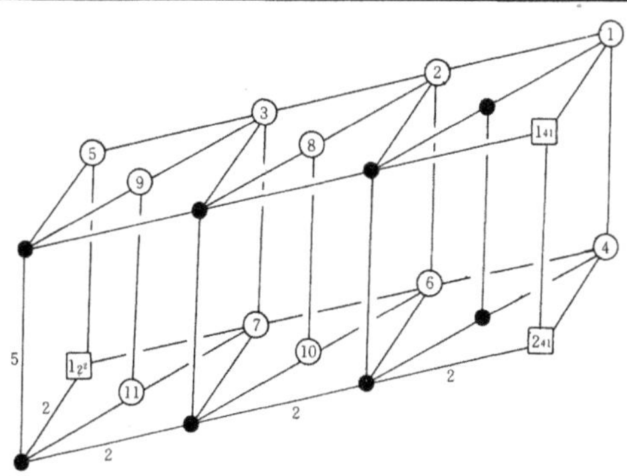

$f=165=3\cdot5\cdot11$					
No.	gen. Character	Type	w	Q	h^*
1	$\chi_3, \chi_5, \chi_{11}$	$(2, 2^2, 2, 5)$	$2\cdot3\cdot5\cdot11$	2 (27)	$2^2\cdot5\cdot11\cdot421$
2	$\chi_3, \chi_5, \chi_{11}^2$	$(2, 2^2, 5)$	$2\cdot3\cdot5$	2 (29)	$5\cdot11$
3	$\chi_3, \chi_5, \chi_{11}^5$	$(2, 2^2, 2)$	$2\cdot3\cdot5$	2 (29)	2^2
4	$\chi_3, \chi_5^2, \chi_{11}$	$(2, 2, 2, 5)$	$2\cdot3\cdot11$	2 (29)	$2\cdot11$
5	$\chi_3, \chi_5^2, \chi_{11}^2$	$(2, 2, 5)$	$2\cdot3$	1 (22)	11
6	$\chi_3, \chi_5^2, \chi_{11}^5$	$(2, 2, 2)$	$2\cdot3$	2 (29)	2
7	$\chi_3, \chi_5\chi_{11}$	$(2, 5, 2^2)$	$2\cdot3$	1 (22)	$2^2\cdot11\cdot421$
8	$\chi_3, \chi_5\chi_{11}^5$	$(2, 2^2)$	$2\cdot3$	1 (22)	2^2
9	$\chi_3, \chi_5^2\chi_{11}$	$(2, 2, 5)$	$2\cdot3$	1 (**)	2
10	$\chi_3, \chi_5^2\chi_{11}^5$	$(2, 2)$	$2\cdot3$	1 (*)	2
11	$\chi_3\chi_{11}, \chi_5$	$(2, 5, 2^2)$	$2\cdot5$	1 (22)	$2^2\cdot5\cdot421$
12	$\chi_3\chi_{11}^5, \chi_5$	$(2, 2^2)$	$2\cdot5$	1 (22)	2^2
13	$\chi_3\chi_5, \chi_{11}$	$(2^2, 2, 5)$	$2\cdot11$	1 (22)	$2^3\cdot421$
14	$\chi_3\chi_5, \chi_{11}^5$	$(2^2, 2)$	2	1 (22)	2^3
15	$\chi_3\chi_5^2, \chi_{11}$	$(2, 2, 5)$	$2\cdot11$	2 (**)	$2\cdot11$
16	$\chi_3\chi_5^2, \chi_{11}^5$	$(2, 2)$	2	2 (*)	2
17	$\chi_3\chi_5\chi_{11}$	$(2^2, 5)$	2	1 (24)	$2^3\cdot421$
18	$\chi_3\chi_5\chi_{11}^5$	2^2	2	1 (24)	2^3
19	$\chi_3\chi_5^2\chi_{11}^2$	$(2, 5)$	2	1 (24)	$2\cdot11$
20	$\chi_3\chi_5^2, \chi_3\chi_{11}$	$(2, 2, 5)$	2	1 (22)	$2^2\cdot11$
21	$\chi_3\chi_5^2, \chi_3\chi_{11}^5$	$(2, 2)$	2	1 (22)	2^2

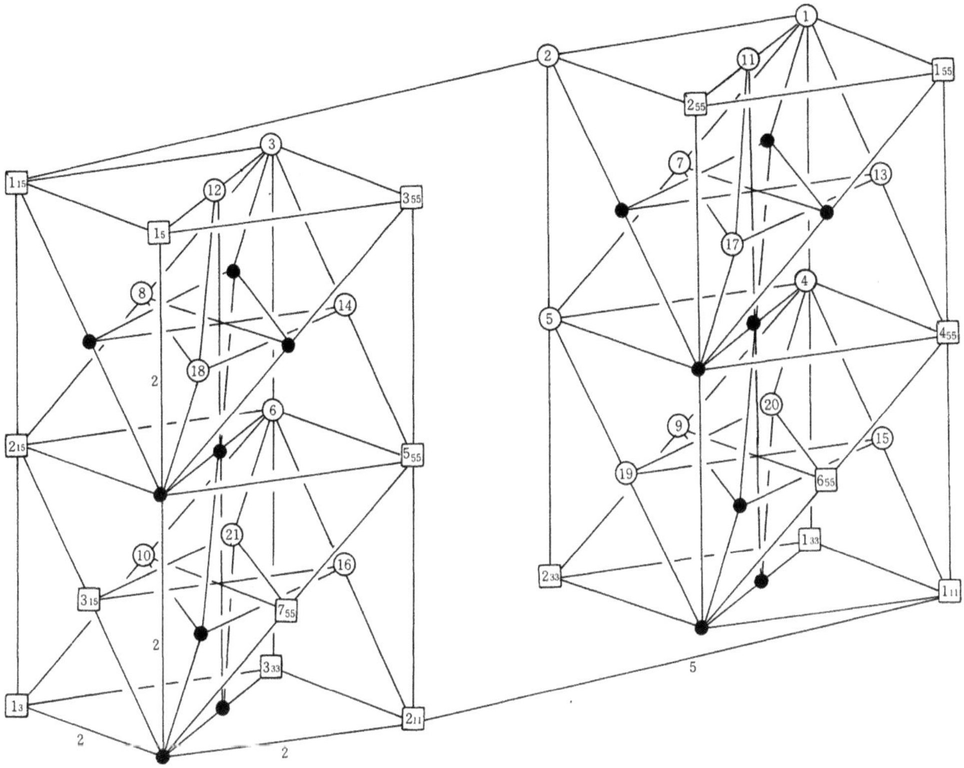

f=167					
No.	gen. Character	Type	w	Q	h^*
1	χ_{167}	(2, 83)	2·167	1 (27)	11·499·51231 89985 48422 90359 47419
2	χ_{167}^{83}	2	2	1 (23)	11

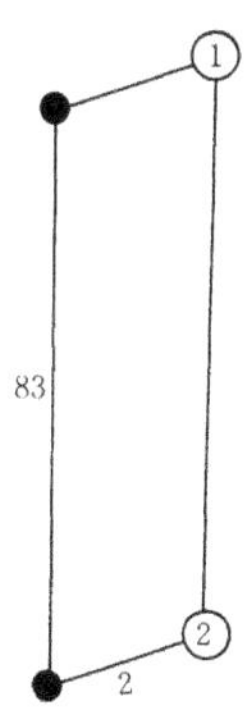

f=168=$2^3 \cdot 3 \cdot 7$					
No.	gen. Character	Type	w	Q	h^*
1	$\chi_{2^2}, \psi_{2^3}; \chi_3, \chi_7$	(2, 2, 2, 2, 3)	$2^3 \cdot 3 \cdot 7$	2 (27)	$2^2 \cdot 3 \cdot 7$
2	$\chi_{2^2}, \psi_{2^3}; \chi_3, \chi_7^2$	(2, 2, 2, 3)	$2^3 \cdot 3$	2 (29)	3
3	$\chi_{2^2}, \psi_{2^3}; \chi_3, \chi_7^3$	(2, 2, 2, 2)	$2^3 \cdot 3$	2 (29)	2^2
4	$\psi_{2^3}, \chi_3, \chi_7$	(2, 2, 2, 3)	$2 \cdot 3 \cdot 7$	2 (29)	$2 \cdot 3$
5	$\psi_{2^3}, \chi_3, \chi_7^2$	(2, 2, 3)	$2 \cdot 3$	1 (22)	3
6	$\psi_{2^3}, \chi_3, \chi_7^3$	(2, 2, 2)	$2 \cdot 3$	2 (29)	2
7	$\chi_{2^2}, \psi_{2^3}; \chi_3\chi_7$	(2, 2, 2, 3)	2^3	1 (22)	$2^2 \cdot 7$
8	$\chi_{2^2}, \psi_{2^3}; \chi_3\chi_7^3$	(2, 2, 2)	2^3	1 (22)	2^2
9	$\chi_{2^2}\psi_{2^3}, \chi_3, \chi_7$	(2, 2, 2, 3)	$2 \cdot 3 \cdot 7$	2 (29)	7
10	$\chi_{2^2}\psi_{2^3}, \chi_3, \chi_7^2$	(2, 2, 3)	$2 \cdot 3$	2 (29)	1
11	$\chi_{2^2}\psi_{2^3}, \chi_3, \chi_7^3$	(2, 2, 2)	$2 \cdot 3$	2 (29)	1
12	$\chi_{2^2}, \chi_3, \psi_{2^3}\chi_7$	(2, 2, 2, 3)	$2^2 \cdot 3$	2 (29)	$2^2 \cdot 7$
13	$\chi_{2^2}, \chi_3, \psi_{2^3}\chi_7^3$	(2, 2, 2)	$2^2 \cdot 3$	2 (29)	2^2
14	$\chi_{2^2}, \psi_{2^3}\chi_3, \chi_7$	(2, 2, 2, 3)	$2^2 \cdot 7$	2 (29)	$2 \cdot 3 \cdot 7$
15	$\chi_{2^2}, \psi_{2^3}\chi_3, \chi_7^2$	(2, 2, 3)	2^2	2 (29)	$2 \cdot 3$
16	$\chi_{2^2}, \psi_{2^3}\chi_3, \chi_7^3$	(2, 2, 2)	2^2	2 (29)	2
17	$\psi_{2^3}, \chi_{2^2}\chi_3, \chi_7$	(2, 2, 2, 3)	$2 \cdot 7$	1 (22)	$2^3 \cdot 7$
18	$\psi_{2^3}, \chi_{2^2}\chi_3, \chi_7^3$	(2, 2, 2)	2	1 (22)	2^3
19	$\psi_{2^3}, \chi_3, \chi_{2^2}\chi_7$	(2, 2, 2, 3)	$2 \cdot 3$	1 (22)	$2^2 \cdot 3 \cdot 7$
20	$\psi_{2^3}, \chi_3, \chi_{2^2}\chi_7^3$	(2, 2, 2)	$2 \cdot 3$	1 (22)	2^2
21	$\chi_3, \psi_{2^3}\chi_7$	(2, 2, 3)	$2 \cdot 3$	1 (**)	2
22	$\chi_3, \psi_{2^3}\chi_7^3$	(2, 2)	$2 \cdot 3$	1 (*)	2
23	$\psi_{2^3}\chi_3, \chi_7$	(2, 2, 3)	$2 \cdot 7$	2 (**)	$2 \cdot 3$
24	$\psi_{2^3}\chi_3, \chi_7^2$	(2, 3)	2	1 (24)	$2 \cdot 3$
25	$\psi_{2^3}\chi_3, \chi_7^3$	(2, 2)	2	2 (*)	2
26	$\chi_{2^2}\psi_{2^3}\chi_7, \chi_3$	(2, 3, 2)	$2 \cdot 3$	1 (22)	$2 \cdot 7$
27	$\chi_{2^2}\psi_{2^3}\chi_7^3, \chi_3$	(2, 2)	$2 \cdot 3$	1 (22)	2
28	$\chi_{2^2}\psi_{2^3}\chi_3, \chi_7$	(2, 2, 3)	$2 \cdot 7$	1 (22)	$2 \cdot 7$
29	$\chi_{2^2}\psi_{2^3}\chi_3, \chi_7^3$	(2, 2)	2	1 (22)	2

30	$\chi_{2^2}, \psi_{2^3}\chi_3\chi_7$	(2, 2, 3)	2^2	1 (∗∗)	2・7
31	$\chi_{2^2}, \psi_{2^3}\chi_3\chi_7^3$	(2, 2)	2^2	1 (∗)	2
32	$\psi_{2^3}, \chi_{2^2}\chi_3\chi_7$	(2, 2, 3)	2	1 (22)	2^3・7
33	$\psi_{2^3}, \chi_{2^2}\chi_3\chi_7^3$	(2, 2)	2	1 (22)	2^3
34	$\chi_{2^2}\psi_{2^3}, \psi_{2^3}\chi_3, \chi_7$	(2, 2, 2, 3)	2・7	2 (29)	2・3
35	$\chi_{2^2}\psi_{2^3}, \psi_{2^3}\chi_3, \chi_7^2$	(2, 2, 3)	2	2 (29)	2・3
36	$\chi_{2^2}\psi_{2^3}, \psi_{2^3}\chi_3, \chi_7^3$	(2, 2, 2)	2	2 (29)	2
37	$\chi_{2^2}\psi_{2^3}, \chi_3, \psi_{2^3}\chi_7$	(2, 2, 2, 3)	2・3	2 (29)	2^2
38	$\chi_{2^2}\psi_{2^3}, \chi_3, \psi_{2^3}\chi_7^3$	(2, 2, 2)	2・3	2 (29)	2^2
39	$\chi_{2^2}, \psi_{2^3}\chi_3, \chi_3\chi_7$	(2, 2, 2, 3)	2^2	2 (29)	2^3・3
40	$\chi_{2^2}, \psi_{2^3}\chi_3, \chi_3\chi_7^3$	(2, 2, 2)	2^2	2 (29)	2^3
41	$\chi_{2^2}\psi_{2^3}, \psi_{2^3}\chi_3, \chi_3\chi_7$	(2, 2, 2, 3)	2	2 (29)	2^3・3・7
42	$\chi_{2^2}\psi_{2^3}, \psi_{2^3}\chi_3, \chi_3\chi_7^3$	(2, 2, 2)	2	2 (29)	2^3
43	$\chi_{2^2}\psi_{2^3}, \chi_3\chi_7$	(2, 2, 3)	2	1 (22)	2・7
44	$\chi_{2^2}\psi_{2^3}, \chi_3\chi_7^3$	(2, 2)	2	1 (22)	2
45	$\chi_{2^2}\psi_{2^3}, \psi_{2^3}\chi_3\chi_7$	(2, 2, 3)	2	1 (22)	2
46	$\chi_{2^2}\psi_{2^3}, \psi_{2^3}\chi_3\chi_7^3$	(2, 2)	2	1 (22)	2
47	$\chi_{2^2}\chi_3, \psi_{2^3}\chi_7$	(2, 2, 3)	2	1 (22)	2^3・7
48	$\chi_{2^2}\chi_3, \psi_{2^3}\chi_7^3$	(2, 2)	2	1 (22)	2^3
49	$\chi_{2^2}\psi_{2^3}\chi_3, \psi_{2^3}\chi_7$	(2, 2, 3)	2	1 (22)	2^3
50	$\chi_{2^2}\psi_{2^3}\chi_3, \psi_{2^3}\chi_7^3$	(2, 2)	2	1 (22)	2^3
51	$\psi_{2^3}\chi_3, \chi_3\chi_7$	(2, 2, 3)	2	1 (22)	2^2・3
52	$\psi_{2^3}\chi_3, \chi_3\chi_7^3$	(2, 2)	2	1 (22)	2^2
53	$\psi_{2^3}\chi_3, \chi_{2^2}\chi_7$	(2, 2, 3)	2	1 (22)	2^2・3・7
54	$\psi_{2^3}\chi_3, \chi_{2^2}\chi_7^3$	(2, 2)	2	1 (22)	2^2
55	$\psi_{2^3}\chi_3, \chi_{2^2}\chi_3\chi_7$	(2, 2, 3)	2	1 (22)	2^2・3
56	$\psi_{2^3}\chi_3, \chi_{2^2}\chi_3\chi_7^3$	(2, 2)	2	1 (22)	2^2
57	$\chi_{2^2}\psi_{2^3}\chi_3\chi_7$	(2, 3)	2	1 (24)	2^2・7
58	$\chi_{2^2}\psi_{2^3}\chi_3\chi_7^3$	2	2	1 (24)	2^2

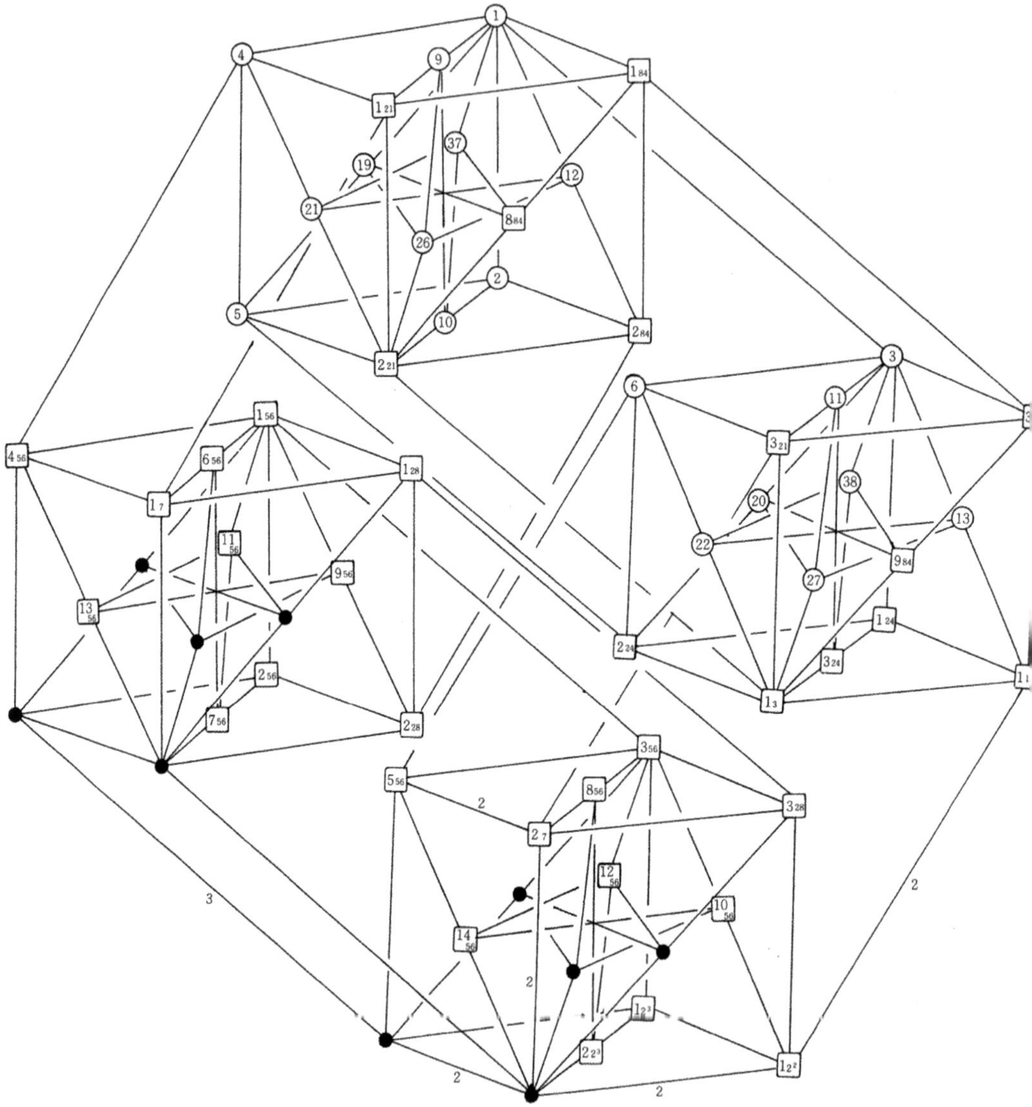
1
4
9
1 84
1 21
37
19
12
21
8 84
26
2
5
10
2 84
2 21
3
6
11
3 21
38
20
13
22
9 84
27
1 24
2 24
3 24
1 3
1 1
1 56
4 56
6 56
1 28
1 7
11 56
9 56
13 56
2 56
7 56
2 28
3 56
5 56
8 56
3 28
2 7
12 56
10 56
14 56
1 2³
2 2³
1 2²
2
3

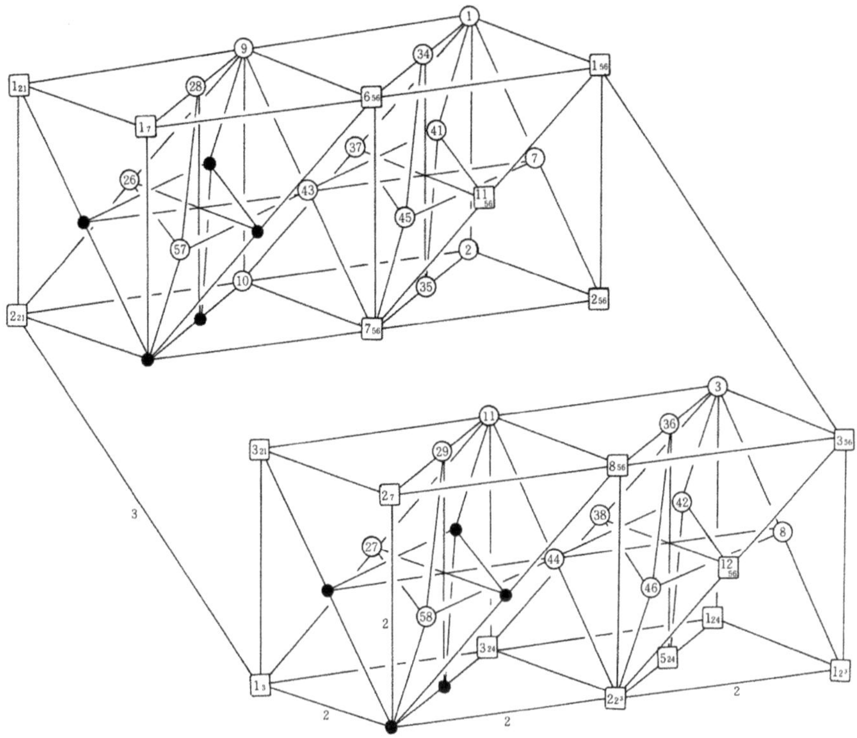

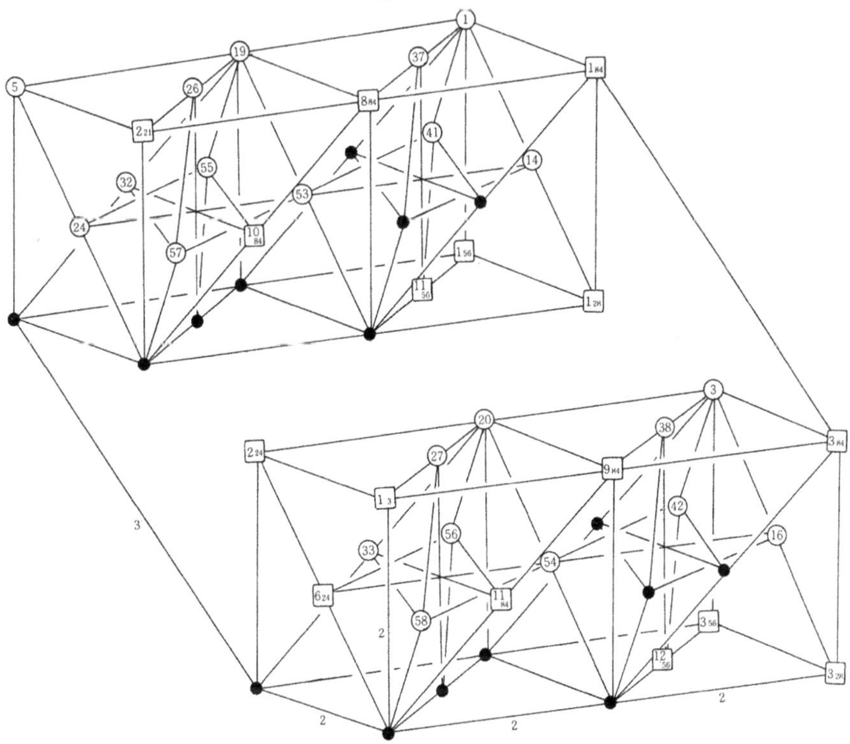

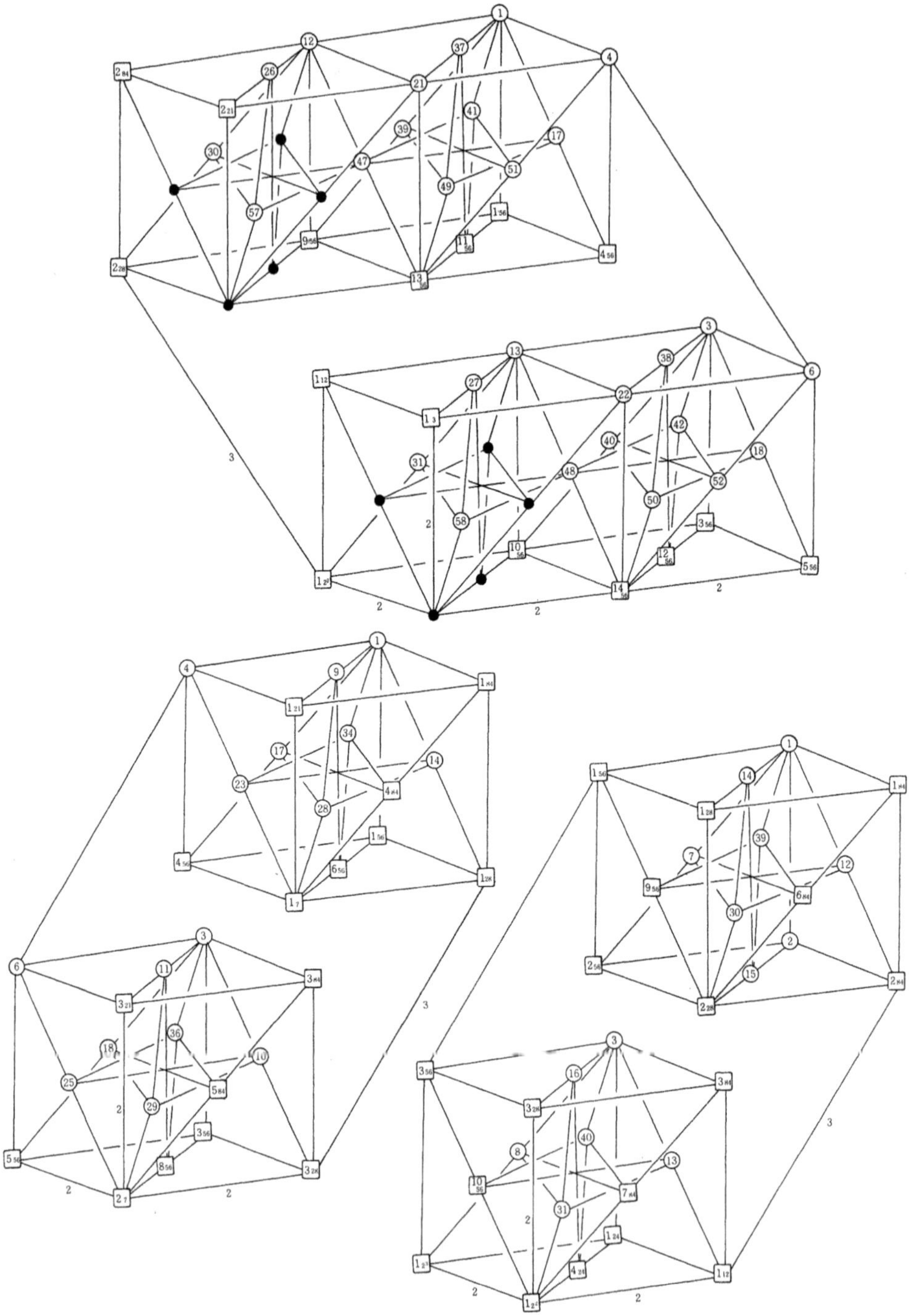

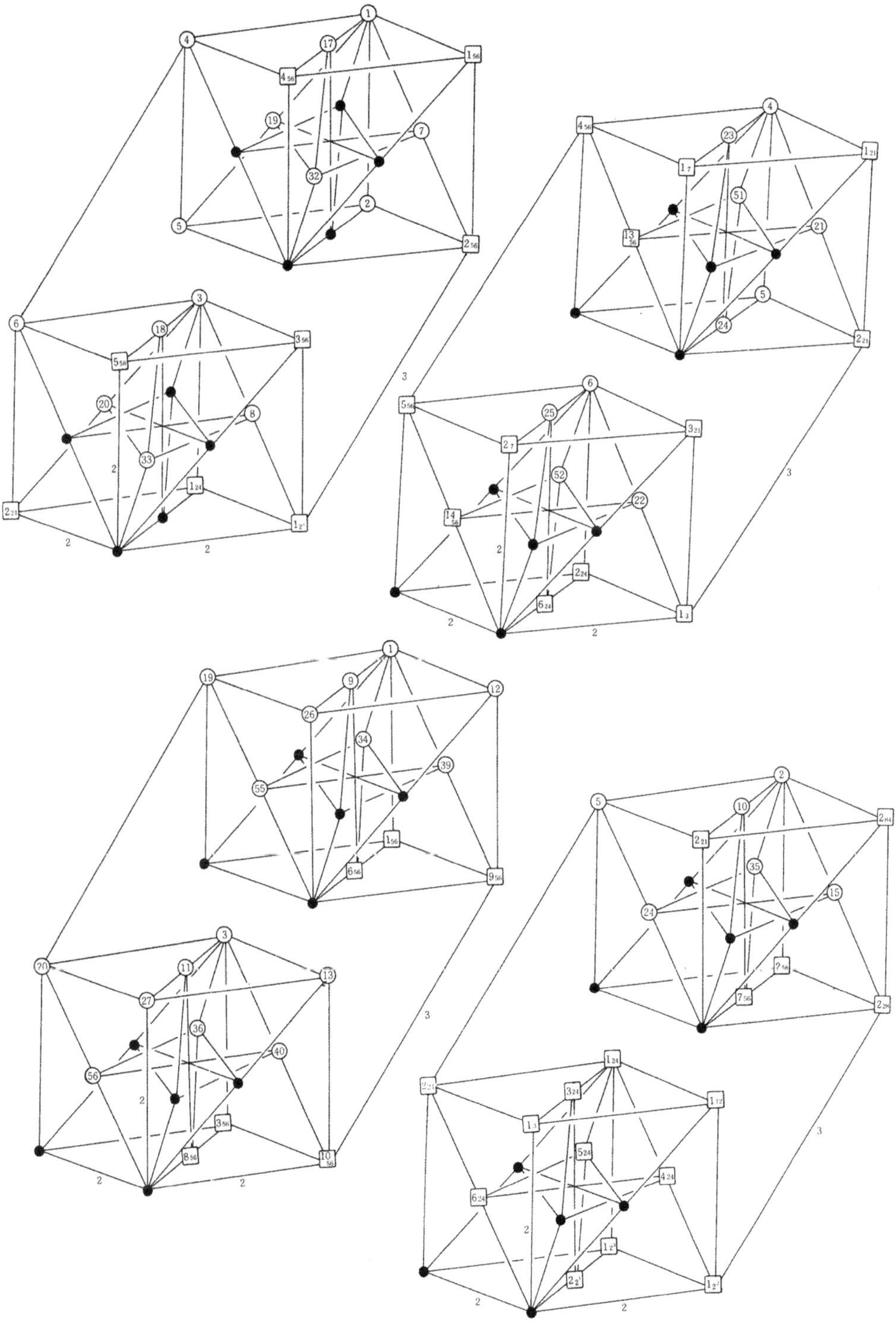

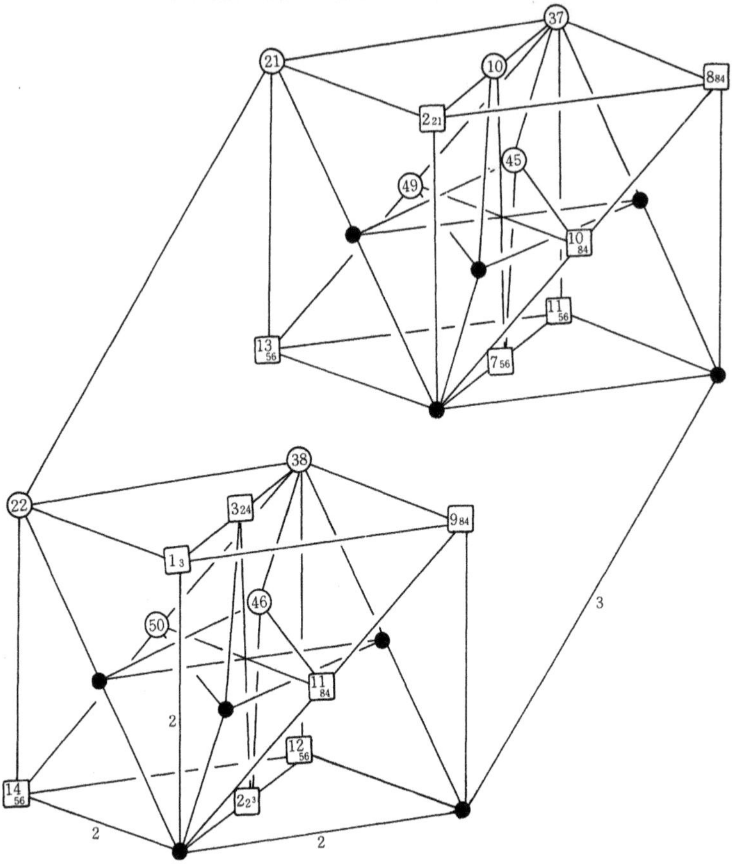

f=169=13²					
No.	gen. Character	Type	w	Q	h^*
1	χ_{13}, ψ_{13^2}	$(2^2, 3, 13)$	$2\cdot 13^2$	1 (27)	313·1873·4733·19695 32962 89361
2	χ_{13}^3, ψ_{13^2}	$(2^2, 13)$	2	1 (23)	1873 · 4733

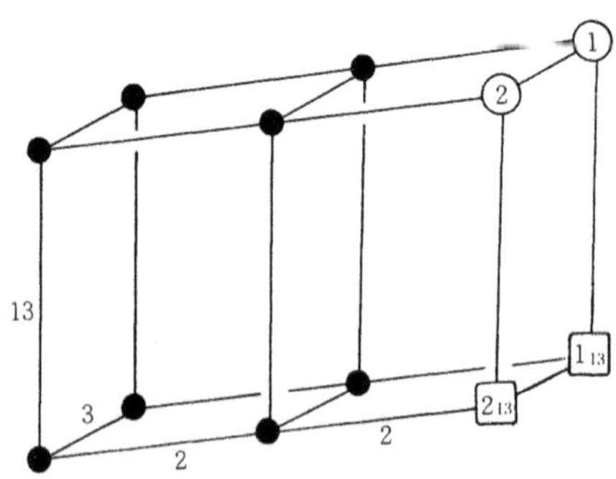

f=171=$3^2\cdot 19$					
No.	gen. Character	Type	w	Q	h^*
1	$\chi_3,\ \psi_{3^2};\ \chi_{19}$	$(2,\ 3,\ 2,\ 3^2)$	$2\cdot 3^2\cdot 19$	2 (27)	$2^2\cdot 3^6\cdot 7\cdot 19\cdot 73\cdot 109\cdot 163$
2	$\chi_3,\ \psi_{3^2};\ \chi_{19}^2$	$(2,\ 3,\ 3^2)$	$2\cdot 3^2$	1 (22)	$3^6\cdot 19\cdot 109$
3	$\chi_3,\ \psi_{3^2};\ \chi_{19}^3$	$(2,\ 3,\ 2,\ 3)$	$2\cdot 3^2$	2 (26)	$2^2\cdot 3^3\cdot 7$
4	$\chi_3,\ \psi_{3^2};\ \chi_{19}^6$	$(2,\ 3,\ 3)$	$2\cdot 3^2$	1 (22)	3^3
5	$\chi_3,\ \psi_{3^2};\ \chi_{19}^9$	$(2,\ 3,\ 2)$	$2\cdot 3^2$	2 (26)	2^2
6	$\psi_{3^2},\ \chi_{19}$	$(3,\ 2,\ 3^2)$	$2\cdot 19$	1 (22)	$2^2\cdot 7\cdot 73\cdot 163$
7	$\psi_{3^2},\ \chi_{19}^3$	$(3,\ 2,\ 3)$	2	1 (22)	$2^2\cdot 7$
8	$\psi_{3^2},\ \chi_{19}^9$	$(3,\ 2)$	2	1 (24)	2^2
9	$\chi_3,\ \psi_{3^2}\chi_{19}^2,\ \chi_{19}^9$	$(2,\ 3^2,\ 2)$	$2\cdot 3$	2 (26)	$3^2\cdot 19\cdot 73$
9′	$\chi_3,\ \psi_{3^2}\bar{\chi}_{19}^2,\ \chi_{19}^9$	$(2,\ 3^2,\ 2)$	$2\cdot 3$	2 (26)	$3^2\cdot 109\cdot 163$
10	$\psi_{3^2}\chi_{19}^2,\ \chi_{19}^9$	$(3^2,\ 2)$	2	1 (24)	73
10′	$\psi_{3^2}\bar{\chi}_{19}^2,\ \chi_{19}^9$	$(3^2,\ 2)$	2	1 (24)	163
11	$\chi_3,\ \psi_{3^2}\chi_{19}^2$	$(2,\ 3^2)$	$2\cdot 3$	1 (24)	$3^2\cdot 19$
11′	$\chi_3,\ \psi_{3^2}\bar{\chi}_{19}^2$	$(2,\ 3^2)$	$2\cdot 3$	1 (24)	$3^2\cdot 109$
12	$\chi_3,\ \psi_{3^2}\chi_{19}^6,\ \chi_{19}^9$	$(2,\ 3,\ 2)$	$2\cdot 3$	2 (26)	3
12′	$\chi_3,\ \psi_{3^2}\bar{\chi}_{19}^6,\ \chi_{19}^9$	$(2,\ 3,\ 2)$	$2\cdot 3$	2 (26)	$3\cdot 7$
13	$\psi_{3^2}\chi_{19}^6,\ \chi_{19}^9$	$(3,\ 2)$	2	1 (24)	1
13′	$\psi_{3^2}\bar{\chi}_{19}^6,\ \chi_{19}^9$	$(3,\ 2)$	2	1 (24)	7
14	$\chi_3,\ \psi_{3^2}\chi_{19}^6$	$(2,\ 3)$	$2\cdot 3$	1 (24)	3
14′	$\chi_3,\ \psi_{3^2}\bar{\chi}_{19}^6$	$(2,\ 3)$	$2\cdot 3$	1 (24)	3

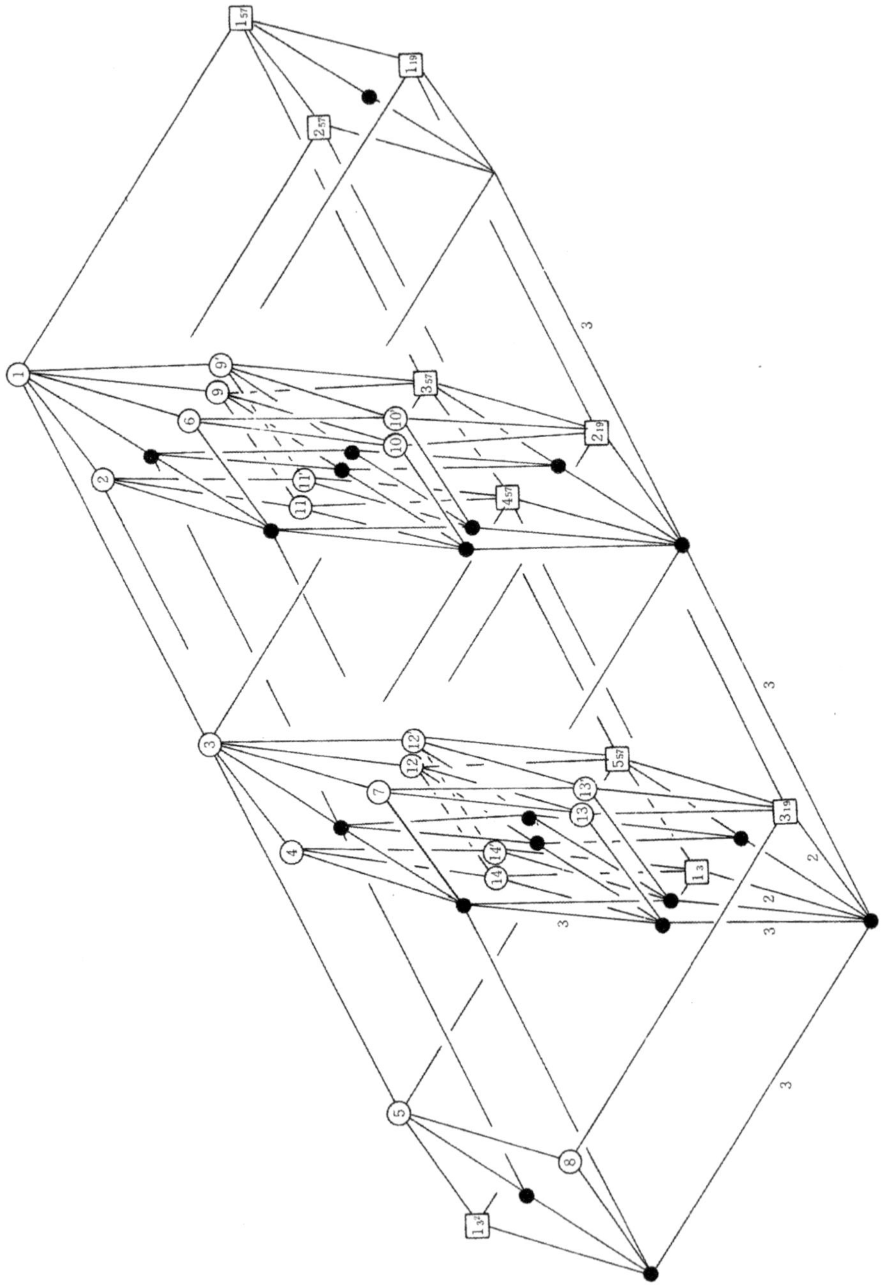

f=172=$2^2\cdot 43$					
No.	gen. Character	Type	w	Q	h^*
1	χ_{2^2}, χ_{43}	(2, 2, 3, 7)	$2^2\cdot 43$	2 (27)	$2^2\cdot 43\cdot 211\cdot 21841$
2	χ_{2^2}, χ_{43}^2	(2, 3, 7)	2^2	1 (24)	$2^2\cdot 43\cdot 21841$
3	χ_{2^2}, χ_{43}^3	(2, 2, 7)	2^2	2 (26)	43
4	χ_{2^2}, χ_{43}^6	(2, 7)	2^2	1 (24)	43
5	χ_{2^2}, χ_{43}^7	(2, 2, 3)	2^2	2 (26)	2^2
6	χ_{2^2}, χ_{43}^{14}	(2, 3)	2^2	1 (24)	2^2
7	χ_{2^2}, χ_{43}^{21}	(2, 2)	2^2	2 (26)	1

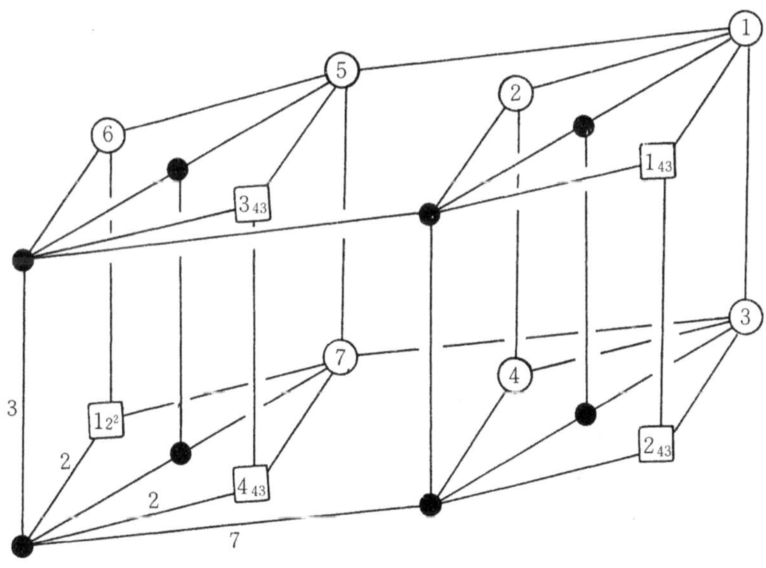

f=173					
No.	gen. Character	Type	w	Q	h^*
1	χ_{173}	(2^2, 43)	$2\cdot 173$	1 (27)	5·20297·2 31169·72571 72936 28518 70621
2	χ_{173}^{43}	2^2	2	1 (23)	5

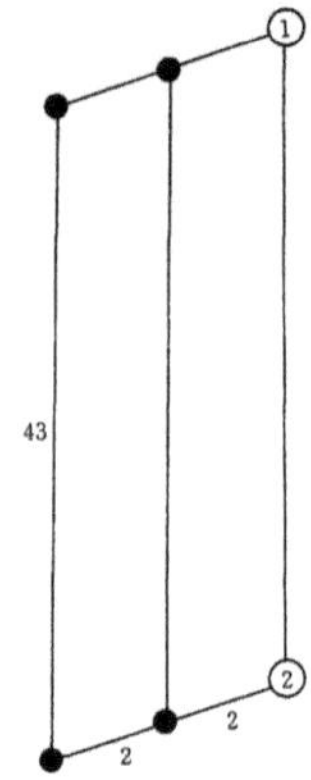

f=175=5²・7					
No.	gen. Character	Type	w	Q	h^*
1	$\chi_5,\ \psi_{5^2}\ ;\ \chi_7$	$(2^2, 5, 2, 3)$	$2\cdot 5^2\cdot 7$	2 (27)	$2^8\cdot 61\cdot 271\cdot 601\cdot 1861$
2	$\chi_5,\ \psi_{5^2}\ ;\ \chi_7^2$	$(2^2, 5, 3)$	$2\cdot 5^2$	1 (24)	$61\cdot 1861$
3	$\chi_5,\ \psi_{5^2}\ ;\ \chi_7^3$	$(2^2, 5, 2)$	$2\cdot 5^2$	2 (26)	2^8
4	$\chi_5^2,\ \psi_{5^2}\ ;\ \chi_7$	$(2, 5, 2, 3)$	$2\cdot 7$	1 (22)	$2^8\cdot 271\cdot 601$
5	$\chi_5^2,\ \psi_{5^2}\ ;\ \chi_7^3$	$(2, 5, 2)$	2	1 (22)	2^8
6	$\psi_{5^2},\ \chi_7$	$(5, 2, 3)$	$2\cdot 7$	1 (24)	$2^4\cdot 601$
7	$\psi_{5^2},\ \chi_7^3$	$(5, 2)$	2	1 (24)	2^4
8	$\psi_{5^2},\ \chi_5^2\chi_7$	$(5, 2, 3)$	2	1 (24)	$2^5\cdot 271$
9	$\psi_{5^2},\ \chi_5^2\chi_7^3$	$(5, 2)$	2	1 (24)	2^5

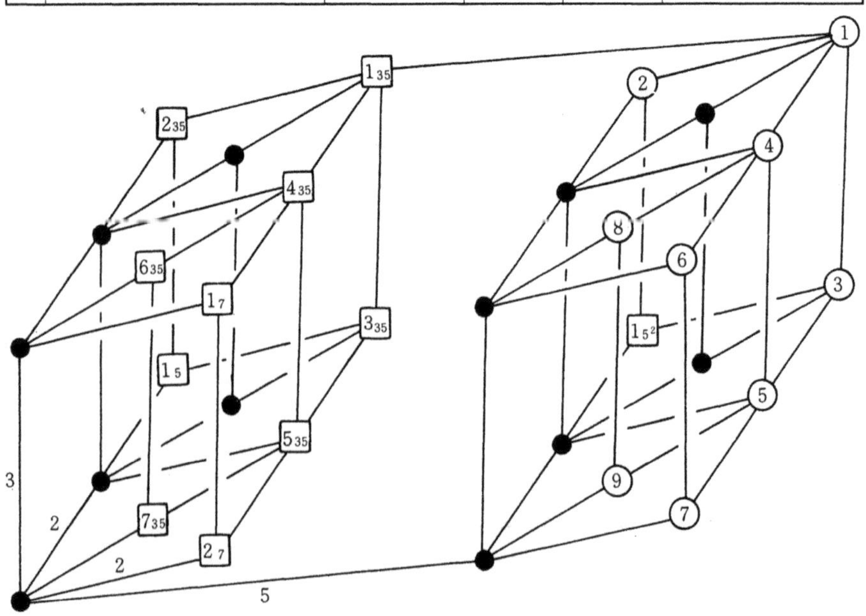

f=176=$2^4 \cdot 11$					
No.	gen. Character	Type	w	Q	h^*
1	$\chi_{2^2},\ \psi_{2^4}\ ;\ \chi_{11}$	(2, 2^2, 2, 5)	$2^4 \cdot 11$	2 (27)	$5^3 \cdot 11 \cdot 41 \cdot 521$
2	$\chi_{2^2},\ \psi_{2^4}\ ;\ \chi_{11}^2$	(2, 2^2, 5)	2^4	1 (22)	$5 \cdot 521$
3	$\chi_{2^2},\ \psi_{2^4}\ ;\ \chi_{11}^5$	(2, 2^2, 2)	2^4	2 (**)	5
4	$\psi_{2^4},\ \chi_{11}$	(2^2, 2, 5)	$2 \cdot 11$	1 (22)	$5^2 \cdot 11 \cdot 41$
5	$\psi_{2^4},\ \chi_{11}^5$	(2^2, 2)	2	1 (22)	5
6	$\chi_{2^2}\psi_{2^4},\ \chi_{11}$	(2^2, 2, 5)	$2 \cdot 11$	2 (26)	$11 \cdot 521$
7	$\chi_{2^2}\psi_{2^4},\ \chi_{11}^2$	(2^2, 5)	2	1 (24)	521
8	$\chi_{2^2}\psi_{2^4},\ \chi_{11}^5$	(2^2, 2)	2	2 (26)	1
9	$\chi_{2^2},\ \psi_{2^4}\chi_{11}$	(2, 2^2, 5)	2^3	2 (26)	$2 \cdot 5^3 \cdot 41$
10	$\chi_{2^2},\ \psi_{2^4}\chi_{11}^5$	(2, 2^2)	2^3	2 (26)	$2 \cdot 5$
11	$\chi_{2^2}\psi_{2^4},\ \chi_{2^2}\chi_{11}$	(2^2, 2, 5)	2	2 (26)	$2 \cdot 5^2 \cdot 41 \cdot 521$
12	$\chi_{2^2}\psi_{2^4},\ \chi_{2^2}\chi_{11}^5$	(2^2, 2)	2	2 (26)	$2 \cdot 5$
13	$\psi_{2^4}\chi_{11}$	(2^2, 5)	2	1 (24)	$2 \cdot 5^2 \cdot 41$
14	$\psi_{2^4}\chi_{11}^5$	2^2	2	1 (24)	$2 \cdot 5$

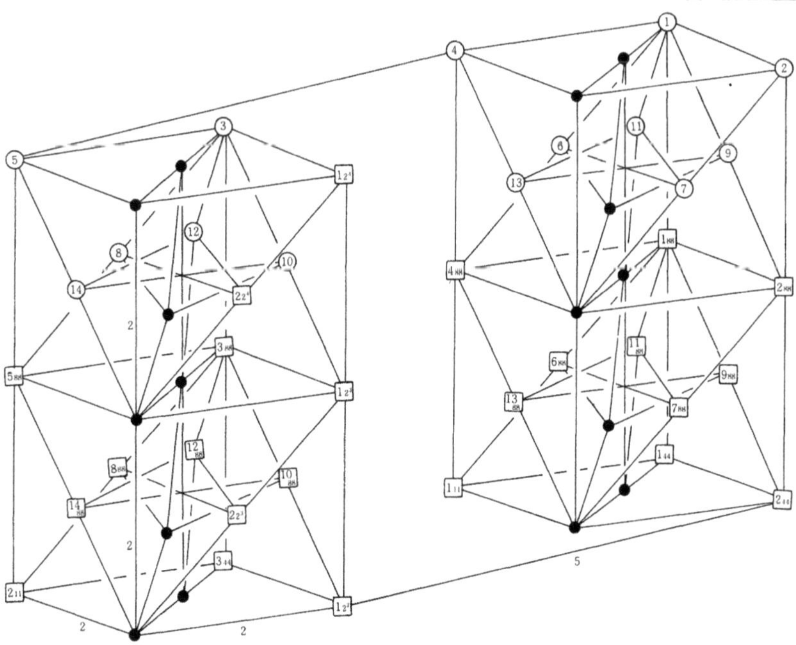

f=177=3・59					
No.	gen. Character	Type	w	Q	h^*
1	χ_3, χ_{59}	(2, 2, 29)	2・3・59	2 (27)	3・59・233・523・37 89257
2	χ_3, χ_{59}^2	(2, 29)	2・3	1 (24)	523・37 89257
3	χ_3, χ_{59}^{29}	(2, 2)	2・3	2 (26)	3

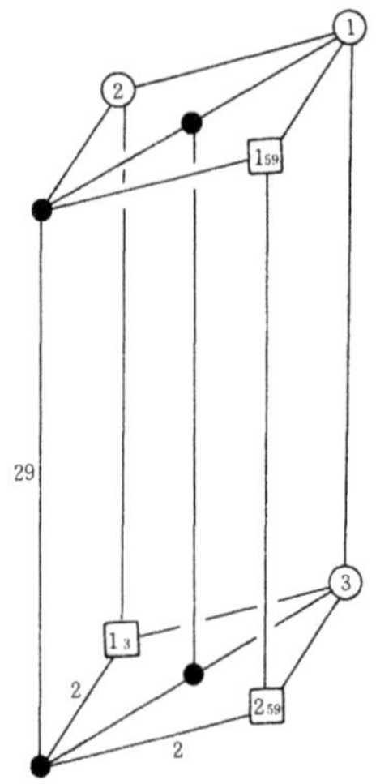

f=179					
No.	gen. Character	Type	w	Q	h^*
1	χ_{179}	(2, 89)	2・179	1 (27)	5・1069・1445 86673 92334 94828 67646 35121
2	χ_{179}^{89}	2	2	1 (23)	5

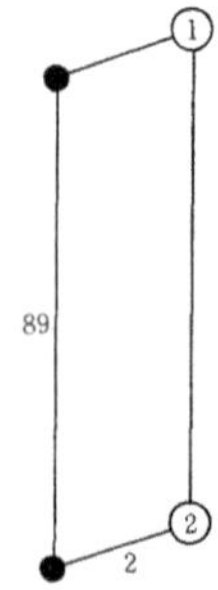

$f=180=2^2\cdot 3^2\cdot 5$					
No.	gen. Character	Type	w	Q	h^*
1	χ_{2^2} ; χ_3, ψ_{3^2} ; χ_5	$(2, 2, 3, 2^2)$	$2^2\cdot 3^2\cdot 5$	2 (27)	$3\cdot 5^2$
2	χ_{2^2} ; χ_3, ψ_{3^2} ; χ_5^2	$(2, 2, 3, 2)$	$2^2\cdot 3^2$	2 (29)	3
3	χ_{2^2}, ψ_{3^2}, χ_5	$(2, 3, 2^2)$	$2^2\cdot 5$	2 (29)	3
4	χ_{2^2}, ψ_{3^2}, χ_5^2	$(2, 3, 2)$	2^2	1 (22)	3
5	$\chi_{2^2}\chi_3$, ψ_{3^2}, χ_5	$(2, 3, 2^2)$	$2\cdot 5$	1 (22)	$2\cdot 5^2$
6	χ_{2^2}, ψ_{3^2}, $\chi_3\chi_5$	$(2, 3, 2^2)$	2^2	1 (22)	$2\cdot 3\cdot 5^2$
7	χ_{2^2}, ψ_{3^2}, $\chi_3\chi_5^2$	$(2, 3, 2)$	2^2	1(**)	1
8	$\chi_{2^2}\chi_5$; χ_3, ψ_{3^2}	$(2^2, 2, 3)$	$2\cdot 3^2$	1 (22)	$2\cdot 5^2$
9	$\chi_{2^2}\chi_5^2$; χ_3, ψ_{3^2}	$(2, 2, 3)$	$2\cdot 3^2$	1(**)	3
10	$\chi_{2^2}\chi_3$, $\chi_3\chi_5^2$, ψ_{3^2}	$(2, 2, 3)$	2	1 (22)	$2\cdot 3$
11	$\chi_{2^2}\chi_3\chi_5$, ψ_{3^2}	$(2^2, 3)$	2	1 (24)	$2^2\cdot 5^2$
12	$\chi_{2^2}\chi_5^2$, ψ_{3^2}	$(2, 3)$	2	1 (24)	$2\cdot 3$

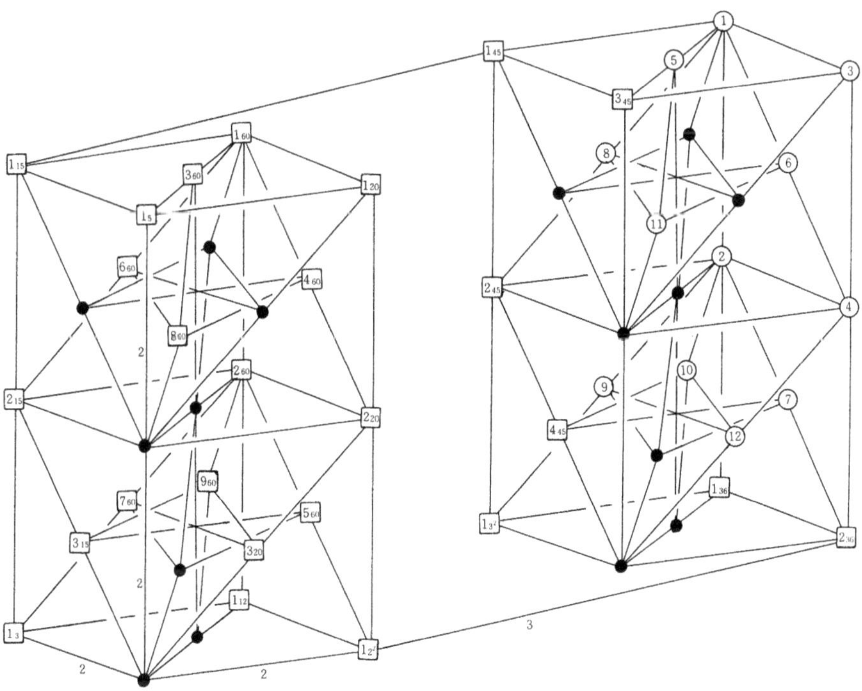

f=181					
No.	gen. Character	Type	w	Q	h^*
1	χ_{181}	$(2^2, 3^2, 5)$	$2 \cdot 181$	1 (27)	$5^3 \cdot 37 \cdot 41 \cdot 61 \cdot 1321 \cdot 2521 \cdot$ 5488 43578 25892 77701
2	χ_{181}^{3}	$(2^2, 3, 5)$	2	1 (23)	$5^3 \cdot 37 \cdot 41 \cdot 61 \cdot 1321$
3	χ_{181}^{5}	$(2^2, 3^2)$	2	1 (23)	$5^2 \cdot 37 \cdot 2521$
4	χ_{181}^{9}	$(2^2, 5)$	2	1 (23)	$5^3 \cdot 41$
5	χ_{181}^{15}	$(2^2, 3)$	2	1 (23)	$5^2 \cdot 37$
6	χ_{181}^{45}	2^2	2	1 (23)	5^2

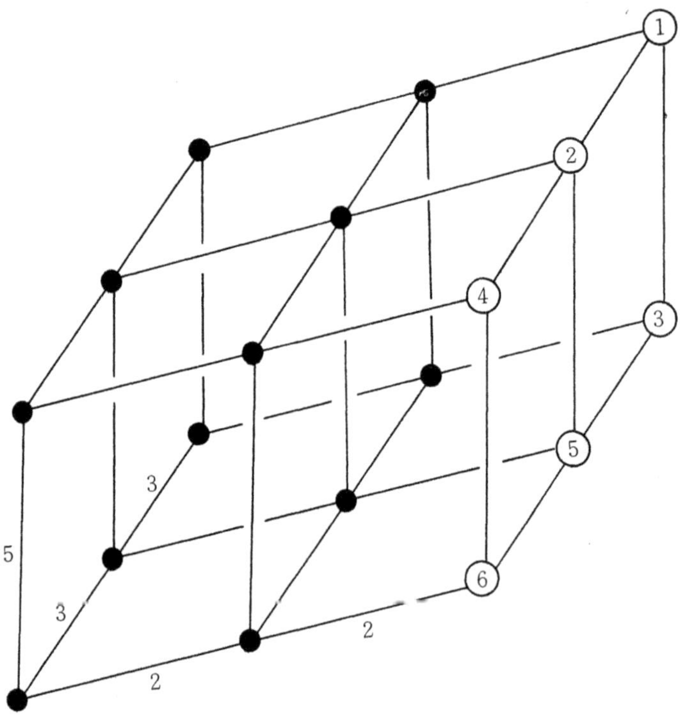

f=183=3 · 61					
No.	gen. Charaster	Type	w	Q	h^*
1	χ_3, χ_{61}	(2, 2^2, 3, 5)	2 · 3 · 61	2 (27)	$2^6\cdot5^2\cdot31^3\cdot41\cdot211\cdot1861$
2	χ_3, χ_{61}^2	(2, 2, 3, 5)	2 · 3	1 (22)	$2^6\cdot5^2\cdot31^3\cdot211$
3	χ_3, χ_{61}^3	(2, 2^2, 5)	2 · 3	2 (26)	$2^2\cdot5^2\cdot31\cdot41$
4	χ_3, χ_{61}^4	(2, 3, 5)	2 · 3	1 (24)	$2^2\cdot5^2\cdot211$
5	χ_3, χ_{61}^5	(2, 2^2, 3)	2 · 3	2 (26)	2^6
6	χ_3, χ_{61}^6	(2, 2, 5)	2 · 3	1 (22)	$2^2\cdot5^2\cdot31$
7	χ_3, χ_{61}^{10}	(2, 2, 3)	2 · 3	1 (22)	2^6
8	χ_3, χ_{61}^{12}	(2, 5)	2 · 3	1 (24)	5^2
9	χ_3, χ_{61}^{15}	(2, 2^2)	2 · 3	2 (26)	2^2
10	χ_3, χ_{61}^{20}	(2, 3)	2 · 3	1 (24)	2^2
11	χ_3, χ_{61}^{30}	(2, 2)	2 · 3	1 (22)	2^2
12	$\chi_3\chi_{61}^2$	(2, 3, 5)	2	1 (24)	$2^5\cdot31^3$
13	$\chi_3\chi_{61}^6$	(2, 5)	2	1 (24)	$2^3\cdot31$
14	$\chi_3\chi_{61}^{10}$	(2, 3)	2	1 (24)	2^5
15	$\chi_3\chi_{61}^{30}$	2	2	1 (24)	2^3

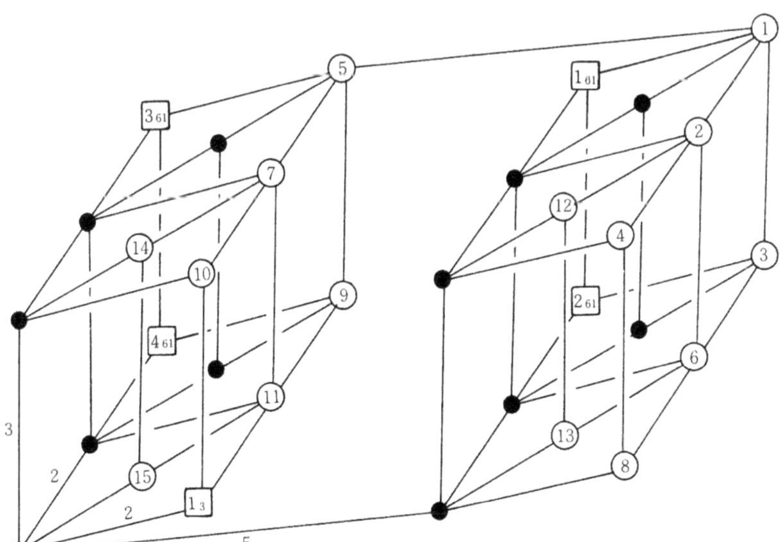

$f=184=2^3\cdot 23$					
No.	gen. Character	Type	w	Q	h^*
1	$\chi_{2^2},\ \psi_{2^3}\ ;\ \chi_{23}$	(2, 2, 2, 11)	$2^3\cdot 23$	2 (27)	$2\cdot 3\cdot 23\cdot 67^2\cdot 2399$
2	$\chi_{2^2},\ \psi_{2^3}\ ;\ \chi_{23}^2$	(2, 2, 11)	2^3	1 (22)	$23\cdot 67^2$
3	$\chi_{2^2},\ \psi_{2^3}\ ;\ \chi_{23}^{11}$	(2, 2, 2)	2^3	2 (**)	$2\cdot 3$
4	$\psi_{2^3},\ \chi_{23}$	(2, 2, 11)	$2\cdot 23$	1 (22)	$2\cdot 3\cdot 2399$
5	$\psi_{2^3},\ \chi_{23}^{11}$	(2, 2)	2	1 (22)	$2\cdot 3$
6	$\chi_{2^2}\psi_{2^3},\ \chi_{23}$	(2, 2, 11)	$2\cdot 23$	2 (26)	$3\cdot 23\cdot 67$
7	$\chi_{2^2}\psi_{2^3},\ \chi_{23}^2$	(2, 11)	2	1 (24)	$23\cdot 67$
8	$\chi_{2^2}\psi_{2^3},\ \chi_{23}^{11}$	(2, 2)	2	2 (26)	3
9	$\chi_{2^2},\ \psi_{2^3}\chi_{23}$	(2, 2, 11)	2^2	2 (26)	$2^2\cdot 67\cdot 2399$
10	$\chi_{2^2},\ \psi_{2^3}\chi_{23}^{11}$	(2, 2)	2^2	2 (26)	2^2
11	$\chi_{2^2}\psi_{2^3},\ \psi_{2^3}\chi_{23}$	(2, 2, 11)	2	2 (26)	$2^2\cdot 23\cdot 67\cdot 2399$
12	$\chi_{2^2}\psi_{2^3},\ \psi_{2^3}\chi_{23}^{11}$	(2, 2)	2	2 (26)	2^2
13	$\psi_{2^3}\chi_{23}$	(2, 11)	2	1 (24)	$2^2\cdot 2399$
14	$\psi_{2^3}\chi_{23}^{11}$	2	2	1 (24)	2^2

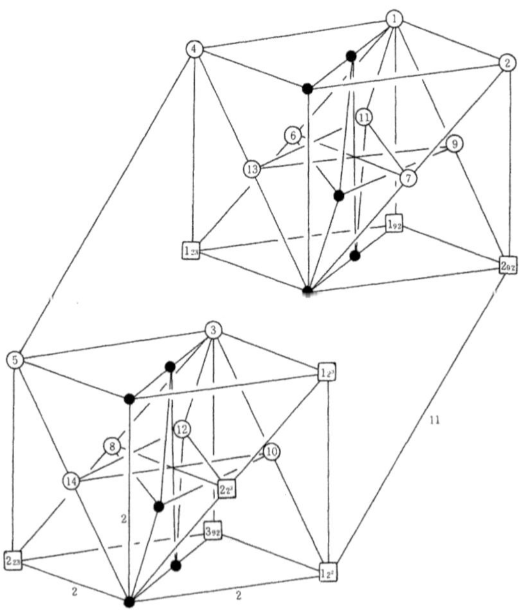

f=185=5 · 37					
No.	gen. Character	Type	w	Q	h^*
1	$\chi_5,\ \chi_{37}$	$(2^2,\ 2^2,\ 3^2)$	$2\cdot5\cdot37$	2 (27)	$5^2\cdot7^2\cdot13\cdot37^2\cdot53^2\cdot9433\cdot23833$
2	$\chi_5,\ \chi_{37}^2$	$(2,\ 2^2,\ 3^2)$	$2\cdot5$	1 (22)	$5\cdot7^2\cdot37\cdot53^2\cdot9433$
3	$\chi_5,\ \chi_{37}^3$	$(2^2,\ 2^2,\ 3)$	$2\cdot5$	2 (26)	$5^2\cdot7^2\cdot13\cdot37$
4	$\chi_5,\ \chi_{37}^4$	$(2^2,\ 3^2)$	$2\cdot5$	1 (24)	$7^2\cdot9433$
5	$\chi_5,\ \chi_{37}^6$	$(2^2,\ 2,\ 3)$	$2\cdot5$	1 (22)	$5\cdot7^2\cdot37$
6	$\chi_5,\ \chi_{37}^9$	$(2^2,\ 2^2)$	$2\cdot5$	2 (26)	5^2
7	$\chi_5,\ \chi_{37}^{12}$	$(2^2,\ 3)$	$2\cdot5$	1 (24)	7^2
8	$\chi_5,\ \chi_{37}^{18}$	$(2^2,\ 2)$	$2\cdot5$	1 (22)	5
9	$\chi_5^2,\ \chi_{37}$	$(2,\ 2^2,\ 3^2)$	$2\cdot37$	1 (22)	$5\cdot13\cdot37\cdot23833$
10	$\chi_5^2,\ \chi_{37}^3$	$(2,\ 2^2,\ 3)$	2	1 (22)	$5\cdot13$
11	$\chi_5^2,\ \chi_{37}^9$	$(2,\ 2^2)$	2	1 (22)	5
12	$\chi_5\chi_{37}^2$	$(2^2,\ 3^2)$	2	1 (24)	$2\cdot5\cdot37\cdot53^2$
13	$\chi_5\chi_{37}^6$	$(2^2,\ 3)$	2	1 (24)	$2\cdot5\cdot37$
14	$\chi_5\chi_{37}^{18}$	2^2	2	1 (24)	$2\cdot5$
15	$\chi_5^2\chi_{37}$	$(2^2,\ 3^2)$	2	1 (24)	$2\cdot5\cdot13\cdot23833$
16	$\chi_5^2\chi_{37}^3$	$(2^2,\ 3)$	2	1 (24)	$2\cdot5\cdot13$
17	$\chi_5^2\chi_{37}^9$	2^2	2	1 (24)	$2\cdot5$

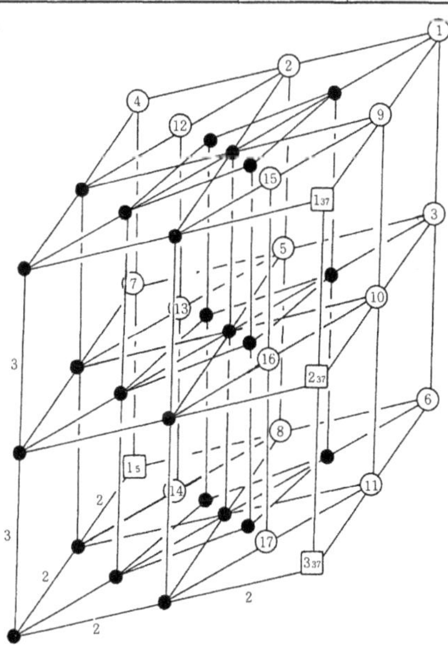

f=187=11 · 17					
No.	gen. Character	Type	w	Q	h^*
1	χ_{11}, χ_{17}	$(2, 5, 2^4)$	2 · 11 · 17	2 (27)	17^2·41·241·4801·2 99681·94 47601
2	χ_{11}, χ_{17}^2	$(2, 5, 2^3)$	2 · 11	1 (22)	17^2·41·241·2 99681
3	χ_{11}, χ_{17}^4	$(2, 5, 2^2)$	2 · 11	1 (22)	17 · 41 · 241
4	χ_{11}, χ_{17}^8	(2, 5, 2)	2 · 11	1 (22)	41
5	χ_{11}^2, χ_{17}	$(5, 2^4)$	2 · 17	1 (24)	4801 · 94 47601
6	χ_{11}^5, χ_{17}	$(2, 2^4)$	2 · 17	2 (26)	17^2
7	χ_{11}^5, χ_{17}^2	$(2, 2^3)$	2	1 (22)	17^2
8	χ_{11}^5, χ_{17}^4	$(2, 2^2)$	2	1 (22)	17
9	χ_{11}^5, χ_{17}^8	(2, 2)	2	1 (22)	1
10	$\chi_{11}\chi_{17}^2$	$(5, 2^3)$	2	1 (24)	2 · 17 · 2 99681
11	$\chi_{11}\chi_{17}^4$	$(5, 2^2)$	2	1 (24)	2 · 17 · 241
12	$\chi_{11}\chi_{17}^8$	(5, 2)	2	1 (24)	2 · 41
13	$\chi_{11}^5\chi_{17}^2$	2^3	2	1 (24)	2 · 17
14	$\chi_{11}^5\chi_{17}^4$	2^2	2	1 (24)	2 · 17
15	$\chi_{11}^5\chi_{17}^8$	2	2	1 (24)	2

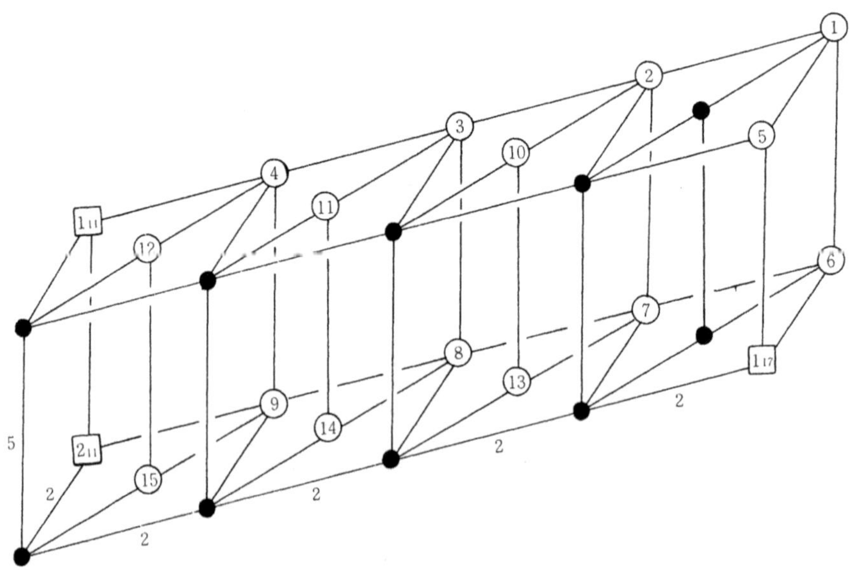

$f=188=2^2\cdot 47$					
No.	gen. Character	Type	w	Q	h^*
1	$\chi_{2^2},\ \chi_{47}$	(2, 2, 23)	$2^2\cdot 47$	2 (27)	5·47·139·7 42717
2	$\chi_{2^2},\ \chi_{47}^2$	(2, 23)	2^2	1 (22)	47 · 7 42717
3	$\chi_{2^2},\ \chi_{47}^{23}$	(2, 2)	2^2	2 (∗∗)	5

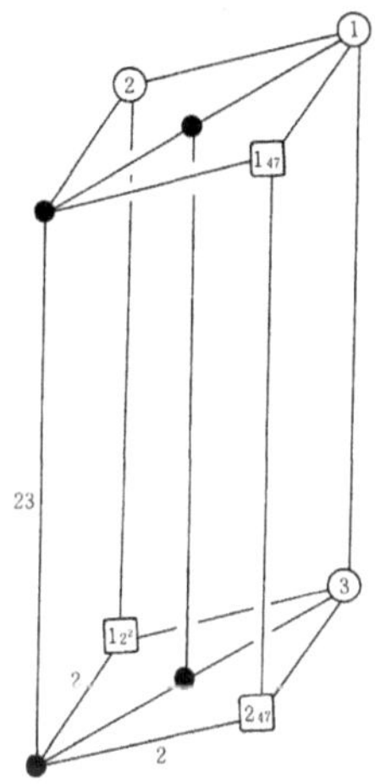

$f=189=3^3\cdot 7$					
No.	gen. Character	Type	w	Q	h^*
1	$\chi_3,\ \psi_{3^3}\ ;\ \chi_7$	(2, 3^2, 2, 3)	$2\cdot 3^3\cdot 7$	2 (27)	7·37·109·127·163·181
2	$\chi_3,\ \psi_{3^3}\ ;\ \chi_7^2$	(2, 3^2, 3)	$2\cdot 3^3$	1 (22)	109 · 181
3	$\chi_3,\ \psi_{3^3}\ ;\ \chi_7^3$	(2, 3^2, 2)	$2\cdot 3^3$	2 (26)	163
4	$\psi_{3^3},\ \chi_7$	(3^2, 2, 3)	$2\cdot 7$	1 (22)	7 · 37 · 127 · 163
5	$\psi_{3^3},\ \chi_7^3$	(3^2, 2)	2	1 (24)	163
6	$\chi_3,\ \psi_{3^3}\chi_7^2,\ \chi_7^3$	(2, 3^2, 2)	$2\cdot 3^2$	2 (26)	127 · 181
6′	$\chi_3,\ \psi_{3^3}\bar{\chi}_7^2,\ \chi_7^3$	(2, 3^2, 2)	$2\cdot 3^2$	2 (26)	37 · 109
7	$\chi_3,\ \psi_{3^3}\chi_7^2$	(2, 3^2)	$2\cdot 3^2$	1 (24)	181
7′	$\chi_3,\ \psi_{3^3}\bar{\chi}_7^2$	(2, 3^2)	$2\cdot 3^2$	1 (24)	109
8	$\psi_{3^3}\chi_7^2,\ \chi_7^3$	(3^2, 2)	2	1 (24)	127
8′	$\psi_{3^3}\bar{\chi}_7^2,\ \chi_7^3$	(3^2, 2)	2	1 (24)	37

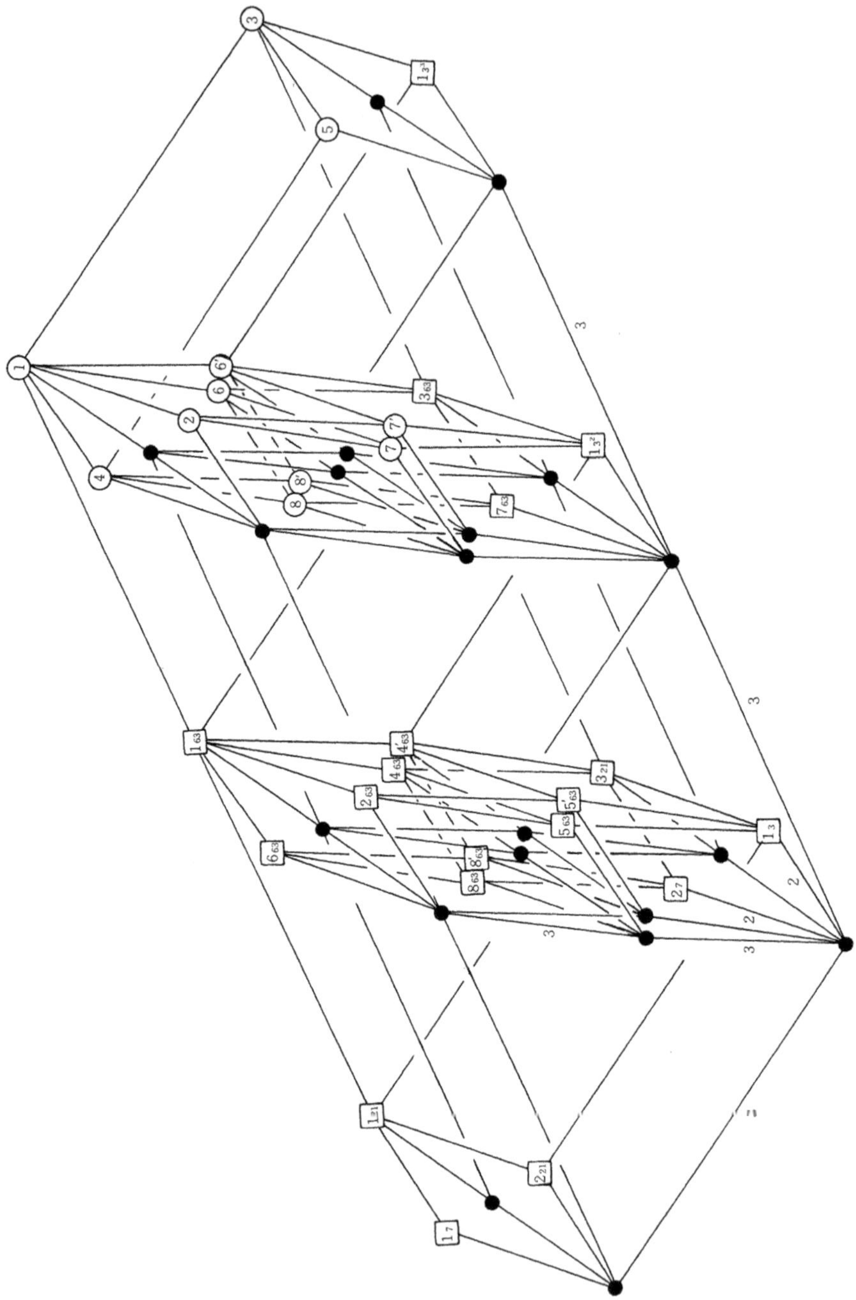

f=191					
No.	gen. Character	Type	w	Q	h^*
1	χ_{191}	(2, 5, 19)	$2\cdot 191$	1 (27)	11·13·51263·6127 71091· 36733 95066 97337 13761
2	χ_{191}^{5}	(2, 19)	2	1 (23)	$13\cdot 51263$
3	χ_{191}^{19}	(2, 5)	2	1 (23)	$11\cdot 13$
4	χ_{191}^{95}	2	2	1 (23)	13

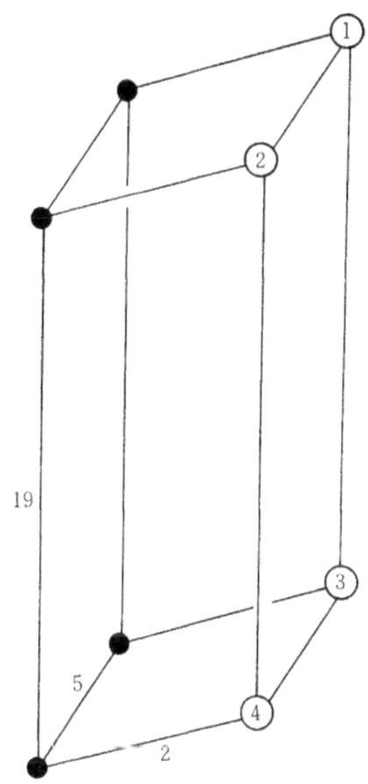

$f=192=2^6\cdot 3$					
No.	gen. Character	Type	w	Q	h^*
1	$\chi_{2^2},\ \psi_{2^6}\ ;\ \chi_3$	$(2,\ 2^4,\ 2)$	$2^6\cdot 3$	2 (27)	$3^2\cdot 17\cdot 401$
2	$\psi_{2^6},\ \chi_3$	$(2^4,\ 2)$	$2\cdot 3$	1 (22)	$3^2\cdot 401$
3	$\chi_{2^2}\psi_{2^6},\ \chi_3$	$(2^4,\ 2)$	$2\cdot 3$	2 (26)	$3^2\cdot 17$
4	$\chi_{2^2},\ \psi_{2^6}\chi_3$	$(2,\ 2^4)$	2^5	2 (26)	$2\cdot 401$
5	$\chi_{2^2}\psi_{2^6},\ \chi_{2^2}\chi_3$	$(2^4,\ 2)$	2	2 (26)	$2\cdot 17\cdot 401$
6	$\psi_{2^6}\chi_3$	2^4	2	1 (24)	$2\cdot 401$

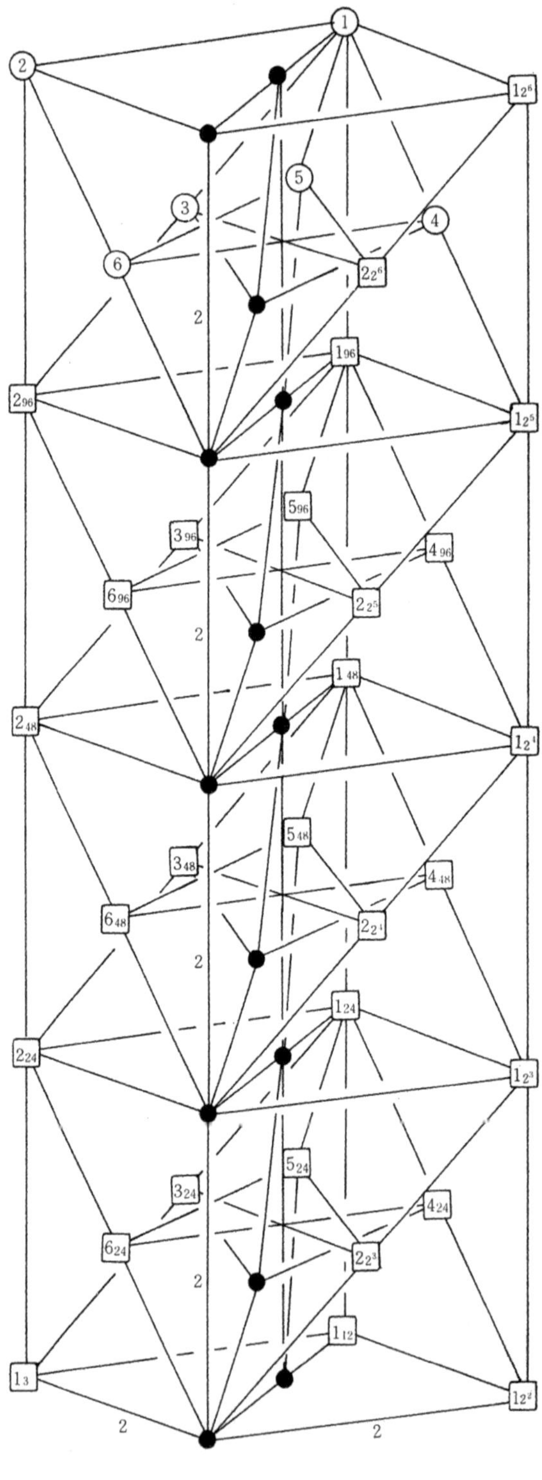
1
2
1_{2^6}
5
3
4
6
2_{2^6}
2
1_{96}
2_{96}
1_{2^5}
5_{96}
3_{96}
4_{96}
6_{96}
2_{2^5}
2
1_{48}
2_{48}
1_{2^4}
5_{48}
3_{48}
4_{48}
6_{48}
2_{2^4}
2
1_{24}
2_{24}
1_{2^3}
5_{24}
3_{24}
4_{24}
6_{24}
2_{2^3}
2
1_{12}
1_3
1_{2^2}
2
2

f=193					
No.	gen. Character	Type	w	Q	h^*
1	χ_{193}	$(2^6,\ 3)$	$2\cdot 193$	1 (27)	6529·15361·29761·91969·103 69729·19 20262 80449
2	χ_{193}^3	2^6	2	1 (23)	19 20262 80449

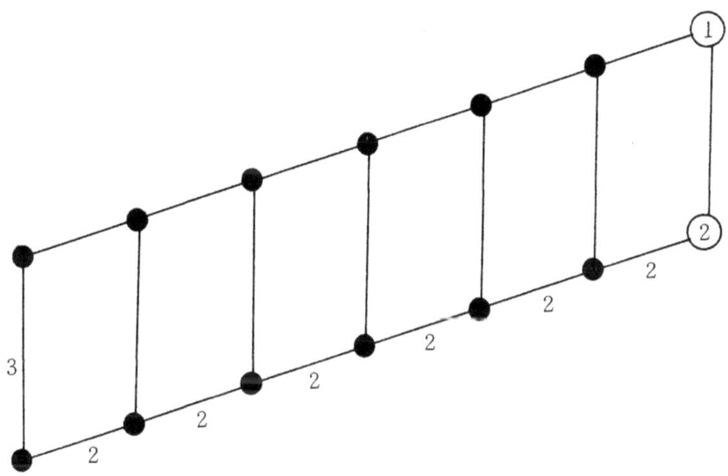

f=195=3・5・13					
No.	gen. Character	Type	w	Q	h^*
1	$\chi_3,\ \chi_5,\ \chi_{13}$	$(2,\ 2^2,\ 2^2,\ 3)$	$2\cdot3\cdot5\cdot13$	2 (27)	$2^{17}\cdot13^2$
2	$\chi_3,\ \chi_5,\ \chi_{13}^2$	$(2,\ 2^2,\ 2,\ 3)$	$2\cdot3\cdot5$	2 (29)	2^{10}
3	$\chi_3,\ \chi_5,\ \chi_{13}^3$	$(2,\ 2^2,\ 2^2)$	$2\cdot3\cdot5$	2 (29)	$2^3\cdot13$
4	$\chi_3,\ \chi_5,\ \chi_{13}^4$	$(2,\ 2^2,\ 3)$	$2\cdot3\cdot5$	2 (29)	2^4
5	$\chi_3,\ \chi_5,\ \chi_{13}^6$	$(2,\ 2^2,\ 2)$	$2\cdot3\cdot5$	2 (29)	2^2
6	$\chi_3,\ \chi_5^2,\ \chi_{13}$	$(2,\ 2,\ 2^2,\ 3)$	$2\cdot3\cdot13$	2 (29)	2^8
7	$\chi_3,\ \chi_5^2,\ \chi_{13}^2$	$(2,\ 2,\ 2,\ 3)$	$2\cdot3$	1 (22)	2^6
8	$\chi_3,\ \chi_5^2,\ \chi_{13}^3$	$(2,\ 2,\ 2^2)$	$2\cdot3$	2 (29)	2^2
9	$\chi_3,\ \chi_5^2,\ \chi_{13}^4$	$(2,\ 2,\ 3)$	$2\cdot3$	1 (22)	2^2
10	$\chi_3,\ \chi_5^2,\ \chi_{13}^6$	$(2,\ 2,\ 2)$	$2\cdot3$	1 (22)	2^2
11	$\chi_3,\ \chi_5\chi_{13},\ \chi_5^2$	$(2,\ 2^2,\ 3,\ 2)$	$2\cdot3$	1 (22)	$2^{12}\cdot13^2$
12	$\chi_3,\ \chi_5\chi_{13}^3,\ \chi_5^2$	$(2,\ 2^2,\ 2)$	$2\cdot3$	1 (22)	$2^4\cdot13$
13	$\chi_3,\ \chi_5\chi_{13}$	$(2,\ 2^2,\ 3)$	$2\cdot3$	1 (22)	$2^6\cdot13$
13′	$\chi_3,\ \chi_5\bar{\chi}_{13}$	$(2,\ 2^2,\ 3)$	$2\cdot3$	1 (22)	$2^6\cdot13$
14	$\chi_3,\ \chi_5\chi_{13}^2$	$(2,\ 2^2,\ 3)$	$2\cdot3$	2(**)	2^5
15	$\chi_3,\ \chi_5\chi_{13}^3$	$(2,\ 2^2)$	$2\cdot3$	1 (22)	$2^2\cdot13$
15′	$\chi_3,\ \chi_5\bar{\chi}_{13}^3$	$(2,\ 2^2)$	$2\cdot3$	1 (22)	2^2
16	$\chi_3,\ \chi_5\chi_{13}^6$	$(2,\ 2^2)$	$2\cdot3$	2 (15)	2
17	$\chi_3,\ \chi_5^2\chi_{13}$	$(2,\ 2^2,\ 3)$	$2\cdot3$	2(**)	2^4
18	$\chi_3,\ \chi_5^2\chi_{13}^2$	$(2,\ 2,\ 3)$	$2\cdot3$	1 (22)	2^3
19	$\chi_3,\ \chi_5^2\chi_{13}^3$	$(2,\ 2^2)$	$2\cdot3$	2 (15)	2^2
20	$\chi_3,\ \chi_5^2\chi_{13}^6$	$(2,\ 2)$	$2\cdot3$	1 (22)	2
21	$\chi_3\chi_5,\ \chi_{13}$	$(2^2,\ 2^2,\ 3)$	$2\cdot13$	1 (22)	$2^8\cdot13^2$
22	$\chi_3\chi_5,\ \chi_{13}^3$	$(2^2,\ 2^2)$	2	1 (22)	$2^2\cdot13$
23	$\chi_3\chi_5^2,\ \chi_{13}$	$(2,\ 2^2,\ 3)$	$2\cdot13$	1(**)	2^5
24	$\chi_3\chi_5^2,\ \chi_{13}^2$	$(2,\ 2,\ 3)$	2	1 (22)	2^6
25	$\chi_3\chi_5^2,\ \chi_{13}^3$	$(2,\ 2^2)$	2	1 (15)	2
26	$\chi_3\chi_5^2,\ \chi_{13}^4$	$(2,\ 3)$	2	1 (22)	2^3
27	$\chi_3\chi_5^2,\ \chi_{13}^6$	$(2,\ 2)$	2	1 (22)	2^2

28	$\chi_5,\ \chi_3\chi_{13}$	$(2^2,\ 2^2,\ 3)$	$2\cdot 5$	1 (22)	$2^{10}\cdot 13^2$
29	$\chi_5,\ \chi_3\chi_{13}^2$	$(2^2,\ 2,\ 3)$	$2\cdot 5$	1 (**)	2^6
30	$\chi_5,\ \chi_3\chi_{13}^3$	$(2^2,\ 2^2)$	$2\cdot 5$	1 (22)	$2^2\cdot 13$
31	$\chi_5,\ \chi_3\chi_{13}^6$	$(2^2,\ 2)$	$2\cdot 5$	1 (15)	2^2
32	$\chi_5^2,\ \chi_3\chi_{13}^2$	$(2,\ 2,\ 3)$	2	1 (22)	2^5
33	$\chi_5^2,\ \chi_3\chi_{13}^6$	$(2,\ 2)$	2	1 (22)	2^3
34	$\chi_5^2,\ \chi_3\chi_5\chi_{13}$	$(2,\ 2^2,\ 3)$	2	1 (22)	$2^7\cdot 13^2$
35	$\chi_5^2,\ \chi_3\chi_5\chi_{13}^3$	$(2,\ 2^2)$	2	1 (22)	$2^3\cdot 13$
36	$\chi_3\chi_5,\ \chi_5\chi_{13}^2$	$(2^2,\ 2,\ 3)$	2	1 (22)	2^7
37	$\chi_3\chi_5,\ \chi_5\chi_{13}^6$	$(2^2,\ 2)$	2	1 (22)	2^3
38	$\chi_3\chi_5^2,\ \chi_5\chi_{13}$	$(2,\ 2^2,\ 3)$	2	1 (22)	$2^7\cdot 13$
38′	$\chi_3\chi_5^2,\ \chi_5\bar{\chi}_{13}$	$(2,\ 2^2,\ 3)$	2	1 (22)	$2^7\cdot 13$
39	$\chi_3\chi_5^2,\ \chi_5\chi_{13}^3$	$(2,\ 2^2)$	2	1 (22)	2^3
39′	$\chi_3\chi_5^2,\ \chi_5\bar{\chi}_{13}^3$	$(2,\ 2^2)$	2	1 (22)	$2^3\cdot 13$
40	$\chi_3\chi_5^2,\ \chi_5^2\chi_{13}$	$(2,\ 2^2,\ 3)$	2	1 (22)	2^8
41	$\chi_3\chi_5^3,\ \chi_5^2\chi_{13}^2$	$(2,\ 2,\ 3)$	2	1 (22)	2^4
42	$\chi_3\chi_5^2,\ \chi_5^2\chi_{13}^3$	$(2,\ 2^2)$	2	1 (22)	2^2
43	$\chi_3\chi_5^2,\ \chi_5^2\chi_{13}^6$	$(2,\ 2)$	2	1 (22)	2^2
44	$\chi_3\chi_5\chi_{13}$	$(2^2,\ 3)$	2	1 (24)	$2^4\cdot 13$
44′	$\chi_3\chi_5\bar{\chi}_{13}$	$(2^2,\ 3)$	2	1 (24)	$2^4\cdot 13$
45	$\chi_3\chi_5\chi_{13}^3$	2^2	2	1 (24)	$2^2\cdot 13$
45′	$\chi_3\chi_5\bar{\chi}_{13}^3$	2^2	2	1 (24)	2^2
46	$\chi_3\chi_5^2\chi_{13}^2$	$(2,\ 3)$	2	1 (24)	2^4
47	$\chi_3\chi_5^2\chi_{13}^6$	2	2	1 (24)	2^2

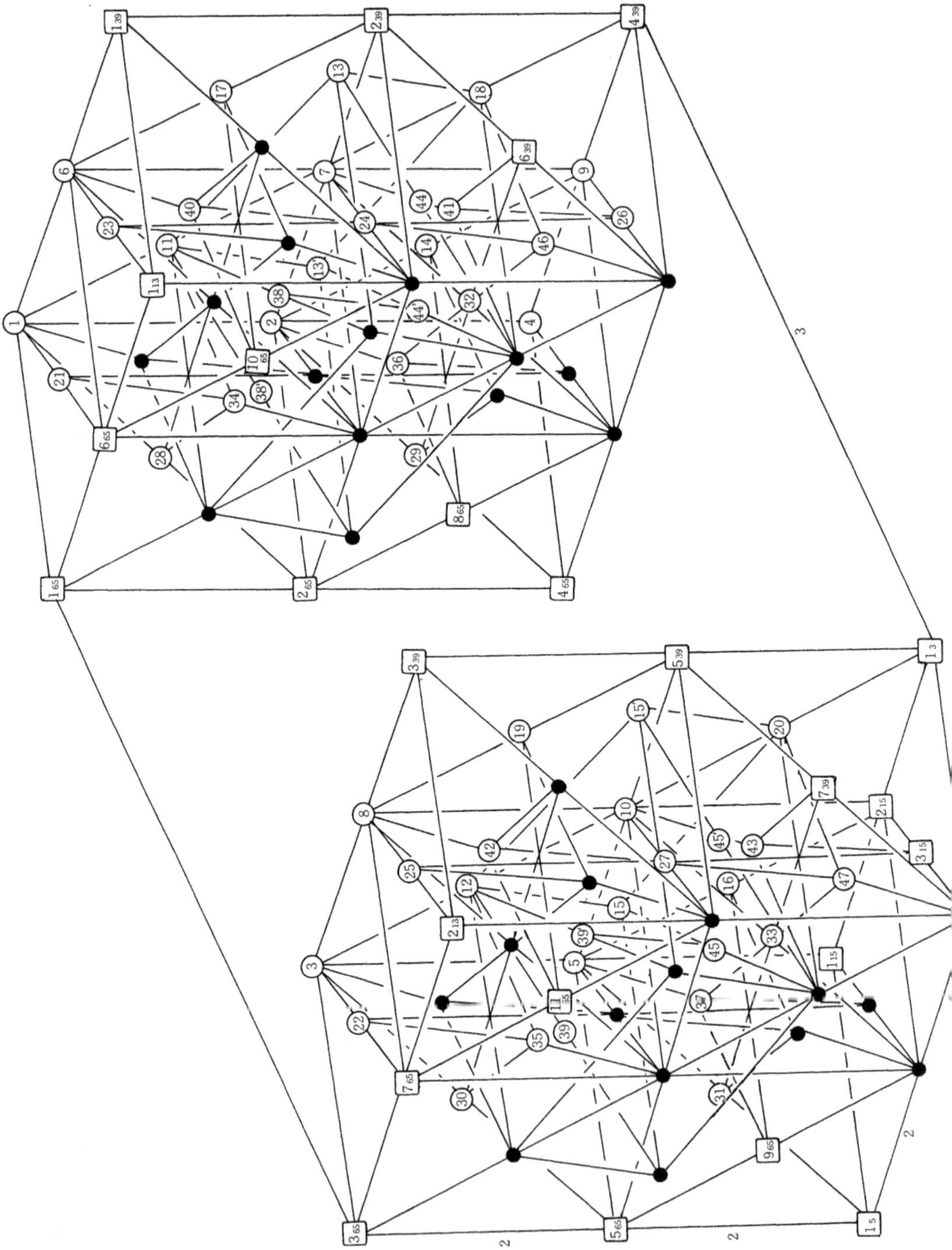

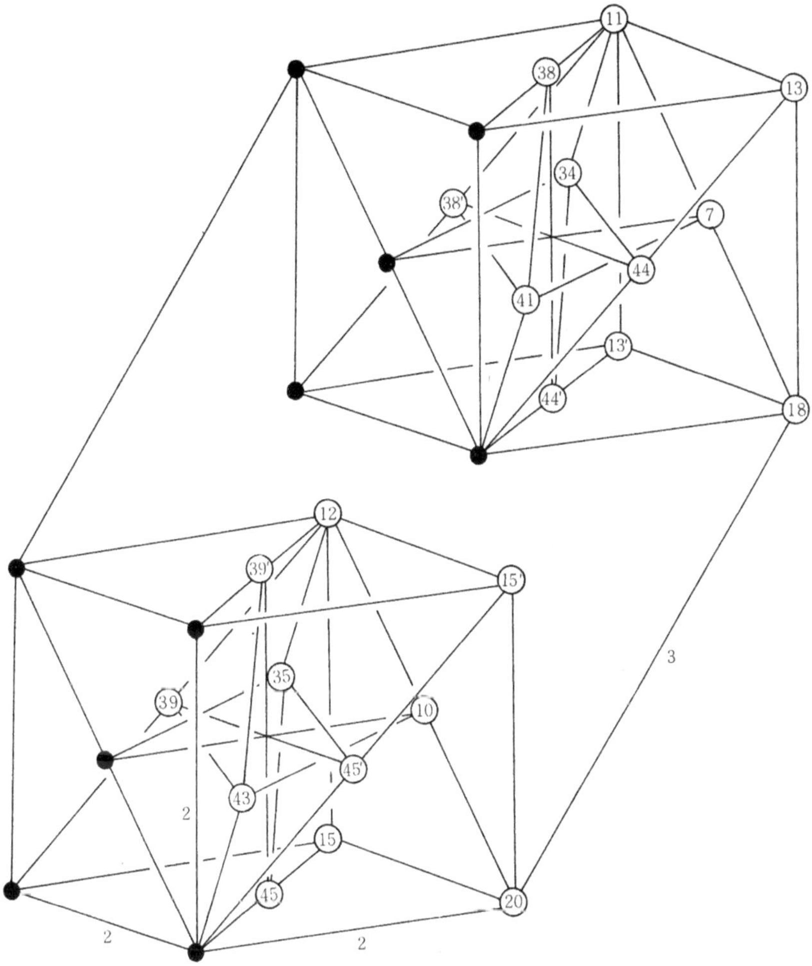

f=196=$2^2 \cdot 7^2$					
No.	gen. Character	Type	w	Q	h^*
1	χ_{2^2} ; χ_7, ψ_{7^2}	(2, 2, 3, 7)	$2^2 \cdot 7^2$	2 (27)	43 · 71 · 27091
2	χ_{2^2} ; χ_7^2, ψ_{7^2}	(2, 3, 7)	2^2	1 (24)	71 · 27091
3	χ_{2^2} ; χ_7^3, ψ_{7^2}	(2, 2, 7)	2^2	2 (29)	71
4	χ_{2^2}, ψ_{7^2}	(2, 7)	2^2	1 (24)	71

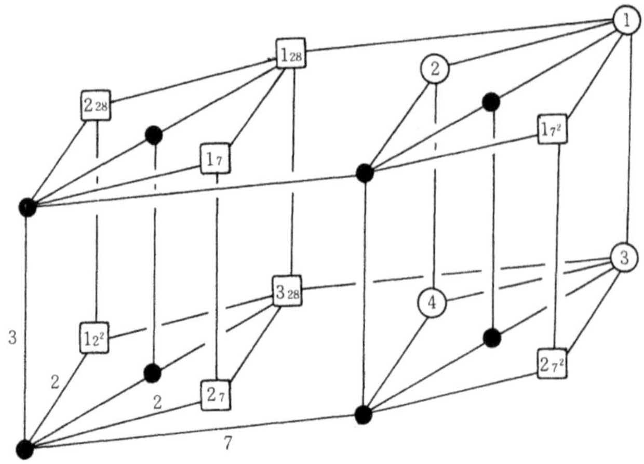

f=197					
No.	gen. Character	Type	w	Q	h^*
1	χ_{197}	$(2^2,\ 7^2)$	$2 \cdot 197$	1 (27)	$2^3 \cdot 5 \cdot 1877 \cdot 7841 \cdot 939\ 83026$ $84870\ 86665\ 62256\ 11549$
2	χ_{197}^{7}	$(2^2,\ 7)$	2	1 (23)	$2^3 \cdot 5 \cdot 1877$
3	χ_{197}^{49}	2^2	2	1 (23)	5

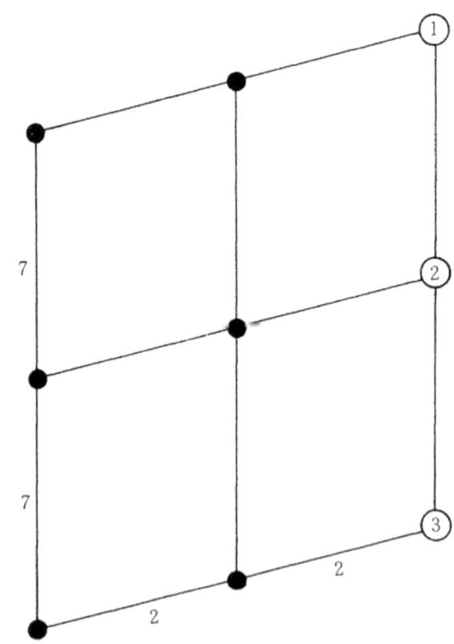

f=199					
No.	gen. Character	Type	w	Q	h^*
1	χ_{199}	$(2,\ 3^2,\ 11)$	$2\cdot 199$	1 (27)	$3^4\cdot 19\cdot 727\cdot 256\ 45093\cdot$ $20\ 72935\ 48177\cdot$ $316\ 81904\ 12839$
2	χ_{199}^{3}	$(2,\ 3,\ 11)$	2	1 (23)	$3^3\cdot 727\cdot 256\ 45093$
3	χ_{199}^{9}	$(2,\ 11)$	2	1 (23)	$3^2\cdot 727$
4	χ_{199}^{11}	$(2,\ 3^2)$	2	1 (23)	$3^4\cdot 19$
5	χ_{199}^{33}	$(2,\ 3)$	2	1 (23)	3^3
6	χ_{199}^{99}	2	2	1 (23)	3^2

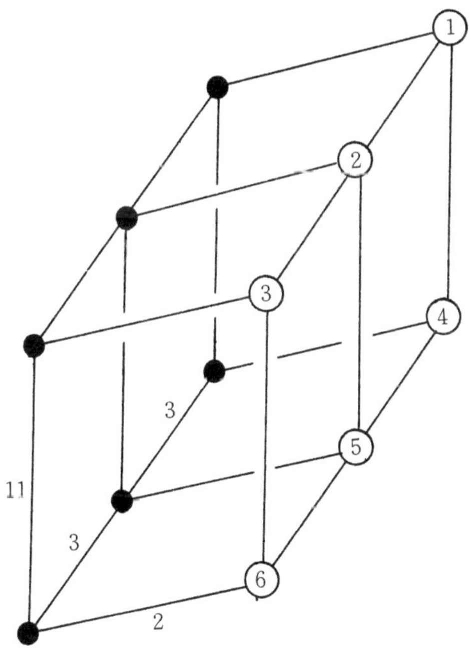

$f=200=2^3\cdot 5^2$					
No.	gen. Character	Type	w	Q	h^*
1	$\chi_{2^2},\ \psi_{2^3}\ ;\ \chi_5,\ \psi_{5^2}$	(2, 2, 2^2, 5)	$2^3\cdot 5^2$	2 (27)	$5\cdot 11^2\cdot 41\cdot 601$
2	$\chi_{2^2},\ \psi_{2^3}\ ;\ \chi_5^2,\ \psi_{5^2}$	(2, 2, 2, 5)	2^3	1 (22)	$5\cdot 11^2\cdot 41$
3	$\chi_{2^2},\ \psi_{2^3}\ ;\ \psi_{5^2}$	(2, 2, 5)	2^3	1 (22)	$5\cdot 11$
4	$\psi_{2^3}\ ;\ \chi_5,\ \psi_{5^2}$	(2, 2^2, 5)	$2\cdot 5^2$	1 (22)	601
5	$\chi_{2^2}\psi_{2^3}\ ;\ \chi_5,\ \psi_{5^2}$	(2, 2^2, 5)	$2\cdot 5^2$	2 (29)	$11\cdot 41$
6	$\chi_{2^2}\psi_{2^3}\ ;\ \chi_5^2,\ \psi_{5^2}$	(2, 2, 5)	2	1 (22)	$11\cdot 41$
7	$\chi_{2^2}\psi_{2^3},\ \psi_{5^2}$	(2, 5)	2	1 (24)	11
8	$\chi_{2^2},\ \psi_{2^3}\chi_5,\ \psi_{5^2}$	(2, 2^2, 5)	2^2	2 (29)	$2\cdot 5\cdot 11\cdot 601$
9	$\chi_{2^2},\ \psi_{2^3}\chi_5^2,\ \psi_{5^2}$	(2, 2, 5)	2^2	1(**)	$5\cdot 41$
10	$\psi_{2^3},\ \chi_{2^2}\chi_5^2,\ \psi_{5^2}$	(2, 2, 5)	2	1 (22)	$2\cdot 11\cdot 41$
11	$\chi_{2^2}\psi_{2^3},\ \chi_{2^2}\chi_5,\ \psi_{5^2}$	(2, 2^2, 5)	2	2 (29)	$2\cdot 11\cdot 41\cdot 601$
12	$\chi_{2^2}\psi_{2^3},\ \chi_{2^2}\chi_5^2,\ \psi_{5^2}$	(2, 2, 5)	2	1(**)	11^2
13	$\psi_{2^3}\chi_5,\ \psi_{5^2}$	(2^2, 5)	2	1 (24)	$2\cdot 601$
14	$\chi_{2^2}\psi_{2^3}\chi_5^2,\ \psi_{5^2}$	(2, 5)	2	1 (24)	$2\cdot 41$

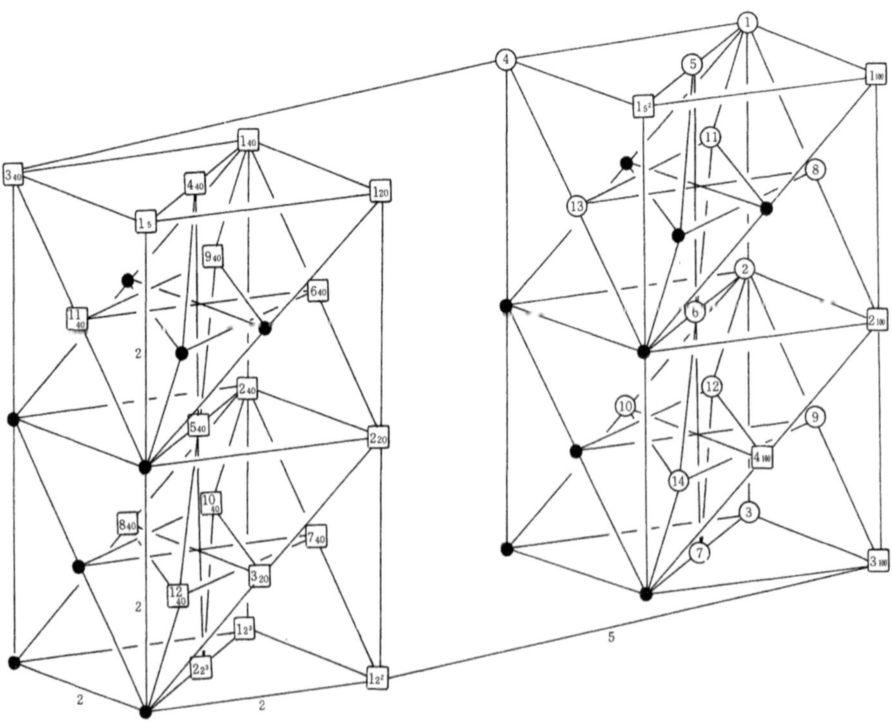

Bibliography

[1] H. Hasse, Über die Klassenzahl abelscher Zahlkörper, Akademie Verlag, 1952.

[2] M. N. Gras, Table numérique du nombre de classes et des unités des extensions cycliques réelles de degré 4 de $\boldsymbol{Q}$, Publ. math. Fac. Sci. Besancon 1977/1978, fbsc. 2.

[3] K. Yoshino and M. Hirabayashi, On the Relative Class Number of the Imaginary Abelian Number Field I, Memoirs of the College of Liberal Arts, Kanazawa Medical University, Vol. 9, 1981.

[4] M. Hirabayashi, On the unit indices of imaginary abelian number fields I, II, Memoirs of Kanazawa Institute of Technology, No. 17, 1982.

[5] M. Hirabayashi and K. Yoshino, Remarks on unit indices of imaginary abelian number fields, (to appear).

(Received August 31, 1982)

Chapter 6
Supplemental Readings

Bibliography on Hasse's Unit Indices of CM-Fields

1. H. Hasse, *Über die Klassenzahl abelscher Zahlkörper,* Mathematische Lehrbucher und Monographien, Band I, Akademie-Verlag, Berlin, 1952; Reprinted with an introduction by J. Martinet : Springer-Verlag: Berlin, 1985.
2. K. Uchida, Imaginary abelian number fields with class number one, *Tôhoku Math. J.* (2) **24** (1972), 487–499.
3. K. Yoshino and M. Hirabayashi, On the relative class number of imaginary abelian number field I, *Mem. College Liberal Arts, Kanazawa Medical Univ.* **9** (1981), 5–53.
4. K. Yoshino and M. Hirabayashi, On the relative class number of imaginary abelian number field II, *Mem. College Liberal Arts, Kanazawa Medical Univ.* **10** (1982), 33–81.
5. M. Hirabayashi and K. Yoshino, The unit indices of imaginary abelian number fields, *Proc. Japan Acad. Ser. A Math. Sci.* **60** (1984), 215–217.
6. M. Hirabayashi and K. Yoshino, Remarks on unit indices of imaginary abelian number fields, *Manuscripta Math.* **60** (1988), 423–436.
7. P. E. Conner and J. Hurrelbrink, *Class Number Parity*, World Scientific, Singapore, 1988.
8. M. Hirabayashi and K. Yoshino, Remarks on unit indices of imaginary abelian number fields II, *Manuscripta Math.* **64** (1989), 235–251.
9. M. Hirabayashi and K. Yoshino, Unit indices of imaginary abelian number fields of type (2, 2, 2), *J. Number Theory* **34** (1990), 346–361.
10. S. Louboutin, On the class number one problem for non-normal quartic CM-fields, *Tôhoku Math. J.* (2) **46** (1994), 1–12.
11. M. Hirabayashi, Unit indices of some imaginary composite quadratic fields, *Pacific J. Math.* **164** (1994), 87–104.
12. F. Lemmermeyer, Ideal class groups of cyclotomic fields I, *Acta Arith.* **72** (1995), 347–359.
13. M. Hirabayashi, Unit indices of some imaginary composite quadratic fields II, *Pacific J. Math.* **173** (1996), 93–104.
14. S. Louboutin, Determination of all quaternion octic CM-fields with class number 2, *J. London Math. Soc.*(2) **54** (1996), 227–238.
15. A. Azizi, Unités de certains corps de nombres imaginaires et abéiens sur **Q**, *Ann. Sci. Math. Québec* **23** (1999), 15–21.
16. R. Okazaki, Inclusion of CM-fields and divisibility of relative class numbers, *Acta Arith.* **92** (2000), 319–338.

© Springer Nature Switzerland AG 2019

H. Hasse, *On the Class Number of Abelian Number Fields*,
https://doi.org/10.1007/978-3-030-01512-1_6

17. S. Louboutin, Hasse unit indices of dihedral octic CM-fields, *Math. Nachr.* **215** (2000), 107–113.
18. R. Okazaki, On parities of relative class numbers of certain CM-extensions. Class field theory – its centenary and prospect (Tokyo, 1998), 419–444, *Adv. Stud. Pure Math.*, **30**, Math. Soc. Japan, Tokyo, 2001.
19. S. Nakajima, Class number of (2, 2)-extensions over the rational number field (Japanese), Proceedings of the 7th Workshop on Number Theory (Kanazawa 2008), (2009), 46–58.
20. K. Yamamura, Bibliography on Hasse's unit indices of CM-fields (Japanese), Proceedings of the 7th Workshop on Number Theory (Kanazawa 2008), (2009), 134–138.
21. A. Azizi, I. Jerrari, A. Zekhnini and M. Talbi, Units of some abelian imaginary number fields of type (2, 4), *Gulf J. Math.* **4** (2016), 166–170.

Bibliography on Calculations of Relative Class Numbers of Imaginary Abelian Number Fields

1. E. E. Kummer, Bestimmung der Anzahl nicht äquivalenter Classen für die aus λten Wurzeln der Einheit gebildeten complexen Zahlen und die idealen Factoren derselben, *J. Reine Angew. Math.* **40** (1850), 93–116.
2. E. E. Kummer, Mémoire sur la théorie des nombres complexes composés de racines de l'unité et de nombres entiers. *J. de math. pures et appl.* **16** (1851), 377–498.
3. H. Hasse, *Über die Klassenzahl abelscher Zahlkörper,* Mathematische Lehrbucher und Monographien, Band I, Akademie-Verlag, Berlin, 1952; Reprinted with an introduction by J. Martinet : Springer-Verlag: Berlin, 1985.
4. G. Schrutka von Rechtenstamm, Tabelle der (Relativ)-Klassenzahlen der Kreiskörper, deren ϕ-Funktion des Wurzelexponenten (Grad) nicht grösser als 256 ist, *Abh. Deutsch. Akad. Wiss. Berlin, Kl. Math. Phys.*, (1964),1–64.
5. T. Metsänkylä, Calculation of the first factor of the class number of the cyclotomic field, *Math. Comp.* **23** (1969), 533–537.
6. M. Newman, A table of the first factor for prime cyclotomic fields, *Math. Comp.* **24** (1970), 215–219.
7. R. Spira, Calculation of the first factor of the cyclotomic class number, Computers in number theory (Proc. Sci. Res. Council Atlas Sympos. No. 2, Oxford, 1969) (A. O. L. Atkin and B. J. Birch, eds.), Academic Press, London and New York, 1971, 149–151.
8. V. M. Galkin, The first factor of the ideal class number of a cyclotomic field (Russian), Uspehi Mat. Nauk **27** (1972), 233–234.
9. D. H. Lehmer, Prime factors of cyclotomic class numbers, *Math. Comp.* **31** (1977), 599–607.
10. D. H. Lehmer and J. M. Masley, Table of the cyclotomic class numbers $h^*(p)$ and their factors for $200 < p < 521$, *Math. Comp.* **32** (1978), 577–582.
11. G. Fung, A. Granville, and H. C. Williams, Computation of the fisrt factor of the class number of cyclotomic fields, *J. Number Theory* **42** (1992), 297–312.
12. V. Jha, Faster computation of the first factor of the class number of $\mathbf{Q}(\zeta_p)$, *Math. Comp.* **64** (1995), 1705–1710.
13. K. Yamamura, Calculation of class numbers by analytic class number formulas (Japanese), Proceedings of the 2nd Symposium on Algebra and Computation (Tokyo 1997).
14. S. Louboutin, Computation of relative class numbers of imaginary abelian number fields, *Experiment. Math.* **7** (1998), 293–303.
15. S. Louboutin, Computation of relative class numbers of imaginary cyclic fields of 2-power degrees, *Algorithmic number theory* (Portland, 1998) (J. P. Buhler, ed.), Lecture Notes in Comput. Sci., vol. 1423, Springer, Berlin, (1998), 475–48.

16. M. A. Shokrollahi, Relative class number of imaginary abelian fields of prime conductor below 10000, *Math. Comp.* **68** (1999), 1717–1728.
17. K. Yamamura, Calculation of absolute norms of integers of cyclotomic fields (Japanese), Proceedings of the 3rd Symposium on Algebra and Computation (Tokyo 1999).
18. T. Taniguchi, On a fast calculation algorithm of the relative class number of cyclotomic fields (Japanese), The 17th Symposium on Algorithmic Number Theory and its Applications (Noda 2007), Japan Society for Industrial and Applied Mathematics (2007).
19. T. Taniguchi, Computation of the relative class number of cyclotomic fields – applications of fast multiprecision polynomial multiplications (Japanese), *Inform. Proc. Soc. Japan*, **50** (2009), 1768–1774.
20. K. Yamamura, Bibliography on calculations of relative class numbers of imaginary abelian number fields (Japanese), Proceedings of the 7th Workshop on Number Theory (Kanazawa 2008) (2009), 139–140.
21. T. Taniguchi, Computation of the minus-part of the class number of the prime-power-th cyclotomic fields (Japanese), Proceedings of the 9th Workshop on Number Theory (Kanazawa 2010), (2011), 47–59.
22. T. Taniguchi, Computation of relative class numbers of cyclotomic fields (Japanese), *Trans. Japan Soc. Ind. Appl. Math.* **25** (2015), 285–307.

Bibliography on Determinantal Expressions of Relative Class Numbers of Imaginary Abelian Number Fields

1. L. Carlitz and F. R. Olson, Maillet's determinant, *Proc. Amer. Math. Soc.* **6** (1955), 265–269.
2. K. Inkeri, Über die Klassenzahl des Kreiskörpers des l^{ten} Einheitswurzeln, *Ann. Acad. Sci. Fenn.* Ser. A I No. 199 (1955), 12 pp.
3. L. Carlitz, A generalization of Maillet's determinant and a bound for the first factor of the class number, *Proc. Amer. Math. Soc.* **12** (1961), 256–261.
4. Z. I. Borevich and I. R. Shafarevich, *Number theory,* Translation from Russian by N. Greenleaf : Academic Press, New York and London, 1966, Chap. 5, Sec. 6, Problem 2.
5. T. Lepistö, On the first factor of the class number of the cyclotomic number field and Dirichlet's L-functions, *Ann. Acad. Sci. Fenn.* Ser. A I, No. 387 (1966), 53 pp.
6. T. Metsänkylä : Bemerkungen über den ersten Faktor Klassenzahl des Kreiskörpers, *Ann. Univ. Turku.* Ser. A I, No. 105 (1967), 15 pp.
7. S. Hyyrö : Über eine Determinantenidentität und den ersten Factor der Klassenzahl des Kreiskörpers, *Ann. Acad. Sci. Fenn.* Ser. A I, No. 398 (1967), 7 pp.
8. T. Metsänkylä, Über den ersten Faktor Klassenzahl des Kreiskörpers, *Ann. Acad. Sci. Fenn.* Ser. A I, No. 416 (1968), 48 pp.
9. T. Metsänkylä, Congruences modulo 2 for class number factors in cyclotomic fields, *Ann. Acad. Sci. Fenn.* Ser. A I, No. 453 (1969), 12 pp.
10. M. Newman, A table of the fisrt factor for prime cyclotomic fields, *Math. Comp.* **24** (1970), 215–219.
11. V. M. Galkin, The first factor of the ideal class number of a cyclotomic field (Russian), U. M. N. **27** (1972), 233–234.
12. J. Kühnová, Maillet's determinant D_{p^n+1}, *Arch. Math. (Brno)* **15** (1979), 209–212.
13. L. Skula, Another proof of Iwasawa's class number formula, *Acta Arith.* **39** (1981), 1–6.
14. K. Tateyama, Maillet's determinant, *Sci. Papers College Gen. Ed. Tokyo* **32** (1982), 97–100.
15. S. Okada, Generalized Maillet determinant, Proceedings of the 29th Symposium on Algebra (Iwate 1983), (1983), 271–279.
16. S. Okada, Generalized Maillet determinant, *Nagoya Math. J.* **94** (1984), 165–170.

17. T. Metsänkylä, Maillet's matrix and irregular primes, *Ann. Univ. Turku.* Ser. A I No. 186 (1984), 72–79.
18. K. Wang, On Maillet's determinant, *J. Number Theory* **18** (1984), 306–312.
19. A. Endô, The relative class numbers of certain imaginary abelian fields, *Abh. Math. Sem. Univ. Hamburg* **58** (1988), 237–243.
20. A. Endô, The relative class numbers of certain imaginary abelian number fields and determinants, *J. Number Theory* **34** (1990), 13–20.
21. T. Funakura, On Kronecker's limit formula for Dirichlet series with periodic coefficients, *Acta Arith.* **55** (1990), 59–73.
22. F. Hazama, Demjanenko matrix, class number, and Hodge group, *J. Number Theory* **34** (1990), 174–177.
23. A. Endô, On an index formula for the relative class numbers of a cyclotomic field, *J. Number Theory* **36** (1990), 332–338.
24. A. Endô, On the Stickelberger ideal of $(2, \cdots, 2)$-extensions of a cyclotomic number field, *Manuscripta Math.* **69** (1990), 107–132.
25. G. Fujisaki, A generalization of Carlitz's determinant, *Sci. Papers College Arts Sci. Univ. Tokyo* **40** (1990), 63–68.
26. K. Girstmair, A recursion formula for the relative class number of the p^n-th cyclotomic field, *Abh. Math. Sem. Univ. Hamburg* **61** (1991), 131–138.
27. K. Girstmair, On the cosets of the $2q$-power group in the unit group modulo p, *Abh. Math. Sem. Univ. Hamburg* **62** (1992), 217–232.
28. K. Girstmair, The relative class numbers of imaginary cyclic fields of degrees 4, 6, 8 and 10, *Math. Comp.* **61** (1993), 881–887, S25–S27.
29. W. Schwarz, Demjanenko matrix and 2-divisibility of class numbers, *Arch. Math. (Basel)* **60** (1993), 154–156.
30. K. Dohmae, Demjanenko matrix for imaginary abelian fields of odd conductors, *Proc. Japan Acad. Ser. A Math. Sci.* **70** (1994), 292–294.
31. J. W. Sands and W. Schwarz, A Demjanenko matrix for abelian fields of prime power conductors, *J. Number Theory* **52** (1995), 85–97.
32. L. Skula, The Stickelberger ideal and the Demjanenko matrix, Algebraic cycles and related topics (Kitasakado, 1994) (F. Hazama, ed.), World Sci. Publishing, River Edge NJ, 1995, 69–76.
33. T. Agoh and L. Skula, Kummer type congruences and Stickelberger subideals, *Acta Arith.* **75** (1996), 235–250.
34. H. Tsumura, On a direct sum decomposition of the Dem'yanenko matrix, *Acta Arith.* **77** (1996), 71–76.
35. H. Tsumura, On Demjanenko's matrix and Maillet's determinant for imaginary abelian number fields, *J. Number Theory* **60** (1996), 70–79.
36. M.-H. Le, A determinant concerning the relative class number of the cyclotomic field $\mathbf{Q}(\zeta_{p^n})$, *Discuss. Math. Algebra Stochastic Methods* **16** (1996), 61–65.
37. C. Hélou, Proof of a conjecture of Terjanian for regular primes, *C. R. Math. Rep. Acad. Sci. Canada* **43** (1996), 193–198.
38. A. Endô, The relative class numbers of certain imaginary abelian number fields of odd conductors, *Proc. Japan Acad. Ser. A Math. Sci.* **72** (1996), 64–68.
39. T. Metsänkylä, Letter to the editor. Comment on : "On Demyanenko's matrix and Maillet's determinant for imaginary abelian number fields" [*J. Number Theory* **60** (1996), no. 1, 70–79] by Tsumura, *J. Number Theory* **64** (1997), 162–163.
40. H. Tsumura, On Demjanenko matrix for general imaginary abelian number fields (Japanese), *Algebraic Number Theory and Fermat's Problem, RIMS (Kyoto 1995).* (K. Komatsu ed.) RIMS Kôkyûroku **971** (1996) 125–133.
41. M. Hirabayashi, A relative class number formula of an imaginary abelian field by means of Demyanenko matrix, Proceedings of the conference on analytic and elementary number theory: a satellite conference of the European Congress on Mathematics '96 (Vienna, 1996) (W. G. Nowak and J. Schoißengeier, eds.), Univ. Wien, 1996, 81–91.

42. F. Hazama, Hodge cycles on the jacobian variety of the Catalan curve, *Compositio Math.* **107** (1997), 339–353.
43. P. Fuchs, Maillet's determinant and a certain basis of Stickelberger ideal, Number theory (Liptovský, Ján, 1995), *Tatra Mt. Math. Publ.* **11** (1997), 121–128.
44. S. Huang, On the representation of the first factor of the class number of the cyclotomic field $\mathbf{Q}(\zeta_p)$ by determinants (Chinese), *J. Zhanjiang Normal College Nat. Sci. Ed.* **18** (1997), 23–24.
45. M. Hirabayashi, A generalization of Maillet and Demyanenko determinants, *Acta Arith.* **83** (1998), 391–397.
46. S. Kanemitsu and T. Kuzumaki, On a generalization of the Maillet determinant, Number theory (Eger, 1996) (K. Győry, A. Pethő and V. T. Sós, eds.), de Gruyter, Berlin, 1998, 271–287.
47. R. Kučera, Formulae for the relative class numbers of an imaginary abelian number field in the form of a determinant, *Nagoya Math. J.* **163** (2001), 167–191.
48. S. Kanemitsu and T. Kuzumaki, On a generalization of the Maillet determinant II, *Acta Arith.* **99** (2001), 343–361.
49. M. Hirabayashi, A generalization of Maillet and Demjanenko determinants for the cyclotomic $\mathbf{Z}_p$-extension, *Abh. Math. Sem. Univ. Hamburg* **71** (2001), 15–28.
50. M. Hirabayashi, Inkeri's determinant for an imaginary abelian number field, *Arch. Math. (Basel)* **79** (2002), 175–181.
51. M. Hirabayashi, A relative class number formula for an imaginary abelian number field by means of Dedekind sum, *Manuscripta Math.* **109** (2002), 223–227.
52. R. Kučera, Formulae for the relative class number of an imaginary abelian number field in the form of a product of determinants, *Acta Math. Inform. Univ. Ostraviensis* **10** (2002), 79–83.
53. A. Endô, A generalization of the Maillet determinant and the Demyanenko determinant, *Abh. Math. Sem. Univ. Hamburg* **73** (2003), 181–193.
54. M.-H. Le, A class of determinants and its applications (Chinese), *J. Wuhan Univ. Sci Engrg.* **16** (2003), 89–91.
55. M. Hirabayashi, A formula of Girstmair and a remark on relative class number formula, Proceedings of the 2003 Nagoya Conference "Yokoi-Chowla Conjecture and Related Problems" (Nagoya, 2003) (S.-I. Katayama, C. Levesque and T. Nakahara, eds.), Saga Univ., Saga, 2004, 11–24.
56. J. Ahn and H. Jung, Determinant formulas and class numbers in cyclotomic fields, Proceedings of the Japan-Korea joint seminar on number theory (Kuju, 2004) (H. K. Kim, ed.), Kyushu Univ., 2004, 1–15.
57. M. Hirabayashi, Generalizations of Girstmair's formulas, *Abh. Math. Sem. Univ. Hamburg* **75** (2005), 83–95.
58. M. Hirabayashi and H. Tsumura, Multiple Dedekind sums and relative class number formulae, *Math. Nachr.* **278** (2005), 1673–1680.
59. M. Hirabayashi, On relative class number formulas for imaginary abelian number fields, Proceedings of the Workshop Number Theory 2005 (Waseda Univ. 2005). (K. Hashimoto and K. Komatsu eds.) 2005, 75–88.
60. T. Agoh and T. Taniguchi, A study of Inkeri's class number formula, Exposition. Math. **24** (2006), 53–79.
61. H. Y. Jung and J. Ahn, Group determinant formulas and class numbers of cyclotomic fields, *J. Korean Math. Soc.* **44** (2007), 499–509.
62. M. Hirabayashi, A determinant formula for the quotient of the relative class numbers of imaginary abelian number fields of relative degree 2, Proceedings of the Symposium on Algebraic Number Theory and Related Topics (K. Hashimoto, Y. Nakajima and H. Tsunogai, eds.), RIMS Kôkyûroku Bessatsu **B4**, Research Institute for Mathematical Sciences, Kyoto Univ., Kyoto, 2007, 21–33.
63. M. Hirabayashi, A generalization of Newmann's formula, *Arch. Math. (Basel)* **90** (2008), 223–229.
64. S. Jakubec, On some new estimates for $h^-(\mathbf{Q}(\zeta_p))$, *Acta Arith.* **137** (2009), 43–50.
65. K. Yamamura, Bibliography on determinantal expressions of relative class numbers of imaginary abelian number fields, Number theory (Higashi-Osaka, 2008) (T. Aoki, S. Kanemitsu, & J. Liu, eds.), Ser. Number Theory Appl., 6, World Sci. Publ., Hackensack, NJ, 2010, 244–250.

66. M. Hirabayashi, A determinant formula for the relative class number formula for an imaginary abelian number field, *Commun. Math.* **22** (2014), 133–140.
67. M. Hirabayashi, A generalization of Jakubec's formula, *Math. Slovaca* **65** (2015), 215–227.
68. S. Jakubec, Connection between multiplication theorem for Bernoulli polynomials and first factor h_p^-, *Math. Slovaca* **67** (2017), 345–348.

Index

© Springer Nature Switzerland AG 2019

H. Hasse, *On the Class Number of Abelian Number Fields*,
https://doi.org/10.1007/978-3-030-01512-1

MIX
Papier aus verantwortungsvollen Quellen
Paper from responsible sources
FSC® C105338

If you have any concerns about our products,
you can contact us on
ProductSafety@springernature.com

In case Publisher is established outside the EU,
the EU authorized representative is:
Springer Nature Customer Service Center GmbH
Europaplatz 3, 69115 Heidelberg, Germany

Printed by Libri Plureos GmbH
in Hamburg, Germany